APPROXIMATION AND COMPUTATION

For more titles in this series, go to
http://www.springer.com/series/7393

Springer Optimization and Its Applications

VOLUME 42

Aims and Scope
Optimization has been expanding in all directions at an astonishing rate during the last few decades. New algorithmic and theoretical techniques have been developed, the diffusion into other disciplines has proceeded at a rapid pace, and our knowledge of all aspects of the field has grown even more profound. At the same time, one of the most striking trends in optimization is the constantly increasing emphasis on the interdisciplinary nature of the field. Optimization has been a basic tool in all areas of applied mathematics, engineering, medicine, economics and other sciences.

The series *Springer Optimization and Its Applications* publishes undergraduate and graduate textbooks, monographs and state-of-the-art expository works that focus on algorithms for solving optimization problems and also study applications involving such problems. Some of the topics covered include nonlinear optimization (convex and nonconvex), network flow problems, stochastic optimization, optimal control, discrete optimization, multiobjective programming, description of software packages, approximation techniques and heuristic approaches.

APPROXIMATION AND COMPUTATION

In Honor of Gradimir V. Milovanović

Edited By

WALTER GAUTSCHI
Purdue University
Department of Computer Science
West Lafayette, Indiana, USA

GIUSEPPE MASTROIANNI
University of Basilicata
Department of Mathematics and Computer Sciences
Potenza, Italy

THEMISTOCLES M. RASSIAS
National Technical University of Athens
Department of Mathematics
Athens, Greece

Springer

Editors
Walter Gautschi
Purdue University
Department of Computer Science
305 N. University Street
West Lafayette, Indiana, 47907
USA
wxg@cs.purdue.edu

Themistocles M. Rassias
National Technical University of Athens
Department of Mathematics
Zografou Campus
15780 Athens
Greece
trassias@math.ntua.gr

Giuseppe Mastroianni
University of Basilicata
Department of Mathematics and
Computer Sciences
Viale dell'Ateneo Lucano, 10
85100 Potenza
Italy
mastroianni.csafta@unibas.it

ISSN 1931-6828
ISBN 978-1-4614-2703-2 ISBN 978-1-4419-6594-3 (eBook)
DOI 10.1007/978-1-4419-6594-3
Springer New York Dordrecht Heidelberg London

Mathematics Subject Classification (2010): 37D40, 65Kxx, 68U35, 68Txx

Cover illustration: "Dynamics of Pebble Energy" – Photo taken by Elias Tyligadas

Printed on acid-free paper

Springer is part of Springer Science+Business Media (www.springer.com)

Preface

The book "APPROXIMATION AND COMPUTATION" deals with recent results in approximation theory, numerical analysis, and various applications of an interdisciplinary character. It contains papers written by experts from 15 countries in their respective subjects. Most of these papers were presented in person during an International Conference, which was held at the University of Niš, Serbia (August 25–29, 2008), to honor the 60th anniversary of the well-known mathematician Professor Gradimir V. Milovanović. Professor Milovanović, next to other distinctions he has received, is a corresponding member of the Serbian Academy of Sciences and Arts.

The book consists of the following five Parts:

1. Introduction
2. Polynomials and Orthogonal Systems
3. Quadrature Formulae
4. Differential Equations
5. Applications

In each part, the papers appear in alphabetical order with respect to the last name of the first-named author, except in Part 1. The latter contains three contributions connected with the scientific work of G.V. Milovanović. Besides biographical data, A. Ivić presents contributions of Milovanović in the field of polynomials and quadrature processes. W. Gautschi describes his collaboration with Milovanović over many years, and Th.M. Rassias gives an account on some major trends in mathematics.

Polynomials, algebraic and trigonometric, and corresponding orthogonal systems are treated as basic constructive elements in Part 2. P. Barry, P.M. Rajković, and M.D. Petković give an application of Sobolev orthogonal polynomials to the computation of a special Hankel determinant. Extremal problems for polynomials in the complex plane, including the well-known conjecture of Sendov about critical points of algebraic polynomials and the mean value conjecture of Smale, as well as certain relations between these two famous open problems, are considered by B. Bojanov. He also formulates a conjecture that seems to be a natural complex analog of Rolle's theorem and contains as a particular case Smale's conjecture. V. Božin and M. Mateljević characterize graphs of maximal energy by means of orthogonal matrices. Their result makes it possible to estimate the energy of graphs without direct computation of eigenvalues. A.S. Cvetković further develops interlacing properties of zeros of shifted Jacobi polynomials, the investigation of which has recently been initiated by K. Driver and K. Jordaan. He, in fact, proves certain improvements of their results. A.S. Cvetković and M.P. Stanić investigate trigonometric polynomials of semi-integer degree with respect to some weight functions on $[-\pi,\pi)$ and orthogonal systems connected with interpolatory quadrature rules with an even maximal trigonometric degree of exactness. W. Gautschi gives an account of computational work in support of conjectured inequalities for zeros of Jacobi polynomials, the sharpness of Bernstein's inequality for Jacobi polynomials, and the positivity of certain quadrature formulae of Newton–Cotes, Gauss–Radau, and Gauss–Lobatto type. The use of symbolic computation is described for generating Gauss quadrature rules with exotic weight functions, specifically weight functions decaying super-exponentially at infinity, and weight functions densely oscillatory at zero. J. Gilewicz and R. Jedynak illuminate the compatibility of continued fraction convergents with Padé approximants. Orthogonal decompositions of fractal sets are considered by Lj.M. Kocić, S. Gegovska-Zajkova, and E. Babače. The last paper in this part by S. Koumandos gives a systematic account of new results on positive trigonometric sums and applications to geometric function theory. His work is related to recent investigations concerning sharpening and generalizations of the celebrated Vietoris inequalities. Far-reaching extensions and results on starlike functions are obtained and new positive sums of Gegenbauer polynomials and a discussion of some challenging conjectures are presented.

Part 3 is dedicated to numerical integration, including quadrature formulas with equidistant nodes and formulas of Gaussian type having multiple nodes. Several quadrature rules for the numerical integration of smooth (nonoscillatory) functions, defined on the real (positive) semiaxis or on the real axis and decaying algebraically at infinity, are examined by G. Monegato and L. Scuderi. Among those considered on the real axis, there are four new alternative numerical approaches. The advantages and the disadvantages of each of them are pointed out through several numerical tests, involving either the computation of a single integral or the numerical solution of some integral equations. G. Nikolov and C. Simian construct Gauss, Lobatto, and Radau quadrature formulae associated with the spaces of parabolic splines with equidistant knots. These quadrature formulae are known to be asymptotically optimal in Sobolev spaces W_p^3. Sharp estimates for the error constant in W_∞^3 are given. Some quadrature rules based on the zeros of Freud polynomials for computing Cauchy principal value integrals on the real line are proposed by I. Notarangelo. M.M. Spalević, and M.S. Pranić give a survey on contour integration methods for estimating the remainder of Gauss–Turán quadrature rules involving analytic functions. Finally, J. Waldvogel proposes to use the trapezoidal rule on the entire real line $\mathbb{R}$ as the standard algorithm for numerical quadrature of analytic functions. Other intervals and slowly decaying integrands may elegantly be handled by means of simple analytic transformations of the integration variable.

Methods for differential equations are considered in Part 4. D.R. Bojović and B.S. Jovanović investigate the convergence of difference schemes for the one-dimensional heat equation with time-dependent operator and the coefficient of the time derivative containing a Dirac delta distribution. An abstract operator method is developed for analyzing this equation. In the paper by S. Burke, Ch. Ortner, and E. Süli, the energy of the Francfort–Marigo model of brittle fracture is approximated, in the sense of Γ-convergence, by the Ambrosio–Tortorelli functional. They formulate and analyze an adaptive finite element algorithm, combining an inexact Newton method with residual-driven adaptive mesh refinement, for the computation of its (local) minimizers. The sequence generated by this algorithm is proved to converge to a critical point. C. Frammartino proposes a Nyström method for solving integral equations equivalent to second-order boundary value problems on the real semiaxis. She proves the stability and convergence of such a procedure and gives some interesting numerical examples. B.S. Jovanović investigates an initial boundary value problem for a one-dimensional hyperbolic equation in two disjoint intervals. A finite difference scheme approximating this initial boundary value problem is proposed and analyzed. P.S. Milojević develops a nonlinear Fredholm alternative theory involving k-ball and k-set perturbations of general homeomorphisms as well as of homeomorphisms that are nonlinear Fredholm maps of index zero. He gives several applications to the unique and finite solvability of potential and semilinear problems with strongly nonlinear boundary conditions and to quasilinear elliptic equations. The work of S. Pilipović, N. Teofanov, and J. Toft shows possible directions for numerically interested mathematicians to approximate different types of singular supports, wave front sets, and pseudodifferential operators in the framework of Fourier-Lebesgue spaces. It contains new results on singular supports in Fourier-Lebesgue spaces and on the continuity properties of certain pseudodifferential operators. Finally, B.M. Piperevski considers a class of matrix differential equations and gives conditions under which this class has a polynomial solution.

Finally, contributions discussing various applications are presented in Part 5. An improved algorithm for Petviashvili's heuristic numerical method for finding solitons in optically induced photonic lattices is presented by R. Jovanović and M. Tuba. An explicit method for the numerical solution of the Fokker–Planck equation of filtered phase noise in modern telecommunication systems is given by D. Milić. Z.S. Nikolić investigates numerically the densification due to gravity-induced skeletal settling during liquid phase sintering. A new methodology is applied for simulation of microstructural evolution of a regular multidomain model. P. Stanimirović, M. Miladinović, and I.M. Jovanović investigate symbolic transformations on "unevaluated" expressions representing objective functions to generate unevaluated composite objective functions required during the implementation of unconstrained nonlinear optimization methods based on the exact line search. N. Stevanović and P.V. Protić deal with Abel-Grassmann's groupoids and introduce the class "root of a band", a generalization of AG-band and AG-3-band. B.T. Todorović, S.R. Rančić, and E.H. Mulalić consider a version of probabilistic supervised machine, learning classifier in named entity recognition. Z. Udovičić deals with interpolating quadratic splines, and finally, Lj.S. Velimirović, S.R. Rančić, and M.Lj. Zlatanović consider infinitesimal bending of curves in E^3.

This book addresses researchers and students in mathematics, physics, and other computational and applied sciences.

As coauthors of Professor Gradimir V. Milovanović, we feel very pleased to have undertaken the preparation of this publication.

Finally, we express our warmest thanks to all of the scientists who contributed to this volume, to the referees for their careful reading of the manuscripts, and to collaborators of Professor Milovanović: Ljubiša Kocić, Aleksandar Cvetković, and Marija Stanić, who helped in the organization of the Conference and in the preparation of this volume.

West Lafayette/Potenza/Athens
August, 2009

Walter Gautschi
Giuseppe V. Mastroianni
Themistocles M. Rassias

Contents

Part I
Introduction

The Scientific Work of Gradimir V. Milovanović

Aleksandar Ivić

1 Introduction

This text is a somewhat changed version of the lecture presented by the author at the conference "Approximation & Computation", held in Niš (Serbia), from 25 to 29 August 2008. The conference was dedicated to the 60th anniversary of Prof. Gradimir V. Milovanović (henceforth GVM for brevity), one of the best mathematicians that Serbia has ever had. The present text is organized as follows. In the next section, the biography of GVM is given. In Sect. 3, a short discussion of the scientific fields in Mathematics where GVM has made his permanent mark will be discussed. Two of the most important of these fields will be discussed in detail. Thus in Sect. 4, we shall present his results involving quadrature processes, and in Sect. 5 those that are connected with polynomials. In Sect. 6, we briefly discuss his recent monograph "*Interpolation Processes: Basic Theory and Applications*", written jointly with G. Mastroianni. Finally, the selected bibliography of GVM will be given at the end of this paper.

2 The Biography of G.V. Milovanović

G.V. Milovanović was born in Zorunovac (Eastern Serbia) on January 2, 1948 to father Vukašin and mother Vukadinka (born Savić). The family's last name was originally Milutinović, but it was changed later to Milovanović, after GVM's great-great-grandfather. He went to high school in the town of Knjaževac (Eastern Serbia), and already there he successfully took part in mathematical competitions.

Aleksandar Ivić
Serbian Academy of Sciences and Arts, Knez Mihailova 35, 11000 Belgrade, Serbia,
e-mail: ivic@rgf.bg.ac.rs

W. Gautschi et al. (eds.), *Approximation and Computation: In Honor of Gradimir V. Milovanović*, Springer Optimization and Its Applications 42,
DOI 10.1007/978-1-4419-6594-3_1, © Springer Science+Business Media, LLC 2011

In 1967, began studies at the Section of Electronics of the Technical Faculty of Niš (Serbia), which soon became independent, as the Faculty of Electronics. Among his professors were Jovan Surutka, Tihomir Aleksić, Petar Madić, Jovan Petrić, and Branko Raković, but it was certainly the well-known mathematician Dragoslav S. Mitrinović who drew him to scientific research. From the beginning of his student days, GVM showed great interest in computers and numerical mathematics.

After obtaining his Bachelor's degree in 1971, he continued graduate studies at the University of Niš, where he obtained his Master's degree in 1974 and his Ph. D. degree in 1976 (in both cases his mentor was Prof. Dragoslav S. Mitrinović). He became an assistant at the University of Niš in 1971, and he passed all the titles until he became a full Professor in 1986 at the Faculty of Electronics. GVM was employed there until July 2008, when he moved to the Megatrend University in Belgrade. In 2006, he was elected as a corresponding member of SANU (Serbian Academy of Sciences and Arts).

During his impressive university career, GVM has taught various courses in Mathematics to undergraduate and graduate students. These include Numerical analysis, Approximation theory, Real analysis, Linear algebra, Special functions, Mathematical programming, etc. This shows his wide mathematical culture and his many mathematical interests. Some of his students (these include Aleksandar Cvetković, Ljubiša Kocić, Milan Kovačević, Marija Stanić, Miodrag Spalević, Nenad Cakić, and Alall Guessab) have become well-established mathematicians of their own.

GVM also held many important functions and offices. He was the Rector of the University of Niš (2004–2006) and he is currently the chairman of the National Council for Science of the Republic of Serbia. He was the vice-rector of the University of Niš (1989–1991), the chairman of the scientific council of the Mathematical Institute of Belgrade, the vice-chairman of the Scientific Society of Serbia, and the dean of the Faculty of Electronics in Niš from 2002 to 2004. He is a member of AMS, GAMM (Gesellschaft für angewandte Mathematik und Mechanik). He was a visiting professor at Purdue University, Université de Pau (France), Università di Basilicata (Potenza, Italy), etc. He participated with talks in many symposia, conferences, and congresses at home and abroad. He supervised 11 doctoral theses, 17 master theses, and many research projects. As a reviewer of scientific projects, he worked for the Ministry of Science of Serbia, Italy (MURST), and Montenegro. He participated in the work of commissions for doctorate theses and promotion of professors in many countries (France, Italy, Cyprus, Australia, India). GVM organized in Niš in 1996 an international mathematical conference dedicated to the memory of Prof. Dragoslav S. Mitrinović. He is married to Dobrila, a former director of "Niteks" from Niš. Their daughter Irena is a medical doctor living with her husband Vladica in Switzerland.

He is the founder of a scientific journal "Facta Universitatis: Series Mathematics and Informatics" at the University of Niš and its editor since its foundation (1986). He is a member of the editorial board of six journals in Serbia, two in Romania and

Bulgaria, and one in Australia, India, Armenia, and Iran, respectively. He is a referee for numerous international mathematical journals and a reviewer of Mathematical Reviews (USA) and Zentralblatt für Mathematik (Germany).

GVM is one of Serbia's most prolific mathematicians of all times. At the time of writing this text, he has published more than 260 works (85 research papers in international periodicals from the SCI list, 80 in other international reviewed journals, 25 chapters in monographs, and 70 in the proceedings of international and domestic conferences). He so far wrote four monographs, 19 textbooks, and numerous technical papers. He published papers in such renowned journals as "Mathematics of Computation", "J. Computational and Applied Mathematics", "SIAM J. Sci. Computation", "Numerische Mathematik", "IMA J. Numerical Analysis", and many others. His latest work, just published by "Springer Verlag", is a comprehensive monograph (written jointly with Prof. Giuseppe Mastroianni) entitled "*Interpolation process - Basic theory and applications*". His earlier monograph (written jointly with D.S. Mitrinović and Th. M. Rassias) "*Topics in Polynomials: Extremal Problems, Inequalities, Zeros*", and published by World Scientific, Singapore, 1994, xiv + 822 pp. is a famous work, called by many the "Bible of Polynomials". His works are quoted in the works of other mathematicians more than 800 times. Of the books in Serbian, the following should be mentioned, his "*Numerička analiza*" ("*Numerical Analysis*") in three volumes, published by Naučna knjiga, Belgrade (first edition 1985). This was the first complete text book on this subject written in Serbian, and it has been widely used by several generations of students throughout former Yugoslavia.

3 Fields of Scientific Work of GVM

Gradimir V. Milovanović worked successfully in several fields of Numerical Analysis and Approximation Theory, where he made important contributions. Broadly speaking, these include

1. Orthogonal polynomials and systems
2. Polynomials (extremal problems, inequalities, and zeros)
3. Approximations by polynomials and splines
4. Interpolation and quadrature processes

It is obviously impossible, in a text such as this one, to cover his results extensively in all the above fields. Thus, in presenting his results a selection of the topics will be made, whose possible shortcomings are due entirely to the author. In presenting his results, it seemed appropriate to provide a short discussion of the topics that are involved. Two of the fields where his most important contributions lie, namely, quadrature processes and polynomials, will be discussed in the next two sections.

4 GVM and Quadrature Processes

Numerical integration begins with Newton's idea (1676)* for finding the weight coefficients $A_1, A_2, \ldots, A_n$ in the so-called *n-point quadrature formula*

$$I(f) = \int_a^b f(t)\,\mathrm{d}t \approx Q_n(f) = A_1 f(\tau_1) + A_2 f(\tau_2) + \cdots + A_n f(\tau_n), \tag{1}$$

for given (usually equidistant) n points (nodes) $\tau_1, \tau_2, \ldots, \tau_n$, such that (1) is exact for all algebraic polynomials of degree at most $n-1$, i.e., for each $f \in \mathcal{P}_{n-1}$. In modern terminology, given n distinct points τ_k and corresponding values $f(\tau_k)$, Newton constructs the unique polynomial $P \in \mathcal{P}_{n-1}$, which at the points τ_k assumes the same values as f, i.e., $P(\tau_k) = f(\tau_k)$, $k = 1,2,\ldots,n$, expressing this interpolation polynomial in terms of divided differences. Evidently, for the interpolation error $r_n(f;t) := f(t) - P(t)$ we have $r_n(f;t) = 0$ for all $f \in \mathcal{P}_{n-1}$. Subsequently, integrating this polynomial over $[a,b]$, Newton obtains (1). Here, we give it in a more convenient (Lagrange) form. Namely, the interpolation polynomial P can be expressed in terms of the so-called fundamental polynomials

$$\ell_k(t) = \frac{\omega_n(t)}{\omega_n'(\tau_k)(t-\tau_k)}, \qquad k = 1,\ldots,n,$$

where $\omega_n(t) = \prod_{k=1}^{n}(t-\tau_k)$, and therefore $A_k = I(\ell_k)$, $k = 1,2,\ldots,n$, and $R_n(f) = I(f) - Q_n(f) = I(r_n(f;\cdot))$. Obviously, the remainder term $R_n(f) = 0$ for each $f \in \mathcal{P}_{n-1}$. The quadrature formula obtained in this way is known as *interpolatory* and it has *degree of exactness* at least $n-1$. We write $d = d(Q_n) \geq n-1$.

Starting from the work of Newton and Cotes and combining it with his earlier work on the hypergeometric series, Gauss (1814) develops his famous method of numerical integration, which dramatically improved the earlier method of Newton and Cotes. Later improvements are due to Jacobi, Christoffel, Markov, Stieltjes, and others. Today, these formulae with maximal degree of precision are known as the *Gauss–Christoffel quadrature formulae*. The nodes τ_k, $k = 1,2,\ldots,n$, are zeros of the polynomial of degree n, which is orthogonal to $\mathcal{P}_{n-1}$ with respect to a given measure $\mathrm{d}\mu(x)$. An extensive survey of Gauss–Christoffel quadrature formulae was written by Gautschi (1981).

4.1 Construction of Gaussian Quadratures

Passing to modern theory, we mention some nonclassical measures: $\mathrm{d}\mu(x) = w(x)\,\mathrm{d}x$ for which the recursion coefficients $\alpha_k(\mathrm{d}\mu)$, $\beta_k(\mathrm{d}\mu)$, $k = 0,1,\ldots,n-1$,

* Works of authors not including GVM are indicated by year only, in order to avoid excessive length.

in the fundamental three-term recurrence relation to the corresponding orthogonal polynomials,

$$\pi_{k+1}(t) = (t - \alpha_k(\mathrm{d}\mu))\pi_k(t) - \beta_k(\mathrm{d}\mu)\pi_{k-1}(t), \quad k = 0, 1, \ldots,$$

with $\pi_0(t) = 1$ and $\pi_{-1}(t) = 0$, have been provided in the literature and used in the construction of Gaussian quadratures.

1. *One-sided Hermite weight* $w(x) = \exp(-x^2)$ on $[0, c]$, $0 < c \leqslant x + \infty$.

2. *Logarithmic weight* $w(x) = x^\alpha \log(1/x)$, $\alpha > -1$ on $(0,1)$.

3. *Airy weight* $w(x) = \exp(-x^3/3)$ on $(0, +\infty)$.

4. *Reciprocal gamma function* $w(x) = 1/\Gamma(x)$ on $(0, +\infty)$.

5. *Einstein's* and *Fermi's weight functions* on $(0, +\infty)$,

$$w_1(x) = \varepsilon(x) = \frac{x}{\mathrm{e}^x - 1} \quad \text{and} \quad w_2(x) = \varphi(x) = \frac{1}{\mathrm{e}^x + 1}.$$

The last functions arise in solid state physics. Integrals with respect to the measure $\mathrm{d}\mu(x) = \varepsilon(x)^r \mathrm{d}x$, $r = 1$ and $r = 2$, are widely used in phonon statistics and lattice specific heats and occur also in the study of radiative recombination processes. The integrals with $\varphi(x)$ are encountered in the dynamics of electrons in metals. For $w_1(x)$, $w_2(x)$, $w_3(x) = \varepsilon(x)^2$ and $w_4(x) = \varphi(x)^2$, Gautschi and Milovanović [21,22] performed the first systematic investigation on the derivation of quadrature rules with high-precision, determined the recursion coefficients α_k and β_k, $k < n = 40$, and gave an application of the corresponding Gauss–Christoffel quadratures to the summation of slowly convergent series, whose general term is expressible in terms of a Laplace transform or its derivative. We call such a summation the *method of Laplace transform.*

In the numerical construction for the measure $\mathrm{d}\mu(x) = [\varepsilon(x)]^r\,\mathrm{d}x$ on $(0, +\infty)$, $r \geqslant 1$, they used the discretized Stieltjes procedure based on the Gauss–Laguerre quadratures, so that

$$\begin{aligned}\int_0^{+\infty} P(x)\,\mathrm{d}\mu(x) &= \frac{1}{r}\int_0^{+\infty} P(x/r)\left(\frac{x/r}{1-\mathrm{e}^{-x/r}}\right)^r \mathrm{e}^{-x}\,\mathrm{d}x \\ &\approx \sum_{k=1}^{N} \frac{A_k^L}{r}\left(\frac{x_k^L/r}{1-\mathrm{e}^{-x_k^L/r}}\right)^r P(x_k^L/r),\end{aligned}$$

where $P \in \mathcal{P}$ and $N \gg n$. The Gauss–Laguerre nodes x_k^L (zeros of the standard Laguerre polynomial $L_N(x)$) and the weights A_k^L can be easily computed for an arbitrary N by the Golub–Welsch algorithm.

6. *The hyperbolic weights* on $(0, +\infty)$,

$$w_1(x) = \frac{1}{\cosh^2 x} \quad \text{and} \quad w_2(x) = \frac{\sinh x}{\cosh^2 x}.$$

The recursion coefficients α_k, β_k, for $k < 40$ were obtained by Milovanović [55–58]. The main application of these quadratures is the summation of slowly convergent series, with the general term $a_k = f(k)$. Such a method known as the *method of contour integration* was given by GVM [55–58].

4.2 Moment-Preserving Spline Approximation and Quadratures

An interesting application of Gaussian–type formulas concerns the so-called *moment-preserving spline approximation* of a given function f on $[0,+\infty)$ (or on a finite interval, e.g. $[0,1]$). Such kind of problems appeared in physics, for example, in the approximation of the Maxwell velocity distribution by a linear combination of Dirac δ-functions or in the corresponding approximation by a linear combination of Heaviside step functions. To get a stable method for this kind of approximation, Gautschi and Milovanović [21, 22] found new applications of Gaussian type of quadratures.

Let f be a given function defined on the positive real line $\mathbb{R}_+ = [0,+\infty)$ and $s_{n,m}$ be a spline of the form

$$s_{n,m}(t) = \sum_{\nu=1}^{n} a_\nu (t_\nu - t)_+^m, \quad 0 \leqslant t < +\infty,$$

where the plus sign on the right is the cutoff symbol, meaning that $u_+ = u$ if $u > 0$ and $u_+ = 0$ if $u \leqslant 0$, $0 < t_1 < \cdots < t_n$, $a_\nu \in \mathbb{R}$. They considered the moment-preserving spline approximation $f(t) \approx s_{n,m}(t)$ such that

$$\int_0^{+\infty} s_{n,m}(t) t^j \,\mathrm{d}V = \int_0^{+\infty} f(t) t^j \,\mathrm{d}V, \quad j = 0,1,\ldots,2n-1,$$

where $\mathrm{d}V$ is the volume element depending on the geometry of the problem. In some concrete applications in physics, up to unimportant numerical factors, $\mathrm{d}V = t^{d-1}\,\mathrm{d}t$, where $d = 1,2,3$ for rectilinear, cylindric, and spherical geometry, respectively.

For fixed $n,m \in \mathbb{N}$, $d \in \{1,2,3\}$ and certain conditions on f, they [21] proved that the spline function $s_{n,m}$ exists uniquely if and only if the measure

$$\mathrm{d}\lambda_m(t) = \frac{(-1)^{m+1}}{m!} t^{m+d} f^{(m+1)}(t)\,\mathrm{d}t \quad \text{on} \quad [0,+\infty)$$

admits an n-point Gauss–Christoffel quadrature formula

$$\int_0^{+\infty} g(x)\,\mathrm{d}\lambda_m(x) = \sum_{\nu=1}^{n} \lambda_\nu^{(n)} g(\tau_\nu^{(n)}) + R_n(g;\mathrm{d}\lambda_m)$$

with distinct positive nodes $\tau_\nu^{(n)}$, where $R_n(g;\mathrm{d}\lambda_m) = 0$ for all $g \in \mathcal{P}_{2n-1}$.

Approximation on a compact interval was considered by Frontini et al. [29].

4.3 Quadratures with Multiple Nodes

One may consider *Quadratures with multiple nodes*, where $\eta_1,\ldots,\eta_m$ $(\eta_1<\cdots<\eta_m)$ are given *fixed* (or *prescribed*) nodes, with multiplicities $m_1,\ldots,m_m$, and $\tau_1,\ldots,\tau_n$ with $(\tau_1<\cdots<\tau_n)$ are *free* nodes, with given odd multiplicities $n_1,\ldots,n_n$, $n_\nu=2s_\nu+1$. The quadrature formula is

$$I(f)=\int_{\mathbb{R}} f(t)\,\mathrm{d}\lambda(t)\cong Q(f),$$

where

$$Q(f)=\sum_{\nu=1}^{n}\sum_{i=0}^{n_\nu-1}A_{i,\nu}f^{(i)}(\tau_\nu)+\sum_{\nu=1}^{m}\sum_{i=0}^{m_\nu-1}B_{i,\nu}f^{(i)}(\eta_\nu),$$

with an algebraic degree of exactness at least $M+N-1$ (D.D. Stancu). He proved (1957) that $\tau_1,\ldots,\tau_n$ are the Gaussian nodes if and only if

$$\int_{\mathbb{R}} t^k Q_N(t)q_M(t)\,\mathrm{d}\lambda(t)=0,\quad k=0,1,\ldots,n-1,$$

where, for $M=\sum_{\nu=1}^{m}m_\nu$ and $N=\sum_{\nu=1}^{n}n_\nu$,

$$q_M(t):=\prod_{\nu=1}^{m}(t-\eta_\nu)^{m_\nu},\quad Q_N(t):=\prod_{\nu=1}^{n}(t-\tau_\nu)^{n_\nu}.$$

Let $\pi_n(t):=\prod_{\nu=1}^{n}(t-\tau_\nu)$. Since $Q_N(t)/\pi_n(t)=\prod_{\nu=1}^{n}(t-\tau_\nu)^{2s_\nu}\geqslant 0$ and assuming $q_M(t)\geq 0$ over the support interval, Milovanović [25] reinterpreted the above "orthogonality conditions" as

$$\int_{\mathbb{R}} t^k\pi_n(t)\,\mathrm{d}\mu(t)=0,\quad k=0,1,\ldots,n-1,$$

where

$$\mathrm{d}\mu(t)=\left(\prod_{\nu=1}^{n}(t-\tau_\nu)^{2s_\nu}\right)\mathrm{d}\hat{\lambda}(t),\quad \mathrm{d}\hat{\lambda}(t)=q_M(t)\,\mathrm{d}\lambda(t).$$

This means that $\pi_n(t)$ is a polynomial orthogonal with respect to the new nonnegative measure $\mathrm{d}\mu(t)$ and, therefore, all zeros $\tau_1,\ldots,\tau_n$ are simple, real, and belong to the support interval. As we see, the measure $\mathrm{d}\mu(t)$ involves the nodes $\tau_1,\ldots,\tau_n$, i.e., the unknown polynomial $\pi_n(t)$, which is implicitly defined. This polynomial $\pi_n(t)$ belongs to the class of so-called *σ-orthogonal polynomials* $\{\pi_{n,\sigma}(t)\}_{n\in\mathbb{N}_0}$, which correspond to the sequence $\sigma=(s_1,s_2,\ldots)$ connected with the multiplicities of the Gaussian nodes, namely, we have $\pi_n(t)=\pi_{n,\sigma}(t)$. If $\sigma=(s,s,\ldots)$, these polynomials reduce to the *s-orthogonal polynomials* Milovanović [88, 89].

Quadratures with only Gaussian nodes $(m=0)$,

$$\int_{\mathbb{R}} f(t)\,\mathrm{d}\lambda(t)=\sum_{\nu=1}^{n}\sum_{i=0}^{2s_\nu}A_{i,\nu}f^{(i)}(\tau_\nu)+R(f),$$

which are exact for all algebraic polynomials of degree at most

$$d_{\max} = 2\sum_{\nu=1}^{n} s_\nu + 2n - 1,$$

are known as the *Chakalov–Popoviciu quadrature formulas*. The case with a weight function $\mathrm{d}\lambda(t) = w(t)\,\mathrm{d}t$ on $[a,b]$ has been investigated by Italian mathematicians Ossicini, Ghizzetti, Guerra, Rosati, and by Chakalov, Stroud, Stancu, Ionescu, Pavel, etc.

At the Third Conference on Numerical Methods and Approximation Theory (Niš 1987), GVM [31] presented a stable method with quadratic convergence for numerically constructing s-orthogonal polynomials, whose zeros are nodes of Turán quadratures. The basic idea of the method to numerically construct s-orthogonal polynomials with respect to the measure $\mathrm{d}\mu(t)$ on the real line $\mathbb{R}$ is a reinterpretation of *s-orthogonality* in terms of implicitly defined standard orthogonality. Further progress in this direction was made by Gautschi and Milovanović [61]. After Milovanović's survey [88], where he gave a connection between quadratures, s- and σ-orthogonality and moment-preserving approximation with defective splines, the interest for this subject rapidly increased. A very efficient method for constructing quadratures with multiple nodes was given recently by Milovanović et al. [109].

4.4 Orthogonality with Respect to a Moment Functional and Corresponding Quadratures

In the previous sections, the inner product was always positive definite, implying the existence of the corresponding orthogonal polynomials with real zeros in the support of the measure. Such zeros appeared as the nodes of the Gaussian formulas. However, there are more general concepts of orthogonality with respect to a given linear moment functional L on the linear space $\mathcal{P}$ of all algebraic polynomials. Because of linearity, the value of the linear functional L at every polynomial is known if the values of L are known at the set of all monomials, i.e., if we know $L(x^k) = \mu_k$, for each $k \in \mathbb{N}_0$. In that case, we can introduce a system of orthogonal polynomials $\{\pi_k\}_{k\in\mathbb{N}_0}$ with respect to the functional L if for all nonnegative integers k and n we have that $\pi_k(x)$ is polynomial of degree k, $L(\pi_k(x)\pi_n(x)) = 0$, if $k \neq n$, $L(\pi_n^2(x)) \neq 0$.

4.4.1 Orthogonality on the Semicircle and Quadratures

Let w be a weight function which is positive and integrable on the open interval $(-1,1)$, though possibly singular at the endpoints, and which can be extended to a function $w(z)$ holomorphic in the half disc $D_+ = \{z \in \mathbb{C} : |z| < 1, \ \mathrm{Im}\, z > 0\}$. Consider the following "inner product":

$$(f,g) = \int_\Gamma f(z)g(z)w(z)(\mathrm{i}z)^{-1}\,\mathrm{d}z = \int_0^\pi f(\mathrm{e}^{\mathrm{i}\theta})g(\mathrm{e}^{\mathrm{i}\theta})w(\mathrm{e}^{\mathrm{i}\theta})\,\mathrm{d}\theta,$$

where Γ is the circular part of ∂D_+ and all integrals are assumed to exist (possibly) as appropriately defined improper integrals. The existence of the corresponding orthogonal polynomials $\{\pi_n\}_{n\in\mathbb{N}_0}$ is not guaranteed.

The case $w = 1$ was considered by Gautschi and Milovanović [21, 22]. The existence and uniqueness of polynomials orthogonal on the unit semicircle were proved via moment determinants. A more general case of the complex weight was considered by Gautschi et al. [28]. Under the condition $\mathrm{Re}\,\mu_0 = \mathrm{Re}\int_0^\pi w(\mathrm{e}^{\mathrm{i}\theta})\,\mathrm{d}\theta \neq 0$, they proved that the orthogonal polynomials $\{\pi_n\}_{n\in\mathbb{N}_0}$ exist uniquely and that they can be represented in the form

$$\pi_n(z) = p_n(z) - \mathrm{i}\theta_{n-1}p_{n-1}(z),$$

where

$$\theta_{n-1} = \theta_{n-1}(w) = \frac{\mu_0 p_n(0) + \mathrm{i}q_n(0)}{\mathrm{i}\mu_0 p_{n-1}(0) - q_{n-1}(0)}.$$

Here the p_k's are standard (real) polynomials orthogonal with respect to the inner product

$$[f,g] = \int_{-1}^{1} f(x)\overline{g(x)}w(x)\,\mathrm{d}x,$$

and the q_k's are the corresponding associated polynomials of the second kind:

$$q_k(z) = \int_{-1}^{1} \frac{p_k(z) - p_k(x)}{z - x} w(x)\,\mathrm{d}x.$$

4.4.2 Orthogonal Polynomials for Oscillatory Weights

Let w be a given weight function on $[-1,1]$ and $\mathrm{d}\mu(x) = xw(x)\mathrm{e}^{\mathrm{i}\zeta x}\,\mathrm{d}x$, where $\zeta \in \mathbb{R}$. In this subsection, we consider the existence of the orthogonal polynomials π_n with respect to the functional

$$L(p) = \int_{-1}^{1} p(x)\,\mathrm{d}\mu(x), \quad \mu_k = L(x^k),\ k \in \mathbb{N}_0.$$

Two cases are intensively studied.

1. *Case* $w(x) = 1$, $\zeta = m\pi$, $m \in \mathbb{Z}\setminus\{0\}$.

Existence and uniqueness of the polynomials π_n were proved via moment determinants by Milovanović and Cvetković [111, 112, 114, 116, 118, 119].

From the Hankel determinants, it is clear that the recursion coefficients are rational functions in $\zeta = m\pi$. Using their software package Cvetković and Milovanović [103], one can generate coefficients even in symbolic form for some reasonable values of n. The complexity of expressions for α_n and β_n increases exponentially with n. However, there is an efficient algorithm for their numerical construction, Milovanović and Cvetković [111, 112, 114, 116, 118, 119].

The corresponding Gaussian quadratures can also be constructed. A possible application of these quadratures is the numerical calculation of integrals involving highly oscillatory integrands, in particular, the calculation of the Fourier coefficients,

$$C_m(f) + \mathrm{i}S_m(f) = \int_{-1}^{1} f(x)\mathrm{e}^{\mathrm{i}m\pi x}\,\mathrm{d}x = \int_{-1}^{1} \frac{f(x)-f(0)}{x}\,\mathrm{d}\mu(x)$$

$$\approx \sum_{\nu=1}^{n} \frac{A_\nu^n}{x_\nu^n}(f(x_\nu^n) - f(0)),$$

where x_ν^n are the zeros of π_n, and A_k^n the corresponding weight coefficients. It is interesting to mention that if m increases, the convergence of this quadrature becomes faster!

2. *Case* $w(x) = 1/\sqrt{1-x^2}$, $\zeta \in \mathbb{R}$.

This was considered recently by Milovanović and Cvetković [111, 112, 114, 116, 118, 119]. They showed that the corresponding moments can be expressed in terms of the Bessel functions J_0 and J_1 as

$$\mu_k = \frac{\mathrm{i}\pi}{(\mathrm{i}\zeta)^k}(P_k(\zeta^2)J_1(\zeta) + \zeta Q_k(\zeta^2)J_0(\zeta)), \quad k \in \mathbb{N}_0,$$

where P_k and Q_k are algebraic polynomials with integer coefficients of degrees $2[k/2]$ and $2[(k-1)/2]$, respectively. They satisfy the following recurrence relation:

$$y_{k+2} = -(k+2)y_{k+1} - \zeta^2 y_k - (k+1)\zeta^2 y_{k-1},$$

with initial conditions $P_0 = 1, P_1 = -1, P_2 = 2 - \zeta^2$ and $Q_0 = 0, Q_1 = 1, Q_2 = -1$. When ζ is a positive zero of J_0, they proved that the polynomials π_n orthogonal with respect to the functional

$$L(p) = \int_{-1}^{1} p(x)\,\mathrm{d}\mu(x), \quad \mu_k = L(x^k), \quad k \in \mathbb{N}_0$$

exist uniquely and satisfy a three-term recurrence relation, for whose coefficients they found asymptotic formulae. They also proved that the essential spectrum of the associated Jacobi operator generated by the three-term recurrence coefficients α_k and β_k, $k \in \mathbb{N}_0$, is $[-1, 1]$.

4.5 Nonstandard Quadratures of Gaussian Type

All previous quadrature rules use the information on the integrand only at some selected points x_k, $k=1,\ldots,n$ (the values of the function f and its derivatives in the cases of rules with multiple nodes). Such quadratures will be called the *standard quadrature formulae*. However, in many cases in physics and technical sciences it is not possible to measure the exact value of the function f at points x_k, so that a standard quadrature cannot be applied. On the contrary, some other information on f can be available, like the average

$$\frac{1}{2h_k}\int_{I_k} f(x)\,\mathrm{d}\mu(x)$$

of this function over some nonoverlapping subintervals I_k, with length of I_k equal to $2h_k$, and their union being a proper subset of $\operatorname{supp}(\mathrm{d}\mu)$. One may also know a fixed linear combination of the values of this function, e.g.

$$af(x-h)+bf(x)+cf(x+h)$$

at some points x_k, where a,b,c are constants and h is a sufficiently small positive number, etc.

The last example comes from the Theory of communications, namely, supposing that we are receiving a signal with an interference. Then the measurements provide linear combinations of the function values rather than a single function value. In other words, for signals (functions) which depend on time in a discrete sense, averaging may be given with respect to a discrete measure, with some jumps at x_k-h, x_k, x_k+h, in our simple example. The same problem appears when a signal f passes through a digital filter.

Suppose now that we want to find the value of the integral of the signal f with respect to some positive measure μ, i.e., to calculate $\int_{\mathbb{R}} f(x)\,\mathrm{d}\mu(x)$, and that we know only the values of the signal after passing through some digital filter, i.e., we know the values $g_k=af(x_k-h)+bf(x_k)+cf(x_k+h)$, $k=1,\ldots,n$.

In this very simple example, using initial conditions, in general, we can state a system of linear equations and calculate the values $f(x_k)$ from the values g_k, and then we apply the quadrature rule. Unfortunately, very often the system of linear equations is ill-conditioned and such a procedure significantly disturbs the final result of computations. Therefore, it is much better to use directly the values g_k instead of $f(x_k)$ if it is possible. This idea leads to the so-called *nonstandard quadratures* (see Milovanović and Cvetković [136]).

Thus, if the information data $\{f(x_k)\}_{k=1}^n$ in the standard quadrature is replaced by $\{(\mathcal{A}^{h_k}f)(x_k)\}_{k=1}^n$, where $\mathcal{A}^h$ is an extension of some linear operator $\mathcal{A}^h:\mathcal{P}\to\mathcal{P}$, $h\geqslant 0$, we get a nonstandard quadrature formula

$$\int_{\mathbb{R}} f(x)\,\mathrm{d}\mu(x)=\sum_{k=1}^n w_k(\mathcal{A}^{h_k}f)(x_k)+R_n(f).$$

This kind of quadratures is based on operator values for a general family of linear operators $\mathcal{A}^h$, acting of the space of algebraic polynomials, such that the degrees of polynomials are preserved.

A typical example for such operators is the *average (Steklov) operator* mentioned before, i.e.,

$$(\mathcal{A}^h p)(x) = \frac{1}{2h}\int_{x-h}^{x+h} p(x)\,\mathrm{d}x, \quad h>0,\ p\in\mathcal{P}.$$

The first idea involves the so-called *interval quadratures*. Let $h_1,\ldots,h_n$ be non-negative numbers such that

$$a < x_1 - h_1 \leqslant x_1 + h_1 < x_2 - h_2 \leqslant x_2 + h_2 < \cdots < x_n - h_n \leqslant x_n + h_n < b, \tag{2}$$

and let $w(x)$ be a given weight function on $[a,b]$. Using these inequalities, it is obvious that we have $2(h_1+\cdots+h_n) < b-a$.

Bojanov and Petrov (2001) proved that the Gaussian interval quadrature rule of the maximal algebraic degree of exactness $2n-1$ exists, i.e.,

$$\int_a^b f(x)w(x)\,\mathrm{d}x = \sum_{k=1}^n \frac{w_k}{2h_k}\int_{x_k-h_k}^{x_k+h_k} f(x)w(x)\,\mathrm{d}x + R_n(f), \tag{3}$$

where $R_n(f)=0$ for each $f\in\mathcal{P}_{2n-1}$. If $h_k=h$, $1\leqslant k\leqslant n$, they also proved the uniqueness of (3). In 2003, they proved the uniqueness of (3) for the Legendre weight ($w(x)=1$) for any set of lengths $h_k\geqslant 0$, $k=1,\ldots,n$, satisfying the condition (2).

Recently Milovanović and Cvetković [104, 107, 110], by using properties of the topological degree of nonlinear mappings, proved that the Gaussian interval quadrature formula is unique for the Jacobi weight function $w(x)=(1-x)^\alpha(1+x)^\beta$, $\alpha,\beta>-1$, on $[-1,1]$ and they proposed an algorithm for numerical construction. For the special case of the Chebyshev weight of the first kind and a special set of lengths, an analytic solution can be given.

Recently, Bojanov and Petrov (2005) proved the existence and uniqueness of the weighted Gaussian interval quadrature formula for a given system of continuously differentiable functions, which constitute an ET system of order two on $[a,b]$.

The cases with interval quadratures on unbounded intervals with the classical generalized Laguerre and Hermite weights have been recently investigated by Milovanović and Cvetković [111, 112, 114, 116, 118, 119, 127, 130, 133].

They also considered the nonstandard quadratures with some special operators of the form

$$(\mathcal{A}^h p)(x) = \frac{1}{2h}\int_{x-h}^{x+h} p(t)\,\mathrm{d}t, \quad (\mathcal{A}^h p)(x) = \sum_{k=-m}^{m} a_k p(x+kh)$$

or

$$(\mathcal{A}^h p)(x) = \sum_{k=-m}^{m-1} a_k p\left(x+(k+\tfrac{1}{2})h\right),$$

and

$$(\mathcal{A}^h p)(x) = \sum_{k=0}^{m} \frac{b_k h^k}{k!} \mathcal{D}^k p(x),$$

where m is a fixed natural number and $\mathcal{D}^k = \mathrm{d}^k/\mathrm{d}x^k$, $k \in \mathbb{N}_0$.

4.6 Gaussian Quadrature for Müntz Systems

Gaussian integration can be extended in a natural way to nonpolynomial functions, taking a system of linearly independent functions

$$\{P_0(x), P_1(x), P_2(x), \ldots\} \quad (x \in [a,b]), \tag{4}$$

usually chosen to be complete in some suitable space of functions. If $\mathrm{d}\sigma(x)$ is a given nonnegative measure on $[a,b]$ and the quadrature rule

$$\int_a^b f(x)\,\mathrm{d}\sigma(x) = \sum_{k=1}^{n} A_k f(x_k) + R_n(f) \tag{5}$$

is such that it integrates exactly the first $2n$ functions in (4), we call the rule (5) *Gaussian with respect to the system* (4). The existence and uniqueness of a Gaussian quadrature rule (5) with respect to the system (4), or shorter a *generalized Gaussian formula*, is always guaranteed if the first $2n$ functions of this system constitute a Chebyshev system on $[a,b]$. Then, all the weights $A_1, \ldots, A_n$ in (5) are positive.

The generalized Gaussian quadratures for Müntz systems go back to Stieltjes in 1884. Taking $P_k(x) = x^{\lambda_k}$ on $[a,b] = [0,1]$, where $0 \leqslant \lambda_0 < \lambda_1 < \cdots$, Stieltjes showed the existence of Gaussian formulae. In his short note, he considered also the Gauss–Radau formulae. A numerical algorithm for constructing the generalized Gaussian quadratures was recently investigated by Ma, Rokhlin and Wandzura (1996), but it is ill conditioned. To obtain the double precision results (REAL*8), the authors have performed the computations in extended precision (Q-arithmetic – REAL*16) for generating Gaussian quadratures up to order 20, and in *Mathematica* (120 digits operations) for generating Gaussian quadratures of higher orders ($n \leqslant 40$).

Milovanović and Cvetković [111, 112, 114, 116, 118, 119] presented an alternative numerical method for constructing the generalized Gaussian quadrature (5) for Müntz polynomials, which is exact for each

$$f \in M_{2n-1}(\Lambda) = \operatorname{span}\{x^{\lambda_0}, x^{\lambda_1}, \ldots, x^{\lambda_{2n-1}}\}.$$

Besides the properties of orthogonal Müntz polynomials on $(0,1)$ and their connection with orthogonal rational functions, GVM gave also a method for numerical evaluation of such generalized polynomials [78]. His method is rather stable and simpler than the previous one, since it is based on orthogonal Müntz systems. It performs calculations in double precision arithmetics to get double precision results.

As it is well known, the Gaussian quadrature rule is unique provided the measure σ has a nonnegative absolutely continuous part and has finitely many atoms on $[0,1]$. Such quadratures possess several properties of the classical Gaussian formulae (for polynomial systems), such as positivity of the weights, rapid convergence, etc. They can be applied to a wide class of functions, including smooth functions, as well as functions with end-point singularities, such as those occurring in the boundary-contact value problems, integral equations, complex analysis, potential theory, and several other fields.

5 GVM and Polynomials

The crowning achievement of GVM in this field is his extensive monograph (written jointly with D.S. Mitrinović and Th. M. Rassias) "*Topics in Polynomials: Extremal Problems, Inequalities, Zeros*", World Scientific, Singapore, 1994, XIV+822 pp. This is a famous work, called by many the "Bible of Polynomials".

The book contains some of the most important results on the analysis of polynomials and their derivatives. Besides the fundamental results which are treated with their proofs, the book also provides an account of the most recent developments concerning extremal properties of polynomials and their derivatives in various metrics with an extensive analysis of inequalities for trigonometric sums and algebraic polynomials, as well as their zeros. The final chapter provides some selected applications of polynomials in approximation theory and computer-aided geometric design. One can also find in this book several new research problems and conjectures with sufficient information concerning the results obtained to date toward the investigation of their solution.

5.1 Classical Orthogonal Polynomials

Let $\mathcal{P}_n$ be the set of all algebraic polynomials of degree not exceeding n. Define the norms

$$\|f\|_\infty := \max_{-1\leqslant t\leqslant 1} |f(t)|$$

and

$$\|f\|_r := \left(\int_{-\infty}^{\infty} |f(t)|^r \mathrm{d}\lambda(t)\right)^{1/r}, \quad r \geqslant 1,$$

where $\mathrm{d}\lambda(t)$ is a given nonnegative measure on the real line (with compact support or otherwise), for which all the moments

$$\mu_k := \int_{-\infty}^{\infty} t^k \mathrm{d}\lambda(t), \quad k = 0, 1, \ldots,$$

exist and $\mu_0 > 0$. When $r = 2$ we have the norm

$$\|f\|_2 = \left(\int_{-\infty}^{\infty} |f(t)|^2 \mathrm{d}\lambda(t)\right)^{1/2},$$

and then the inner product is defined by

$$(f,g) := \int_{-\infty}^{\infty} f(t)\overline{g(t)}\, \mathrm{d}\lambda(t).$$

A standard case of orthogonal polynomials is when

$$\mathrm{d}\lambda(t) = w(t)\,\mathrm{d}t,$$

where $w(t)$ is nonnegative, all its moments exist and $\mu_0 > 0$. An important case are the *classical orthogonal polynomials* on an interval $(a,b) \in \mathbb{R}$ of orthogonality. One can take the following intervals:

$$(a,b) = (-1,1),\ (0,+\infty) \text{ or } (-\infty,+\infty)$$

with the inner product

$$(f,g) = \int_a^b f(t)\overline{g(t)}w(t)\,\mathrm{d}t.$$

Orthogonal polynomials $\{Q_n(t)\}$ on (a,b) with this inner product are called *classical orthogonal polynomials* if $w(t)$ satisfies the differential equation

$$\frac{\mathrm{d}}{\mathrm{d}t}\Big(A(t)w(t)\Big) = B(t),$$

where B is a polynomial of the first degree and

$$A(t) = \begin{cases} 1-t^2, & \text{if } (a,b) = (-1,1), \\ t, & \text{if } (a,b) = (0,\infty), \\ 1, & \text{if } (a,b) = (-\infty,\infty). \end{cases}$$

The classical orthogonal polynomial $Q_n(t)$ is a solution of $A(t)y'' + B(t)y' + \lambda_n y = 0$ with an explicit constant λ_n.

5.2 *Extremal Problems of Markov–Bernstein Type for Polynomials*

There are many results on extremal problems and inequalities for the class $\mathcal{P}_n$ of all algebraic polynomials of degree at most n. Markov (1889) solved the extremal problem of determining

$$A_n = \sup_{P \in \mathcal{P}_n} \frac{\|P'\|_\infty}{\|P\|_\infty}.$$

The best constant is $A_n = n^2$ and the extremal polynomial is $P^*(t) = cT_n(t)$, where $T_n(t) = \cos(n \arccos t)$ is the Chebyshev polynomial of the first kind of degree n and c is an arbitrary constant, alternatively $A_n = T_n'(1)$,

$$\|P'\|_\infty \leqslant n^2 \|P\|_\infty \quad (P \in \mathcal{P}_n).$$

Markov (1892) proved

$$\|P^{(k)}\|_\infty \leqslant T_n^{(k)}(1) \|P\|_\infty \quad (P \in \mathcal{P}_n),$$

and Bernstein (1912)

$$\|P'\|_\infty \leqslant n\|P\| \quad \Big(\|f\| = \max_{|z|\leqslant 1} |f(z)|,\ P \in \mathcal{P}_n\Big).$$

When combined, these results yield

$$|P'(t)| \leqslant \min\left\{n^2, \frac{n}{\sqrt{1-t^2}}\right\} \quad (-1 \leqslant t \leqslant 1). \tag{6}$$

Excluding only certain particular cases, the best constant $A_n(k,p)$ in a general Markov type inequality

$$\|P^{(k)}\|_p \leqslant A_n(k,p)\|P\|_p \quad (1 \le k \le n)$$

in the L^p-norm $(p \geqslant 1)$ is still not known, even for $p = 2$ and $w(t) = 1$ on $[-1,1]$. The case $p = 2$ was independently investigated by Dörfler (1987) and GVM [25,30]. GVM obtained the best constant in terms of eigenvalues of a five-diagonal matrix and reduced it to two sequences of certain orthogonal polynomials.

Guessab and Milovanović [48, 50] considered a weighted L^2-analog of Bernstein's inequality, which can be stated as

$$\|\sqrt{1-t^2}\,P'(t)\|_\infty \leqslant n\|P\|_\infty.$$

Using the norm $\|f\|^2 = (f,f)$ with a classical weight $w(t)$, they determined the best constant $C_{n,m}(w)$ $(1 \leqslant m \leqslant n)$ in the inequality

$$\|A^{m/2}P^{(m)}\| \leqslant C_{n,m}(w)\|P\|, \tag{7}$$

where A is defined above, namely, for all polynomials $P \in \mathcal{P}$ there holds (7), with the best constant

$$C_{n,m}(w) = \sqrt{\lambda_{n,0}\lambda_{n,1}\cdots\lambda_{n,m-1}},$$

where $\lambda_{n,k} = -(n-k)\left(\frac{1}{2}(m+k-1)A''(0) + B'(0)\right)$. Equality holds in (7) if and only if P is a multiple of the classical polynomial $Q_n(t)$ orthogonal with respect to the weight function $w(t)$.

Agarwal and Milovanović [40] proved that for all polynomials $P \in \mathcal{P}$,

$$(2\lambda_n + B'(0))\,\|\sqrt{A}\,P'\|^2 \leqslant \lambda_n^2\|P\|^2 + \|AP''\|^2,$$

with equality again if and only if P is a multiple of the classical polynomial $Q_n(t)$ orthogonal with respect to the weight function $w(t)$. Here, $\lambda_n = \lambda_{n,0}$.

5.2.1 L^2–Inequalities with Laguerre Measure for Nonnegative Polynomials

Varma (1981) investigated the problem of determining the best constant $C_n(\alpha)$ in the L^2–inequality

$$\|P'\|^2 \leqslant C_n(\alpha)\|P\|^2$$

for polynomials with nonnegative coefficients, with respect to the generalized Laguerre weight function $w(t) = t^\alpha \mathrm{e}^{-t}$ $(\alpha > -1)$ on $[0, \infty)$. If P_n is an algebraic polynomial of degree n with nonnegative coefficients, then for $\alpha \geqslant \frac{1}{2}(\sqrt{5}-1)$ Varma proved that

$$\int_0^\infty (P_n'(t))^2 w(t)\,\mathrm{d}t \leqslant \frac{n^2}{(2n+\alpha)(2n+\alpha-1)} \int_0^\infty (P_n(t))^2 w(t)\,\mathrm{d}t.$$

The equality holds for $P_n(t) = t^n$. For $0 \leqslant \alpha \leqslant \frac{1}{2}$

$$\int_0^\infty (P_n'(t))^2 w(t)\,\mathrm{d}t \leqslant \frac{1}{(2+\alpha)(1+\alpha)} \int_0^\infty (P_n(t))^2 w(t)\,\mathrm{d}t. \tag{8}$$

Moreover, (8) is best possible in the sense that for $P_n(t) = t^n + \lambda t$ the expression on the left-hand side of (8) can be made arbitrarily close to the one on the right-hand side if λ is sufficienlty large. The ranges $\alpha < 0$ and $\frac{1}{2} < \alpha < \frac{1}{2}(\sqrt{5}-1)$ are not covered by Varma's results. This gap was filled by GVM [16]. He determined

$$C_n(\alpha) = \sup_{P \in W_n} \frac{\|P'\|^2}{\|P\|^2} \tag{9}$$

for all $\alpha \in (-1, \infty)$, where

$$W_n := \left\{ P \;\middle|\; P(t) = \sum_{\nu=0}^{n} a_\nu t^\nu,\ a_0 \geqslant 0,\ a_1 \geqslant 0,\ \ldots,\ a_{n-1} \geqslant 0,\ a_n > 0 \right\}.$$

Namely, taking

$$\alpha_n = \frac{1}{2(n+1)}\left(\sqrt{17n^2+2n+1} - 3n + 1\right), \quad n = 1,2,\ldots,$$

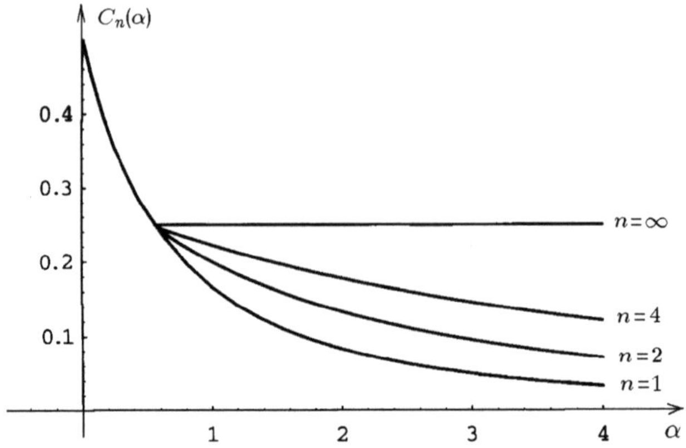

Fig. 1 The constant $C_n(\alpha)$ for $n = 1,2,4$ and $n = \infty$

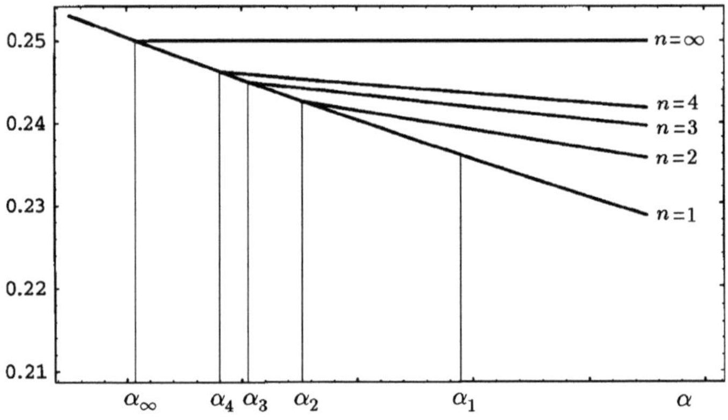

Fig. 2 Enlarged nontrivial part in Fig. 1

GVM [16] constructs $C_n(\alpha)$ in (9) as

$$C_n(\alpha) = \begin{cases} \dfrac{1}{(2+\alpha)(1+\alpha)}, & -1 \le \alpha \le \alpha_n, \\ \dfrac{n^2}{(2n+\alpha)(2n+\alpha-1)}, & \alpha_n \le \alpha < \infty, \end{cases}$$

which is presented in Figs. 1 and 2.

5.2.2 Extremal Problems for the Lorentz Class of Polynomials

Extremal problems for the Lorentz class of polynomials with respect to the Jacobi weight $w(t) = (1-t)^\alpha (1+t)^\beta$, $\alpha, \beta > -1$ were investigated by Milovanović and Petković [32]. Let L_n be the Lorentz class of polynomials

$$P(t) := \sum_{\nu=0}^{n} b_\nu (1-t)^\nu (1+t)^{n-\nu} \quad (b_\nu \geqslant 0 \text{ for } \nu = 0,1,\ldots,n).$$

In their work, they determined the best constant

$$C_n^{(k)}(\alpha,\beta) := \sup \frac{\|P'\|^2}{\|P\|^2},$$

where the supremum is over the polynomials from L_n for which $P^{(i-1)}(\pm 1) = 0$ for $i = 1,\ldots,k$. They proved that

$$C_n^{(0)}(\alpha,\beta) := \frac{n^2(2n+\alpha+\beta)(2n+\alpha+\beta+1)}{4(2n+\lambda)(2n+\lambda-1)},$$

where $\lambda = \min(\alpha,\beta)$. In particular,

$$C_n^{(0)}(1,1) := \frac{n(n+1)(2n+3)}{4(2n+1)},$$

a result already obtained by Erdős and Varma (1986).

5.3 Orthogonal Polynomials on Radial Rays

A topic which had absorbed GVM for a long time is the "introduction of new concepts of orthogonality". For example, one can mention orthogonality on the semicircle, on a circular arc, on the radial rays, orthogonality of Müntz polynomials, multiple orthogonality, s- and σ-orthogonality, etc.

Milovanović [66–68,70,92] studied orthogonal polynomials on radial rays in the complex plane and gave several applications of his results. Let $M \in \mathbb{N}$ and

$$a_s > 0, \quad s = 1,2,\ldots,M, \quad 0 \leqslant \theta_1 < \theta_2 < \cdots < \theta_M < 2\pi.$$

Consider points $z_s = a_s \varepsilon_s, \varepsilon_s = \mathrm{e}^{\mathrm{i}\theta_s}$ $(s = 1,2,\ldots,M)$ and define the inner product

$$(f,g) := \sum_{s=1}^{M} \mathrm{e}^{-\mathrm{i}\theta_s} \int_{\ell_s} f(z)\overline{g(z)}|w_s(z)|\,\mathrm{d}z,$$

where ℓ_s are the radial rays in the complex plane which connect the origin and the points z_s, while $w_s(z)$ are suitable complex weights. The case $M = 6$ is shown in Fig. 3. Precisely, $\omega_s(x) = |w_s(z)| = |w_s(x\varepsilon_s)|$ are weight functions on $(0,a_s)$. One can write this as

$$(f,g) := \sum_{s=1}^{M} \int_0^{a_s} f(x\varepsilon_s)\overline{g(x\varepsilon_s)}\omega_s(x)\,\mathrm{d}x,$$

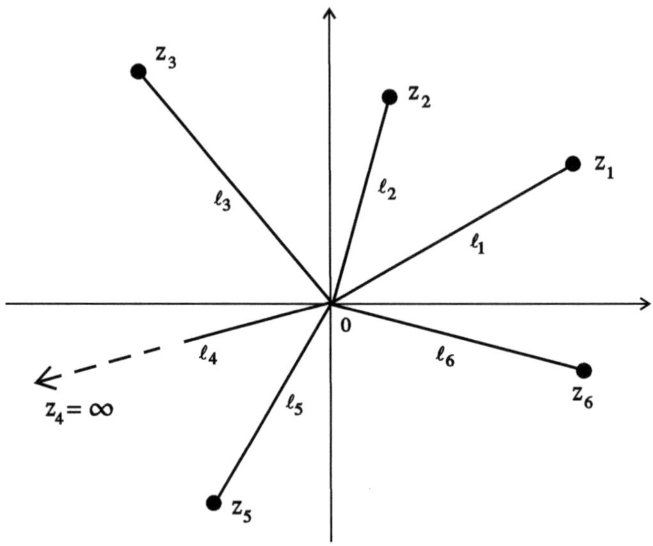

Fig. 3 The rays in the complex plane ($M = 6$)

and for $M = 2, \theta_1 = 0, \theta_2 = \pi$,

$$(f,g) = \int_0^{a_1} f(x)\overline{g(x)}\omega_1(x)\,\mathrm{d}x + \int_0^{a_2} f(-x)\overline{g(-x)}\omega_2(x)\,\mathrm{d}x,$$

that is,

$$(f,g) = \int_a^b f(x)\overline{g(x)}\omega(x)\,\mathrm{d}x$$

with $a = -a_2,\ b = a_1$ and

$$\omega(x) = \begin{cases} \omega_1(x), & \text{if } 0 < x < b, \\ \omega_2(-x), & \text{if } a < x < 0. \end{cases}$$

GVM proves the existence and uniqueness of orthogonal polynomials on radial rays $\pi_N(z)$ $(N = 0, 1, 2, \ldots)$. He considers the numerical construction of these polynomials, recurrence relations, and connections with standard polynomials orthogonal on $\mathbb{R}$.

Some interesting classes of orthogonal polynomials with rays of equal lengths, distributed equidistantly in the complex plane, and with same weights on the rays, are also considered.

Two applications of these polynomials in physics and electrostatics are discussed. The first one is a physical problem connected with a nonlinear diffusion equation, where the equations for the dispersion of a buoyant contaminant can be approximated by the Erdogan–Chatwin equation

$$\partial_t c = \partial_y \Big\{ \Big[D_0 + \big(\partial_y c\big)^2 D_2 \Big] \partial_y c \Big\},$$

where D_0 is the dispersion coefficient appropriate for neutrally buoyant contaminants, and the D_2 term represents the increased rate of dispersion associated with the buoyancy-driven currents. Smith (1982) obtained analytic expressions for the similarity solutions of this equation in the limit of strong nonlinearity ($D_0 = 0$), i.e.,

$$\partial_t c = D_2 \partial_y \left[(\partial_y c)^3 \right],$$

both for a concentration jump and for a finite discharge. He also investigated the asymptotic stability of these solutions. It is interesting that the stability analysis for the finite discharge involves a family of orthogonal polynomials $Y_N(z)$, such that

$$(1 - z^4) Y_N'' - 6z^3 Y_N' + N(N+5) z^2 Y_N = 0.$$

The degree N is restricted to the values $0, 1, 4, 5, 8, 9, \ldots$, so that the first few (monic) polynomials are:

$$1,\ z,\ z^4 - \frac{1}{3},\ z^5 - \frac{5}{11} z,\ z^8 - \frac{14}{17} z^4 + \frac{21}{221},\ z^9 - \frac{18}{19} z^5 + \frac{3}{19} z,\ \ldots .$$

These polynomials are a special case of Milovanović's polynomials orthogonal on the radial rays in the complex plane for $M = 4$ and

$$\omega(x) = (1 - x^4)^{1/2} x^2.$$

The second application is an electrostatic interpretation of the zeros of polynomials $\pi_N(z)$. It is a nontrivial generalization of the first electrostatic interpretation of the zeros of Jacobi polynomials given by Stieltjes in 1885.

Namely, an electrostatic system of M positive point charges all of strength q, which are placed at fixed points ξ_s given by

$$\xi_s = \exp\left(\frac{2(s-1)\pi \mathrm{i}}{M} \right) \quad (s = 1, 2, \ldots, M)$$

and a charge of strength p ($> -(M-1)/2$) at the origin $z = 0$, as well as N positive free unit charges, positioned at $z_1, z_2, \ldots, z_N$, is in electrostatic equilibrium if these points z_k are zeros of the polynomial $\pi_N(z)$ orthogonal with respect to the inner product

$$(f, g) = \int_0^1 \left(\sum_{s=1}^{M} f(x\varepsilon_s) \overline{g(x\varepsilon_s)} \right) \omega(x)\, \mathrm{d}x,$$

with the weight function

$$\omega(x) = (1 - x^M)^{2q-1} x^{M+2(p-1)}.$$

This polynomial can be expressed in terms of the monic Jacobi polynomials

$$\pi_N(z) = 2^{-n} z^{\nu} \hat{P}_n^{(2q-1,(2p+2\nu-1)/M)}(2z^M - 1),$$

where $N = Mn + \nu$, $n = [N/M]$.

6 GVM and Interpolation Processes

We conclude by giving an account of the recent comprehensive monograph "*Interpolation processes: Basic Theory and Applications*", written jointly by GVM and Giuseppe Mastroianni. This work is a crowning achievement of GVM's work in this expanding field. Interpolation of functions is one of the basic parts of Approximation Theory. There are many books on approximation theory, yet only a few of them are devoted only to interpolation processes. The classical books on interpolation address numerous negative results, i.e., results on divergent interpolation processes, usually constructed over some equidistant system of nodes. The new book of GVM and Mastroianni deals mainly with new results on convergent interpolation processes in uniform norm, for algebraic and trigonometric polynomials, not yet published in other textbooks and monographs on approximation theory and numerical mathematics. Basic tools in this field (orthogonal polynomials, moduli of smoothness, K-functionals, etc.), as well as some selected applications in numerical integration, integral equations, moment-preserving approximation, and summation of slowly convergent series are also given.

The first chapter provides an account of basic facts on approximation by algebraic and trigonometric polynomials introducing the most important concepts regarding the approximation of functions.

The second chapter is devoted to orthogonal polynomials on the real line and weighted polynomial approximation. For polynomials orthogonal on the real line they give the basic properties and introduce and discuss the associated polynomials, functions of the second kind, Stieltjes polynomials, as well as the Christoffel functions and numbers. The *classical orthogonal polynomials* as the most important class of orthogonal polynomials on the real line are treated in Sect. 2.3, and new results on *nonclassical orthogonal polynomials*, including methods for their numerical construction, are studied in Sect. 2.4. Introducing the weighted functional spaces, moduli of smoothness and K-functionals, the weighted best polynomial approximations on $(-1,1)$, $(0,+\infty)$ and $(-\infty,+\infty)$ are treated in Sect. 2.5, as well as the weighted polynomial approximation of functions having interior isolated singularities.

Trigonometric approximation is considered in Chap. 3. Approximations by sums of Fourier and Fejér and de la Vallée Poussin means are presented. Their discrete versions and the Lagrange trigonometric operator are also investigated. As a basic tool for studying approximating properties of the Lagrange and de la Vallée Poussin operators they consider the so-called Marcinkiewicz inequalities. Besides the uniform approximation, they also investigate the Lagrange interpolation error in the L^p-norm $(1 < p < +\infty)$ and give some estimates in the L^1-Sobolev norm, including some weighted versions.

Chapter 4 treats algebraic interpolation processes $\{L_n(\mathcal{X})\}_{n\in\mathbb{N}}$ in the uniform norm, starting with the so-called optimal system of nodes $\mathcal{X}$, which provides Lebesgue constants of order $\log n$ and the convergence of the corresponding interpolation processes. Moreover, the error of these approximations is near to the error of the best uniform approximation. Beside two classical examples of the well-known optimal system of nodes (zeros of the Jacobi polynomials $P_n^{(\alpha,\beta)}(x)$ $(-1 < \alpha, \beta \leqslant -1/2)$ and the so-called Clenshaw's abscissas), they introduce more general results for constructing interpolation processes at nodes with an *arc sine distribution* having Lebesgue constants of order $\log n$.

The final chapter provides some selected applications in numerical analysis. In the first section on quadrature formulae, they present some special Newton–Cotes rules, the Gauss–Christoffel, Gauss–Radau, and Gauss–Lobatto quadratures, the so-called product integration rules, as well as a method for the numerical integration of periodic functions on the real line with respect to a rational weight function. Also, they include the error estimates of Gaussian rules for some classes of functions. The second section is devoted to methods for solving the Fredholm integral equations of the second kind. The methods are based on the so-called *Approximation and Polynomial Interpolation Theory* and lead to the construction of polynomial sequences converging to the exact solutions in some weighted uniform norms. In the third section, they consider some kinds of moment-preserving approximations by polynomials and splines. In the last section of this chapter, the authors consider two recent methods of summation of slowly convergent series based on integral representations of series and an application of Gaussian quadratures. In the first method, they assume that the general term of the series is expressible in terms of the Laplace transform (or its derivative) of a known function. It leads to the Gaussian quadrature formulas with respect to the Einstein and Fermi weight functions on $(0,+\infty)$. The second method is based on contour integration over a rectangle in the complex plane, thus reducing the summation of a slowly convergent series to a problem of Gaussian quadrature on $(0,+\infty)$ with respect to some hyperbolic weight functions.

7 The Selected Bibliography of GVM

What follows is the selected bibliography of GVM [1– 150], covering all periods of his scientific activity. The full bibliography can be found at the site:

http://gauss-megatrend.edu.rs/publ0.html

References

1. D.Đ. Tošić, G.V. Milovanović: *An application of Newton's method to simultaneous determination of zeros of a polynomial*, Univ. Beograd. Publ. Elektrotehn. Fak. Ser. Mat. Fiz. No 412–No 460 (1973), 175–177.
2. G.V. Milovanović: *A method to accelerate iterative processes in Banach space*, Univ. Beograd. Publ. Elektrotehn. Fak. Ser. Mat. Fiz. No 461–No 479 (1974), 67–71.
3. R.Ž. Đorđević, G.V. Milovanović: *A generalization of E. Landau's theorem*, Univ. Beograd. Publ. Elektrotehn. Fak. Ser. Mat. Fiz. No 498–No 541 (1975), 97–106.
4. G.V. Milovanović: *On some integral inequalities*, Univ. Beograd. Publ. Elektrotehn. Fak. Ser. Mat. Fiz. No 498–No 541 (1975), 119–124.

5. P.M. Vasić, G.V. Milovanović: *On an inequality of Iyengar*, Univ. Beograd. Publ. Elektrotehn. Fak. Ser. Mat. Fiz. No 544–No 576 (1976), 18–24.
6. G.V. Milovanović, J.E. Pečarić: *On generalization of the inequality of A. Ostrowski and some related applications*, Univ. Beograd. Publ. Elektrotehn. Fak. Ser. Mat. Fiz. No 544–No 576 (1976), 155–158.
7. G.V. Milovanović, J.E. Pečarić: *Some considerations of Iyengar's inequality and some related applications*, Univ. Beograd. Publ. Elektrotehn. Fak. Ser. Mat. Fiz. No 544–No 576 (1976), 166–170.
8. G.V. Milovanović: *On some functional inequalities*, Univ. Beograd. Publ. Elektrotehn. Fak. Ser. Mat. Fiz. No 599 (1977), 1–59 (Serbian, English summary).
9. G.V. Milovanović, I.Ž. Milovanović: *On discrete inequalities of Wirtinger's type*, J. Math. Anal. Appl. **88** (1982), 378–387.
10. G.V. Milovanović, R.Ž. Đorđević: *A generalization of modified BirkhoffYoung quadrature formula*, Univ. Beograd. Publ. Elektrotehn. Fak. Ser. Mat. Fiz. No 735–No 762 (1982), 130–134.
11. G.V. Milovanović, M.S. Petković: *On the convergence order of a modified method for simultaneous finding polynomial zeros*, Computing **30** (1983), 171–178.
12. M.S. Petković, G.V. Milovanović: *A note on some improvements of the simultaneous method for determination of polynomial zeros*, J. Comput. Appl. Math. **9** (1983), 65–69.
13. G.V. Milovanović, I.Ž. Milovanović: *Some discrete inequalities of Opial's type*, Acta Sci. Math. (Szeged) **47** (1984), 413–417.
14. W. Gautschi, G.V. Milovanović: *On a class of complex polynomials having all zeros in a half disc*, In: Numerical methods and approximation theory (Niš, 1984), pp. 49–53, Univ. Niš, Niš, 1984.
15. W. Gautschi, G.V. Milovanović: *Polynomials orthogonal on the semicircle*, Rend. Sem. Mat. Univ. Politec. Torino, Special Issue (July 1985), 179–185.
16. G.V. Milovanović: *An extremal problem for polynomials with nonnegative coefficients*, Proc. Amer. Math. Soc. **94** (1985), 423–426.
17. W. Gautschi, G.V. Milovanović: *Gaussian quadrature involving Einstein and Fermi functions with an application to summation of series*, Math. Comp. **44** (1985), 177–190; Supplement to Gaussian quadrature involving Einstein and Fermi functions with an application to summation of series, Math. Comp. **44** (1985), S1–S11.
18. G.V. Milovanović, M.S. Petković: *On computational efficiency of the iterative methods for simultaneous approximation of polynomial zeros*, ACM Trans. Math. Software **12** (1986), 295–306.
19. G.V. Milovanović, B.P. Acharya, T.N. Pattnaik: *On numerical evaluation of double integrals of an analytic function of two complex variables*, BIT **26** (1986), 521–526.
20. M.S. Petković, G.V. Milovanović, L.V. Stefanović: *Some higher-order methods for the simultaneous approximation of multiple polynomial zeros*, Comput. Math. Appl. **12A** (1986), 951–962.
21. W. Gautschi, G.V. Milovanović: *Spline approximations to spherically symmetric distributions*, Numer. Math. **49** (1986), 111–121.
22. W. Gautschi, G.V. Milovanović: *Polynomials orthogonal on the semicircle*, J. Approx. Theory **46** (1986), 230–250.
23. G.V. Milovanović, I.Ž. Milovanović: *A generalization of a result of A. Meir for nondecreasing sequences*, Rocky Mountain J. Math. **16** (1986), 237–239.
24. G.V. Milovanović, S. Wrigge: *Least squares approximation with constraints*, Math. Comp. **46** (1986), 551–565.
25. G.V. Milovanović: *Various extremal problems of Markov's type for algebraic polynomials*, Facta Univ. Ser. Math. Inform. **2** (1987), 7–28.
26. W. Gautschi, M.A Kovačević, G.V. Milovanović: *The numerical evaluation of singular integrals with coth-kernel*, BIT **27** (1987), 389–402.
27. G.V. Milovanović, G. Djordjević: *On some properties of Humbert's polynomials*, Fibonacci Quart. **25** (1987), 356–360.

28. W. Gautschi, H. Landau, G.V. Milovanović: *Polynomials orthogonal on the semicircle II*, Constr. Approx. **3** (1987), 389–404.
29. M. Frontini, W. Gautschi, G.V. Milovanović: *Moment-preserving spline approximation on finite intervals*, Numer. Math. **50** (1987), 503–518.
30. G.V. Milovanović: *Some applications of the polynomials orthogonal on the semicircle*, In: Numerical Methods (Miskolc, 1986), pp. 625–634, Colloquia Mathematica Societatis Janos Bolyai, Vol. 50, North-Holland, Amsterdam-New York, 1987.
31. G.V. Milovanović: *Construction of s-orthogonal polynomials and Turán quadrature formulae*, In: Numerical methods and approximation theory III (Niš, 1987), pp. 311–328, Univ. Niš, Niš, 1988.
32. G.V. Milovanović, M.S. Petković: *Extremal problems for Lorentz classes of nonnegative polynomials in L_2 metric with Jacobi weight*, Proc. Am. Math. Soc. **102** (1988), 283–289.
33. G.V. Milovanović, M.A. Kovačević: *Moment-preserving spline approximation and Turán quadratures*, In: Numerical mathematics (Singapore, 1988), pp. 357–365, ISNM, Vol. 86, Birkhäuser, Basel, 1988.
34. G.V. Milovanović: *Complex orthogonality on the semicircle with respect to Gegenbauer weight: Theory and applications*, In: Topics in Mathematical Analysis (Th.M. Rassias, ed.), pp. 695–722, Ser. Pure Math., 11, World Sci. Publishing, Teaneck, NJ, 1989.
35. G.V. Milovanović, P.M. Rajković: *Geronimus concept of orthogonality for polynomials orthogonal on a circular arc*, Rend. Mat. Appl. **10** (1990), 383–390.
36. R.P. Agarwal, G.V. Milovanović: *On an inequality of Bogar and Gustafson*, J. Math. Anal. Appl. **146** (1990), 207–216.
37. G.V. Milovanović, D.S. Mitrinović, Th.M. Rassias: *On some extremal problems for algebraic polynomials in L^r norm*, In: Generalized Functions and Convergence (Katowice, 1988), pp. 343–354, World Scientific, Singapore, 1990.
38. F. Calió, M. Frontini, G.V. Milovanović: *Numerical differentiation of analytic functions using quadratures on the semicircle*, Comput. Math. Appl. **22** (1991), 99–106.
39. G.V. Milovanović, Lj.M. Kocić: *Integral spline operators in CAGD*, Atti Sem. Mat. Fis. Univ. Modena **39** (1991), 433–454.
40. R.P. Agarwal, G.V. Milovanović: *A characterization of the classical orthogonal polynomials*, Progress in Approximation Theory (P. Nevai, A. Pinkus, eds.), pp. 1–4, Academic, New York, 1991.
41. L. Gori, M.L. Lo Cascio, G.V. Milovanović: *The s-orthogonal polynomials: a method of construction*, In: IMACS Annals on Computing and Applied Mathematics, Vol. 9: Orthogonal Polynomials and Their Applications (C. Brezinski, L. Gori, A. Ronveaux, eds.), pp. 281–285, Baltzer, Basel, 1991.
42. G.V. Milovanović, Th.M. Rassias: *Inequalities connected with trigonometric sums*, In: Constantin Carathéodory: An International Tribute, Vol. II (Th.M. Rassias, ed.), pp. 875–941, World Sci. Publishing, Teaneck, NJ, 1991.
43. G.V. Milovanović: *On polynomials orthogonal on the semicircle and applications*, J. Comput. Appl. Math. **49** (1993), 193–199.
44. G.V. Milovanović: *Recurrence relation for a class of polynomials associated by the generalized Hermite polynomials*, Publ. Inst. Math. (Beograd) (N.S.) **54** (68) (1993), 35–37.
45. G.V. Milovanović, D.S. Mitrinović, Th.M. Rassias: *On some Turán's extremal problems for algebraic polynomials*, In: Topics in Polynomials of One and Several Variables and Their Applications: A Mathematical Legaci of P.L. Chebyshev (1821–1894) (Th.M. Rassias, H.M. Srivastava, A. Yanushauskas, eds.), pp. 403–433, World Scientific, Singapore, 1993.
46. G.V. Milovanović, I.Ž. Milovanović, L.Z. Marinković: *Extremal problems for polynomials and their coefficients*, In: Topics in Polynomials of One and Several Variables and Their Applications: A Mathematical Legaci of P.L. Chebyshev (1821–1894) (Th.M. Rassias, H.M. Srivastava, A. Yanushauskas, eds.), pp. 435–455, World Scientific, Singapore, 1993.

47. G.V. MILOVANOVIĆ, D.S. MITRINOVIĆ, TH.M. RASSIAS: *Topics in Polynomials: Extremal Problems, Inequalities, Zeros*, World Scientific Publ. Co., Singapore, 1994, xiv+822 pp.
48. A. GUESSAB, G.V. MILOVANOVIĆ: *Extremal problems of Markov's type for some differential operators*, Rocky Mt. J. Math. **24** (1994), 1431–1438.
49. G.V. MILOVANOVIĆ, P.M. RAJKOVIĆ: *On polynomials orthogonal on a circular arc*, J. Comput. Appl. Math. **51** (1994), 1–13.
50. A. GUESSAB, G.V. MILOVANOVIĆ: *Weighted L_2-analogues of Bernstein's inequality and classical orthogonal polynomials*, J. Math. Anal. Appl. **82** (1994), 244–249.
51. G.V. MILOVANOVIĆ, A. GUESSAB: *An estimate for coefficients of polynomials in L_2 norm*, Proc. Am. Math. Soc. **120** (1994), 165–171.
52. G.V. MILOVANOVIĆ: *Summation of series and Gaussian quadratures*, In: Approximation and Computation (R.V.M. Zahar, ed.), pp. 459–475, ISNM Vol. 119, Birkhäuser, Basel, 1994.
53. G.V. MILOVANOVIĆ: *Summation of slowly convergent series via quadratures*, In: Advances in Numerical Methods and Applications (I.T. Dimov, Bl. Sendov, P.S. Vassilevski, eds.), pp. 154–161, World Scientific, Singapore, 1994.
54. G.V. MILOVANOVIĆ: *Extremal problems for polynomials: old and new results*, In: Proceedings Open Problems in Approximation Theory (B. Bojanov, ed.), pp. 138–155, SCT Publishing, Singapore, 1994.
55. G.V. MILOVANOVIĆ: *Some nonstandard types of orthogonality (A survey)*, Filomat **9** (1995), 517–542.
56. G.V. MILOVANOVIĆ: *Expansions of the Kurepa function*, Publ. Inst. Math. (Beograd) (N.S.) **57** (71) (1995), 81–90.
57. G.V. MILOVANOVIĆ: *Summation of series and Gaussian quadratures, II*, Numer. Algorithms **10** (1995), 127–136.
58. G.V. MILOVANOVIĆ: *Generalized Hermite polynomials on the radial rays in the complex plane*, In: Theory of Functions and Applications, Collection of Works Dedicated to the Memory of M.M. Djrbashian (H.B. Nersessian, ed.), pp. 125–129, Yerevan, Louys Publishing House, 1995.
59. M.A. KOVAČEVIĆ, G.V. MILOVANOVIĆ: *Spline approximation and generalized Turán quadratures*, Portugal. Math. **53** (1996), 355–366.
60. G.V. MILOVANOVIĆ: *A sequence of Kurepa's functions*, Scientific Review, Ser. Sci. Eng. **19–20** (1996), 137–146.
61. W. GAUTSCHI, G.V. MILOVANOVIĆ: *S-orthogonality and construction of Gauss-Turán type quadrature formulae*, J. Comput. Appl. Math. **86** (1997), 205–218.
62. G.V. MILOVANOVIĆ, I.Ž. MILOVANOVIĆ: *Discrete inequalities of Wirtinger's type for higher differences*, J. Inequal. Appl. **1** (1997), 301–310.
63. G.V. MILOVANOVIĆ, A.K. VARMA: *On Birkhoff* $(0,3)$ *and* $(0,4)$ *quadrature formulae*, Numer. Funct. Anal. Optim. **18** (1997), 427–433.
64. LJ.M. KOCIĆ, G.V. MILOVANOVIĆ: *Shape preserving approximations by polynomials and splines*, Comput. Math. Appl. **33** (1997), 59–97.
65. G.V. MILOVANOVIĆ, P.S. STANIMIROVIĆ: *On Moore-Penrose inverse of block matrices and full-rank factorization*, Publ. Inst. Math. (Beograd) (N.S.) **62** (76) (1997), 26–40.
66. G.V. MILOVANOVIĆ: *A class of orthogonal polynomials on the radial rays in the complex plane*, J. Math. Anal. Appl. **206** (1997), 121–139.
67. G.V. MILOVANOVIĆ: *Integral inequalities for algebraic polynomials*, In: General Inequalities 7 (C. Bandle, ed.), pp. 17–25, ISNM Vol. 123, Birkhäuser, Basel, 1997.
68. G.V. MILOVANOVIĆ: *Orthogonal polynomial systems and some applications*, In: Inner Product Spaces and Applications (Th.M. Rassias, ed.), pp. 115–182, Pitman Res. Notes Math. Ser. 376, Longman, Harlow, 1997.
69. B. DANKOVIĆ, G.V. MILOVANOVIĆ, S.LJ. RANČIĆ: *Malmquist and Müntz orthogonal systems and applications*, In: Inner Product Spaces and Applications (Th.M. Rassias, ed.), pp. 22–41, Pitman Res. Notes Math. Ser. 376, Longman, Harlow, 1997.

70. G.V. MILOVANOVIĆ: *S-orthogonality and generalized Turán quadratures: Construction and applications*, In: Approximation and Optimization, Vol. I (Cluj-Napoca, 1996) (D.D. Stancu, Gh. Coman, W.W. Breckner, P. Blaga, eds.), pp. 91–106, Transilvania, Cluj-Napoca, Romania, 1997.
71. G.V. MILOVANOVIĆ: *Orthogonal polynomials on the radial rays and an electrostatic interpretation of zeros*, Publ. Inst. Math. (Beograd) (N.S.) **64** (78) (1998), 53–68.
72. G.V. MILOVANOVIĆ, B. DANKOVIĆ, S.LJ. RANČIĆ: *Some Müntz orthogonal systems*, J. Comput. Appl. Math. **99** (1998), 299–310.
73. G.V. MILOVANOVIĆ: *Numerical calculation of integrals involving oscillatory and singular kernels and some applications of quadratures*, Comput. Math. Appl. **36** (1998), 19–39.
74. G.V. MILOVANOVIĆ: *Life and inequalities: D. S. Mitrinović (1908–1996)*, In: Recent Progress in Inequalities (G.V. Milovanović, ed.), Mathematics and Its Applications, Vol. 430, pp. 1–10, Kluwer, Dordrecht, 1998.
75. G.V. MILOVANOVIĆ, I.Ž. MILOVANOVIĆ: *Discrete inequalities of Wirtinger's type*, In: Recent Progress in Inequalities (G.V. Milovanovic, ed.), Mathematics and Its Applications, Vol. 430, pp. 289–308, Kluwer, Dordrecht, 1998.
76. G.V. MILOVANOVIĆ: *Extremal problems for restricted polynomial classes in L_p norm*, In: Approximation Theory: In Memory of A.K. Varma (N.K. Govil, R.N. Mohapatra, Z. Nashed, A. Sharma, J. Szabados, eds.), Monogr. Textbooks Pure Appl. Math., Vol. 212, pp. 405–432, Marcel Dekker, New York, 1998.
77. G.V. MILOVANOVIĆ, TH.M. RASSIAS: *New developments on Turán's extremal problems for polynomials*, In: Approximation Theory: In Memory of A.K. Varma (N.K. Govil, R.N. Mohapatra, Z. Nashed, A. Sharma, J. Szabados, eds.), Monogr. Textbooks Pure Appl. Math., Vol. 212, pp. 433–447, Marcel Dekker, New York, 1998.
78. G.V. MILOVANOVIĆ: *Müntz orthogonal polynomials and their numerical evaluation*, In: Applications and Computation of Orthogonal Polynomials (W. Gautschi, G.H. Golub, and G. Opfer, eds.), pp. 179–202, ISNM, Vol. 131, Birkhäuser, Basel, 1999.
79. G.V. MILOVANOVIĆ: *A note to the paper "On a new class of polynomials" by C.L. Parihar and S.K. Harne*, J. Indian Acad. Math. **21** (1999), 277–278.
80. G.V. MILOVANOVIĆ, D.DJ. TOŠIĆ: *An extremal problem for the length of algebraic polynomials*, Univ. Beograd. Publ. Elektrotehn. Fak. Ser. Mat. **10** (1999), 31–36.
81. G.V. MILOVANOVIĆ: *Extremal problems and inequalities of Markov-Bernstein type for polynomials*, In: Analytic and Geometric Inequalities and Applications (Th.M. Rassias, H.M. Srivastava, eds.), Mathematics and Its Applications, Vol. 478, pp. 245–264, Kluwer, Dordrecht, 1999.
82. X. LI, I. GUTMAN, G.V. MILOVANOVIĆ: *The β-polynomials of complete graphs are real*, Publ. Inst. Math. (Beograd) (N.S.) **67** (81) (2000), 1–6.
83. G.V. MILOVANOVIĆ: *Dragoslav S. Mitrinović (1908–1995)*, In: Lives and Work of the Serbian Scientists, Biographies and Bibliographies, Vol. 6 (M. Sarić, ed.), pp. 519–581, Serbian Academy of Sciences and Arts, Belgrade, 2000 (Serbian).
84. G.V. MILOVANOVIĆ: *Some generalized orthogonal systems and their connections*, In: Proceedings of the Symposium "Contemporary Mathematics" (Belgrade, 1998) (N. Bokan, ed.), pp. 181–200, Faculty of Mathematics, University of Belgrade, 2000.
85. G.V. MILOVANOVIĆ, TH.M. RASSIAS: *Inequalities for polynomial zeros*, In: Survey on Classical Inequalities (Th. M. Rassias, ed.), Mathematics and Its Applications, Vol. 517, pp. 165–202, Kluwer, Dordrecht, 2000.
86. G.V. MILOVANOVIĆ, TH.M. RASSIAS: *Distribution of zeros and inequalities for zeros of algebraic polynomials*, In: Functional Equations and Inequalities (Th. M. Rassias, ed.), Mathematics and Its Applications, Vol. 518, pp. 171–204, Kluwer, Dordrecht, 2000.
87. G.V. MILOVANOVIĆ, A.S. CVETKOVIĆ: *Note on a construction of weights in Gauss-type quadrature rule*, Facta Univ. Ser. Math. Inform. **15** (2000), 69–83.
88. G.V. MILOVANOVIĆ: *Quadrature with multiple nodes, power orthogonality, and moment-preserving spline approximation*, Numerical analysis 2000, Vol. V, Quadrature and orthogonal polynomials (W. Gautschi, F. Marcellán, and L. Reichel, eds.), J. Comput. Appl. Math. **127** (2001), 267–286.

89. G.V. MILOVANOVIĆ: *Contribution and influence of S.B. Prešić to numerical factorization of polynomials*, In: A Tribute to S. B. Prešić (Papers Celebrating his 65th Birthday), pp. 47–56, Mathematics Institut SANU, Belgrade, 2001.
90. R.P. AGARWAL, G.V. MILOVANOVIĆ: *Extremal problems, inequalities, and classical orthogonal polynomials*, Appl. Math. Comput. **128** (2002), 151–166.
91. G. MASTROIANNI, G.V. MILOVANOVIĆ: *Weighted integration of periodic functions on the real line*, Appl. Math. Comput. **128** (2002), 365–378.
92. G.V. MILOVANOVIĆ: *Orthogonal polynomials on the radial rays in the complex plane and applications*, Rend. Circ. Mat. Palermo, Serie II, Suppl. **68** (2002), 65–94.
93. G.V. MILOVANOVIĆ, M.M. SPALEVIĆ: *Quadrature formulae connected to σ-orthogonal polynomials*, J. Comput. Appl. Math. **140** (2002), 619–637.
94. G.V. MILOVANOVIĆ, A. PETOJEVIĆ: *Generalized factorial functions, numbers and polynomials*, Math. Balkanica (N.S.) **16** (2002), 113–130.
95. G.V. MILOVANOVIĆ, A.S. CVETKOVIĆ: *Numerical integration of functions with logarithmic end point singularity*, Facta Univ. Ser. Math. Inform. **17** (2002), 57–74.
96. G. MASTROIANNI, G.V. MILOVANOVIĆ: *Weighted interpolation of functions with isolated singularities*, In: Approximation Theory: A volume dedicated to Blagovest Sendov (B. Bojanov, ed.), pp. 310–341, DARBA, Sofia, 2002.
97. G.V. MILOVANOVIĆ, M.M. SPALEVIĆ: *Error bounds for Gauss-Turán quadrature formulae of analytic functions*, Math. Comp. **72** (2003), 1855–1872.
98. G.V. MILOVANOVIĆ, M. STANIĆ: *Construction of multiple orthogonal polynomials by discretized Stieltjes-Gautschi procedure and corresponding Gaussian quadrature*, Facta Univ. Ser. Math. Inform. **18** (2003), 9–29.
99. G.V. MILOVANOVIĆ, A.S. CVETKOVIĆ: *An application of little $1/q$-Jacobi polynomials to summation of certain series*, Facta Univ. Ser. Math. Inform. **18** (2003), 31–46.
100. L.N. DJORDJEVIĆ, D.M. MILOŠEVIĆ, G.V. MILOVANOVIĆ, H.M. SRIVASTAVA: *Some finite summation formulas involving multivariable hypergeometric polynomials*, Integral Transform. Spec. Funct. **14** (2003), 349–361.
101. G.V. MILOVANOVIĆ, A.S. CVETKOVIĆ: *Complex Jacobi matrices and quadrature rules*, Filomat **17** (2003), 117–134.
102. G.V. MILOVANOVIĆ, A.S. CVETKOVIĆ: *Numerical construction of the generalized Hermite polynomials*, Univ. Beograd. Publ. Elektrotehn. Fak. Ser. Mat. **14** (2003), 49–63.
103. A.S. CVETKOVIĆ, G.V. MILOVANOVIĆ: *The Mathematica Package "OrthogonalPolynomials"*, Facta Univ. Ser. Math. Inform. **19** (2004), 17–36.
104. G.V. MILOVANOVIĆ, A.S. CVETKOVIĆ: *Uniqueness and computation of Gaussian interval quadrature formula for Jacobi weight function*, Numer. Math. **99** (2004), 141–162.
105. N.P. CAKIĆ, G.V. MILOVANOVIĆ: *On generalized Stirling numbers and polynomials*, Math. Balkanica (N.S.) **18** (2004), 241–248.
106. G.V. MILOVANOVIĆ, M. STANIĆ: *Multiple orthogonal polynomials on the semicircle and corresponding quadratures of Gaussian type*, Math. Balkanica (N.S.) **18**(2004), 373–387.
107. G.V. MILOVANOVIĆ, A.S. CVETKOVIĆ: *Convergence of Gaussian quadrature rules for approximation of certain series*, East J. Approx. **10** (2004), 171–187.
108. G.V. MILOVANOVIĆ, M.M. SPALEVIĆ: *Error analysis in some Gauss-Turán-Radau and Gauss-Turán-Lobatto quadratures for analytic functions*, J. Comput. Appl. Math. **164/165** (2004), 569–586.
109. G.V. MILOVANOVIĆ, M.M. SPALEVIĆ, A.S. CVETKOVIĆ: *Calculation of Gaussian type quadratures with multiple nodes*, Math. Comput. Model. **39** (2004), 325–347.
110. G.V. MILOVANOVIĆ, A.S. CVETKOVIĆ: *Standard and non-standard quadratures of Gaussian type*, In: Approximation Theory: A volume dedicated to Borislav Bojanov (D. K. Dimitrov, G. Nikolov, R. Uluchev, eds.), pp. 186–200, Marin Drinov Academic Publishing House, Sofia, 2004.
111. G.V. MILOVANOVIĆ, A.S. CVETKOVIĆ: *Orthogonal polynomials related to the oscillatory-Chebyshev weight function*, Bull. Cl. Sci. Math. Nat. Sci. Math. **30** (2005), 47–60.
112. G.V. MILOVANOVIĆ, A.S. CVETKOVIĆ: *Gaussian type quadrature rules for Müntz systems*, SIAM J. Sci. Comput. **27** (2005), 893–913.

113. G.V. MILOVANOVIĆ, M. STANIĆ: *Multiple orthogonality and quadratures of Gaussian type*, Rend. Circ. Mat. Palermo, Serie II, Suppl. **76** (2005), 75–90.
114. G.V. MILOVANOVIĆ, A.S. CVETKOVIĆ: *Some inequalities for symmetric functions and an application to orthogonal polynomials*, J. Math. Anal. Appl. **311** (2005), 191–208.
115. G.V. MILOVANOVIĆ, M.M. SPALEVIĆ: *An error expansion for some Gauss-Turán quadratures and L^1-estimates of the remainder term*, BIT **45** (2005), 117–136.
116. G.V. MILOVANOVIĆ, A.S. CVETKOVIĆ: *Remarks on "Orthogonality of some sequences of the rational functions and Müntz polynomials"*, J. Comput. Appl. Math. **173** (2005), 383–388.
117. G.V. MILOVANOVIĆ, M.M. SPALEVIĆ: *Bounds of the error of Gauss–Turán-type quadratures*, J. Comput. Appl. Math. **178** (2005), 333–346.
118. G.V. MILOVANOVIĆ, A.S. CVETKOVIĆ: *Orthogonal polynomials and Gaussian quadrature rules related to oscillatory weight functions*, J. Comput. Appl. Math. **179** (2005), 263–287.
119. G.V. MILOVANOVIĆ, A.S. CVETKOVIĆ: *Gauss-Laguerre interval quadrature rule*, J. Comput. Appl. Math. **182** (2005), 433–446.
120. G.V. MILOVANOVIĆ, A.S. CVETKOVIĆ, M.M. MATEJIĆ: *On positive definiteness of some linear functionals*, Stud. Univ. Babeş-Bolyai Math. **51** (2006), no. 4, 157–166.
121. G.V. MILOVANOVIĆ, M.M. SPALEVIĆ: *Gauss-Turán quadratures of Kronrod type for generalized Chebyshev weight functions*, Calcolo **43** (2006), 171–195.
122. G.V. MILOVANOVIĆ, M.M. SPALEVIĆ: *Quadrature rules with multiple nodes for evaluating integrals with strong singularities*, J. Comput. Appl. Math. **189** (2006), 689–702.
123. G.V. MILOVANOVIĆ, A.S. CVETKOVIĆ: *Gauss-Radau and Gauss-Lobatto interval quadrature rules for Jacobi weight function*, Numer. Math. **102** (2006), 523–542.
124. G.V. MILOVANOVIĆ, A.S. CVETKOVIĆ, M.P. STANIĆ: *Two conjectures for integrals with oscillatory integrands*, Facta Univ. Ser. Math. Inform. **22** (2007), 77–90.
125. G. MASTROIANNI, G.V. MILOVANOVIĆ: *Polynomial approximation on unbounded intervals by Fourier sums*, Facta Univ. Ser. Math. Inform. **22** (2007), 155–164.
126. D. DOLIĆANIN, G.V. MILOVANOVIĆ, V.G. ROMANOVSKI: *Linearizablity conditions for a cubic system*, Appl. Math. Comput. **190** (2007), 937–945.
127. G.V. MILOVANOVIĆ, A.S. CVETKOVIĆ: *Gauss-Hermite interval quadrature rule*, Comput. Math. Appl. **54** (2007), 544–555.
128. G.V. MILOVANOVIĆ, A.S. CVETKOVIĆ, M.P. STANIĆ: *Gaussian quadratures for oscillatory integrands*, Appl. Math. Lett. **20** (2007), 853–860.
129. G.V. MILOVANOVIĆ, M.M. SPALEVIĆ: *Monotonicity of the error term in Gauss-Turán quadratures for analytic functions*, ANZIAM J. **48** (2007), 567–581.
130. G.V. MILOVANOVIĆ, A.S. CVETKOVIĆ: *A note on three-step iterative methods for nonlinear equations*, Stud. Univ. Babeş-Bolyai Math. **52** (2007), no. 3, 137–146.
131. G.V. MILOVANOVIĆ, A.S. CVETKOVIĆ, Z.M. MARJANOVIĆ: *Connection of semi-integer trigonometric orthogonal polynomials with Szegő polynomials*, 394–401, Lecture Notes in Comput. Sci., 4310, Springer, Berlin, 2007.
132. G.V. MILOVANOVIĆ, A.S. CVETKOVIĆ, M.P. STANIĆ: *Trigonometric orthogonal systems and quadrature formulae with maximal trigonometric degree of exactness*, 402–409, Lecture Notes in Comput. Sci., 4310, Springer, Berlin, 2007.
133. G.V. MILOVANOVIĆ, A.S. CVETKOVIĆ: *Logarithmic modification of the Jacobi weight function*, Stud. Univ. Babeş-Bolyai Math. **52** (2007), no. 4, 143–153.
134. G.V. MILOVANOVIĆ, M.M. SPALEVIĆ: *A note on the bounds of the error of Gauss-Turán-type quadratures*, J. Comput. Appl. Math. **200** (2007), 276–282.
135. G. MASTROIANNI, G.V. MILOVANOVIĆ: *Interpolation Processes – Basic Theory and Applications*, Springer Monographs in Mathematics, Springer, Berlin, 2008, xiv+444 pp.
136. G.V. MILOVANOVIĆ, A.S. CVETKOVIĆ, M.P. STANIĆ: *Trigonometric orthogonal systems and quadrature formulae*, Comput. Math. Appl. **56** (2008), 2915–2931.
137. G.V. MILOVANOVIĆ: *Quadrature processes – development and new directions*, Bull. Cl. Sci. Math. Nat. Sci. Math. **33** (2008), 11–41.
138. A.S. CVETKOVIĆ, G.V. MILOVANOVIĆ: *Positive definite solutions of some matrix equations*, Linear Algebra Appl. **429** (2008), 2401–2414.

139. G.V. Milovanović, A.S. Cvetković, M.P. Stanić: *Explicit formulas for five-term recurrence coefficients of orthogonal trigonometric polynomials of semi-integer degree*, Appl. Math. Comput. **198** (2008), 559–573.
140. G.V. Milovanović, M.M. Spalević, M.S. Pranić: *Maximum of the modulus of kernels in Gauss-Turán quadratures*, Math. Comp. **77** (2008), 985–994.
141. G.V. Milovanović, M.M. Spalević, M.S. Pranić: *On the remainder term of Gauss-Radau quadratures for analytic functions*, J. Comput. Appl. Math. **218** (2008), 281–289.
142. G.V. Milovanović, A.S. Cvetković, M.P. Stanić: *Orthogonal polynomials for modified Gegenbauer weight and corresponding quadratures*, Appl. Math. Letters **22** (2009), 1189–1194.
143. G.V. Milovanović, A.S. Cvetković, M.P. Stanić: *Quadrature formulae with multiple nodes and a maximal trigonometric degree of exactness*, Numer. Math. **112** (2009), 425–448.
144. G.V. Milovanović, A.S. Cvetković: *Numerical integration with complex Jacobi weight function*, 20–31, Lecture Notes in Comput. Sci., 5434, Springer, Berlin, 2009.
145. G.V. Milovanović, M.M. Spalević, M.S. Pranić: *Error estimates for Gauss-Turán quadratures and their Kronrod extensions*, IMA J. Numer. Anal. **29** (2009), 486–507.
146. G.V. Milovanović, M.M. Spalević, M.S. Pranić: *Error estimates for Gaussian quadratures of analytic functions*, J. Comput. Appl. Math. **233** (2009), 802–807.
147. G. Mastroianni, G.V. Milovanović: *Some numerical methods for second kind Fredholm integral equation on the real semiaxis*, IMA J. Numer. Anal. **29** (2009), 1046–1066.
148. G.V. Milovanović, A.S. Cvetković: *Nonstandard Gaussian quadrature formulae based on operatror values*, Adv. Comput. Math. **32** (2010), 431–486.
149. G. Mastroianni, G.V. Milovanović: *Well-conditioned matrices for numerical treatment of Fredholm integral equations of the second kind*, Numer. Linear Algebra Appl. **16** (2009), 995–1011.
150. G.V. Milovanović, M.M. Spalević, M.S. Pranić: *Bounds of the error of Gauss-Turn-type quadratures, II*, Appl. Numer. Math. **60** (2010), 1–9.

My Collaboration with Gradimir V. Milovanović

Walter Gautschi

1. Collaborative efforts, and joint publications resulting therefrom, are much more prevalent in the physical sciences than they are in mathematics. The reason is that research in the physical sciences usually requires team work involving a number of scientists with specialized skills, whereas research in mathematics is a much more individual and solitary enterprise. Nevertheless, even in mathematics, collaboration between different mathematicians may come about through a variety of circumstances. In my own experience, most of my collaboration originated in my attending mathematical conferences, visiting other institutions, or entertaining guests at my own institution. Another not insignificant group of collaborators comes from Ph.D. or postdoctoral students. In all these cases, an important aspect is interpersonal communication and oral exchange of ideas. Not so in the case of Gradimir! Here, collaboration started anonymously, almost ghostlike, during a process of refereeing, exactly 25 years ago. (I may be permitted to divulge information that normally is held confidential!) That is when I received a manuscript from the editor of *Mathematics of Computation* authored by some Gradimir Milovanović, a name I had never heard of before. I was asked to referee it for the journal, whose editor-in-chief I was to become shortly thereafter.

2. The topic of the manuscript looked interesting enough: It was a matter of computing integrals that frequently occur in solid state physics, e.g. the total energy of thermal vibration of a crystal lattice, which is expressible as an integral

$$\int_0^\infty f(t)\frac{t}{\mathrm{e}^t - 1}\,\mathrm{d}t, \tag{1}$$

Walter Gautschi
Department of Computer Sciences, Purdue University, West Lafayette, IN 47907-2066, USA
e-mail: wxg@cs.purdue.edu

W. Gautschi et al. (eds.), *Approximation and Computation: In Honor of Gradimir V. Milovanović*, Springer Optimization and Its Applications 42,
DOI 10.1007/978-1-4419-6594-3_2,

where $f(t)$ is related to the phonon density of states, or the crystal lattice heat capacity at constant volume, which is

$$\int_0^\infty g(t)\left(\frac{t}{e^t-1}\right)^2 dt,$$

with $g(t)=e^t f(t)$. Gradimir's idea was to compute these integrals, and similar ones with $t/(e^t-1)$ replaced by $1/(e^t+1)$, by Gaussian quadrature, treating $t/(e^t-1)$, or its square, as a weight function. This is a neat way of dealing with the poles of this function at $\pm 2k\pi i$, $k=0,1,2,\ldots$, which otherwise would adversely interfere with more standard integration techniques.

Another simple, but interesting observation of Gradimir was this: Integrals of the type (1) can be used to sum infinite series,

$$\sum_{k=1}^\infty a_k = \int_0^\infty h(t)\frac{t}{e^t-1}\,dt, \tag{2}$$

if the general term of the series, $a_k = -F'(k)$, is the (negative) derivative of the Laplace transform $F(p)=\int_0^\infty e^{-pt}h(t)\,dt$ evaluated at $p=k$ of some known function h. Since series of this kind are typically slowly convergent, the representation (2) offers a useful summation procedure, the sequence of n-point Gaussian quadrature rules, $n=1,2,3,\ldots$, applied to the integral on the right converging rapidly if h is sufficiently smooth.

This is all very nice, but how do we generate Gaussian quadrature rules with such unusual weight functions? Classically, there is an approach via orthogonal polynomials and the moments of the weight function,

$$\mu_k = \int_0^\infty t^k \frac{t}{e^t-1}\,dt, \quad k=0,1,2,\ldots.$$

In fact, this is the road Gradimir took in his manuscript, noting that the moments are expressible in terms of the Riemann zeta function,

$$\mu_k = (k+1)!\,\zeta(k+2), \quad k=0,1,2,\ldots.$$

It was at this point where I felt I had to exercise my prerogatives as a referee: I criticized the highly ill-conditioned nature of this approach and proposed more stable alternative methods that I developed just a year or two earlier. In the process, I rewrote a good portion of the manuscript and informed the editor that the manuscript so revised would be an appropriate and interesting contribution to computational mathematics. I suggested, subject to the author's approval, to publish the work as a joint paper. The approval was forthcoming, and that is how our first joint publication [6] came about.

In retrospect, Gradimir's original approach via moments has regained some viability since software has become available inthe last few years that allows generating the required orthogonal polynomials in variable-precision arithmetic. One such program is the Matlab symbolic Chebyshev algorithm `schebyshev.m` (downloadable from

http://www.cs.purdue.edu/archives/2002/wxg/codes/SOPQ.html),

which generates the required recurrence coefficients directly from the moments. Table 1 in [6], and similarly Tables 2–4 (cf. 4 [Sects. 4–5]), can thus be produced very simply using the following Matlab script:

```
syms mom ab
digits(65); dig=65;
for k=1:80
   mom(k)=vpa(gamma(vpa(k+1))*zeta(vpa(k+1)));
end
ab=schebyshev(dig,40,mom);
ab=vpa(ab,25)
```

True, it takes 65-decimal-digit arithmetic to overcome the severe ill-conditioning and obtain the first 40 recursion coefficients (in the array `ab`) of the orthogonal polynomials to 25 decimal digits. But this is a one-time shot; once these coefficients are available, one can revert to ordinary arithmetic to compute the desired Gaussian quadratures and the integrals in question.

3. In March of 1984, on a visit to Niš, I had the opportunity to finally meet my collaborator in person. He invited me to dinner at his home (my compliments to Dobrila for her culinary art!), after which Gradimir and I retired to his study, where we engaged in a most lively brainstorming session. I was astonished how well he knew earlier work of mine. He must have read my short 1984 paper on spherically symmetric distributions and their approximation by step functions matching as many moments of the distribution as possible. Because he obviously had thought about extending this type of approximation to more general spline approximations. Another idea that surfaced during this discussion was orthogonality on the semicircle and related (complex-valued) orthogonal polynomials. We agreed to pursue these topics further, which provided enough material to keep us busy for several years to come. It so happened that it was the second of these problems that received our attention first, but soon enough we worked on both problems concurrently.

4. Polynomials that are orthogonal on curves Γ in the complex plane have a long history in the case where the underlying inner product is Hermitian, i.e., of the form $(u,v) = \int_\Gamma u(z)\overline{v(z)}\,\mathrm{d}\sigma(z)$, $\mathrm{d}\sigma$ being a positive measure on Γ; see, e.g., [14, Chaps. 11 and 16]. The case most studied, by far, is the unit circle, $\Gamma = \{\mathrm{e}^{\mathrm{i}\theta}, 0 \le \theta < 2\pi\}$, which gives rise to Szegö's theory of orthogonal polynomials on the unit circle. The question we asked ourselves is this: what happens if the second factor in the inner product is not conjugated? We decided to begin our study with a prototype inner product, namely, $\mathrm{d}\sigma$ the Lebesgue measure, and Γ the upper half of the unit circle (the whole unit circle being ruled out by Cauchy's theorem). Thus, we began looking at

$$(u,v) = \int_0^\pi u(e^{i\theta})v(e^{i\theta})\,d\theta, \tag{3}$$

postponing for later the study of more general weight functions.

The moment functional associated with (3) is $\mathcal{L}z^k = (1,z^k)$, $k = 0,1,2,\ldots$; it is well known that a sequence of monic polynomials $\{\pi_n\}$ orthogonal with respect to the inner product (3) exists uniquely if the moment sequence $\{\mu_k\}$, $\mu_k = \mathcal{L}z^k$, is quasi-definite, i.e., $\Delta_n \neq 0$ for all $n \geq 1$, where

$$\Delta_n = \det \begin{bmatrix} \mu_0 & \mu_1 & \cdots & \mu_{n-1} \\ \mu_1 & \mu_2 & \cdots & \mu_n \\ \cdots & \cdots\cdots & & \cdots \\ \mu_{n-1} & \mu_n & \cdots & \mu_{2n-2} \end{bmatrix}. \tag{4}$$

We were able to prove quasi-definiteness by explicit computation of the moments and the determinant in (4).

Since $(zu,v) = (u,zv)$, there must exist a three-term recurrence relation to the polynomials $\{\pi_n\}$; we found it to be of the form

$$\begin{gathered} \pi_{k+1}(z) = (z - i\alpha_k)\pi_k(z) - \beta_k\pi_{k-1}(z), \quad k = 0,1,2,\ldots, \\ \pi_{-1}(z) = 0, \quad \pi_0(z) = 1, \end{gathered} \tag{5}$$

where

$$\alpha_0 = \vartheta_0, \quad \alpha_k = \vartheta_k - \vartheta_{k-1},\ \beta_k = \vartheta_{k-1}^2,\ k = 1,2,\ldots, \tag{6}$$

and

$$\vartheta_k = \frac{2}{2k+1}\left[\frac{\Gamma((k+2)/2)}{\Gamma((k+1)/2)}\right]^2, \quad k \geq 0. \tag{7}$$

As $k \to \infty$, one finds $\alpha_k \to 0$, $\beta_k \to \frac{1}{4}$, familiar from Szegö's class of polynomials orthogonal on the interval $[-1,1]$.

Interestingly, the polynomials π_n are closely connected to Legendre polynomials,

$$\pi_n(z) = \hat{P}_n(z) - i\vartheta_{n-1}\hat{P}_{n-1}(z), \quad n \geq 1, \tag{8}$$

where $\hat{P}_k$ is the monic Legendre polynomial of degree k. This allowed us to derive a linear second-order differential equation for π_n, which, like the differential equation for Legendre polynomials, has regular singular points at 1, -1, and ∞, but unlike Legendre polynomials, an additional singular point on the negative imaginary axis, which depends on n and approaches the origin monotonically as $n \uparrow \infty$.

All zeros of π_n are contained in the half disk $D_+ = \{z \in \mathbb{C} : |z| < 1, \operatorname{Im} z > 0\}$ and located symmetrically with respect to the imaginary axis. They are all simple and can be computed as the eigenvalues of the real, nonsymmetric, tridiagonal matrix having the first n of the coefficients α_k on the diagonal, the first $n-1$ of the ϑ_k on the upper side diagonal and their negatives on the lower side diagonal.

There is a Gaussian quadrature formula for integrals over the semicircle,

$$\int_0^\pi g(e^{i\theta})\,d\theta = \sum_{\nu=1}^n \sigma_\nu g(\zeta_\nu), \quad g \in \mathbb{P}_{2n-1}, \tag{9}$$

where ζ_ν are the zeros of π_n and σ_ν the (complex) Christoffel numbers. The latter can be computed by an adaptation of the well-known Golub/Welsch procedure.

All these results are briefly announced in [7] and fully developed in [8], where one also finds applications of the Gauss formula (9) to numerical differentiation and the evaluation of Cauchy principal value integrals.

Partial results for Gegenbauer weight functions had already been obtained, when new impulses were received through collaboration with Henry J. Landau, cf. [12]. This resulted in a considerable simplification of the existence and uniqueness theory. Indeed, if the inner product is

$$(u,v) = \int_0^\pi u(e^{i\theta})v(e^{i\theta})w(e^{i\theta})\,d\theta, \tag{10}$$

where w is positive on $(-1,1)$ and holomorphic in D_+, then the (monic) polynomials $\{\pi_n\}$ orthogonal with respect to (10) exist uniquely if

$$\mathrm{Re}\,(1,1) = \mathrm{Re} \int_0^\pi w(e^{i\theta})\,d\theta \neq 0.$$

This is always true for symmetric weight functions,

$$w(-z) = w(z) \quad \text{and} \quad w(0) > 0, \tag{11}$$

for example, the Gegenbauer weight $w(z) = (1-z^2)^{\lambda-1/2}$, $\lambda > -1/2$, and also for the Jacobi weight function $w(z) = (1-z)^\alpha(1+z)^\beta$, $\alpha > -1$, $\beta > -1$.

There are interesting interrelations between the (monic) complex polynomials $\{\pi_n\}$ orthogonal with respect to the inner product (10), the (monic) real polynomials $\{p_n\}$ orthogonal with respect to the inner product $[u,v] = \int_{-1}^1 u(x)v(x)w(x)\,dx$, and the associated polynomials of the second kind,

$$q_n(z) = \int_{-1}^1 \frac{p_n(z)-p_n(x)}{z-x} w(x)\,dx, \quad n = 0,1,2,\ldots; \quad q_{-1}(z) = -1.$$

Thus, for example (cf. (8)),

$$\pi_n(z) = p_n(z) - i\vartheta_{n-1}p_{n-1}(z), \quad n = 0,1,2,\ldots,$$

where

$$\vartheta_{n-1} = \frac{\mu_0\, p_n(0) + iq_n(0)}{i\mu_0\, p_{n-1}(0) - q_{n-1}(0)}, \quad \mu_0 = (1,1),$$

or, alternatively,

$$\vartheta_n = \mathrm{i}a_n + \frac{b_n}{\vartheta_{n-1}}, \quad n = 0,1,2,\dots; \quad \vartheta_{-1} = \mu_0, \tag{12}$$

where a_k, b_k are the recursion coefficients for the real orthogonal polynomials $\{p_n\}$. For symmetric weight functions (11), one can prove $\mu_0 = \pi w(0) > 0$, so that by (12), since $a_n = 0$ and $b_n > 0$, all ϑ_n are positive.

Moreover, the three-term recurrence relation for the π_n again has the form (5), where now

$$\alpha_0 = \vartheta_0 - \mathrm{i}a_0, \quad \alpha_k = \vartheta_k - \vartheta_{k-1} - \mathrm{i}a_k \quad (k \geq 1),$$
$$\beta_0 = \mu_0, \quad \beta_k = \vartheta_{k-1}(\vartheta_{k-1} - \mathrm{i}a_{k-1}) \quad (k \geq 1).$$

For symmetric weight functions ($a_k = 0$), this reduces to (6), and for Gegenbauer weight functions, one finds

$$\vartheta_0 = \frac{\Gamma(\lambda+1/2)}{\sqrt{\pi}\,\Gamma(\lambda+1)}, \quad \vartheta_k = \frac{1}{\lambda+k}\,\frac{\Gamma((k+2)/2)\Gamma(\lambda+(k+1)/2)}{\Gamma((k+1)/2)\Gamma(\lambda+(k/2))}, \; k \geq 1,$$

generalizing (7).

With regard to the location of the zeros of π_n, we showed for symmetric weight functions that they are contained in D_+, with the possible exception of a single (simple) zero on the positive imaginary axis outside the unit disk. (For a related result, see also [5]). The exception cannot occur for Gegenbauer weights, at least not when $n \geq 2$, and all zeros in this case can be shown to be simple. For Gegenbauer weights, one can also obtain the linear second-order differential equation for π_n, which has properties analogous to those stated above for Legendre weight functions.

5. Our second joint venture deals with a problem of spline approximation on the half line $\mathbb{R}_+ = \{t : t \geq 0\}$. Given a function f on $\mathbb{R}_+$ having finite moments, we want to approximate f by a spline function s of degree $m \geq 0$ that also has finite moments; in fact, we want f and s to have the same successive moments up to an order as high as possible.

Now any spline function of degree m is the sum of a polynomial of degree m and a linear combination of truncated mth powers. If this is to have finite moments on $\mathbb{R}_+$, then the polynomial part must be identically zero, and the spline s therefore is of the form

$$s_{n,m}(t) = \sum_{\nu=1}^{n} a_\nu (\tau_\nu - t)_+^m, \tag{13}$$

where u_+^m are the truncated powers

$$u_+^m = \begin{cases} u^m & \text{if } u \geq 0, \\ 0 & \text{if } u < 0, \end{cases} \qquad m = 0,1,2,\dots.$$

The coefficients a_ν are real and the "knots" τ_ν mutually distinct and positive, say $0 < \tau_1 < \tau_2 < \cdots < \tau_n$, but otherwise can be freely chosen. Since there are $2n$ unknowns, we can impose $2n$ moment conditions,

$$\int_{\mathbb{R}_+} t^j s_{n,m}(t)\,\mathrm{d}t = \mu_j, \quad j = 0,1,2,\ldots,2n-1, \tag{14}$$

where $\mu_j = \int_{\mathbb{R}_+} t^j f(t)\,\mathrm{d}t$ are the (given) moments of f. The problem thus amounts to solving the system (14) of $2n$ nonlinear equations in the $2n$ unknowns a_ν, τ_ν, $\nu = 1,2,\ldots,n$.

The problem is reminiscent of Gaussian quadrature and in fact can be solved by constructing a suitable n-point Gaussian quadrature rule [9]. Indeed, if f is such that

(a) $f \in C^{m+1}(\mathbb{R}_+)$
(b) The moments $\mu_j = \int_{\mathbb{R}_+} t^j f(t)\,\mathrm{d}t$, $j = 0,1,2,\ldots,2n-1$ exist
(c) $f^{(\mu)}(t) = o(t^{-2n-\mu})$ as $t \to \infty$, $\mu = 0,1,\ldots,m$

then the equations (14) have a unique solution if and only if the measure

$$\mathrm{d}\lambda_m(t) = \frac{(-1)^{m+1}}{m!} t^{m+1} f^{(m+1)}(t)\,\mathrm{d}t \quad \text{on } \mathbb{R}_+ \tag{15}$$

admits an n-point Gaussian quadrature formula

$$\int_{\mathbb{R}_+} g(t)\,\mathrm{d}\lambda_m(t) = \sum_{\nu=1}^{n} \lambda_\nu^G g(t_\nu^G), \quad g \in \mathbb{P}_{2n-1}, \tag{16}$$

satisfying $0 < t_1^G < t_2^G < \cdots < t_n^G$. If so, then the solution to (14) is

$$\tau_\nu = t_\nu^G;\ a_\nu = \frac{\lambda_\nu^G}{[t_\nu^G]^{m+1}}, \quad \nu = 1,2,\ldots,n. \tag{17}$$

In general, of course, $\mathrm{d}\lambda_m$ is not a positive measure, and therefore the existence of the Gauss formula (16) with positive nodes is by no means guaranteed. However, when f is completely monotone, i.e., $(-1)^k f^{(k)}(t) > 0$ on $\mathbb{R}_+$ for $k = 0,1,2,\ldots$, then the measure (15) is obviously positive and under the assumptions (a)–(b) can be shown to have finite moments of orders up to $2n-1$. In this case, the quadrature formula (16) exists uniquely and has distinct positive nodes t_ν^G. Moreover, by (17), the coefficients a_ν are all positive, so that $s_{n,m}$ is also completely monotone, at least in the weak sense that $(-1)^k s_{n,m}^{(k)}(t) \geq 0$ for all $k \geq 0$ a.e. on $\mathbb{R}_+$.

In case the spline approximation $s_{n,m}(t)$ exists, its error $f(t) - s_{n,m}(t)$ at $t = x$ can be expressed in terms of the error of the Gauss formula (16) for a special spline function $g(t) = t^{-(m+1)}(t-x)_+^m$ (cf. [10, Theorem 2.3]).

Similar problems of moment-preserving spline approximation can be considered on a finite interval, say $[0,1]$. In this case, we can add to (13) a polynomial of degree m, which increases the degree of freedom by $m+1$. We may use this increased degree of freedom either to add $m+1$ more moment conditions, or to impose $m+1$

boundary conditions of the form $s_{n,m}^{(k)}(1) = f^{(k)}(1)$, $k = 0, 1, \ldots, m$. The relevant measure then becomes

$$d\lambda_m(t) = \frac{(-1)^{m+1}}{m!} f^{(m+1)}(t)\, dt \quad \text{on } [0,1], \tag{18}$$

and the solution of the two problems can be given (if it exists) in terms of generalized Gauss–Lobatto formulae for the first problem, and generalized Gauss–Radau formulae for the other, both for integration with respect to the measure (18); cf. [1]. The numerical construction of such formulae, however, is rather more complicated, and has been considered by the author only recently in [2, 3]. Gradimir, together with M.A. Kovačević [13], also studied moment-preserving approximation on $\mathbb{R}_+$ by defective splines, which gives rise to Gauss–Turán quadrature rules for the measure (15). Undoubtedly, this led Gradimir to wonder about how to compute these quadratures effectively.

6. Gauss–Turán quadrature fromulae are of Gaussian type, i.e., have maximum algebraic degree of exactness, and involve not only values of the integrand function, but also values of its successive derivatives up to an even order $2s$, all evaluated at a common set of n nodes. Since there are $(2s+1)n$ coefficients (n for each derivative) and n nodes to be determined, the maximum degree of exactness is expected to be $2(s+1)n-1$, and the formula thus has the form

$$\int_{\mathbb{R}} f(t)\, d\lambda(t) = \sum_{i=0}^{2s} \sum_{\nu=1}^{n} \lambda_{i,\nu} f^{(i)}(\tau_\nu), \quad f \in \mathbb{P}_{2(s+1)n-1}, \tag{19}$$

where $d\lambda$ is a given positive measure. It is known that the nodes τ_ν must be the zeros of the (monic) polynomial $\pi_n = \pi_{n,s}$ of degree n whose $(2s+1)$st power is orthogonal (relative to the measure $d\lambda$) to all polynomials of degree $<n$. In other words, π_n is the nth-degree polynomial orthogonal with respect to the positive measure

$$d\mu(t) = \pi_n^{2s}(t)\, d\lambda(t). \tag{20}$$

We have here a case of implicit orthogonality–also called s-orthogonality–since the polynomial π_n to be determined appears also in the measure of orthogonality. The problem of computing Gauss–Turán quadrature rules (19), considered in [10], thus will in some way come down to a problem of solving a system of nonlinear equations.

Gradimir's idea was to embed π_n in a sequence of $n+1$ polynomials $\pi_0, \pi_1, \ldots, \pi_n$, namely, the polynomials orthogonal with respect to the measure (20). As such, they must satisfy a three-term recurrence relation

$$\begin{aligned} &\pi_{k+1}(t) = (t-\alpha_k)\pi_k(t) - \beta_k \pi_{k-1}(t), \quad k = 0, 1, \ldots, n-1, \\ &\pi_{-1}(t) = 0, \;\; \pi_0(t) = 1. \end{aligned} \tag{21}$$

Although the coefficients $\alpha_0, \alpha_1, \ldots, \alpha_{n-1}$; $\beta_0, \beta_1, \ldots, \beta_{n-1}$ are not known, in fact, need to be determined, we know from the general theory of orthogonal polynomials that they are expressible in terms of inner products involving the polynomials $\pi_0, \pi_1, \ldots, \pi_{n-1}$; specifically,

$$\alpha_k = \frac{\int_{\mathbb{R}} t\pi_k^2(t)\,\mathrm{d}\mu(t)}{\int_{\mathbb{R}} \pi_k^2(t)\,\mathrm{d}\mu(t)}, \quad k = 0, 1, \ldots, n-1;$$

$$\beta_k = \frac{\int_{\mathbb{R}} \pi_k^2(t)\,\mathrm{d}\mu(t)}{\int_{\mathbb{R}} \pi_{k-1}^2(t)\,\mathrm{d}\mu(t)}, \quad k = 1, 2, \ldots, n-1,$$

and $\beta_0 = \int_{\mathbb{R}} \mathrm{d}\mu(t)$ by convention. If we insert here the definition (20) of the measure $\mathrm{d}\mu$ and clear all fractions, we arrive at the following system of $2n$ equations:

$$\begin{aligned}
&\beta_0 - \int_{\mathbb{R}} \pi_n^{2s}(t)\,\mathrm{d}\lambda(t) = 0,\\
&\int_{\mathbb{R}} (\alpha_k - t)\pi_k^2(t)\pi_n^{2s}(t)\,\mathrm{d}\lambda(t) = 0, \quad k = 0, 1, \ldots, n-1;\\
&\int_{\mathbb{R}} (\beta_k\pi_{k-1}^2(t) - \pi_k^2(t))\pi_n^{2s}(t)\,\mathrm{d}\lambda(t) = 0, \quad k = 1, 2, \ldots, n-1.
\end{aligned} \tag{22}$$

We note that (22) represents a system of $2n$ nonlinear equations in the $2n$ unknowns α_k, β_k. Indeed, each of the polynomials π_r appearing in (22) can be thought of as a function of $\alpha_0, \alpha_1, \ldots, \alpha_{r-1}$; $\beta_0, \beta_1, \ldots, \beta_{r-1}$ by virtue of the relations in (21). Furthermore, each integrand in (22) is a polynomial of degree $\leq 2(s+1)n-2$, and therefore all integrals in (22) can be evaluated exactly by an N-point Gaussian quadrature rule relative to the measure $\mathrm{d}\lambda$, where $N = (s+1)n$. The same is true for the integrals appearing in the Jacobian matrix of the system (22). We were able, therefore, to apply to (22) the Newton–Kantorovich method, which could be made to converge by a careful choice of the initial approximations $\alpha_k^{(0)}$, $\beta_k^{(0)}$ to the unknowns α_k, β_k. Once the α_k and β_k have been obtained, the zeros τ_ν of π_n, i.e., the nodes in the quadrature rule (19), can be computed by the well-known Golub/Welsch procedure.

All that remains is to compute the coefficients $\lambda_{i,\nu}$ in (19). By inserting in (19) suitably selected polynomials of degree $\leq 2(s+1)n-1$, the coefficients $\lambda_{i,\nu}$ for each fixed ν, $1 \leq \nu \leq n$, can be found by solving an upper triangular system of $2s+1$ linear algebraic equations (cf. [10, Theorem 3.3]).

7. Conformal maps in fluid mechanics often require Cauchy principal value integrals of the form

$$(I_{[\alpha,\beta]}\phi)(\xi) = ⨍_{\alpha}^{\beta} \phi(\tau)\coth\frac{\tau-\xi}{2}\,\mathrm{d}\tau, \quad \alpha < \xi < \beta, \tag{23}$$

which are notoriously difficult to evaluate numerically. Some possible approaches are discussed in [11]. If the interval $[\alpha,\beta]$ is finite, (23) can be transformed to

$$(I_a f)(x) = \frac{1}{a} \fint_{-1}^{1} \frac{f(t)}{t-x} w(t)\,\mathrm{d}t, \quad -1 < x < 1, \tag{24}$$

where $x = [2\xi - (\alpha+\beta)]/(\beta-\alpha)$, $a = (\beta-\alpha)/4$, and

$$w(t) = \omega(a(t-x)), \quad \omega(u) = u \coth u.$$

The difficulty here is caused by the poles of $w(t)$ at $t = x \pm ki\pi/a$, $k = 1,2,\dots$, which can be close to the interval $[-1,1]$ if a is large. Following standard procedure, (24) is first written in the form

$$(I_a f)(x) = \frac{1}{a}\left\{ f(x) \ln \frac{\sinh a(1-x)}{\sinh a(1+x)} + \int_{-1}^{1} g(t) w(t)\,\mathrm{d}t \right\}, \tag{25}$$

where $g(t) = [x,t]f$ is the divided difference of f at x and t. The integral in (25) can then be computed either by Gauss quadrature relative to the (nonstandard) weight function w, which gives excellent results but is somewhat expensive, or by the less expensive Gauss–Legendre quadrature, which, however, works well only for a relatively small. An alternative procedure is interpolatory quadrature of (24) on the zeros of the respective orthogonal polynomials, which cirumvents the need of computing a divided difference. Error analyses are provided, either in real variable form, involving derivatives, or in terms of contour integration in the complex plane.

Cauchy principal value integrals (23) with infinite interval $[\alpha,\beta] = [-\infty,\infty]$ can be rendered accessible to the same approaches after suitable truncation of the interval.

8. In looking back on my collaboration with Gradimir, I can only marvel at the spontaneity and originality of his input, which often reduced my own role to one of implementor and organizer. It has been truly a pleasure to work together with Gradimir, and I am sure I am sharing this feeling with the many other individuals who have had the privilege of collaborating with Gradimir. I wish him many more years of good health and continued excellence and success in his research.

References

1. Frontini, M., Gautschi, W., Milovanović, G.V.: Moment-preserving spline approximation on finite intervals. Numer. Math. **50** (5), 503–518 (1987)
2. Gautschi, W.: Generalized Gauss–Radau and Gauss–Lobatto formulae. BIT **44** (4), 711–720 (2004)
3. Gautschi, W.: High-order generalized Gauss–Radau and Gauss–Lobatto formulae for Jacobi and Laguerre weight functions. Numer. Algorithm **51** (2), 143–149 (2009)
4. Gautschi, W.: Variable-precision recurrence coefficients for nonstandard orthogonal polynomials. Numer. Algorithms **52** (3), 409–418 (2010)

5. Gautschi, W., Milovanović, G.V.: On a class of complex polynomials having all zeros in a half disc. Numerical methods and approximation theory (Niš, 1984), 49–53, Univ. Niš, Niš (1984)
6. Gautschi, W., Milovanović, G.V.: Gaussian quadrature involving Einstein and Fermi functions with an application to summation of series. Math. Comp. **44** (169), 177–190 (1985)
7. Gautschi, W., Milovanović, G.V.: Polynomials orthogonal on the semicircle. Rend. Sem. Mat. Univ. e Politec. Torino. Special issue (1985), 179–185
8. Gautschi, W., Milovanović, G.V.: Polynomials orthogonal on the semicircle. J. Approx. Theory **46** (3), 230–250 (1986)
9. Gautschi, W., Milovanović, G.V.: Spline approximations to spherically symmetric distributions. Numer. Math. **49** (1), 111–121 (1986)
10. Gautschi, W., Milovanović, G.V.: s-orthogonality and construction of Gauss–Turán-type quadrature formulae. J. Comput. Appl. Math. **86** (1), 205–218 (1997)
11. Gautschi, W., Kovačević, M.A., Milovanović, G.V.: The numerical evaluation of singular integrals with coth-kernel. BIT **27** (3), 389–402 (1987)
12. Gautschi, W., Landau, H.J., Milovanović, G.V.: Polynomials orthogonal on the semicircle, II. Constr. Approx. **3** (4), 389–404 (1987)
13. Milovanović, G.V., Kovačević, M.A.: Moment-preserving spline approximation and Turán quadratures. In: Numerical Mathematics, Singapore 1988, 357–365, Internat. Schriftenreihe Numer. Math., vol. 86, Birkhäuser, Basel (1988)
14. Szegö, G.: Orthogonal Polynomials. 4th ed., American Mathematical Society, Colloquium Publications, vol. 23, Amer. Math. Soc., Providence, RI (1975)

On Some Major Trends in Mathematics

Themistocles M. Rassias

Dedicated to Gradimir V. Milovanović on the occasion of his 60th birthday

An attempt will be given to present some ideas regarding the present state and the near future of Mathematics and some of its applications. Since assessments and any predictions in this field of science are necessarily subjective, I shall communicate to you the opinions of renowned contemporary mathematicians with some of whom I have recently come into contact. In addition, I will mention to you some of the work of Professor Gradimir V. Milovanović that fits the context of present and future Mathematics.

I will mention three basic directions of current mathematical activity, which I personally find exciting.

1. The realization that difficult problems require deep study of nonlinearity, non convexity, group theory, Discrete Mathematics, as well as relationships between deterministic and probabilistic ideas (Stochastic as it is also called). Specifically, we are witnessing an increased interest in the interplay between probabilistic ideas and deterministic ones, in particular, in such classical branches of Pure and Applied Mathematics, such as Partial Differential Equations and Number Theory.

2. Discrete Mathematics fueled by the computer revolution and rekindled interest in Number Theory, notably the Riemann Hypothesis, as well as Goldbach Conjecture and Coding Theory.

3. The recognition that "good" solutions of many applied problems from other Sciences and Technology, such as Cosmology, Quantum Mechanics and Biology,

Themistocles M. Rassias
Department of Mathematics, National Technical University of Athens, Zografou Campus, 15780 Athens, Greece
e-mail: trassias@math.ntua.gr

W. Gautschi et al. (eds.), *Approximation and Computation: In Honor of Gradimir V. Milovanović*, Springer Optimization and Its Applications 42,
DOI 10.1007/978-1-4419-6594-3_3,

often call upon the panorama of Mathematics. This fact has contributed to the recent popularization of the strength of Mathematics. One may note here the newly established Gauss prize in Mathematics.

Particular problems of mathematical interest belonging to the above directions are those which were proposed by the Clay Mathematics Institute in the United States of America led by its Founding Scientific Board consisting of Alain Connes (Fields Medal, 1982), Arthur Jaffe (Harvard University), Edward Witten (Fields Medal, 1990), and Andrew Wiles (Princeton University, the man who proved Fermat's Last Theorem), which decided to establish a set of seven prize problems. The aim of these problems was "to record some of the most difficult issues with which mathematicians were struggling at the beginning of the third millennium; to recognize achievement in Mathematics of historical dimension; to elevate in the consciousness of the general public the fact that, in Mathematics the frontier is still open, indeed widening and abounding in significant unsolved problems; and to emphasize the importance of working toward solutions of the deepest, most difficult problems".

One of these seven problems exemplifies the unexpected difficulties for their solution.

In the year 1900, Poincaré announced a proof of the general n-dimensional case of what subsequently has been called the Poincaré Conjecture. Four years later in the year 1904, Poincaré found a mistake in his own n-dimensional proof. Poincaré formulated his Conjecture for 3-manifolds in 1904 as follows:

"*Any closed, simply connected 3-manifold is homeomorphic to the 3-sphere*".

This problem was solved recently by the Russian mathematician Grigori Perelman, who declined to accept the Fields Medal at the International Congress of Mathematicians that was held in Madrid in the year 2006.

The paper in which Poincaré posed this problem in the year 1904 contributed to the founding of Topology as an independent branch of Pure Mathematics. Solving the Poincaré Conjecture is a big honor for Perelman, but it is also a great achievement for all Mathematicians, for it gives a measure of how far our understanding of the nature of space has advanced. To paraphrase Newton, Perelman has seen far, but to do so he stood on the shoulders of giants, who came before him. One giant is Richard Hamilton. Without Hamilton's work, Perelman's work would not have been possible. One of the most interesting aspects of the resolution of the Poincaré Conjecture is the nature of the solution. While the problem is purely topological in its formulation, the proof is NOT. The proof uses deep techniques and results from other areas of Mathematics, namely, Mathematical Analysis and Differential Geometry.

In fact most, if not all, of the outstanding problems in Mathematics have gone outside their domain of definition for their solutions. Of the above-mentioned three directions, I wish to single out two areas of future Mathematics.

One is computers with which most of us are more or less acquainted and the other is Biology that most of us are less familiar with.

The mathematical logician Alan Turing is by many considered to be the father of modern computer science and the most important inventor of the modern computer, which as we know has revolutionized society. Thus, the predominant theoretical branch in Mathematics, mathematical logic, resulted in the most applied tool of our civilization.

The digital world is still ahead of us.

Roger Penrose in the year 1991 pointed out some limitations of artificial intelligence. His argumentation brings in the interesting question of whether the Mandelbrot set is decidable and implications of the Gödel Incompleteness Theorem. However, Stephen Smale proposes that a broader study is called for, one which involves deeper models of the brain, and of the computer, in a search of what artificial and human intelligence have in common and how they differ. Smale would look in a direction where learning, problem-solving, and game theory play a substantial role, together with the mathematics of real numbers, approximations, probability, and geometry.

It is interesting that the most astonishing things in the world can be studied mathematically.

Although Mathematics often deals with things that change in a very predictable way, much in life is quite unpredictable.

We are currently witnessing a remarkable period of renewed collaboration between mathematicians and scientists of several other new fields.

I believe, I am not oversimplifying if I predict that a good portion of Science and Technology of the twenty-first century will deal with

Number 1: Discrete Mathematics (because of computers)

Number 2: Mathematics stemming from Biology

Number 3: Nonlinear Dynamics and the interaction between Algebra, Number Theory, Analysis and Geometry

Thus, there is an attempt toward the algebraization of Geometry, the geometrization of Algebra, the algebraization of Analysis, and so on.

We now present the opinions of living scientists regarding some of the trends mentioned above.

Thomas Banchoff (ex-President of the Mathematical Association of America). He mentioned: "One marvelous modern development for unification in Mathematics is scientific visualization. Computer graphics makes it possible to communicate ideas from all parts of Mathematics to all people: researchers, artists, students, and the general public".

Timothy Gowers (University of Cambridge, Fields Medal, 1998). He also mentioned: "Several current areas of Mathematics are less tightly defined than the traditional areas such as algebraic topology or functional analysis. For example, Arithmetic Combinatorics is a fusion of techniques and insights from Analysis, Probability, Combinatorics and Number Theory. This tendency, to mix ideas from several branches of Mathematics, was clearly visible at the ICM in Madrid, 2006".

Sergei Konyagin (University of Moscow). He replied: "The process of unification of Mathematics in Russia is slow, and there are no evident dramatic changes in the last years. But in my opinion (I am writing about subjects I am familiar with) the most interesting results are based on a combination of ideas from different mathematics areas".

Frank Morgan (Massachusetts, USA). He mentioned: "Mathematics is finally learning to deal with the singularities which encode the mysteries of the universe".

Alexei Sossinsky (Independent University of Moscow, Institute for Problems in Mechanics – Russian Academy of Sciences). Similarly, he mentioned that "One of the crucial developments in recent years has been the disappearance of any boundary between Mathematics and Theoretical Physics. The ICM in Madrid, 2006, has strikingly demonstrated a new interaction between Probability Theory and Analysis, between structure and randomness".

Stephen Smale (University of California at Berkeley, Fields Medal, 1966), during the International Congress of Mathematicians that was held in Berkeley in 1986, predicted the value of future important branches of Mathematics by stating: "This subject (i.e., algorithms, complexity, etc.) is now likely to change Mathematics itself. Algorithms become a subject of study, not just a means of solving problems".

At the ICM in Madrid, 2006, the following quotes were written in the Daily News of the Congress regarding the new field of Mathematics Art that was created by Benoit Mandelbrot. "His discovery of the so-called Mandelbrot set consisted in a number of visual observations phrased into conjectures demanding rigorous mathematical treatment" ... "Mathematics has always remained isolated and has concentrated on simple figures. I've been very lucky to work with the Mathematics of roughness. In a certain sense, the circle has been completed: in many cultures the "rough" has always been considered Art. Then this concept arrived to Mathematics and afterwards to other sciences, then back to Mathematics again, and finally to Art once more through fractal art".

Some of the work of Professor Gradimir V. Milovanović fits the context of present and future Mathematics. Professor Milovanović made essential contributions in the following subjects:

(a) Extremal problems for polynomials (mainly Markov, Bernstein and Turán type on restricted and unrestricted polynomial classes). We mention here an elegant solution of Markov's extremal problem, stated originally by A.K. Varma in 1981, for a restricted class of polynomials with nonnegative coefficients with respect to the generalized Laguerre measure on the open interval of strictly positive real numbers. This result obtained by Milovanović was published in the year 1985 in the Proceedings of the American Mathematical Society after several attempts that had been made earlier by many other mathematicians.

(b) Concepts of orthogonality and corresponding numerical construction (orthogonality on the unit semicircle and a circular arc with respect to a non-Hermitian inner product, orthogonality with respect to certain complex "weight functions" and oscillatory weights, orthogonality on the radial rays in the complex plane, s-and σ-orthogonality and a connection with quadratures with multiple nodes, multiple orthogonality, orthogonal Müntz systems, and other nonpolynomial systems and corresponding quadratures of Gaussian type, etc.). Especially, in various important cases Gradimir V. Milovanović introduced a new type of orthogonality that was connected to a number of applications in numerical mathematics or in some applied and computational sciences (Physics, Electrotechnics, Telecommunications, etc.).

(c) Quadratures with multiple nodes. Milovanović gave efficient algorithms for numerical construction and methods for the error estimations in different classes of functions including spline approximation, finite elements, etc.

(d) Nonstandard Gaussian quadratures. Recently, Gradimir Milovanović and his collaborator Aleksandar S. Cvetković have developed the theory of so-called nonstandard Gaussian quadrature formulae based on operator values for a general family of linear operators, acting on the space of algebraic polynomials, such that the degrees of polynomials are preserved. They proposed a stable numerical algorithm for constructing such quadrature formulae. In particular, for some special classes of linear operators they obtained interesting explicit results connected with the theory of orthogonal polynomials. This is a new direction of the famous Gaussian method of integration.

Finally, I would like to mention that together with Dragoslav S. Mitrinović , we greatly enjoyed the writing of our 821 – page book "*Topics in Polynomials: Extremal Problems, Inequalities, Zeros*", World Scientific Publishing Co., Singapore – New Jersey – London, 1994, which took seven years of scientific interaction. My admiration and friendship for Gradimir V. Milovanović and a number of colleagues from the Serbian academic community is based on long-standing association and indeed collaboration. In conclusion, it is a great privilege for me to be participating at the Conference among eminent mathematicians to honor the 60th anniversary of Professor Gradimir V. Milovanović.

Part II
Polynomials and Orthogonal Systems

An Application of Sobolev Orthogonal Polynomials to the Computation of a Special Hankel Determinant

Paul Barry, Predrag M. Rajković, and Marko D. Petković

Dedicated to Professor Gradimir V. Milovanović
on the occasion of his 60th birthday

1 Introduction

The *Hankel transform* of a given number sequence A is the sequence of Hankel determinants H given by

$$A = \{a_n\}_{n\in\mathbb{N}_0} \;\rightarrow\; H = \{h_n\}_{n\in\mathbb{N}_0} : \; h_n = |a_{i+j-2}|_{i,j=1}^{n}.$$

This term was first used by J.W. Layman [6]. The closed-form computation of Hankel determinants is of great combinatorial interest related to partitions and permutations. A long list of known determinant evaluations can be found in [5]. We also obtained a few interesting results given in [3] and [8]. Many methods are known for the evaluation of these determinants.

Here, we will consider the sequence A whose general element is

$$a_n = \sum_{k=0}^{n} T_{n,k} r^k, \quad r \in \mathbb{Z}, \tag{1}$$

Paul Barry
School of Science, Waterford Institute of Technology, Ireland, e-mail: pbarry@wit.ie

Predrag M. Rajković
Faculty of Mechanical Engineering, University of Niš, Serbia, e-mail: pedja.rajk@masfak.ni.ac.rs

Marko D. Petković
Faculty of Sciences and Mathematics, University of Niš, Serbia, e-mail: dexterofnis@gmail.com

W. Gautschi et al. (eds.), *Approximation and Computation: In Honor of Gradimir V. Milovanović*, Springer Optimization and Its Applications 42,
DOI 10.1007/978-1-4419-6594-3_4, © Springer Science+Business Media, LLC 2011

where

$$T_{n,k} = \binom{2n}{n-k} - \binom{2n}{n-k-1} = \frac{2k+1}{n+k+1}\binom{2n}{n-k}, \quad 0 \le k \le n;\ n \in \mathbb{N}_0. \quad (2)$$

We will define $T_{n,k} = 0$ in all other cases.

The first members of A and its Hankel transform are:

$$A = \{1, 1+r, 2+3r+r^2, 5+9r+5r^2+r^3, \ldots\},$$
$$H = \{1, 1+r, (1+r)^2, (1+r)^3, \ldots\}.$$

Our purpose is to prove that, in general, $h_n = (1+r)^{n-1}$.

2 Preliminaries

The ordinary generating function of the sequence a_n is the power series $f(x) = \sum_{n=0}^{\infty} a_n x^n$, with $a_n = [x^n]f(x)$, where the operator $[x^n]$ extracts the coefficient of x^n. Sequences are often referred to by their 'A' number in the On-Line Encyclopedia of Integer Sequences [9].

Let us recall that the sequence of Catalan numbers $C_n = \frac{1}{n+1}\binom{2n}{n}$, $n \in \mathbb{N}_0$, can be written in the integral form

$$C_n = \frac{1}{2\pi}\int_0^4 x^n \sqrt{\frac{4-x}{x}}\,\mathrm{d}x, \quad n \in \mathbb{N}_0.$$

This sequence has the generating function

$$c(x) = \sum_{n=0}^{\infty} C_n x^n = \frac{1-\sqrt{1-4x}}{2x}.$$

Definition 1. The *Riordan array* is an infinite lower-triangular matrix $\mathcal{M} = [m_{j,k}]$ defined by a pair of generating functions $g(x) = \sum_{k=0}^{\infty} g_k x^k$ and $f(x) = \sum_{k=1}^{\infty} f_k x^k$ where $f_1 \neq 0$, whose k-th column is generated by $g(x)f(x)^k$ (the first column being indexed by 0), i.e. $m_{j,k} = [x^j]\left(g(x)f(x)^k\right)$, $j,k \in \mathbb{N}_0$. The matrix $\mathcal{M}$ corresponding to the pair of functions f and g is denoted by (g, f) or $\mathcal{R}(g, f)$.

The *Riordan group* is a set of infinite lower-triangular matrices, where each matrix is defined by a pair of generating functions.

We will consider the transform defined by the Riordan array $\left(c(x), xc^2(x)\right)$. This matrix is the inverse of the Riordan matrix $\left(1/(1+x), x/(1+x)^2\right)$, which is the coefficient array of the Morgan Voyce polynomials. They are closely linked to the Chebyshev polynomials of the second kind.

The general element $T_{n,k}$ of $\left(c(x), xc^2(x)\right)$ is given by (2). This is the sequence $A039599$ and represents the number of lattice paths from $(0,0)$ to (n,n) with steps

$E = (1,0)$ and $N = (0,1)$ which touch but do not cross the line $x - y = k$ and only situated above this line. Also, this sequence appears in the definition of the "Ballot" transform (see [1]).

By the theory of Riordan arrays, the generating function of a_n can be evaluated from

$$\left(c(x), xc^2(x)\right) \cdot \frac{1}{1-rx} = \frac{c(x)}{1-rxc^2(x)},$$

i.e.,

$$G(x) = \sum_{n=0}^{\infty} a_n x^n = \frac{(r+1)\sqrt{1-4x}+r-1}{2r-(r+1)^2 x}.$$

The function

$$F(x) = \frac{1}{x} G\left(\frac{1}{x}\right) = \frac{(r+1)\sqrt{1-4/x}+r-1}{2rx-(r+1)^2}$$

will be very useful in the next sections.

3 The Properties of Number Sequences

Notice

$$T_{n,0} = C_n, \;\; T_{n,n} = 1, \quad T_{n,k} = 0, \; k < 0 \vee k > n. \tag{3}$$

The sequence $\{T_{n,k}\}$ satisfies the following recurrence relationo

$$T_{n,k} = T_{n-1,k-1} + 2T_{n-1,k} + T_{n-1,k+1}, \quad k = 0, 1, \ldots, n. \tag{4}$$

Let us denote $\zeta = (r+1)^2/r$.

Lemma 1. *For the sequence $\{a_n\}$, the following recurrence relation is valid,*

$$a_n = \zeta \, a_{n-1} - \frac{r+1}{r} C_{n-1}. \tag{5}$$

Proof. We can write

$$\begin{aligned}
a_n - \frac{(r+1)^2}{r} a_{n-1} &= a_n - r\, a_{n-1} - 2\, a_{n-1} - \frac{1}{r} a_{n-1} \\
&= \sum_{k=0}^{n} T_{n,k} r^k - \sum_{k=0}^{n-1} T_{n-1,k} r^{k+1} - 2 \sum_{k=0}^{n-1} T_{n-1,k} r^k - \sum_{k=0}^{n-1} T_{n-1,k} r^{k-1} \\
&= -\frac{1}{r} T_{n-1,0} + \left(T_{n,0} - 2T_{n-1,0} - T_{n-1,1}\right) \\
&\quad + \sum_{k=1}^{n-2} \left(T_{n,k} - T_{n-1,k-1} - 2T_{n-1,k} - T_{n-1,k+1}\right) r^k \\
&\quad + \left(T_{n,n-1} - 2T_{n-1,n-1} - T_{n-1,n-2}\right) r^{n-1} + \left(T_{n,n} - T_{n-1,n-1}\right) r^n.
\end{aligned}$$

By using (3) and (4), we finish the proof. □

Theorem 1. *The sequence $\{a_n\}$ can be represented in the integral form*

$$a_n = \frac{r+1}{2\pi} \int_0^4 \frac{x^n}{(r+1)^2 - rx} \sqrt{\frac{4-x}{x}}\, \mathrm{d}x + \frac{r-1}{r} \zeta^n. \tag{6}$$

Proof. We will apply mathematical induction. Since

$$\frac{r+1}{2\pi} \int_0^4 \frac{1}{(r+1)^2 - rx} \sqrt{\frac{4-x}{x}}\, \mathrm{d}x + \frac{r-1}{r} = 1,$$

AND $a_0 = 1$, we have that the formula is valid for $n = 0$.

Suppose that the formula is true for n. Hence

$$\zeta\, a_n = \frac{r+1}{2\pi} \int_0^4 \frac{\zeta x^n}{(r+1)^2 - rx} \sqrt{\frac{4-x}{x}}\, \mathrm{d}x + \frac{r-1}{r} \zeta^{n+1}.$$

Since

$$\frac{r+1}{r} C_n = \frac{r+1}{2\pi r} \int_0^4 x^n \sqrt{\frac{4-x}{x}}\, \mathrm{d}x,$$

we can write

$$\zeta\, a_n - \frac{r+1}{r} C_n = \frac{r+1}{2\pi} \int_0^4 x^n \left(\frac{\zeta}{(r+1)^2 - rx} - \frac{1}{r} \right) \sqrt{\frac{4-x}{x}}\, \mathrm{d}x + \frac{r-1}{r}\, \zeta^{n+1}.$$

By using (5), we get

$$a_{n+1} = \frac{r+1}{2\pi} \int_0^4 \frac{x^{n+1}}{(r+1)^2 - rx} \sqrt{\frac{4-x}{x}}\, \mathrm{d}x + \frac{r-1}{r}\, \zeta^{n+1},$$

wherefrom we complete the proof. □

4 The Hankel Determinants and Orthogonal Polynomials

Among the few methods for evaluating Hankel determinants, we concentrate on the method based on the theory of distributions and orthogonal polynomials.

Namely, the Hankel determinant h_n of the sequence $\{a_n\}_{n\geq 0}$ equals

$$h_n = a_0^n \beta_1^{n-1} \beta_2^{n-2} \cdots \beta_{n-2}^2 \beta_{n-1}, \tag{7}$$

where $\{\beta_n\}_{n\geq 1}$ is the sequence given by:

$$\mathcal{G}(x) = \sum_{n=0}^{\infty} a_n x^n = \cfrac{a_0}{1 - \alpha_0 x - \cfrac{\beta_1 x^2}{1 - \alpha_1 x - \cfrac{\beta_2 x^2}{1 - \alpha_2 x - \cdots}}}.$$

The above sequences $\{\alpha_n\}_{n\geq 0}$ and $\{\beta_n\}_{n\geq 1}$ are the coefficients in the recurrence relation

$$Q_{n+1}(x) = (x-\alpha_n)Q_n(x) - \beta_n Q_{n-1}(x),$$

where $\{Q_n(x)\}_{n\geq 0}$ is the monic polynomial sequence orthogonal with respect to the functional $\mathcal{U}$ determined by $a_n = \mathcal{U}[x^n]$, $n = 0,1,2,\ldots$. In some cases, there exists a weight function $w(x)$ such that the functional $\mathcal{U}$ can be expressed by

$$\mathcal{U}[f] = \int_{\mathbb{R}} f(x)w(x)\,\mathrm{d}x, \quad f(x) \in C(\mathbb{R});\ w(x) \geq 0.$$

So, we can associate with every weight $w(x)$ two sequences of coefficients, i.e., $w(x) \mapsto \{\alpha_n, \beta_n\}_{n\in\mathbb{N}_0}$, by

$$\alpha_n = \frac{\mathcal{U}\left[xQ_n^2(x)\right]}{\mathcal{U}\left[Q_n^2(x)\right]}, \quad \beta_n = \frac{\mathcal{U}\left[Q_n^2(x)\right]}{\mathcal{U}\left[Q_{n-1}^2(x)\right]}, \quad \beta_0 = \mathcal{U}(1), \quad n \in \mathbb{N}_0.$$

Finding the weight function can be started by the function $F(z) = 1/zG(1/z)$. From the theory of distribution functions (see Chihara [2]), we have Stieltjes's inversion function

$$\psi(t) - \psi(s) = -\frac{1}{\pi}\int_s^t \operatorname{Im} F(x+\mathrm{i}y)\,\mathrm{d}x.$$

hence we find the distribution function $\psi(t)$. After differentiation of $\psi(t)$ and simplification of the resulting expression, we finally have $w(x) = \psi'(x)$.

The following lemma will be very useful in our further discussion. The majority of statements is proven in [4].

Lemma 2. *Let* $w(x) \mapsto \{\alpha_n, \beta_n\}_{n\in\mathbb{N}_0}$, $\widetilde{w}(x) \mapsto \{\widetilde{\alpha}_n, \widetilde{\beta}_n\}_{n\in\mathbb{N}_0}$. *Then*

(i) $\widetilde{w}(x) = Cw(x) \Rightarrow \{\widetilde{\alpha}_n = \alpha_n,\ \widetilde{\beta}_0 = C\beta_0,\ \widetilde{\beta}_n = \beta_n,\ n \in \mathbb{N}\}$;

(ii) $\widetilde{w}(x) = w(ax+b) \Rightarrow \left\{\widetilde{\alpha}_n = \dfrac{\alpha_n - b}{a},\ \widetilde{\beta}_0 = \dfrac{\beta_0}{|a|},\ \widetilde{\beta}_n = \dfrac{\beta_n}{a^2},\ n \in \mathbb{N}\right\}$;

(iii) *If* $w_c(x) = \widetilde{w}(x)/(x-c)$, $c \notin \operatorname{supp}(\widetilde{w})$, *then*

$$\alpha_{c,0} = \widetilde{\alpha}_0 + r_0,\ \alpha_{c,k} = \widetilde{\alpha}_k + r_k - r_{k-1},\ \beta_{c,0} = -r_{-1},\ \beta_{c,k} = \widetilde{\beta}_{k-1}\frac{r_{k-1}}{r_{k-2}}, \quad k \in \mathbb{N},$$

where

$$r_{-1} = -\int_{\mathbb{R}} w_c(x)\,\mathrm{d}x, \quad r_n = c - \widetilde{\alpha}_n - \frac{\widetilde{\beta}_n}{r_{n-1}}, \quad n = 0,1,\ldots.$$

(iv) *If* $\widehat{w}_d(x) = \widetilde{w}(x)/(d-x)$, $d > x$, $\forall x \in \operatorname{supp}(\widetilde{w})$, *then*

$$\widehat{\alpha}_{d,0} = \widetilde{\alpha}_0 + r_0,\ \widehat{\alpha}_{d,k} = \widetilde{\alpha}_k + r_k - r_{k-1},\ \widehat{\beta}_{d,0} = r_{-1},\ \widehat{\beta}_{d,k} = \widetilde{\beta}_{k-1}\frac{r_{k-1}}{r_{k-2}}, \quad k \in \mathbb{N},$$

where

$$r_{-1} = \int_{\mathbb{R}} \widehat{w}_d(x)\,\mathrm{d}x, \quad r_n = d - \widetilde{\alpha}_n - \frac{\widetilde{\beta}_n}{r_{n-1}}, \quad n = 0, 1, \ldots.$$

5 Connections with Classical Orthogonal Polynomials

From the formula (6), we conclude that the computation of Hankel determinants is directly connected with the monic polynomial sequence $\{Q_n(x)\}$, which is orthogonal with respect to the discrete Sobolev inner product

$$\widetilde{\varphi}(f,g) = \frac{r+1}{2\pi} \int_0^4 \frac{f(x)g(x)}{(r+1)^2 - rx} \sqrt{\frac{4-x}{x}}\,\mathrm{d}x + \frac{r-1}{r} f(\zeta)g(\zeta), \quad \zeta = \frac{(r+1)^2}{r}.$$

We will start with a special Jacobi polynomial $P_n(x) = P_n^{(1/2,-1/2)}(x)$, $n \in \mathbb{N}_0$, which is also known as *the Chebyshev polynomial of the fourth kind*. The sequence of these polynomials is orthogonal with respect to

$$w^*(x) = w^{(1/2,-1/2)}(x) = \sqrt{\frac{1-x}{1+x}}, \quad x \in (-1,1).$$

They satisfy the three-term recurrence relation (Chihara [2]):

$$P_{n+1}(x) = (x - \alpha_n^*)P_n(x) - \beta_n^* P_{n-1}(x), \quad n \in \mathbb{N}_0, \quad P_{-1}(x) = 0, \; P_0(x) = 1,$$

where

$$\alpha_0^* = -\frac{1}{2}, \; \alpha_n^* = 0, \quad \beta_0^* = \pi, \; \beta_n^* = \frac{1}{4}, \quad n \in \mathbb{N}.$$

For the weight function

$$\widetilde{w}(x) = w^*\left(\frac{x}{2} - 1\right) = \sqrt{\frac{4-x}{x}}, \quad x \in (0,4),$$

applying Lemma 2 (ii) with $a = 1/2$ and $b = -1$, we find the coefficients $\widetilde{\alpha}_0 = 1$, $\widetilde{\alpha}_n = 2$, $n \geq 1$, $\widetilde{\beta}_0 = 2\pi$, $\widetilde{\beta}_n = 1$, $n \geq 1$. Furthermore, we will define the weight function

$$\widehat{w}(x) = \frac{\widetilde{w}(x)}{(r+1)^2/r - x} = \frac{1}{(r+1)^2/r - x} \sqrt{\frac{4-x}{x}}, \quad x \in (0,4).$$

Since $d = (r+1)^2/r > 4$ for $r > 0$, we can apply the case (iv) from Lemma 2. So, we find $r_{-1} = 2\pi/(r+1)$, $r_n = 1/r$, $n \in \mathbb{N}_0$. Hence

$$\widehat{\alpha}_0 = \frac{r+1}{r}, \; \widehat{\alpha}_k = 2, \quad k \in \mathbb{N}, \quad \widehat{\beta}_0 = \frac{2\pi}{r+1}, \; \widehat{\beta}_1 = \frac{r+1}{r}, \; \widehat{\beta}_k = 1, \quad k \in \mathbb{N}; \; k \geq 2.$$

Finally, let us denote with $\{S_n(x)\}$ the sequence of monic polynomials orthogonal with respect to the inner product $\varphi(f,g) = \int_{\mathbb{R}} f(x)g(x)w(x)\,\mathrm{d}x$, where the weight $w(x)$ is defined by

$$w(x) = \frac{r+1}{2\pi r}\widehat{w}(x) = \frac{r+1}{2\pi r} \cdot \frac{1}{(r+1)^2 - rx}\sqrt{\frac{4-x}{x}}, \quad x \in (0,4).$$

Applying the case (i) from Lemma 2, we find

$$\alpha_0 = \frac{r+1}{r}, \ \alpha_k = 2, \quad k \in \mathbb{N}, \quad \beta_0 = \frac{1}{r}, \ \beta_1 = \frac{r+1}{r}, \ \beta_k = 1, \quad k \in \mathbb{N};\ k \geq 2.$$

Their squared norms are

$$\varphi(S_0,\ S_0) = \frac{1}{r}, \quad \varphi(S_n,\ S_n) = \beta_n\beta_{n-1}\ldots\beta_0 = \frac{r+1}{r^2}, \quad n \in \mathbb{N}.$$

6 The Connection with Polynomials Orthogonal with Respect to a Discrete Sobolev Inner Product

Here, we will recall the results from the paper [7] for $\lambda = r/(r-1)$ and $c = \zeta$.

The sequence of monic polynomials $\{Q_n(x)\}$ orthogonal with respect to the inner product

$$\widetilde{\varphi}(f,g) = \varphi(f,g) + \frac{1}{\lambda} f(c)g(c),$$

is quite determined by $\{S_n(x)\}$, λ and c.

Lemma 3. *The polynomials $\{Q_n(x)\}$ satisfy the three-term recurrence relation*

$$Q_{n+1}(x) = (x - \sigma_n)Q_n(x) - \tau_n Q_{n-1}(x), \quad n \in \mathbb{N}, \quad Q_{-1}(x) = 0,\ Q_0(x) = 1.$$

The first few members of the sequence $\{Q_n(x)\}$ are:

$$Q_0(x) = 1, \quad Q_1(x) = x - (r+1), \quad Q_2(x) = x^2 - (r+3)x + (r+1).$$

Hence $\tau_0 = \mu_0 = 1$ and $\tau_1 = r+1$.

Lemma 4. *The polynomials $\{S_n(x)\}$ at the point ζ have the following values: $S_0(\zeta) = 1,\ S_n(\zeta) = (r+1)r^{n-1},\ n \in \mathbb{N}$.*

Proof. The proof can be done by mathematical induction. □

Let us denote by

$$K_m(c,d) = \sum_{j=0}^{m} \frac{S_j(c)S_j(d)}{\varphi(S_j,S_j)}, \quad \lambda_m = 1 + \frac{K_m(c,c)}{\lambda}, \quad m \in \mathbb{N}.$$

Here, one has $K_m(\zeta,\zeta) = r(r^{2m+1}-1)$, $\lambda_m = r^{2m+1}$, $m \in \mathbb{N}$. Also, in the paper [7], one has proven that

$$\widetilde{\varphi}(Q_n,Q_n) = \varphi(S_n,S_n)\frac{\lambda_n}{\lambda_{n-1}}, \quad n \in \mathbb{N};\ n \geq 2.$$

Hence $\widetilde{\varphi}(Q_0,Q_0) = 1$, $\widetilde{\varphi}(Q_1,Q_1) = r+1$, $\widetilde{\varphi}(Q_n,Q_n) = r+1$, $n \in \mathbb{N}$; $n \geq 2$. Since $\widetilde{\varphi}(Q_n,Q_n) = \tau_n\tau_{n-1}\ldots\tau_1\tau_0$, we have $\tau_0 = 1$, $\tau_1 = r+1$, $\tau_n = 1, n \in \mathbb{N}; n \geq 2$.

Now, we have all elements for formula (7) and we can compute h_n by $h_n = \mu_0^n\tau_1^{n-1}\tau_2^{n-2}\cdots\tau_{n-2}^2\tau_{n-1}$.

Theorem 2. *The Hankel transform of the sequence* $\{a_n\}$ *defined in* (1) *is given by* $h_n = (r+1)^{n-1}$, $n \in \mathbb{N}$.

Acknowledgements This research was supported by the Science Foundation of Republic Serbia, Project No. 144023 and Project No. 144011.

References

1. Barry, P.: A Catalan transform and related transformations on integer sequences. J. Integer Seq. **8**, Article 05.4.5 (2005)
2. Chihara, T.S.: An Introduction to Orthogonal Polynomials. Gordon and Breach, New York (1978)
3. Cvetković, A., Rajković, P., Ivković, M.: Catalan numbers, the Hankel transform and Fibonacci numbers. J. Integer Seq. **5**, Article 02.1.3 (2002)
4. Gautschi, W.: Orthogonal Polynomials: Computation and Approximation. Clarendon, Oxford (2004)
5. Krattenthaler, C.: Advanced determinant calculus: A complement. Linear Algebra Appl. **411** (2005) 68–166
6. Layman, J.W.: The Hankel transform and some of its properties. J. Integer Seq. **4**, Article 01.1.5 (2001)
7. Marcellán, F., Ronveaux, A.: On a class of polynomials orthogonal with respect to a discrete Sobolev inner product. Indag. Mathem. N.S. **1**, 451–464 (1990)
8. Rajković, P.M., Petković, M.D., Barry, P.: The Hankel transform of the sum of consecutive generalized Catalan numbers. Integral Transform. Spec. Funct. **18** No. 4, 285–296 (2007)
9. Sloane, NJA: The On-Line Encyclopedia of Integer Sequences. Published electronically at http://www.research.att.com/~njas/sequences/, 2007

Extremal Problems for Polynomials in the Complex Plane

Borislav Bojanov

Dedicated to Gradimir V. Milovanović on the occasion of his 60th birthday

1 Introduction

We shall use the notation

$$\mathcal{P}_n := \left\{a_n z^n + a_{n-1} z^{n-1} + \cdots + a_1 z + a_0\right\}$$

for the class of all algebraic polynomials of degree n with complex coefficients.

According to the Fundamental Theorem of Algebra, every polynomial $P(z)$ of degree exactly n (i.e., with nonzero leading coefficient $a_n \neq 0$) has exactly n zeros $z_1, \ldots, z_n$ in the complex plane $\mathbb{C}$. It can be represented in the form

$$P(z) = a_n(z - z_1) \cdots (z - z_n)$$

and thus, up to a constant factor, every polynomial of degree exactly n can be identified with the set of n points in the plane, its zeros. The zeros of the derivative $P'(z)$ are called *critical points* of the polynomial. This name reflects the fact that every algebraic polynomial $P(z)$ is locally univalent in any domain that does not contain zeros of its derivative, i.e., if $P'(\zeta) \neq 0$, then $P(z_1) \neq P(z_2)$ for every two distinct points z_1, z_2 from the disk $\{z : |z - \zeta| < \delta\}$, for any sufficiently small δ.

The set of zeros $\{z_k\}_1^n$ defines uniquely the set of critical points $\{\xi_j\}_1^{n-1}$ of a polynomial $P \in \mathcal{P}_n$. However, a small perturbation of the zeros may provoke a serious replacement of the critical points. That is why problems about the geometry of zeros and critical points of algebraic polynomials are difficult to resolve. They have been in the scope of interest of great masters in classical analysis and algebra.

Borislav Bojanov
University of Sofia, Department of Mathematics, Blvd. James Boucher 5, 1164 Sofia, BULGARIA, Deceased April 8, 2009

W. Gautschi et al. (eds.), *Approximation and Computation: In Honor of Gradimir V. Milovanović*, Springer Optimization and Its Applications 42,
DOI 10.1007/978-1-4419-6594-3_5,

Fifty years ago (in the spring of 1958), the Bulgarian mathematician Blagovest Sendov made the following conjecture which, at first sight, looks like a student exercise in an introductory course on analytic functions. But it is not. Using the standard notations

$$D(a;\rho) := \{z : |z-a| \leq \rho\}, \quad D := D(0;1),$$

the conjecture reads.

Sendov's conjecture. *Let $P(z) = (z-z_1)\cdots(z-z_n)$ be a polynomial of degree $n \geq 2$ with all its zeros in the unit disk D. Then, for each zero z_k of P, the disk $D(z_k;1)$ contains at least one zero of the derivative P'.*

The conjecture stays still open despite the effort of many mathematicians. More than 100 papers have been published verifying the conjecture, in particular, cases or dealing with certain modifications of the original problem. Sendov made this conjecture trying to find an analogue of Rolle's theorem in the complex plane. Another open problem, raised in 1981 by Smale [38], can be considered also as analog of Rolle's theorem. There is much common between these two famous problems in the analytic theory of polynomials and usually they are treated together in recent publications (see [28, 32]). The main purpose of this paper is to call the attention of more mathematicians problem solvers to this half-a-century-old question, to present the basic ideas used up to now in the works on both problems, and to formulate another conjecture that seems to us to be the most natural analog of Rolle's theorem in the complex plane. It contains as a particular case Smale's conjecture.

The second subject of the paper is the Bernstein majorization theorem. We use here the notion "majorization" to describe the situation when two functions satisfy the inequality $|Q(z)| \leq |P(z)|$ over a certain region Ω in the complex plane. In such a case, we say that P *majorizes* Q over Ω.

Bernstein [2] (see also de Bruijn [8]) proved the following remarkable result which is often used in the study of complex polynomial problems.

As is usual, the boundary of a domain G in the plane is denoted by ∂G and

$$\|P\|_G := \sup_{z\in G} |P(z)|$$

is the uniform norm of P on the set G.

Theorem A. *Assume that G is a convex domain in the complex plane. Let P and Q be any polynomials of degree n with complex coefficients such that*

$$|Q(z)| \leq |P(z)| \quad \textit{for every } z \in \partial G.$$

Then

$$|Q'(z)| \leq |P'(z)| \quad \textit{for every } z \in \partial G,$$

provided all the n zeros of P lie in G.

Theorem A was proved by Bernstein [2] for the unit disk D. Later, it was noted by de Bruijn [8] that it holds for every convex domain G. The celebrated Bernstein inequality

$$|Q'(z)| \le n\|Q\|_D, \quad z \in D,\ Q \in \mathcal{P}_n,$$

is an immediate consequence of Theorem A. Indeed, take $P(z) = z^n$ and any polynomial $Q \in \mathcal{P}_n$ with $\|Q\|_D = 1$. Then, clearly, $|Q(z)| \le |P(z)|$ on ∂D, and by the theorem, for every $z \in \partial D$,

$$|Q'(z)| \le |P'(z)| = n,$$

which is the Bernstein inequality.

We are going to give here some further applications of Theorem A.

2 Gauss–Lucas Theorem and Extensions

The critical points of a polynomial P have the following interesting interpretation. Let us introduce the set

$$\Gamma(P,\varepsilon) := \{z : |P(z)| = \varepsilon\}.$$

This is the so-called *lemniscate* of the polynomial. The value of $|P(z)|$ is less than ε inside $\Gamma(P,\varepsilon)$ and bigger than ε outside $\Gamma(P,\varepsilon)$. When ε is a very small number, $\Gamma(P,\varepsilon)$ consists of closed curves each of them containing a zero of $P(z)$. Moreover, Walsh [43] proved that for sufficiently small ε these curves are convex. When increasing ε, the components (*branches*) of the lemniscate become larger and at a certain moment two or more of them will touch each other. The point of contact is namely a zero of the derivative of $P(z)$. Every critical point is a point of contact of two (or more) branches of the lemniscate for a certain $\varepsilon > 0$ or it coincides with a zero z_k of the polynomial P, in case z_k is a multiple zero of P. To know more about lemniscates, one can learn from the paper of Walsh [42].

Another interpretation, a physical one (see, for example [18, 28]) that associates with each zero z_k a force F_k acting to any point z in the plane (i.e., a force that repels z) with a magnitude inversely proportional to the distance from z_k to z, leads to the conclusion that the critical points are points of equilibrium of the forces $\{F_k\}$. Consequently, if a half-plane $\mathcal{H}$ contains all the zeros of a polynomial P, it contains also all points of equilibrium, that is, all critical points. This basic fact in the geometry of polynomials is stated in the following.

Gauss–Lucas Theorem. *All critical points of an algebraic polynomial lie in the convex hull of its zeros.*

In [11] Dimitrov gave an improvement of this classical result cutting additionally the vertexes of the convex hull corresponding to simple zeros of the polynomial. The following interesting extension of Gauss–Lucas theorem was obtained recently by Pereira [27] and, independently, by Malamud [17].

We give first some definitions.

Definition 1. Let $S = \{s_{ij}\}$ be an m by n real matrix. We say that S is a doubly stochastic rectangular matrix if and only if:

(i) $s_{ij} \geq 0$ for all i, j;
(ii) $s_{i1} + \cdots + s_{in} = 1$ for $i = 1, \ldots, m$;
(iii) $s_{1j} + \cdots + s_{mj} = m/n$ for $j = 1, \ldots, n$.

Definition 2. Let $(x_1, \ldots, x_m)$ and $(y_1, \ldots, y_n)$ be two sets of complex numbers. We say that $(x_1, \ldots, x_m)$ is *majorized* by $(y_1, \ldots, y_n)$ if there exists a doubly stochastic m by n matrix S such that $x_i = s_{i1}y_1 + \cdots + s_{in}y_n$ for $i = 1, \ldots, m$.

The Gauss–Lucas theorem says that for every $P \in \mathcal{P}_n$ there exists a stochastic (by rows) n by n matrix $C = \{c_{ij}\}$ such that the critical points $\xi_1, \ldots, \xi_{n-1}$ of P can be represented in terms of the zeros $z_1, \ldots, z_n$ of P in the form

$$\xi_i = c_{i1}y_1 + \cdots + c_{in}y_n \quad \text{for } i = 1, \ldots, n,$$

where we have set $\xi_n := (z_1 + \cdots + z_n)/n$. The next improvement due to Pereira [27] and Malamud [17] consists of the existence of a representation with a doubly stochastic matrix S. The following is true.

Theorem B. *Let P be an arbitrary polynomial of degree n. Then the critical points $\xi_1, \ldots, \xi_{n-1}$ of P are majorized by the zeros $z_1, \ldots, z_n$ of P.*

One can find in [27] and [17] some interesting consequences from this majorization result, one of them being a proof of the De Bruijn–Springer conjecture [9] from 1947: *Let $\Phi : \mathbb{C} \to \mathbb{R}$ be a convex function. Let P be a polynomial of degree n with zeros $z_1, \ldots, z_n$ and critical points $\xi_1, \ldots, \xi_{n-1}$. Then*

$$\frac{1}{n-1} \sum_{j=1}^{n-1} \Phi(\xi_j) \leq \frac{1}{n} \sum_{k=1}^{n} \Phi(z_k).$$

With the same notations as in the above and under the additional restriction that $z_1 + \cdots + z_n = 0$ Schoenberg [34] conjectured that

$$\frac{1}{n-2} \sum_{j=1}^{n-1} |\xi_j|^2 \leq \frac{1}{n} \sum_{k=1}^{n} |z_k|^2.$$

This also was proved by Pereira [27] and Malamud [17]. Other interesting analogs of Rolle's theorem were considered recently by Sendov [36] where he constructed polynomials P with large domains around two fixed zeros of P that are void of critical points.

The Gauss–Lucas theorem and its extensions show that if we know all the zeros of a polynomial P, we can describe a finite domain containing all the zeros of its derivative – namely, the convex hull of the zeros of P. In the case of a polynomial with real coefficients Rolle's theorem gives much more – it says that between any two distinct zeros of P there is at least one zero of P'. A similar statement is true also in the complex case, but with a much larger domain (determined by two zeros of P), which contains at least one zero of the derivative. Such a proposition is the following.

Grace–Heawood Theorem. *Let P be a polynomial of degree $n \geq 2$. If z_1, z_2 are zeros of P, then the disk*

$$\left\{ z : \left| z - \frac{z_1 + z_2}{2} \right| \leq \left| \frac{z_1 - z_2}{2} \right| \cot \frac{\pi}{n} \right\}$$

contains at least one zero of P'.

For example, if $P \in \mathcal{P}_n$ and we know that it vanishes at 0 and 1, then we can assert that the disk centered at $1/2$ and with a radius of order n will contain a zero of the derivative. Moreover, the Grace–Heawood theorem cannot be improved, it does not hold with a smaller radius of the disk. In other words, the Grace–Heawood theorem says that having fixed two zeros z_1 and z_2, the worst choice of the remaining $n-2$ zeros of a polynomial of degree n, in the whole plane $\mathbb{C}$, is to place them uniformly spaced, together with z_1 and z_2, on the circle that passes through z_1 and z_2. By "worst" we mean a choice that will produce the biggest region around the segment $[z_1, z_2]$ void of critical points.

3 Sendov's Conjecture

Let us denote by

$$\mathcal{P}_n(D) := \{P(z) = (z - z_1) \cdots (z - z_n) \, : \, z_k \in D, \, \forall k\}$$

the subset of polynomials of degree n that have all their zeros in the unit disk D. With any of these polynomials, we associate the radius

$$r(P) := \max_{1 \leq k \leq n} \min_{\xi} \{|z_k - \xi| \, : \, P'(\xi) = 0\}$$

and define

$$r^* := \max_{P \in \mathcal{P}_n(D)} r(P).$$

Sendov's conjecture is equivalent to the statement that $r^* = 1$. The estimates

$$1 \leq r^* \leq 2$$

are elementary. Indeed, $D \subset D(z_k; 2)$ for every zero z_k of $P \in \mathcal{P}_n(D)$. Then, by the Gauss–Lucas theorem, $r^* \leq 2$.

On the contrary, all zeros of the polynomial $g(z) = z^n - 1$ lie on the boundary ∂D of the unit disk, while its derivative $g'(z) = nz^{n-1}$ has only one zero (of multiplicity $n-1$), which lies in the origin. Therefore, $r(g) = 1$ and evidently $r^* \geq 1$.

We shall say that the polynomial P from $\mathcal{P}_n(D)$ is *extremal*, if $r(P) = r^*$. It is not difficult to conjecture that the polynomial $z^n - 1$ is the unique (up to rotation) extremal polynomial of degree n for Sendov's problem. That is why sometimes Sendov's conjecture is given in the following stronger version.

Stronger form of Sendov's conjecture. *For every natural n, the equality $r^* = 1$ holds, and $r(P) = r^* = 1$ only for the polynomial $P(z) = z^n - 1$ (up to rotation).*

In this form, Sendov's conjecture was proved for polynomials of degree less than or equal to 8. For $n = 2$, the conjecture is obviously true since the only zero of the derivative lies at the middle of the linear segment connecting the two zeros of the polynomial. According to Sendov [35], he verified his conjecture for $n = 3$ but he did not publish the proof, hoping to solve the problem in the general case. The first signs of interest in the conjecture have occurred in 1968–1969 when the problem was solved for $n = 3,4,5$ by several authors. The case $n = 4$ and $n = 5$ was treated by Meir and Sharma [22]. The conjecture is now verified in the strong form for all $n \leq 8$. A complete bibliography on this subject and, in particular, complete references to results for small n are given in [23, 28] and in the survey paper by Sendov [35]. Morris Marden was fascinated by the beauty and the simplicity of the conjecture and did a lot to make it known to the mathematical community (see [19, 20]).

Finding good upper estimates for the radius r^* is one of the main directions in which the mathematicians working on Sendov's problem put their effort. As we noticed already, the Gauss–Lucas theorem yields the bound $r^* \leq 2$. Using the stronger statement of Theorem B, we can easily derive the estimate: For every zero z_k of P, there is a zero ξ of P' such that

$$|z_k - \xi| \leq \left(1 - \frac{1}{n}\right)(1 + |z_k|).$$

To see this, we simply take the equation $\xi_i = s_{i1}z_1 + \cdots + s_{ik}z_k + \cdots + s_{in}z_n$ with the maximal coefficient s_{ik}, subtract z_k from the both sides and estimate

$$|\xi_i - z_k| \leq (1 - s_{ik})|z_k| + 1 - s_{ik} \leq \left(1 - \frac{1}{n}\right)(1 + |z_k|).$$

Taking now $z_k = 0$ we get the proposition: *If $P(0) = 0$ and all other zeros of P lie in the unit disk, then the disk $D(0; 1 - 1/n)$ contains at least one zero of P'.* This is a trivial consequence from Theorem B but it suggests an interesting problem, a specification of Sendov's question for particular subclass of polynomials:

Problem 1. For any fixed natural n, find the smallest r_n such that the disk $D(0; r_n)$ contains at least one zero of P' for every polynomial $P \in \mathcal{P}_n(D)$ that vanishes at $z = 0$.

The polynomial $P(z) = (z-1/2)^n - 1/2^n$ evidently vanishes at $z=0$ and all zeros of its derivative lie at the point $z=1/2$. This example implies $1/2 \le r_n$. However, another polynomial, namely $P(z) = z(z^{n-1}-1)$, yields a much better lower estimate. Clearly, all critical points of P lie on the circle of radius $(1/n)^{1/(n-1)}$ and center in the origin. This, together with the above-mentioned upper estimate, implies

$$\left(\frac{1}{n}\right)^{1/(n-1)} \le r_n \le 1 - \frac{1}{n}.$$

Therefore, r_n tends to 1 as $n \to \infty$. It would be interesting to find the extremal polynomial and the exact value of r_n, at least for small n. Note that a result of Meir and Sharma [22] (see also [23, pp. 228–229]) gives the inequalities

$$r_3 \le \frac{\sqrt{2}}{\sqrt{3}}, \quad r_4 \le \frac{\sqrt{2}}{2}.$$

More precise estimates are induced by the bound (7.3.17) in [28], namely,

$$r_3 \le \frac{1}{\sqrt{3}}, \quad r_4 \le \frac{\sqrt{5}}{2\sqrt{3}}, \quad r_5 \le \frac{\sqrt{2}}{2}.$$

In the particular case $n=3$, we obtain the exact value: $r_3 = 1/\sqrt{3}$.

Let us return now to the original problem of Sendov.

In 1971, Schmeisser [31] proved that $r^* \le 1.568$. Next we present the bound for r^* obtained in [5].

Theorem 1. *Let $P(z) = (z-z_1)\cdots(z-z_n)$ be a polynomial of degree $n \ge 2$ with all its zeros in the unit disk D. Then*

$$r(P) \le (1 + |z_1 \cdots z_n|)^{1/n}.$$

Let us first note an immediate consequence from the above theorem.

Since $|z_k| \le 1$ for all k, we have

$$r^* \le (1 + |z_1 \cdots z_n|)^{1/n} \le 2^{1/n} < 1 + \frac{1}{n}$$

which means that Sendov's conjecture is asymptotically true. Besides using the fact that the conjecture was verified for all $n \le 8$, we obtain an absolute bound $r^* \le 2^{1/8} < 1.0905$ which holds for every n. With more care, one can get $r^* \le 1.055767$ (see [28, Theorem 7.3.17]). Any proof of Sendov's conjecture for further $n = 9, 10, \ldots$ would lead to an improvement of the absolute upper bound.

Sketch of the proof of Theorem 1. We start with the following

Definition 3. The polynomials p and q of degree n are *apolar*, if

$$\sum_{k=0}^{n} (-1)^k p^{(k)}(0) q^{(n-k)}(0) = 0.$$

If p and q are given in the form

$$p(x)=\sum_{k=0}^{n} a_k x^k, \quad q(x)=\sum_{k=0}^{n} b_k x^k,$$

then the above relation can be rewritten as

$$\sum_{k=0}^{n} \frac{(-1)^k}{\binom{n}{k}} a_k b_{n-k}=0.$$

Thus, the apolarity is just an orthogonality relation. If two polynomials are apolar with p, then each linear combination of them is also apolar with p. In conclusion, all polynomials of degree n whose vectors of coefficients $\mathbf{c} := (c_0,\ldots,c_n)$ lie in the hyperplane that is orthogonal to the vector $\mathbf{a}^\circ := (a_0^\circ,\ldots,a_n^\circ)$ with

$$a_k^\circ := (-1)^k \binom{n}{k}^{-1} a_k, \quad k=0,\ldots,n,$$

are apolar with p. Therefore, the next theorem that reveals an interesting connection between the zeros of two apolar polynomials is indeed a deep and important result in the analytic theory of polynomials.

Grace's Theorem. *Let p and q be apolar polynomials. Then every circular domain which contains all zeros of one of these polynomials contains at least one zero of the other polynomial.*

By a circular domain, we mean any domain whose boundary is a circle or a line.

The proof of Grace's theorem can be found in most of the books treating zeros of algebraic polynomials. We recommend the excellent work by Obrechkoff [26] where several proofs of this result, as well as many important applications, are given.

The proof of Theorem 1 is based on the fact that the mid-perpendicular of the segment connecting any two distinct zeros of a polynomial separates the zeros of its derivative. We derived this fact in [3] as a corollary from Grace's theorem, as demonstrated in Lemma 1 below, and showed by simple geometric arguments that it yields the estimate $r^* < (1+\sqrt{5})/2$. As we have been informed later, Lemma 1 was known much earlier. It was mentioned by Szegö [40] in 1922 but has remained unnoticed and was not presented in none of the monographs and surveys on the distribution of zeros of polynomials, published prior to the publication of [3]. So, it is worth recalling this nice observation here and attach the proof.

Lemma 1. *Let p be a polynomial of degree $n \ge 2$. If z_1, z_2 are any two distinct zeros of p, then each of the closed half-planes, whose boundary is the mid-perpendicular of the line segment $[z_1,z_2]$, connecting z_1 and z_2, contains a zero of the derivative p' of the polynomial.*

Proof. Without loss of generality, we may assume that $z_1=0$, $z_2=1$. Then the assumption $p(0)=0$ shows that the polynomial p is of the form

$$p(z)=a_1 z + a_2 z^2 + \cdots + a_n z^n,$$

whereas $p(1) = 0$ implies the relation $a_1 + \cdots + a_n = 0$. Then we conclude that the polynomials $p'(z)$ and $q(z) := (z-1)^n - z^n$ are apolar. Indeed,

$$\sum_{k=0}^{n-1} (-1)^k p^{(k+1)}(0) q^{(n-1-k)}(0) = -n! \sum_{k=0}^{n-1} \frac{p^{(k+1)}(0)}{(k+1)!} = -n!(a_1 + \cdots + a_n) = 0.$$

Besides, it is easy to show that the zeros of $q(z)$ are situated on the line $x = 1/2$. Then by Grace's theorem, each of the half-planes $x \geq 1/2$, $x \leq 1/2$ contains a zero of $p'(z)$. □

As an immediate consequence of Theorem 1 we obtain: *Sendov's conjecture is true for polynomials that vanish at the origin.* The proof is just one line:

$$P(0) = 0 \implies z_1 \cdots z_n = 0 \implies r(P) \leq 1.$$

Actually, the assertion that the Sendov conjecture is true for polynomials P such that $P(0) = 0$ follows directly from Lemma 1. Indeed, if z_k is any zero of P, distinct from 0, then the mid-perpendicular ℓ of the segment $[0, z_k]$ will separate the zeros of P', i.e., both parts of the unit disk, cut along the line ℓ, will contain at least one zero of the derivative $P'(z)$. But clearly the part that contains the zero z_k lies in the disk $D(z_k; 1)$ and consequently there is a critical point in this disk. The claim is proved.

As we mentioned already, Lemma 1 is the basic tool in the proof of Theorem 1. The idea is to show that in a small neighborhood of any fixed zero z_k of P there exists a point ζ at which the polynomial P attains the value $P(0)$. Then the polynomial $Q(z) := P(z) - P(0)$ would vanish for $z = 0$ and $z = \zeta$ and according to the claim just proved, Q' would have at least one zero in the disk $D(\zeta; 1)$. But evidently $Q' \equiv P'$. Therefore, P will have a critical point in the disk $D(\zeta; 1)$. If ζ is sufficiently close to z_k, say in a distance δ, then the disk $D(z_k; 1+\delta)$ centered at z_k and with a radius $1+\delta$ will contain $D(\zeta; 1)$. Therefore, Theorem 1 would be proved if we succeed to show that δ is a sufficiently small number. But this is seen (roughly speaking) from the fact that $|P(0)|$ is small, i.e., $|P(0)| = |z_1 \cdots z_n| \leq 1$, while $|P'(z_k)| = n|(z_k - \xi_1) \cdots (z_k - \xi_{n-1})|$ and the assumption that all critical points $\xi_1, \ldots, \xi_{n-1}$ of the polynomial P lie outside the disk $D(z_k; 1)$ will imply that $|P'(z_k)|$ is of order n, i.e., quite big. This implies that in a small neighborhood of z_k (in a disk of radius $1/n$) the polynomial $P(z)$ will take all complex values smaller or equal to 1 in absolute value. In particular, this small neighborhood will contain also a point ζ at which $P(\zeta) = P(0)$ what was to be shown to end the proof of Theorem 1. □

Sendov's conjecture was proved also for some special classes of polynomials. For example, the following proposition, given in [28, Corollary 7.3.13] (improving a result due to Schmeisser [30]), is true: *Let*

$$P(z) = (z - z_1) \cdots (z - z_n) = \sum_{j=1}^{m} a_j z^{n_j},$$

where $0 \leq n_1 \leq \cdots \leq n_m \leq n$ *and* $m < (n+2)/3$. *Then* $r(P) \leq 1$.

It is rather surprising that Sendov's conjecture is not proved yet even for polynomials with real coefficients and real critical points.

Now we shall give some results that treat the existence of a critical point in a circular domain around a fixed zero z_k of the polynomial P. If z_k lies on the boundary of the unit disk, then P has a critical point in a smaller subdomain of the disk $D(z_k;1)$. This was established by Goodman et al. [16], and independently by Schmeisser [31].

Theorem C. *Let $P(z)$ be a polynomial from $\mathcal{P}_n(D)$ with a zero z_1 such that $|z_1|=1$. Then the disk $D(z_1/2;1/2)$ contains a critical point of the polynomial P.*

The proof is based on the classical Laguerre Theorem. Let us recall it.

Laguerre's Theorem. *Let p be a polynomial of degree $n \geq 2$ and let $\alpha \in \mathbb{C}$ be such that $p(\alpha) \neq 0$, $p'(\alpha) \neq 0$. Then every circle Γ which passes through the points α and β,*

$$\beta := \alpha - \frac{np(\alpha)}{p'(\alpha)},$$

separates at least two zeros of p or all zeros lie on the circle Γ.

We refer to [18, 26, 28] for the proof.

Proof of Theorem C. Let us represent the polynomial P in the form $P(z) = (z-z_1)Q(z)$, where Q is the corresponding polynomial of degree $n-1$. One easily verifies that $P'(z_1) = Q(z_1)$, $P''(z_1) = 2Q'(z_1)$. If $P'(z_1) = 0$, there is nothing to prove. Let us assume that $P'(z_1) \neq 0$. This yields $Q(z_1) \neq 0$.

Let us apply first Laguerre's theorem to the polynomial $P'(z)$ and the point $\alpha = z_1$. We have

$$\beta = z_1 - \frac{(n-1)Q(z_1)}{2Q'(z_1)}.$$

If β is in the disk $D(z_1/2;1/2)$, then $P'(z)$ has there a zero and the theorem would be proved. Let us assume that $\beta \notin D(z_1/2;1/2)$, i.e.,

$$\left|\beta - \frac{z_1}{2}\right| > \frac{1}{2}. \tag{1}$$

Let us apply now Laguerre's theorem to the polynomial $Q(z)$ and the point $\alpha = z_1$. For the associated point β_1, we obtain

$$\beta_1 = z_1 - \frac{(n-1)Q(z_1)}{Q'(z_1)}.$$

Therefore, $2\beta - \beta_1 = z_1$. This, together with assumption (1) gives $|\beta_1| = |2\beta - z_1| > 1$ and therefore one can draw a circle Γ through the points β_1 and z_1, which contains the unit disk D. This is a contradiction since in such a case Γ would not separate the zeros of Q. The proof is complete. □

Corollary 1. *If all the zeros of the polynomial P lie on the boundary of the unit disk, then $r(P) \leq 1$.*

Corollary 2. *Let $P(z) = (z - z_1) \cdots (z - z_n)$ be a polynomial of degree $n \geq 2$. Let us assume that the vertexes of the convex hull $\mathcal{K}(\mathcal{P})$ of the zeros $z_1, \ldots, z_n$ of P lie on the unit circle ∂D. Then every disk $D(z_k; 1)$ contains a critical point of P.*

The last assertion follows from Theorem C and the fact that the disks $D(z_1/2; 1/2)$, $D(z_2/2; 1/2)$ corresponding to two adjacent vertexes of $\mathcal{K}(\mathcal{P})$ cover completely the line segment $[z_1, z_2]$ connecting the two vertexes z_1 and z_2, and thus, the triangle with vertexes z_1, z_2, 0. This observation implies that all disks, corresponding to the vertexes of $\mathcal{K}(\mathcal{P})$ cover the domain $\mathcal{K}(\mathcal{P})$ and therefore, each zero z_j of P which is not a vertex will lie in some of the covering disks. But all of these disks are of radius $1/2$ and each of them contains a critical point. Then for every zero z_j, there will be a critical point in a distance not exceeding 1.

The following extremal problem occurs naturally in the study of Sendov's conjecture. For a fixed point w in the complex plane, let us denote by $\mathcal{P}_n(D, w)$ the subset of polynomials of degree n which vanish at the point w and the rest of their $n-1$ zeros lie in the unit disk. With every polynomial P from $\mathcal{P}_n(D, w)$ we associate the minimal radius $\rho(P; w)$ of the disk centered at w, which has a critical point on its boundary, but no one in its interior. Sendov's conjecture actually asserts that $\rho(P; w) \leq 1$ for every point w of the unit disk D and every polynomial P from $\mathcal{P}_n(D, w)$. Let

$$\rho(w) := \max\{\rho(P; w) : P \in \mathcal{P}_n(D, w)\}.$$

The polynomial P^*, for which the above maximum is attained, is called *the maximal polynomial* for the point w. The radius $\rho(P^*; w)$ will be called *critical radius for* w, and the boundary of the disk $D(w; \rho(P^*; w))$ – *critical circle*. The description of the maximal polynomials for a given point w is a difficult problem. The maximal polynomial with biggest critical radius is *extremal* for Sendov's conjecture.

Because of preservation of the set of polynomials $\mathcal{P}_n(D)$ after a rotation of the zeros, it suffices to show that the critical radius does not exceed 1 only for the real points from the interval $[0, 1]$. We have already shown that this inequality is valid for the end-points $z = 0$ and $z = 1$. Let us assume now that $w \in (0, 1)$. Let P^* be the maximal polynomial for the point $w = w_0$ and assume that there is exactly one critical point ξ of P^* on the critical circle of w_0. In a small neighborhood of w_0, the critical point ξ is an analytic function of w. Consider the difference $w - \xi(w)$. It is also an analytic function of w in a small disk U around w_0. Let us choose U so small that it is contained totally in the unit disk D. By the Maximum-Modulus Principle, the distance $|w - \xi(w)|$ attains its maximal value over U at a certain point w_1 on the boundary of U. This shows that the maximal polynomial of w_0 is not extremal since we found a point w_1 in the unit disk D (recall that w_0 is an interior point of D), whose critical radius is bigger than that of w_0. Such an argument would lead us to the conclusion that the extremal polynomial must be associated with a point from the boundary of D and then, in view of Corollary 1, to a proof of Sendov's conjecture. The problem is that more than one critical point may lie on the critical circle. Then the above reasoning does not work since one of the distances (between the zero and a critical point) may increase while some other could decrease when w varies in U. For example, in the case of the polynomial $z^n - 1$ all critical points lie on the critical circle. Moreover, they all coalesce at a single point. Thus, it is not likely that by varying only one zero (i.e., w) of the corresponding maximal polynomial, we would achieve an increase of the critical radius, if more than one critical point lie on the critical circle. Recently, Schmieder suggested in his electronic publication [33] an interesting way of variation of the maximal polynomial by Blaschke transform, varying simultaneously all zeros so that they remain in the unit disk. For given complex parameters $\mathbf{h} = (h_1, \dots, h_n)$, he considers the functions

$$\varphi(z, h_j) := \frac{z + h_j}{1 + \overline{h}_j z}, \quad j = 1, \dots, n.$$

For $|h_j| < 1$ they represent conformal mappings of the unit disk into itself so that the interior points go to interior points, and the boundary points go to points on the boundary. Schmieder suggested a variation of the corresponding maximal polynomial P by means of the parameters $t\mathbf{h}$, so that it goes to the polynomial $Q(z, t, \mathbf{h})$ with zeros $z_j(t) = z_j(th_j) := \varphi(z_j, th_j)$, where t is a real parameter. Let z_1 be a zero of P for which $P'(z_1) \neq 0$. Consider the behavior of $z_1(t)$ in a neighborhood of $t = 0$. By definition

$$z_1(t) = \frac{z + th_1}{1 + t\overline{h}_1 z}$$

and hence $z_1'(0) = h_1 - \overline{h}_1 z_1^2$. Assume that ξ is a zero of P' but it is not a zero of P and P''. Then, by the Implicit Function Theorem, there exists a differentiable curve $\xi(t)$ satisfying the conditions

$$\xi(0) = \xi, \quad \frac{\partial Q}{\partial z}(\xi(t), t) \equiv 0.$$

For small t, the curves $\xi(t)$ and $z_1(t)$ can be represented in the following way

$$z_1(t) = z_1 + t z_1'(0) + O(t^2), \quad \xi(t) = \xi + t\xi'(0) + O(t^2).$$

Therefore,

$$|z_1(t) - \xi(t)| = \left| 1 + t\frac{z_1'(0) - \xi'(0)}{z_1 - \xi} + O(t^2) \right|.$$

Now it is seen that if

$$\operatorname{Re} \frac{z_1'(0) - \xi'(0)}{z_1 - \xi} > 0, \tag{2}$$

then we will obtain $|z_1(t) - \xi(t)| > |z_1 - \xi|$ for sufficiently small t. Hopefully, the study of the expression on the left side of (2) could lead us to finding appropriate parameters $\mathbf{h}$ which secure inequalities (2) for every zero ξ, lying on the critical circle, of the derivative of the maximal polynomial for z_1. This will prove that all zeros of the extremal polynomial should lie on the boundary of the unit disk, and then, by already known results, would yield a proof of Sendov's conjecture. We must mention, however, that the expression in (2) was derived only in the case when the zeros ξ of the derivative, which lie on the critical circle, are simple. The main difficulty occurs in the case of multiple critical points on the critical circle. Then one should use variations of higher order and the corresponding expressions become more complicated. Recently, Borcea [7] showed that Schmieder's approach, as any other approach that uses only variation of the first order, cannot lead to the proof of the claim that all zeros of the extremal polynomial lie on the boundary of the unit disk.

A comprehensive review on the results devoted to Sendov's conjecture, with simplified proofs, extensions, and improvements, is given in the recently published book by Rahman and Schmeisser [28].

4 The Conjecture of Smale

Studying Newton's method for finding the zeros of algebraic polynomials, Steve Smale [38] has arrived at a problem of estimating the derivative of a polynomial in terms of divided differences of a special form. More precisely, it was necessary to find an absolute constant K, as small as possible, such that for any given $z \in \mathbb{C}$ for which $f'(z) \neq 0$ there exists at least one critical point ξ of the polynomial f so that

$$|f'(z)| \geq \frac{1}{K} \left| \frac{f(\xi) - f(z)}{\xi - z} \right|.$$

Clearly, it is sufficient to prove the above inequality only for the point $z = 0$ and a polynomial f for which $f(0) = 0$ and $f'(0) \neq 0$. That is why the conjecture is usually formulated in the following form.

Smale's conjecture. *Let f be a polynomial of degree n and such that $f(0) = 0$ and $f'(0) \neq 0$. Then*

$$\min \left\{ \left| \frac{f(\xi)}{\xi f'(0)} \right| : f'(\xi) = 0 \right\} \leq K$$

where $K = 1$ or most probably $K = (n-1)/n$.

Smale [38] proved the inequality with the constant $K = 4$. On the contrary, the polynomial $f(z) = z^n - z$ is an example that K cannot be a number smaller than $(n-1)/n$. Therefore, the following bounds hold for the best constant

$$\frac{n-1}{n} \leq K \leq 4.$$

Recently, the upper bound has been diminished to $4 - 2/(n+1)$ in [10]. The conjecture with the best constant $K = (n-1)/n$ was verified for polynomials of degree $n \leq 10$ by Marinov and Sendov [37] using a powerful computer to estimate numerically a certain multivariate function of complex variables. There are also proofs for special classes of polynomials.

Next we present a result of Tischler [41] which treats polynomials with zeros on the unit circle.

Theorem 2. *Let $f(z) = (z - z_1)\cdots(z - z_n)$ be a polynomial for which $f(0) = 0$ and all remaining $n-1$ zeros are situated on the unit circle ∂D. Then*

$$\min\left\{\left|\frac{f(\xi)}{\xi f'(0)}\right| : f'(\xi) = 0\right\} \leq \frac{n-1}{n}.$$

The equality is attained if and only if $f(z) = a_1 z + a_n z^n$, where $a_1 a_n \neq 0$.

Proof. We shall derive this result from a note of Szegö [31], concerning the following problem possed by Paul Erdös and U. Fush: *Let*

$$p(z) = (z - w_1)\cdots(z - w_m), \quad |w_k| \leq 1,\ k = 1,\ldots,m.$$

Prove that the set $\{z : |p(z)| \leq 1\}$ consists of at most $m-1$ components.

The solution of Szegö follows: It is known that the components of the lemniscate $\Gamma(p,C)$, C being a constant, touch each other only at points ξ for which $p'(\xi) = 0$. Let $\xi_1,\ldots,\xi_{m-1}$ be the zeros of the derivative $p'(z)$. It is easily verified that

$$\min_j |p(\xi_j)|^{m-1} \leq |p(\xi_1)\cdots p(\xi_{m-1})| = m^{-m}|p'(w_1)\cdots p'(w_m)| = m^{-m}\prod_{i\neq k}|w_i - w_k|.$$

The maximum of the last product is attained only for the polynomial $p_*(z) = z^m - 1$, up to rotation (see I. Schur, Mathematische Zeitschrift, 1918, p. 385). And in this extremal case we have $|p'(w_1)| = \cdots = |p'(w_m)| = m$. Therefore,

$$\min_j |p(\xi_j)| \leq 1,$$

with equality only for p_*. This shows that there is at least one critical point ξ_j in the interior of the lemniscate $|p(z)| = 1$, and consequently, the components of this lemniscate are at most $m-1$. The Szegö proof is completed.

We shall make use of the inequality

$$|p'(w_1)\cdots p'(w_m)| \leq m^m, \tag{3}$$

obtained in the process of the proof above. It easily implies Tischler's result. Indeed, let $z_1 = 0$, and let us denote by $\xi_1,\ldots,\xi_{n-1}$ the zeros of $f'(z)$. Assume that

$$\left|\frac{f(\xi_j)}{\xi_j f'(0)}\right| > \frac{n-1}{n}, \quad \text{for all } j = 1,\ldots,n-1.$$

Since $\xi_1\cdots\xi_{n-1} = f'(0)/n$, the last assumption implies

$$\left(\frac{n-1}{n}\right)^{n-1} < \frac{n|f(\xi_1)\cdots f(\xi_{n-1})|}{|f'(0)|^n}. \tag{4}$$

But

$$|f(\xi_1)\cdots f(\xi_{n-1})| = |f'(z_1)\cdots f'(z_n)|/n^n = |f'(0)f'(z_2)\cdots f'(z_n)|/n^n$$
$$= |f'(0)z_2\cdots z_n\, p'(z_2)\cdots p'(z_n)|/n^n,$$

where $p(z) = (z-z_2)\cdots(z-z_n)$. Let us note that $|f'(0)| = |z_2\cdots z_n| = 1$. Now, applying the Szegö inequality (3) to the polynomial p, we obtain

$$|p'(z_2)\cdots p'(z_n)| \leq (n-1)^{n-1}.$$

Then (4) leads to the contradiction:

$$\left(\frac{n-1}{n}\right)^{n-1} < n\cdot\frac{(n-1)^{n-1}}{n^n} = \left(\frac{n-1}{n}\right)^{n-1}.$$

The proof is complete. □

Both conjectures, those of Sendov and Smale, have been verified in similar particular cases (see the survey by Schmeisser [32] on this subject). Is there some connection between them? We continue our presentation with another formulation of Smale's conjecture, which shows that both conjectures are actually claims about the existence of a critical point of a polynomial in a domain, described in a certain way in terms of the zeros of the polynomial. In Sendov's case, this is a disk of radius 1 around a fixed zero z_k of the polynomial f, while in Smale's case this is a certain lemniscate of the polynomial $f(z)/(z-z_k)$. In other words, both conjectures are complex analogs of Rolle's theorem.

Let $z_1,\ldots,z_n$ be the zeros of the polynomial $f \in \mathcal{P}_n$ and assume that $f'(z_k) \neq 0$ for a certain fixed k. Recall that

$$\ell_{nk}(z) := \frac{f(z)}{(z-z_k)f'(z_k)}$$

is the basic polynomial for Lagrange interpolation associated with the node z_k. Smale's conjecture can be formulated in the following equivalent way: $|\ell_{nk}(\xi)| \leq K$ for at least one zero ξ of f'. Indeed, if $z_k = 0$ the expression $\ell_{nk}(\xi)$ is the same as that in the original conjecture. In particular, for $K = 1$, the above conjecture asserts that there exists at least one zero ξ of the derivative of f at which the basic Lagrange polynomial ℓ_{nk} is bounded by 1 in absolute value. Formulated in this way, it is easily verified in the case of polynomials f with only real zeros. Indeed, let $x_1 < \cdots < x_n$ be any given real points and $f(x) := (x-x_1)\cdots(x-x_n)$. Let $\xi_1 < \cdots < \xi_{n-1}$ be the zeros of $f'(x)$. Clearly, the points $\{x_j\}$ and $\{\xi_k\}$ interlace, i.e.,

$$x_1 < \xi_1 < x_2 < \cdots < \xi_{k-1} < x_k < \xi_k < x_{k+1} < \cdots < \xi_{n-1} < x_n.$$

Since, by construction, $\ell_{nk}(x_k) = 1$ and $\ell_{nk}(x_{k-1}) = \ell_{nk}(x_{k+1}) = 0$, then evidently $|\ell_{nk}(\xi)| < 1$ at least at one of the critical points ξ_{k-1}, ξ_k, which are closest to x_k. Otherwise, it would follow that $\ell_{nk}(x)$ has at least three local extrema in the interval (x_{k-1}, x_{k+1}), which leads to contradiction with the fact that f' has exactly $n-1$ zeros. Thus, the claim is obviously true in the real case.

The proof of Smale's conjecture with a constant $K = (n-1)/n$ for polynomials with only real critical points was given by Tischler [41] and presented also in [28]. Interpretations of the problem from the point of view of electrostatics were discussed in [12].

Next we formulate Smale's conjecture in the following more appealing form.

Alternative formulation of Smale's conjecture. *Let $z_1, \ldots, z_{n-1}$ be any fixed points in the complex plane. Set $q(z) = (z-z_1)\cdots(z-z_{n-1})$. There is a constant K ($K = 1$ or most possibly $K = (n-1)/n$) such that for any given $\varepsilon \geq 0$ and each point z_n on the boundary of the lemniscate $\Gamma(q, \varepsilon)$, there is at least one critical point of the polynomial $f(z) = (z-z_1)\cdots(z-z_n)$ in the interior of $\Gamma(q, K\varepsilon)$.*

It is easily seen that both conjectures are equivalent. Indeed, assume that the ACS (alternative conjecture of Smale) is true. Take any polynomial $f(z) = \prod_{j=1}^{n}(z-z_j)$ with $f'(z_n) \neq 0$. Let $|q(z_n)| = \varepsilon$. Then, by the ASC, there exists a zero ξ of f' inside the lemniscate $\Gamma(q; K\varepsilon)$. Hence $|q(\xi)| < K\varepsilon$ and consequently,

$$|\ell_{nn}(\xi)| = \frac{1}{|q(z_n)|}|q(\xi)| = \frac{1}{\varepsilon}|q(\xi)| < K.$$

Thus, SC (Smale's conjecture) follows from ASC.

Assume now that SC is true. For given $z_1, \ldots, z_{n-1}$ and $\varepsilon > 0$, take any point z_n on the lemniscate $\Gamma(q, \varepsilon)$. Then

$$\ell_{nn}(z) = \frac{1}{|q(z_n)|}q(z) = \frac{1}{\varepsilon}q(z)$$

and thus, by SC, $|\ell_{nn}(\xi)| \leq K$ for some critical point ξ of f. But the later means that $|q(\xi)| < K\varepsilon$ and therefore ξ is interior to the lemniscate $\Gamma(q, K\varepsilon)$, which is exactly the claim in the ASC.

The alternative form of Smale's conjecture and some computer experiments suggest even a stronger assertion which we state here with a constant $K = 1$.

Conjecture 1. For any given $\varepsilon \geq 0$ and every point z_n from any particular branch γ of the lemniscate $\Gamma(q, \varepsilon)$, the domain surrounded by γ contains at least one critical point of the polynomial $f(z) = (z-z_1)\cdots(z-z_n)$.

This would be a nice analog of Rolle's theorem in the complex plane, if true. The conjecture is true for $\varepsilon = 0$ and for sufficiently big ε. It was actually verified for polynomials with real zeros in the corresponding proof above. Besides, it is also evidently true for polynomials of degree 2. Let us sketch the proof of the conjecture for $n = 3$. To do this, we may fix two of the zeros, say $z_1 = 0$ and $z_2 = 1$, and define $q(z) := z(z-1)$. For any given $\varepsilon \geq 0$ and $0 \leq t \leq \varepsilon$, we consider the lemniscate

$\Gamma(q,t)$ and take a particular branch of it, call it $\gamma(t)$. Remember that $|q(z)| = t$ for every $z \in \gamma(t)$. Next we choose an arbitrary point z_3 on $\gamma(t)$. Set

$$f(z) := (z - z_3)q(z) = (z - z_3)z(z-1)$$

and introduce the function $m(t) := \min\{|f(z)| : z \in \gamma(t)\}$. Clearly, the minimum of $|f(z)|$ on $\gamma(t)$ is attained at a point $z(t)$ from $\gamma(t)$ which is in a minimal distance from the third zero z_3 amid all points from $\gamma(t)$. Then $z(t)$ is a point of contact of $\gamma(t)$ and the circle $C(t)$ with center at z_3 and radius $|z(t) - z_3|$.

Assume first that z_3 does not lie on the line $x = 1/2$ and let, for the sake of definiteness, z_3 be in the half plane $x < 1/2$. Then there is only one point of contact $z(t)$ of $C(t)$ and $\gamma(t)$.

Consider now the lemniscate $\{z : |f(z)| = m(t)\}$. It has one component $g_1(t)$ around $z = 0$ and lying inside $\gamma(t)$ and another one $g_2(t)$ around z_3 and lying inside the circle $C(t)$. The components lie inside the corresponding curves $C(t)$ and $\gamma(t)$ since $|f(z)| \geq |f(z(t))| = m(t)$ on these curves with equality only for $z = z(t)$. Thus, the component $g_1(t)$ is inside $\gamma(t)$ and touches its boundary at $z(t)$. The other component $g_2(t)$ is inside the circle $C(t)$. Next we choose a special t. Since $m(0) = m(\varepsilon) = 0$, there is a point $t_* \in (0, \varepsilon)$ for which $m(t)$ attains its maximum on $[0, \varepsilon]$. It is seen that for $t = t_*$ the second component $g_2(t_*)$ should also touch the circle $C(t_*)$ at the point $z(t_*)$ since $z(t_*)$ is the only point z on $C(t_*)$ at which $|f(z)| = m(t_*)$. Then $z(t_*)$ is a point of contact of two distinct components of the lemniscate $|f(z)| = m(t_*)$ and therefore $z(t_*)$ is a critical point of f. Since $0 < t_* < \varepsilon$, the point $z(t_*)$, being on $\gamma(t_*)$, lies inside $\gamma(\varepsilon)$. This is what we wanted to show.

Consider now the case when z_3 lies on the line $x = 1/2$. Then there are two points of contact z' and z'', but because of symmetry, the component $g_2(t_*)$ should touch $C(t_*)$ at both points and in this case there are even 2 critical points of f in $\gamma(\varepsilon)$. This completes the proof.

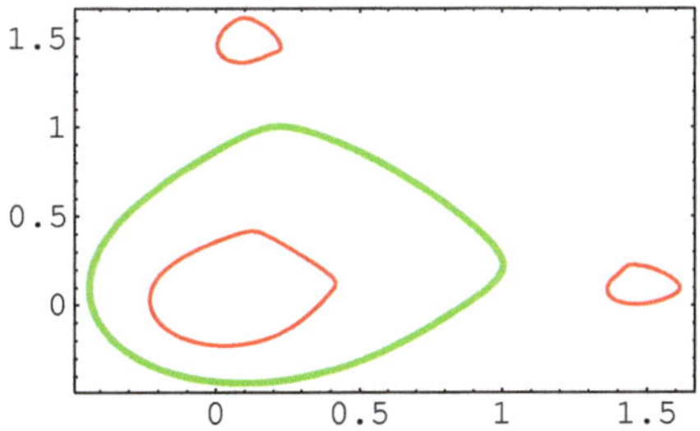

Fig. 1 The curves δ (*thin*) and γ (*thick*) for $f(z) = (z-a)z(z-1)(z-2)$

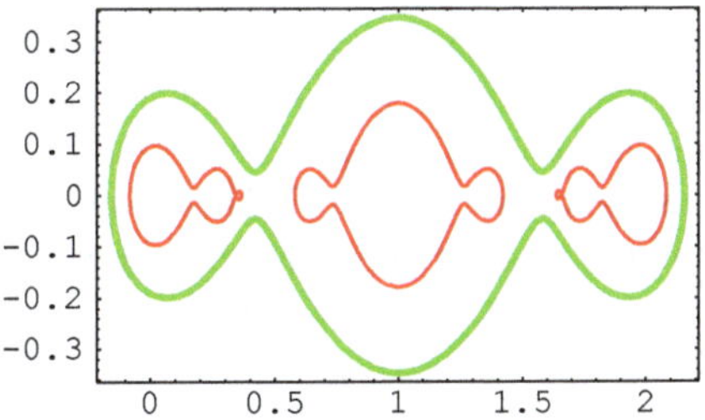

Fig. 2 The curves δ (*thin*) and γ (*thick*) for $f(z) = (z-a)z(z-2)(z-2i)$

Computer experiments indicate that the conjecture holds also for polynomials of degree 4 but we have not yet a rigorous proof. We show here on Fig. 1 the curve δ (the thin one) described by the zero ξ of the derivative of the polynomial $f(z) = (z-a)z(z-1)(z-2)$ of degree 4 when the point $z_4 = a$ is traversing the boundary (the tick line) of the component γ of the lemniscate of $\{z : |q(z)| = |q(z_4)|\}$ with $q(z) = z(z-1)(z-2)$ and $z_4 = 0.19\mathrm{i}$.

In the second example, (Fig. 2) we take $f(z) = (z-a)z(z-2)(z-2\mathrm{i})$, $q(z) = z(z-2)(z-2\mathrm{i})$ and $z_4 = 1+0.2\mathrm{i}$.

Note that in these examples there is always a component of δ which lies in the interior of γ, which means that for every a on γ there is a corresponding critical point ξ of f which is inside γ, as stated in our conjecture.

We finish this section with a simple observation concerning Smale's conjecture with a constant $K = 1$. Let $P(z)$ be a polynomial of degree n with zeros $z_1,\dots,z_n$ and critical points $\xi_1,\dots,\xi_{n-1}$. Fix one zero, say z_n, and consider the lemniscate $\Gamma := \{z : |P'(z)| = t\}$ such that z_n lies on Γ. If z_n can be connected with a critical point ξ so that the line segment $[z_n,\xi]$ lies inside the lemniscate Γ, then we would have

$$|P(\xi)| = \left|\int_{\xi}^{z_n} P'(z)\,\mathrm{d}z\right| \le \int_{\xi}^{z_n} |P'(z)|\cdot|\mathrm{d}z| \le |P'(z_n)|\cdot|z_n-\xi|$$

and thus

$$\frac{|P(\xi)|}{|(z_n-\xi)P'(z_n)|} = \frac{1}{|(z_n-\xi)P'(z_n)|}\left|\int_{\xi}^{z_n} P'(z)\,\mathrm{d}z\right| \le 1.$$

This is just the assertion of Smale's conjecture with a constant $K = 1$. Therefore, the following problem, if solved, will imply Smale's conjecture.

Problem 2. Let $P(z)$ be a polynomial of degree n with zeros $z_1,\dots,z_n$ and critical points $\xi_1,\dots,\xi_{n-1}$. Prove that for each zero z_k of P there is a critical point ξ_j such that the line segment $[z_k,\xi_j]$ connecting z_k and ξ_j lies in the lemniscate $\{z : |P'(z)| = |P'(z_k)|\}$.

In other words, what we need to show is that from every zero of P, lying on the boundary of a lemniscate of P', one can "see" directly at least one critical point of the polynomial P.

According to a result of Walsh [43, Theorem 3], any Jordan curve Γ can be approximated by a single branch that is disjoint from the other branches of a lemniscate of an algebraic polynomial P such that $P(z)$ has a simple zero in Γ and no other zeros interior to Γ. This shows that the claim in Problem 2 is not true for any point z_k on the lemniscate of the derivative. But it could be true for those particular points z_k that are zeros of P.

5 Majorization of Polynomials

Let P and Q be algebraic polynomials of degree n and let S be a point set in the complex plane. Assume that Ω is a region in the plane that does not contain the set S. We shall consider in this section majorization theorems of the following form: *If $|Q(z)| \le |P(z)|$ for every $z \in S$, then P majorizes Q on Ω, i.e., $|Q(z)| \le |P(z)|$ for all $z \in \Omega$.*

A famous example is the following observation due to Bernstein [1] (and rediscovered later by Erdös [15]). We give it here in a slightly more general form than its usual presentation in the literature.

Let us denote by π_n the class of all real algebraic polynomials of degree less than or equal to n.

Theorem D. *Let P be any polynomial from π_n with n distinct zeros $x_1 < \cdots < x_n$ in $(-1,1)$. Let $\{t_k\}_0^n$ be any points such that*

$$-1 = t_0 < x_1 < t_1 < x_2 < \cdots < t_{n-1} < x_n < t_n = 1.$$

Assume that $Q \in \pi_n$ and $|Q(t_j)| \le |P(t_j)|$ for $j = 0,\ldots,n$. Then $|Q(z_0)| \le |P(z_0)|$ for every z_0 with $|z_0| \ge 1$.

As an immediate consequence from the last theorem we can derive from Theorem A the following result: *Under the conditions of Theorem D, we have*

$$|Q^{(k)}(z_0)| \le |P^{(k)}(z_0)|, \quad k = 1,\ldots,n, \tag{5}$$

for every z_0 with $|z_0| \ge 1$.

To show this we just take a circle Γ passing through the point z_0 and containing the unit disk D. Then, by Theorem D, $|Q(z)| \le |P(z)|$ for $z \in \Gamma$. Now, an application of Theorem A to P and Q gives $|Q'(z)| \le |P'(z)|$ for $z \in \Gamma$ and the assertion follows by induction for higher derivatives.

Furthermore, under the conditions of Theorem A, an application of the classical Rouché Theorem yields the following.

Proposition 1. *Assume that G is a convex domain in the complex plane. Let P and Q be any polynomials of degree n with complex coefficients such that $|Q(z)| \le |P(z)|$ for $z \in \partial G$. Then*

$$|Q^{(k)}(z)| \le |P^{(k)}(z)|, \quad k = 0,\ldots,n, \tag{6}$$

for any z lying outside G, provided all n zeros of P lie in G.

Proof. In view of (5), we need to prove inequality (6) only for $k = 0$. Assume that $|Q(z_0)| > |P(z_0)|$ for some $z_0 \notin G$. Then $Q(z_0) = cP(z_0)$ with an appropriate constant c such that $|c| > 1$. Therefore, z_0 is a zero of the polynomial $cP(z) - Q(z)$. But by Rouché's Theorem, all zeros of $cP(z) - Q(z)$ lie in G since $|cP(z)| > |Q(z)|$ on Γ, a contradiction. □

The above-mentioned examples are direct and simple consequences from well-known facts in the analytic theory of polynomials. Next, we present recent results on majorization which reveal further extremal properties of the famous Tchebycheff polynomials $T_n(x) := \cos n \arccos x$ for $x \in [-1,1]$.

Let us denote by $\mathcal{H}_R$ the closed half-plane $\{z \in \mathbb{C} : \operatorname{Re} z \ge R\}$. The following majorization theorem was proved in [13].

Theorem E. *Let $\eta_j = \cos(n-j)\pi/n$ for $j = 0,1,\ldots,n$, and let $t_0,t_1,\ldots,t_n$ be any sequence of $n+1$ numbers in $[-1,\infty)$ satisfying the separation condition*

$$t_{j+1} - t_j \geq \eta_{j+1} - \eta_j, \quad j = 0,1,\ldots,n-1.$$

Suppose that f is a real polynomial of degree at most n such that $|f(t_j)| \leq 1$, $j = 0,1,\ldots,n$. Then, for all $R \geq t_n$, we have

$$|f^{(k)}(z)| \leq |T_n^{(k)}(z)|, \quad z \in \mathcal{H}_R,\ k = 0,1,\ldots,n,$$

where equality holds for any $z \neq 1$ if and only if $f(z) \equiv \pm T_n(z)$.

Actually, this slightly more general formulation of the original publication [13] was given in [28] (see Theorems 12.4.15 and 15.2.10 there).

Note that the points $\{\eta_k\}$ in the last theorem are the extremal points of the Tchebycheff polynomial $T_n(x)$ in $[-1,1]$ and thus the assumption $|f(t_j)| \leq 1$ means $|f(t_j)| \leq |T_n(\eta_j)|$. Nikolov showed in [24] that similar assumptions in the Duffin–Schaeffer theorem [14] can be replaced by the requirement that f be bounded at any other $n+1$ points $\{t_j\}$ that interlace with the zeros of $T_n(x)$. Following this hint, one can see that Theorem E remains true in the following extended form.

Proposition 2. *Let P be any polynomial from π_n with n distinct zeros $x_1 < \cdots < x_n$ in $(-1,1)$ and $\{\eta_j\}_{j=0}^n$ be arbitrary points such that*

$$-1 = \eta_0 < x_1 < \eta_1 < x_2 < \cdots < \eta_{n-1} < x_n < \eta_n = 1.$$

Let $t_0,\ldots,t_n$ be any sequence of $n+1$ numbers in $[-1,\infty)$ satisfying the separation condition

$$t_{j+1} - t_j \geq \eta_{j+1} - \eta_j, \quad j = 0,1,\ldots,n-1.$$

Suppose that f is a real polynomial of degree at most n such that

$$|f(t_j)| \leq |P(\eta_j)|, \quad j = 0,1,\ldots,n.$$

Then, for all $R \geq t_n$, we have

$$|f^{(k)}(z)| \leq |P^{(k)}(z)|, \quad z \in \mathcal{H}_R,\ k = 0,1,\ldots,n,$$

where equality holds for any $z \neq 1$ if and only if $f = \pm P$.

The proof goes along the same lines as in [13]. One has to prove the inequality only for $k = 0$, then the result follows for any k by Theorem A.

Next we discuss the polynomial inequality established by the brothers Markov.

V. Markov's inequality. *Let P be any polynomial of degree n with real coefficients. Then, for every $k = 1,\ldots,n$,*

$$|P^{(k)}(x)| \leq \|P\|_{[-1,1]} T_n^{(k)}(1), \quad -1 \leq x \leq 1.$$

The following result from [6] makes it possible to derive Markov's inequality using the technique of majorization.

Theorem 3. *Let $\xi_0 + \mathrm{i}\eta_0$ be any point in the strip $-1 \le \operatorname{Re} z \le 1$. Then, for every polynomial $f \in \pi_n$ such that $|f(t)| \le 1$ on $[-1,1]$, there holds the estimate*

$$|f(\xi_0 + \mathrm{i}\eta_0)| \le \max_{x \in [-1,1]} |T_n(x + \mathrm{i}\eta_0)|.$$

Now using the known property (see, for example [29, Lemma 2.7.5]) of the Tchebycheff polynomial T_n, established by Duffin and Schaeffer [14],

$$|T_n(x+\mathrm{i}y)| < |T_n(1+\mathrm{i}y)|, \quad -1 < x < 1, \; -\infty < y < \infty, \tag{7}$$

we arrive at the important conclusion: *For every polynomial $f \in \pi_n$ with $|f(t)| \le 1$ on $[-1,1]$ and for any given point $\xi_0 \in [-1,1]$, we have*

$$|f(\xi_0 + \mathrm{i}y)| \le |T_n(1+\mathrm{i}y)|, \quad y \in (-\infty, \infty).$$

In other words, every polynomial from π_n, bounded by 1 on $[-1,1]$, is majorized on the line $x = \xi_0$ by the values of $T_n(z)$ on the line $x = 1$. Then Theorem A implies

$$|f^{(k)}(\xi_0 + \mathrm{i}y)| \le |T_n^{(k)}(1+\mathrm{i}y)|, \quad y \in (-\infty, \infty),$$

for every $k = 1, \ldots, n$, which is the Duffin–Schaeffer extension (see [14], or [4] for further references) of the inequality proved by Vladimir Markov [21].

We conjectured in [4] that inequality (7) holds for every symmetric polynomial (i.e., even or odd) which is bounded by 1 on $[-1,1]$, attains its maximal absolute value there at the end-points, and has all its zeros in $[-1,1]$. Nikolov [25] proved this property for the class of ultraspherical polynomials (i.e., orthogonal on $[-1,1]$ with respect to the weight $(1-x^2)^{\lambda-1/2}$, $\lambda \ge 0$).

6 Bernstein's Inequality on Lemniscates

For a given positive number c and a polynomial P, we shall denote by $E(P,c)$ the set

$$E(P,c) := \{z : |P(z)| \le c\}.$$

Clearly, the boundary of $E(P,c)$ is the lemniscate $\Gamma(P,c)$ and all zeros of P are interior to $E(P,c)$. The following is true.

Proposition 3. *Let $P \in \mathcal{P}_n$. Assume that c is a positive number such that $E(P,c)$ is a connected domain. Then P is the polynomial of least deviation from zero on $E(P,c)$ among all polynomials Q from $\mathcal{P}_n$ that have the same leading coefficient as P, that is,*

$$\max_{z \in E(P,c)} |P(z)| \le \max_{z \in E(P,c)} |Q(z)|$$

with equality only if $Q = P$.

Proof. Assume the contrary. Then there is a polynomial $Q \in \mathcal{P}_n$, distinct from P, and with the same leading coefficient as that of P, such that

$$\max_{z \in E(P,c)} |Q(z)| \le \max_{z \in E(P,c)} |P(z)|.$$

Because of the Maximum-Modulus Principle for analytic functions, the last inequality implies

$$\max_{z \in \Gamma(P,c)} |Q(z)| \le \max_{z \in \Gamma(P,c)} |P(z)|$$

and thus, for any sufficiently small $\varepsilon > 0$,

$$\max_{z \in \Gamma(P,c)} |Q(z)| < \max_{z \in \Gamma(P,c)} (1+\varepsilon)|P(z)|$$

Then, by Rouché's Theorem, P and $(1+\varepsilon)P - Q$ must have the same number of zeros in $E(P,c)$. But, on the one side, P has n zeros in $E(P,c)$ by assumption, whereas

$$(1+\varepsilon)P(z) - Q(z) = \varepsilon z^n + Q_{n-1}(z)$$

with a certain polynomial Q_{n-1} of degree $n-1$. Now it is seen that one of the zeros of $(1+\varepsilon)P(z) - Q(z)$ tends to infinity as ε approaches zero. Thus, for small ε, it has at most $n-1$ zeros in $E(P,c)$, a contradiction. The proof is complete. □

The last proposition says that any polynomial is the only polynomial of least deviation on its own lemniscate. An immediate consequence of this is the following. Assume that c is such that $E(P,c)$ is connected. Let $|Q(z)| \le |P(z)|$ on $\Gamma(P,c)$ for some $Q \in \mathcal{P}_n$. Then the leading coefficient of Q is smaller than the leading coefficient of P.

The following question is quite natural:

Question 1. Let G be a convex bounded domain and $P_n(G,z)$ be the monic polynomial of least uniform deviation on G of degree n. Is the polynomial $P_n(G,z)$ extremal for the Bernstein inequality on G?

In other words, is it true that

$$\|Q'\|_G \le \|P_n'(G,\cdot)\|_G \cdot \frac{\|Q\|_G}{\|P_n(G,\cdot)\|_G}$$

for every $Q \in \mathcal{P}_n$? The answer is positive in the important cases when G is $[-1,1]$ or D. This question is somewhat relevant to a negative result of Schaeffer and Szegö [39] (see formula (9) in the Appendix of their paper).

Note that all zeros of $P_n(G,z)$ must lie in G. Indeed, if a certain zero z_k lies outside G, then a small replacement of z_k toward G would improve the deviation from zero on G of the polynomial.

Another question of similar nature is

Question 2. Is $P_n(E,z)$ extremal for the Bernstein inequality on its own lemniscate E, provided E is connected?

The next proposition is a simple observation concerning the last question.

Proposition 4. *For a given $P \in \mathcal{P}_n$, let c be such that $E(P,c)$ is connected. Assume that $|Q(z)| \le |P(z)|$ on $\Gamma(P,c)$ for some $Q \in \mathcal{P}_n$. Let Ω be the convex hull of $E(P,c)$, and let $\mathcal{A}$ be the intersection of $\Gamma(P,c)$ and $\partial\Omega$. Then, for each $k = 0,\dots,n$,*

$$|Q^{(k)}(z)| \le |P^{(k)}(z)| \quad \textit{for every } z \in \mathcal{A}.$$

Proof. By Rouché's Theorem, all zeros of $P(z) - Q(z)$ lie in $E(P,c)$. This implies that $|Q(z)| \le |P(z)|$ outside $E(P,c)$, and thus on $\partial\Omega$. Then Theorem A implies that $|Q^{(k)}(z)| \le |P^{(k)}(z)|$ for every $z \in \partial\Omega$ and consequently, on $\mathcal{A}$. This finishes the proof. □

Acknowledgements The author is grateful to his colleagues Lozko Milev and Nikola Naidenov for their help in performing computer calculations confirming Conjecture 1 for polynomials of small degree.

This work was supported by the Sofia University Research Grant # 135/2008 and by Swiss-NSF Scopes Project IB7320-111079.

References

1. Bernstein, S.N.: Sur une propriété de polinômes. Comm. Soc. Math. Kharkow Sér. 2. **14**, 1-2, 1–6 (1913) In: Collected Works, Volume 1, Izd. AN SSSR, Moscow, 146–150 (1952)
2. Bernstein, S.N.: Sur la limitation des dérivées des polynomes. C. R. Acad. Sci. Paris **190**, 338–341 (1930)
3. Bojanov, B.: The conjecture of Sendov about the critical points of polynomials. Fiz.-Mat. Spisanie, 140–150 (1984)
4. Bojanov, B.: Markov-type inequalities for polynomials and splines. 31–90. In: Approximation Theory X: Abstract and Classical Analysis, Charles Chui, L.L. Schumaker, and J. Stöckler (eds.), Vanderbilt University Press, Nashville, TN (2002)
5. Bojanov, B., Rahman, Q.I., Szynal, J.: On a conjecture of Sendov about the critical points of a polynomial. Math. Z. **190**, No 2, 281–186 (1985)
6. Bojanov, B., Naidenov, N.: Majorization of polynomials on the plane. II. East J. Math. **12**, No 2, 189–202 (2006)
7. Borcea, J.: Maximal and inextensible polynomials. Preprint: *arXiv math.CV/0601600vl*, May 29, 2006.
8. de Bruijn, N.G.: Inequalities concerning polynomials in the complex domain. Nederl. Akad. Wetensch. Indag. Math. **9**, 591–598 (1947)
9. de Bruijn, N.G., Springer, T.A.: On the zeros of a polynomial and of its derivative II. Nederl. Akad. Wetensch. Indag. Math. **9**, 264–270 (1947)
10. Conte, A., Fujikawa, B., Lakic, N.: Smale's mean value conjecture and the coefficients of univalent functions. Proc. Am. Math. Soc. **135**, No 10, 3295–3300 (2007)

11. Dimitrov, D.K.: A refinement of the Gauss-Lucas theorem. Proc. Am. Math. Soc. **126**, 2065–2070 (1998)
12. Dimitrov, D.K.: Smale's conjecture on mean values of polynomials and electrostatics. Serdica Math. J. **33**, No 4, 399–410 (2007)
13. Dryanov, D.P., Rahman, Q.I.: On a polynomial inequality of E. J. Remez. Proc. Am. Math. Soc. **128**, 1063–1070 (2000)
14. Duffin, R.J., Schaeffer, A.C.: A refinement of an inequality of the brothers Markoff. Trans. Am. Math. Soc. **50**, 517–528 (1941)
15. Erdös, P.: Some remarks on polynomials. Bull. Am. Math. Soc. **53**, 1169–1176 (1947)
16. Goodman, A.W., Rahman, Q.I., Ratti, J.S.: On the zeros of a polynomial and its derivative. Proc. Am. Math. Soc. **21**, 273–274 (1969)
17. Malamud, S.M.: Inverse spectral problem for normal matrices and generalization of the Gauss-Lucas theorem. arXiv:math.CV/0304158v3 6Jul2003.
18. Marden, M.: Geometry of Polynomials. Am. Math. Soc. Providence (1966)
19. Marden, M.: Much ado about nothing. Am. Math. Mon. **83**, 788–789 (1976)
20. Marden, M.: Conjectures on the critical points of a polynomial. Am. Math. Mon. **90**, 267–276 (1983)
21. Markov, V.A.: On the functions of least deviation from zero in a given interval. St. Petersburg (1892) (in Russian); German translation with shortenings: W. Markoff: Über Polynome, die in einem gegebenen Intervalle möglichst wenig von Null abweichen. Math. Ann. **77**, 213–258 (1916)
22. Meir, A., Sharma, A.: On Ilief's conjecture. Pacific J. Math. **31**, 459–467 (1969)
23. Milovanović, G.V., Mitrinović, D.S., Rassias, Th.M.: Topics in Polynomials: Extremal Problems, Inequalities, Zeros. World Scientific, Singapore (1994)
24. Nikolov, G.: Inequalities of Duffin-Schaeffer type. SIAM J. Math. Anal. **33**, 686–698 (2001)
25. Nikolov, G.: An extension of an inequality of Duffin and Schaeffer. Constr. Approx. **21**, 181–191 (2005)
26. Obrechkoff, N.: Zeros of Polynomials. Bulgarian Academic Monographs (7), Marin Drinov Academic Publishing House, Sofia (2003) (English translation of the original text, published in Bulgarian in 1963).
27. Pereira, R.: Differentiators and the geometry of polynomials. J. Math. Anal. Appl. **285**, 336–348 (2003)
28. Rahman, Q.I., Schmeisser, G.: Analytic Theory of Polynomials. Oxford Science Publications, Clarendon, Oxford (2002)
29. Rivlin, T.J.: Chebyshev Polynomials: From Approximation Theory to Algebra and Number Theory. Second edition, Wiley, New York (1990)
30. Schmeisser, G.: On Ilieff's conjecture. Math. Z. **156**, 165–175 (1977),
31. Schmeisser, G.: Bemerkungen zu einer Vermutung von Ilieff. Math. Z. **111**, 121–125 (1969)
32. Schmeisser, G.: The conjectures of Sendov and Smale. pp. 353–369. In: Approximation Theory: A volume dedicated to Blagovest Sendov (B. Bojanov, Ed.), DARBA, Sofia (2002)
33. Schmieder, G.: A proof of Sendov's conjecture. Preprint: arXiv:math.CV/0206173 v6 27 May 2003
34. Schoenberg, I.J.: A conjectured analogue of Rolle's theorem for polynomials with real and complex coefficients. Am. Math. Mon. **93**, 8–13 (1986)
35. Sendov, Bl.: Hausdorff geometry of polynomials. East J. Approx. **7**, No 2, 123–178 (2001)
36. Sendov, Bl.: Complex analogues of the Rolle's theorem. Serdica Math. J. **33**, No 4, 387–398 (2007)
37. Sendov, Bl., Marinov, P.: Verification of Smale's mean value conjecture for $n \leq 10$. C. R. Acad. Bulg. Sci. **60**, No 11, 1151–1156 (2007)
38. Smale, S.: The fundamental theorem of algebra and complexity theory. Bull. Am. Math. Soc. (New Series) **4**, 1–36 (1981)
39. Schaeffer, A.C., Szegö, G.: Inequalities for harmonic polynomials in two and three dimensions. Trans. Am. Math. Soc. **50**, 187–225 (1941)
40. Szegö, G.: Components of the set $|(z-z_1)\cdots(z-z_n)| \leq 1$. Am. Math. Mon. **58**, No 9, p. 639 (1951)

41. Tischler, D.: Critical points and values of polynomials. J. Complex. **5**, 438–456 (1989)
42. Walsh, J.L.: Lemniscates and equipotential curves of Green's function. Am. Math. Mon. **42**, 1–17 (1935)
43. Walsh, J.L.: On the convexity of the ovals of lemniscates. pp. 419-423. In: Studies in Mathematical Analysis and Related Topics, Stanford University Press, Stanford, Calif. (1962)

Energy of Graphs and Orthogonal Matrices

V. Božin and M. Mateljević

Dedicated to Professor Gradimir V. Milovanović
on the occasion of his 60th birthday

1 Introduction and Notation

The energy of a graph is a concept that first arose in theoretical chemistry (see [3,6]) as an approximation to total π-electron energy of a molecule. For further chemistry applications, see book [4]. Research on graph energy is nowadays very active (see [1,2,5,7–13,16] and recent editions of MATCH, in particular MATCH **59** (2008), no. 2).

From the mathematician's point of view, an interesting problem is to characterize graphs having the maximum energy among all graphs with n vertices. An early conjecture [3] was that the complete graphs have maximum energy. However, in a series of recent papers various families of hyperenergetic graphs (which have an energy larger than the complete graphs) were constructed. Koolen and Moulton in [10] and [11] provided an upper bound for the energy, which is sharp for some, very special, values of n. It was conjectured in [8] (see also [17]) that this bound is achieved for all even squares, and this conjecture has been proved in [9]. For n that is not a square of an even number, this bound is not sharp, and the problem of maximal energy remains open in the general case. The method developed here provides a way to improve the upper bound for energy for arbitrary n.

V. Božin
Faculty of Mathematics, University of Belgrade, Studentski Trg 16, Belgrade, Serbia,
e-mail: bozinv@mi.sanu.ac.rs

M. Mateljević
Faculty of Mathematics, University of Belgrade, Studentski Trg 16, Belgrade, Serbia,
e-mail: miodrag@matf.bg.ac.rs

W. Gautschi et al. (eds.), *Approximation and Computation: In Honor of Gradimir V. Milovanović*, Springer Optimization and Its Applications 42,
DOI 10.1007/978-1-4419-6594-3_6,

We will denote by $\mathcal{G}_n$ the family of all simple graphs with n vertices $v_1,\dots,v_n$.

The adjacency matrix $A = A(G)$ of the graph $G \in \mathcal{G}_n$ is a square matrix of order n whose entry in the i-th row and j-th column is defined as

$$a_{ij} = \begin{cases} 1 & \text{if the vertices } v_i \text{ and } v_j \text{ are adjacent,} \\ 0 & \text{otherwise.} \end{cases}$$

The characteristic polynomial of a graph is the characteristic polynomial of its adjacency matrix: $\varphi(G) = \varphi(G,\lambda) = \det(\lambda I - A)$.

Its roots are called eigenvalues of the graph.

For a graph $G \in \mathcal{G}_n$ with adjacency matrix $A = A(G)$ and eigenvalues $\lambda_1, \lambda_2, \dots, \lambda_n$, the energy is defined as [3,5]

$$E = E(G) = E(A) = \sum_{k=1}^{n} |\lambda_k|.$$

Note that because the sum of all eigenvalues of a graph is zero, $E = 2E^+$ where E^+ denotes the sum of positive eigenvalues.

We say that a square matrix P (or corresponding operator) is a projector if $P^2 = P$.

We say that a projector $P : \mathbb{R}^n \to \mathbb{R}^n$ is orthogonal if there is a subspace M of $\mathbb{R}^n$ such that the range of P is M and $(x - Px) \perp M$ for every $x \in \mathbb{R}^n$. A projector is orthogonal if and only the corresponding matrix is symmetric.

Recall that a square matrix A of type $n \times n$ is orthogonal if and only if its columns form an orthonormal basis of $\mathbb{R}^n$. It corresponds to a linear transformation A of $\mathbb{R}^n$ to $\mathbb{R}^n$ which preserves the length of vectors.

We will denote by $\mathcal{O}^s(n)$ the family of all orthogonal symmetric matrices of type $n \times n$.

For two matrices $A = [a^i_j]$ and B the tensor product is the block matrix

$$A \otimes B = \begin{bmatrix} a^1_1 B & \cdots & a^1_n B \\ \vdots & & \\ a^m_1 B & \cdots & a^m_n B \end{bmatrix}.$$

Let a^+ denote $\max\{a, 0\}$ for $a \in \mathbb{R}$. For a matrix $A = [a_{ij}]$, we define $|A|_1 = \sum |a_{ij}|$. We define $E(G,A) = \sum_{i \neq j, (i,j) \in G} a_{ij}$.

We say that an orthogonal symmetric matrix $O = [o_{ij}]$ is extremal for a graph $G \in \mathcal{G}_n$ if $E(G) = E(G,O) = \sum_{i \neq j, (i,j) \in G} o_{ij}$.

2 Characterization by Projectors and Orthogonal Matrices

The standard basis for $\mathbb{R}^n$ is $E_1, \dots, E_n$, where E_j is the element of $\mathbb{R}^n$ with 1 in the j-th place and zeros on all other places. If $T : \mathbb{R}^n \to \mathbb{R}^m$ is a linear transformation, the $m \times n$ matrix of T is the matrix $\mathbf{t}$ whose j-th column is $T(E_j)$ for $j = 1, \dots, n$, and vice versa if $\mathbf{t} = [t_{ij}]$ of $m \times n$, then the operator T of the matrix $\mathbf{t}$ is defined by $T(E_j) = (t_{1j}, \dots, t_{mj}) = \sum_{k=1}^{m} t_{kj} E_k$.

If $T:\mathbb{R}^n \to \mathbb{R}^n$ is a linear transformation and β a base in $\mathbb{R}^n$, by $[T]_\beta$ we denote the matrix with respect to the base β.

Let A be a real symmetric matrix of type $n \times n$ and e_k be orthonormal eigenvectors for A, that is $Ae_k = \lambda_k e_k$, $k = 1,\dots,n$ and E_k be the standard basis in $\mathbb{R}^n$. If Q is a linear operator defined by $Q(E_k) = e_k$, then $Q^{-1}AQ = D$, where $D = \operatorname{diag}(\lambda_1,\lambda_2,\dots,\lambda_n)$.

A real matrix A is symmetric if and only if there is an orthogonal matrix Q such that $Q^{-1}AQ = D$, where $D = \operatorname{diag}(\lambda_1,\lambda_2,\dots,\lambda_n)$.

In linear algebra, the trace of an $n \times n$ square matrix A is defined to be the sum of the elements on the main diagonal (the diagonal from the upper left to the lower right) of A, i.e., $\operatorname{tr}(A) = a_{11} + \cdots + a_{nn} = \sum_i a_{ii}$, where a_{ij} represents the entry on the i-th row and j-th column of A. Equivalently, the trace of a matrix is the sum of its eigenvalues, making it an invariant with respect to a change of basis.

The trace is similarity-invariant, which means that A and $Q^{-1}AQ$ have the same trace. This is because trace is independent of the order of multiplication and it is invariant under cyclic permutation of matrices in a product: $\operatorname{tr}(Q^{-1}AQ) = \operatorname{tr}(AQQ^{-1}) = \operatorname{tr} A$.

Theorem 1. *Suppose that A is a real symmetric matrix with* $\operatorname{tr} A = 0$. *If P is an orthogonal projector matrix then $E = E(A) \geq 2\operatorname{tr}(AP)$; and $E = 2\max\operatorname{tr}(AP)$, where* max *is taken over all orthogonal projectors P.*

Proof. Let $(\lambda_1,\lambda_2,\dots,\lambda_n)$ be eigenvalues of A. Suppose that $\lambda_1 > \lambda_2 > \dots \lambda_p \geq 0$ and $\lambda_k < 0$ for $k > p$. Let $e = (e_1,\dots,e_n)$ be the orthonormal basis of eigenvectors of the matrix A and P_0 be a projector onto the p dimensional space corresponding to nonnegative eigenvalues.

Then, since $[A]_e = \operatorname{diag}(\lambda_1,\lambda_2,\dots,\lambda_n)$ and $\operatorname{tr}(AP_0) = \operatorname{tr}([A]_e[P_0]_e)$, it is clear that $E^+(A) = \operatorname{tr}(AP_0)$.

We first verify that for any vector $x \in \mathbb{R}^n$

$$(x,Ax) \leq (x,AP_0x), \tag{1}$$

where $(\cdot,\cdot)$ denotes the usual scalar product in $\mathbb{R}^n$. Namely, if $x = \sum_{k=1}^n t_k e_k$, then $Ax = \sum_{k=1}^n t_k\lambda_k e_k$, $P_0x = \sum_{k=1}^p t_k e_k$ and $AP_0x = \sum_{k=1}^p t_k\lambda_k e_k$. Hence $(x,Ax) = \sum_{k=1}^n t_k^2\lambda_k$ and $(x,AP_0x) = \sum_{k=1}^p t_k^2\lambda_k$; therefore, we have (1).

Let $\alpha = (f_1,\dots,f_n)$ be an orthonormal basis and $Af_k = \sum_{j=1}^n a_{kj}f_j$. Then

$$\operatorname{tr} A = \sum_{i=1}^n a_{ii} = \sum_{i=1}^n (f_i,Af_i). \tag{2}$$

Now let P be an arbitrary orthogonal projection. Let $\beta_1 = (b_1,\dots,b_r)$ be an orthonormal basis for $\operatorname{Im} P$ and $\beta_2 = (b_{r+1},\dots,b_n)$ an orthonormal basis for $\operatorname{Ker} P$. Then $\beta = \beta_1 \cup \beta_2$ is a basis and

$$P = [P]_\beta = \begin{bmatrix} I_r & 0 \\ 0 & 0 \end{bmatrix},$$

where I_r is a unit matrix.

Since $Pb_i = b_i$, $i = 1,\dots,r$ and $Pb_i = 0$, $i = r+1,\dots,n$, we find $APb_i = Ab_i$, $i = 1,\dots,r$ and $APb_i = 0$, $i = r+1,\dots,n$. Then, by (1) and (2),

$$\operatorname{tr}(AP) = \sum_{i=1}^{r}(b_i, Ab_i) \le \sum_{i=1}^{r}(b_i, AP_0 b_i) \le \operatorname{tr}(AP_0) = E^+. \quad \square$$

Lemma 1. *P is an orthogonal projector if and only if $P = (I+O)/2$, where O is an orthogonal symmetric matrix.*

Proof. Suppose that O is an orthogonal symmetric matrix and $P = (I+O)/2$. Then P is symmetric, $O = O^t$, $OO^t = I$ and therefore $O^2 = I$. Hence $(I+O)(I+O) = (I+O)+O+O^2 = 2(I+O)$, that is $P^2 = P$; thus, since P is symmetric, we conclude that P is an orthogonal projector.

Conversely, if P is an orthogonal projector and $O = 2P - I$, then $O^2 = (2P-I)^2 = 4P - 4P + I = I$, hence $O^2 = I$; moreover, since P is symmetric, O is symmetric, and hence $O^t = O = O^{-1}$, that is O an orthogonal matrix. $\square$

Theorem 2. *Let G be a simple graph with n vertices and $\mathcal{O}^s(n)$ be the family of orthogonal symmetric matrices of type $n \times n$. Then*

$$E(G) \le \frac{n}{2} + \max_{O} \frac{1}{2}|O|_1 \le \frac{n}{2} + \frac{1}{2}n^{3/2},$$

where max *is taken over all $O \in \mathcal{O}^s(n)$.*

It is clear that

$$\max_{G} E(G) \le \frac{n}{2} + \max_{O} \frac{1}{2}|O|_1 \le \frac{n}{2} + \frac{1}{2}n^{3/2},$$

where max is taken over all $G \in \mathcal{G}_n$.

Proof. If O_n is an orthogonal symmetric matrix with $|o_{ij}| = \sqrt{1/n}$, $o_{ii} = -\sqrt{1/n}$ and $\sum o_{ij} = n$, then O_n is extremal.

Step 1. If $O = [o_{ij}]$ is an orthogonal matrix, then $|O|_1 \le n^{3/2}$. Since $O = [o_{ij}]$ is an orthogonal matrix, $\sum_i o_{ik}^2 = 1$, and $\sum_{i,j}|o_{ij}|^2 = n$. Hence $(\sum_{i,j}|o_{ij}|)^2 \le n^2 \sum_{i,j}|o_{ij}|^2 = n^3$. The equality holds if and only if $|o_{ij}| = c_n = \text{const}$, i.e., if and only if $|o_{ij}| = 1/\sqrt{n}$.

Step 2. If $O = [o_{ij}]$ is an orthogonal matrix, then

$$\sum o_{ij}^+ \le \frac{n}{2} + \frac{1}{2}|O|_1. \tag{3}$$

If $P = (I+O)/2$ is a projector and $v_0 = (1,\dots,1)$, then $|Pv_0|^2 = \sum p_{ij} = n/2 + 1/2\sum o_{ij}$. Since $|Pv_0|^2 \le |v_0|^2 = n$, we find $\sum o_{ij} \le n$. Hence, using $\sum o_{ij}^+ = 1/2(\sum|o_{ij}| + o_{ij})$, we find (3). If A is the symmetric $(0,1)$-adjacency matrix of G and $AP = [b_{ij}]$, then $b_{ij} = \sum_k a_{ik}p_{jk}$ and therefore $\operatorname{tr}(AP) = \sum_i b_{ii} = \sum_{i,k=1}^{n} a_{ik}p_{ik}$. Hence, since A is the symmetric $(0,1)$-adjacency matrix, $\operatorname{tr}(AP) \le \sum_{i \ne j} p_{ij}^+$.

By Lemma 1, P is an orthogonal projector if and only if $P = (I+O)/2$, where O is an orthogonal symmetric matrix. If $i \neq j$, $p_{ij} = o_{ij}/2$,

$$\operatorname{tr}(AP) \leq \frac{1}{2}\sum_{i\neq j} o_{ij}^{+} \leq \frac{1}{4}\sum_{i\neq j} o_{ij} + |o_{ij}| \leq \frac{n}{4} + \frac{1}{4}|O|_1 \leq \frac{n}{4} + \frac{1}{4}n^{3/2}$$

and therefore $\max E(G) \leq n/2 + 2\max|O|_1/4 \leq n/2 + 1/2n^{3/2}$. □

By Theorem 1 and Lemma 1, we get:

Theorem 3. *For a graph $G \in \mathcal{G}_n$, $E(G) = \max\sum_{i\neq j,(i,j)\in G} o_{ij}$, where* max *is taken over all orthogonal symmetric matrices $O = [o_{ij}]$.*

There is a maximal orthogonal symmetric matrix such that $E(G) = \sum_{i\neq j,(i,j)\in G} o_{ij}$.

Proof. Since $\mathcal{O}^s(n)$ is compact, there is a maximal orthogonal symmetric matrix $O = [o_{ij}]$ such that $E(G) = \sum_{i\neq j,(i,j)\in G} o_{ij}$. □

3 Some Applications

Let us denote

$$o_i^{+} = \sum_{j,i\neq j,(i,j)\in G} o_{ij}^{+},$$

and let d_k denote the index of a vertex $v_k \in G$.

Proposition 1. *Let m be number of edges. Then $E(G) \leq \sum_{k\in G}\sqrt{d_k} \leq \sqrt{2mn}$.*

Proof. By Theorem 3, there is a maximal orthogonal symmetric matrix $O = [o_{ij}]$ such that $E(G) = \sum_{i\neq j,(i,j)\in G} o_{ij}$. Let

$$o_i^{+} = \sum_{j,i\neq j,(i,j)\in G} o_{ij}^{+}.$$

By Cauchy–Schwarz, $(o_i^{+})^2 \leq d_i \sum_{j,i\neq j,(i,j)\in G}(o_{ij}^{+})^2 \leq d_i$. Hence $o_i^{+} \leq \sqrt{d_i}$ and $E(G) \leq \sum_i o_i^{+} \leq \sum_i \sqrt{d_i}$. Since $\sum_k d_k = 2m$, we get the McClelland inequality $E(G) \leq \sqrt{2mn}$. □

Let $E_n = \max_{G\in\mathcal{G}_n} E(G)$. We say that an orthogonal symmetric matrix $O = [o_{ij}]$ is extremal if $E_n = \sum_{i\neq j} o_{ij}^{+}$ and for $G \in \mathcal{G}_n$ that it is an extremal energy graph if $E_n = E(G)$.

For a given real matrix $M = [m_{ij}]$, we define a graph $G = G_M$ by $(v_i, v_j) \in G$ if and only if $m_{ij} > 0$.

Since $E(G,O) = \sum_{i\neq j,(i,j)\in G} o_{ij} \leq \sum_{i\neq j} o_{ij}^{+}$ and $E(G_O,O) = \sum_{i\neq j} o_{ij}^{+}$, by Theorem 3, we get:

Proposition 2. *For every integer $n \geq 1$,*

$$E_n = \max_{O \in \mathcal{O}^s(n)} \sum_{i \neq j} o_{ij}^+ = \frac{1}{2} \max_{O \in \mathcal{O}^s(n)} \left(\sum_{i \neq j} o_{ij} + |o_{ij}| \right),$$

and there is an orthogonal symmetric extremal matrix $O = [o_{ij}]$ and an extremal energy graph $G \in \mathcal{G}_n$.

Since $O \in \mathcal{O}^s(n)$ if and only if $-O \in \mathcal{O}^s(n)$, we have $|\sum o_{ij}| \leq n$ for $O \in \mathcal{O}^s(n)$.

Proposition 3. *Suppose that a matrix $O = [o_{ij}]$ of type $n \times n$ is an orthogonal symmetric extremal matrix. Then $E_n = E_n^1 := \frac{n}{2} + \frac{1}{2} n^{3/2}$ if and only if*

(a) $\sum_j o_{ij} = 1$
(b) $|o_{ij}| = 1/\sqrt{n}$
(c) $o_{ii} \leq 0$

Proof. Let $P = (I + O)/2$. If $E_n = E_n^1$, then $\sum o_{ij} = n$. Hence $|Pv|^2 = n$ and therefore $Pv = v$; so we get (a). The condition (a) can be replaced with $\sum o_{ij} = n$. □

We can verify that $E_n = n/2 + 1/2 n^{3/2}$ for $n = 4^k$ using orthogonal matrices. We construct orthogonal matrices $\mathbf{O}_n$, $n = 4^k$, of type $n \times n$ with elements $o_{ij} = \pm 1/\sqrt{n}$ by induction (we first construct $\mathbf{A}_{4^k}$):

$$\mathbf{A}_4 = \begin{bmatrix} -1 & 1 & 1 & 1 \\ 1 & -1 & 1 & 1 \\ 1 & 1 & -1 & 1 \\ 1 & 1 & 1 & -1 \end{bmatrix}.$$

We also need $\mathbf{B}_4$:

$$\mathbf{B}_4 = \begin{bmatrix} 1 & 1 & 1 & -1 \\ 1 & 1 & -1 & 1 \\ 1 & -1 & 1 & 1 \\ -1 & 1 & 1 & 1 \end{bmatrix}.$$

Using the tensor product we define $\mathbf{A}_{4^{k+1}} = \mathbf{B}_4 \otimes \mathbf{A}_{4^k}$:

$$\mathbf{A}_{4^{k+1}} = \begin{bmatrix} \mathbf{A}_{4^k} & \mathbf{A}_{4^k} & \mathbf{A}_{4^k} & -\mathbf{A}_{4^k} \\ \mathbf{A}_{4^k} & \mathbf{A}_{4^k} & -\mathbf{A}_{4^k} & \mathbf{A}_{4^k} \\ \mathbf{A}_{4^k} & -\mathbf{A}_{4^k} & \mathbf{A}_{4^k} & \mathbf{A}_{4^k} \\ -\mathbf{A}_{4^k} & \mathbf{A}_{4^k} & \mathbf{A}_{4^k} & \mathbf{A}_{4^k} \end{bmatrix}.$$

Let $\mathbf{O}_n = 1/\sqrt{n}\mathbf{A}_n$. Using Proposition 3, we can verify that the matrices $\mathbf{O}_n$ are extremal for $n = 4^k$. For $n = 4$ the complete graph is extremal, $E_4 = 2 \cdot 3 = 6$; the matrix $\mathbf{O}_4 = \frac{1}{2}\mathbf{A}_4$ is extremal. For $n = 4^2 = 16$, the matrix $\mathbf{O}_{16} = \frac{1}{4}\mathbf{B}_4 \otimes \mathbf{A}_4$ is extremal: $|\mathbf{O}_{16}^+| = \sum_{i \neq j} o_{ij}^+ = 12 \cdot 12/4 + 4 \cdot 4/4 = 36 + 4 = 40 = E_{16}$; $|\mathbf{O}_{16}|/2 = 16$ $\sqrt{16}/2 = 32$.

J. Koolen and V. Moulton, using different methods, proved the following: The energy of a graph on n vertices is at most $n(1+\sqrt{n})/2$. Equality holds if and only if G is a strongly regular graph with parameters $(n,(n+\sqrt{n})/2,(n+2\sqrt{n})/4,(n+2\sqrt{n})/4)$. Graphs of this type correspond to Hadamard matrices of a special kind, or equivalently, orthogonal matrices from Proposition 3. Since the parameters of a strongly regular graph are integer, n must be an even square, and W. Haemers and Q. Xiang have recently proved that this necessary condition for existence is also sufficient [9]. In the case of general n, the problem of maximal energy graph with n vertices remains open.

4 Conference Matrices and Asymptotic Behavior of Maximal Energy

A conference matrix C of order n, $n \geq 1$, is a square matrix of order n with entries $+1$ and -1 off the diagonal, and zeros on the diagonal, such that $CC^T = (n-1)I$ (see [14, 15]). Note that the columns of a conference matrix form an orthogonal basis. Conference matrices are normalized so that the first row and the first column have entries $+1$ except for the zero on the diagonal.

For a symmetric conference matrix $C = [c_{ij}]$, normalized with $c_{1i} = c_{i1} = 1$ for $i > 1$, we consider $G \in \mathcal{G}_n$ with vertices v_i and v_j joined by an edge whenever $c_{ij} = 1$. Let $O = 1/\sqrt{n-1}C$, then O is orthogonal and we have

$$E(G) \geq E(G,O) = \sum_{i \neq j,(i,j) \in G} o_{ij} = \frac{1}{2\sqrt{n-1}}(n^2+n-2) = \left(\frac{n}{2}+1\right)\sqrt{n-1}.$$

This is asymptotically the same as the maximal energy predicted by the inequality $E(G) \leq n/2 + 1/2 n^{3/2}$.

For a symmetric conference matrix of order n to exist, it is necessary that $n-2$ is divisible by 4, and that $n-1$ is the sum of two squares (see [14]). It is known that symmetric conference matrices exist for $n = p+1$, where p is a prime number with $p-1$ divisible by 4. Namely, we can check that the matrix

$$c_{11} = 0, \quad c_{1i} = c_{i1} = 1, \; i > 1, \quad c_{ij} = c_{ji} = \left(\frac{i-j}{p}\right), \; i,j > 1, \; i \neq j,$$

is a symmetric normalized conference matrix.

Here, $\left(\frac{i-j}{p}\right)$ stands for the Legendre symbol (cf. Appendix). Because $p-1$ is divisible by 4, we have $\left(\frac{-1}{p}\right) = 1$, and hence

$$\left(\frac{i-j}{p}\right) = \left(\frac{-1}{p}\right)\left(\frac{j-i}{p}\right) = \left(\frac{j-i}{p}\right),$$

so the matrix will indeed be symmetric.

We proceed to check the orthogonality relations. For $2 \le j,k \le p+1$ let

$$A_{jk} = \sum_{i=2}^{p+1} \left(\frac{i-j}{p}\right)\left(\frac{i-k}{p}\right).$$

The scalar product of two columns C_j and C_k of our matrix C will be $A_{jk}+1$, and we have to check that $A_{jk} = -1$ for $j \neq k$. First note that for $j \neq k$ there is a residue l such that $l(j-k)-1$ is divisible by p, and so, summing over all residues modulo p, we have

$$A_{jk} = \sum_{i=2}^{p+1} \left(\frac{i-j}{p}\right)\left(\frac{i-k}{p}\right) = \sum_i \left(\frac{i}{p}\right)\left(\frac{i+j-k}{p}\right) = \sum_i \left(\frac{l^2}{p}\right)\left(\frac{i}{p}\right)\left(\frac{i+j-k}{p}\right)$$

and therefore

$$A_{jk} = \sum_i \left(\frac{li}{p}\right)\left(\frac{li+1}{p}\right) = \sum_i \left(\frac{i}{p}\right)\left(\frac{i+1}{p}\right), \quad j \neq k.$$

Hence, for $k \neq j$, all these sums are equal to some constant a. Now

$$\sum_{j=2}^{p+1} A_{jk} = \sum_j \left(\frac{i-j}{p}\right)\left(\frac{i-k}{p}\right) = \left(\frac{i-k}{p}\right)\sum_j \left(\frac{i-j}{p}\right) = \left(\frac{i-k}{p}\right)\sum_j \left(\frac{j}{p}\right) = 0,$$

where all sums go over all residues modulo p. Thus, we have

$$0 = \sum_j A_{jk} = \sum_{j \neq k} A_{jk} + A_{kk} = (p-1)a + A_{kk}$$

and since

$$A_{kk} = \sum_{i=2}^{p+1} \left(\frac{i-k}{p}\right)\left(\frac{i-k}{p}\right) = p-1,$$

we conclude $0 = (p-1)a + A_{kk} = (p-1)a + p - 1 = (p-1)(a+1)$ and therefore that the constant a is -1. Hence, we conclude that the columns C_j and C_k of C must be orthogonal for $k, j > 1$, and the first column is orthogonal to all other columns by virtue of the relation

$$\sum_i \left(\frac{i-j}{p}\right) = 0.$$

The graphs obtained from conference matrices we just described are related to the regular graphs called Paley graphs of order p, namely, they are obtained from Paley graphs by adding one vertex and joining it to all the other vertices.

Nikiforov [13] has showed that the maximal energy of graphs behaves asymptotically as $1/2n^{3/2}$ when n goes to infinity. He was using an estimate for eigenvalues of Paley graphs and the fact that for large n there is always a prime p such that $p-1$ is divisible by 4, and p is between n and $n + O(n^{11/20})$. We can obtain the asymptotic behavior from the graphs described above in the same way, using the graph which we obtained from the conference matrix and the characterization of graph energy by means of orthogonal matrices instead of an estimate for eigenvalues.

Acknowledgements We would like to thank Professors D. Cvetković, I. Gutman and M. Marković for their comments on the subject.

Appendix

By $\mathbb{N} = \{0, 1, 2, \dots\}$, we denote natural numbers. The Legendre symbol $\left(\frac{n}{p}\right)$, for $n \in \mathbb{N}$ and p an odd prime number, is defined to be 0 when n is divisible by p, one if there exists $q \in \mathbb{N}$ such that $n \equiv q^2 \pmod{p}$ and n is not divisible by p, and -1 otherwise.

The following formulas then hold:

$$\left(\frac{nm}{p}\right) = \left(\frac{n}{p}\right)\left(\frac{m}{p}\right), \quad \left(\frac{-1}{p}\right) = (-1)^{(p-1)/2}, \quad \sum_{n=0}^{p-1}\left(\frac{n}{p}\right) = 0.$$

References

1. Cvetković, D., Doob, M., Sachs, H.: Spectra of Graphs. Johann Ambrosius Barth, Heidelberg (1995)
2. Cvetković, D., Grout, J.: Maximal energy graphs should have a small number of distinct eigenvalues. Bull. Acad. Serbe Sci. Arts, Cl. Sci. Math. Nat. Sci. Math. **134**, No. 32, 43–57 (2007)
3. Gutman, I.: The energy of a graph. Ber. Math.-Stat. Sekt. Forschungszent. Graz **103**, 1–22 (1978)
4. Gutman, I., Polansky, O.E.: Mathematical Concepts in Organic Chemistry. Springer, Berlin (1986)
5. Gutman, I.: The energy of a graph: Old and new results. Springer, Berlin (2001)
6. Gutman, I.: Uvod u Hemijsku Teoriju Grafova. Kragujevac (2003)
7. Gutman, I., Mateljević, M.: Note on the Coulson integral formula. J. Math. Chem. **39**, 259–266 (2006)
8. Haemers, W.: Strongly regular graphs with maximal energy. Linear Algebra Appl. **429**, 2710–2718 (2008)
9. Haemers, W., Xiang, Q.: Strongly regular graphs with parameters $(4m^4, 2m^4 + m^2, m^4 + m^2, m^4 + m^2)$ exist for all $m > 1$. CentER Discussion paper series Nr.: 2008-86, Tilburg University.
10. Koolen, J., Moulton, V.: Maximal energy graphs. Adv. Appl. Math. **26**, 47–52 (2001)
11. Koolen, J., Moulton, V.: Maximal energy bipartite graphs. Graphs Combinator. **19**, 131–135 (2003)
12. Nikiforov, V.: Graphs and matrices with maximal energy. J. Math. Anal. Appl. **327**, 735–738 (2007) doi:10.1016/j.jmaa.2006.03.089. Page 2. 736.
13. Nikiforov, V.: The energy of graphs and matrices. J. Math. Anal. Appl. **326**, 1472–1475 (2007) doi: 10.1016/j.jmaa.2006.03.072. Page 2.
14. van Lint, J.H., Seidel, J.J.: Equilateral point sets in elliptic geometry. Indagat. Math.. **28**, 335–348 (1966)

15. Goethals, J.M., Seidel, J.J.: Orthogonal matrices with zero diagonal. Can. J. Math. **19**, 1001–1010 (1967)
16. Liu, J., Liu, B.: A Laplacian-energy-like invariant of a graph. MATCH **59**, 355–372 (2008)
17. Muzychuk, M., Xiang, Q.: Symmetric Bush-type Hadamard matrices of order $4m^2$ exist for all odd m. Proc. Am. Math. Soc. **134**, 2197–2204 (2006)

Interlacing Property of Zeros of Shifted Jacobi Polynomials

Aleksandar S. Cvetković

Dedicated to Professor Gradimir V. Milovanović on his 60th birthday

1 Introduction

The interlacing property of zeros of orthogonal polynomials is a classical problem with many applications; see, e.g., [3,4]. Here, we mention only a few of them in numerical integration. For example, it is important for obtaining positive interpolatory quadrature rules [5] for developing methods for the integration of Hilbert transforms [7], etc. A related problem is also recently treated in [2].

In this note, we focus our attention to the case of Jacobi polynomials. To be able to present results, we repeat some facts about Jacobi polynomials.

Let $P_n^{\alpha,\beta}$, $n \in \mathbb{N}_0$, denote monic Jacobi polynomials of degree n with parameters $\alpha, \beta > -1$. It is well known that these polynomials satisfy the orthogonality relation

$$\int_{-1}^{1} P_n^{\alpha,\beta}(x) P_k^{\alpha,\beta}(x)(1-x)^\alpha (1+x)^\beta \, \mathrm{d}x = \|P_n^{\alpha,\beta}\|^2 \delta_{n,k}, \quad n,k \in \mathbb{N}_0,$$

and, as a consequence, the following three-term recurrence relation

$$P_{k+1}^{\alpha,\beta}(x) = \left(x - a_k^{\alpha,\beta}\right) P_k^{\alpha,\beta}(x) - b_k^{\alpha,\beta} P_{k-1}^{\alpha,\beta}(x), \quad k \in \mathbb{N}_0,$$

with $P_{-1}^{\alpha,\beta} = 0$ and $P_0^{\alpha,\beta} = 1$. The recursion coefficients are given by

$$a_k^{\alpha,\beta} = \frac{\beta^2 - \alpha^2}{(2k+\alpha+\beta+2)(2k+\alpha+\beta)}, \quad b_k^{\alpha,\beta} = \frac{4k(k+\alpha)(k+\beta)(k+\alpha+\beta)}{(2k+\alpha+\beta)^2[(2k+\alpha+\beta)^2 - 1]},$$

A.S. Cvetković
Faculty of Sciences and Mathematics, University of Niš, P.O. Box 224, 18000 Niš, Serbia, e-mail: aleksandarcvetkovic@pmf.ni.ac.rs

W. Gautschi et al. (eds.), *Approximation and Computation: In Honor of Gradimir V. Milovanović*, Springer Optimization and Its Applications 42, DOI 10.1007/978-1-4419-6594-3_7,

where b_0 can be chosen arbitrary, but is chosen to be $b_0 = \mu^{\alpha,\beta}(\mathbb{R})$. The Jacobi measure $\mu^{\alpha,\beta}$ is absolutely continuous and is given by $\mathrm{d}\mu^{\alpha,\beta}(x) = w^{\alpha,\beta}(x)\chi_{[-1,1]}(x)\mathrm{d}x = (1-x)^{\alpha}(1+x)^{\beta}\chi_{[-1,1]}(x)\mathrm{d}x$. It is well known that $\|P_n^{\alpha,\beta}\|^2 = \prod_{k=0}^{n} b_k^{\alpha,\beta}$ (cf. [1], [6, Chap. 2], [8]). In particular, we are going to need the identity

$$P_n^{\alpha,\beta}(x) = (-1)^n P_n^{\beta,\alpha}(-x). \tag{1}$$

Now we can state our main result.

Theorem 1. *Let $x_k^{\alpha,\beta}$, $k = 1,\ldots,n$, be the zeros of the Jacobi polynomial $P_n^{\alpha,\beta}$, with the stipulation that $x_0^{\alpha,\beta} = -1$ and $x_{n+1}^{\alpha,\beta} = 1$. For every $t \in (0,1)$, we have*

$$x_k^{\alpha+2,\beta} < x_k^{\alpha+2t,\beta} < x_k^{\alpha,\beta} < x_{k+1}^{\alpha+2,\beta}, \quad k = 1,\ldots,n,$$

$$x_{k-1}^{\alpha,\beta+2} < x_k^{\alpha,\beta} < x_k^{\alpha,\beta+2t} < x_k^{\alpha,\beta+2}, \quad k = 1,\ldots,n.$$

Let $\delta_n^{\alpha,\beta} = (\alpha-\beta)/(2+\alpha+\beta+2n)$ and let $j \in \{0,1,\ldots,n\}$ be such that $x_j^{\alpha,\beta} < \delta_n^{\alpha,\beta} < x_{j+1}^{\alpha,\beta}$. Then $x_j^{\alpha+1,\beta+1} < \delta_n^{\alpha,\beta} < x_{j+1}^{\alpha+1,\beta+1}$ and

$$x_k^{\alpha,\beta} < x_k^{\alpha+1,\beta+1} < x_{k+1}^{\alpha,\beta}, \quad k = 1,\ldots,j,$$

$$x_k^{\alpha+1,\beta+1} < x_k^{\alpha,\beta} < x_{k+1}^{\alpha+1,\beta+1}, \quad k - j+1,\ldots,n.$$

2 Proof of Theorem 1

We need some auxiliary results.

Lemma 1. *Let $\gamma_n^{\alpha,\beta} = \|P_n^{\alpha+1,\beta}\|^2/\|P_n^{\alpha,\beta}\|^2$ and $\eta_n^{\alpha,\beta} = 2(1+\alpha)/(2+\alpha+\beta+2n)$. For $n \in \mathbb{N}$, we have*

$$\begin{aligned}
(1-x)P_n^{\alpha+1,\beta}(x) &= -P_{n+1}^{\alpha,\beta}(x) + \gamma_n^{\alpha,\beta}P_n^{\alpha,\beta}(x),\\
(1+x)P_n^{\alpha,\beta+1}(x) &= P_{n+1}^{\alpha,\beta}(x) + \gamma_n^{\beta,\alpha}P_n^{\alpha,\beta}(x),\\
(1-x)^2P_n^{\alpha+2,\beta}(x) &= \left(x-1-\eta_n^{\alpha,\beta}\right)P_{n+1}^{\alpha,\beta}(x) + (\gamma_n^{\alpha+1,\beta}\gamma_n^{\alpha,\beta} - b_{n+1}^{\alpha,\beta})P_n^{\alpha,\beta}(x),\\
(1+x)^2P_n^{\alpha,\beta+2}(x) &= \left(x+1+\eta_n^{\beta,\alpha}\right)P_{n+1}^{\alpha,\beta}(x) + (\gamma_n^{\beta+1,\alpha}\gamma_n^{\beta,\alpha} - b_{n+1}^{\alpha,\beta})P_n^{\alpha,\beta}(x),\\
(1-x^2)P_n^{\alpha+1,\beta+1}(x) &= \left(-x+\delta_n^{\alpha,\beta}\right)P_{n+1}^{\alpha,\beta}(x) + (\gamma_n^{\alpha,\beta+1}\gamma_n^{\beta,\alpha} + b_{n+1}^{\alpha,\beta})P_n^{\alpha,\beta}(x),
\end{aligned}$$

where $\delta_n^{\alpha,\beta}$ has the same meaning as in Theorem 1.

Proof. Using orthogonality conditions for Jacobi polynomials, we have

$$\int_{-1}^{1}(1-x)P_n^{\alpha+1,\beta}(x)P_{n+k}^{\alpha,\beta}(x)\,\mathrm{d}\mu^{\alpha,\beta}(x)=0,\quad k\in\mathbb{Z}\setminus\{0,1\},$$

$$\int_{-1}^{1}(1-x)P_n^{\alpha+1,\beta}(x)P_n^{\alpha,\beta}(x)\mathrm{d}\mu^{\alpha,\beta}(x)$$
$$=\int_{-1}^{1}P_n^{\alpha+1,\beta}(x)P_n^{\alpha+1,\beta}(x)\mathrm{d}\mu^{\alpha+1,\beta}(x)=\|P_n^{\alpha+1,\beta}\|^2,$$

$$\int_{-1}^{1}(1-x)P_n^{\alpha+1,\beta}(x)P_{n+1}^{\alpha,\beta}(x)\mathrm{d}\mu^{\alpha,\beta}(x)$$
$$=-\int_{-1}^{1}P_{n+1}^{\alpha,\beta}(x)P_{n+1}^{\alpha,\beta}(x)\mathrm{d}\mu^{\alpha,\beta}(x)=-\|P_{n+1}^{\alpha,\beta}\|^2.$$

Then from these identities and (1), we easily get the first two expressions from the statement. The last three identities are obtained by the successive application of the first two identities. □

We are going to need the following Markov Theorem (see [8, p. 116]).

Theorem 2. *Let w_1 and w_2 be two positive weight functions on $(-1,1)$ and assume that w_2/w_1 is an increasing function on $(-1,1)$. If x_k^1, x_k^2, $k=1,\dots,n$, are the zeros of the orthogonal polynomials with respect to the weights w_1 and w_2, then $x_k^1<x_k^2$.*

Proof of Theorem 1. Using the results of Lemma 1, for the zeros $x_k^{\alpha,\beta}$, $k=1,\dots,n$, of the polynomial $P_n^{\alpha,\beta}$, we have

$$\left(1-x_k^{\alpha,\beta}\right)^2\left(1-x_{k+1}^{\alpha,\beta}\right)^2P_n^{\alpha+2,\beta}(x_k^{\alpha,\beta})P_n^{\alpha+2,\beta}(x_{k+1}^{\alpha,\beta})$$
$$=\left(x_k^{\alpha,\beta}-1-\eta_n^{\alpha,\beta}\right)\left(x_{k+1}^{\alpha,\beta}-1-\eta_n^{\alpha,\beta}\right)P_{n+1}^{\alpha,\beta}(x_k^{\alpha,\beta})P_{n+1}^{\alpha,\beta}(x_{k+1}^{\alpha,\beta}).$$

Using this equality and the fact that the zeros of $P_{n+1}^{\alpha,\beta}$ and $P_n^{\alpha,\beta}$ are interlaced, we conclude that

$$P_n^{\alpha+2,\beta}(x_k^{\alpha,\beta})P_n^{\alpha+2,\beta}(x_{k+1}^{\alpha,\beta})<0,\quad k=1,\dots,n-1.$$

Since the function $w^{\alpha,\beta}(x)/w^{\alpha+2,\beta}(x)=(1-x)^{-2}$ is increasing on $(-1,1)$, we have that $x_k^{\alpha+2,\beta}<x_k^{\alpha,\beta}$, $k=1,\dots,n$, according to Theorem 2. Combining the previous two results we conclude that

$$x_k^{\alpha+2,\beta}<x_k^{\alpha,\beta}<x_{k+1}^{\alpha+2,\beta},\quad k=1,\dots,n.$$

Using again Theorem 2, we conclude that

$$x_k^{\alpha+2,\beta}<x_k^{\alpha+2t,\beta}<x_k^{\alpha,\beta},\quad k=1,\dots,n,$$

since the functions

$$\frac{w^{\alpha,\beta}(x)}{w^{\alpha+2t,\beta}(x)} = (1-x)^{-2t} \quad \text{and} \quad \frac{w^{\alpha+2t,\beta}(x)}{w^{\alpha+2,\beta}(x)} = (1-x)^{-2(1-t)}$$

are increasing on $(-1,1)$. Combining these results we get the first statement of this theorem.

The second type of inequalities can be derived using identity (1), according to which $x_k^{\alpha,\beta} = -x_{n+1-k}^{\beta,\alpha}$, $k=1,\ldots,n$.

To prove the second part of this theorem, we use the last identity from Lemma 1. Namely, if we put $x = \delta_n^{\alpha,\beta}$, we get

$$\operatorname{sign} P_n^{\alpha+1,\beta+1}(\delta_n^{\alpha,\beta}) = \operatorname{sign} P_n^{\alpha,\beta}(\delta_n^{\alpha,\beta}) = (-1)^{n-j}. \tag{2}$$

Choosing $x = x_k^{\alpha,\beta}$ and $x = x_{k+1}^{\alpha,\beta}$, we get

$$\begin{aligned}\left(1-(x_k^{\alpha,\beta})^2\right)\left(1-(x_{k+1}^{\alpha,\beta})^2\right) P_n^{\alpha+1,\beta+1}(x_k^{\alpha,\beta}) P_n^{\alpha+1,\beta+1}(x_{k+1}^{\alpha,\beta}) \\ = \left(-x_k^{\alpha,\beta}+\delta_n^{\alpha,\beta}\right)\left(-x_{k+1}^{\alpha,\beta}+\delta_n^{\alpha,\beta}\right) P_{n+1}^{\alpha,\beta}(x_k^{\alpha,\beta}) P_{n+1}^{\alpha,\beta}(x_{k+1}^{\alpha,\beta}).\end{aligned}$$

This equality gives

$$\operatorname{sign}\left[P_n^{\alpha+1,\beta+1}(x_k^{\alpha,\beta}) P_n^{\alpha+1,\beta+1}(x_{k+1}^{\alpha,\beta})\right] = \operatorname{sign}\left[P_{n+1}^{\alpha,\beta}(x_k^{\alpha,\beta}) P_{n+1}^{\alpha,\beta}(x_{k+1}^{\alpha,\beta})\right] = -1, \tag{3}$$

for $k=1,\ldots,j-1,j+1,\ldots,n-1$. As we can see, every interval $(x_k^{\alpha,\beta}, x_{k+1}^{\alpha,\beta})$, $k = 1,\ldots,j-1,j+1,\ldots,n-1$, contains at least one zero of the polynomial $P_n^{\alpha+1,\beta+1}$. As in the statement of this theorem, we assume that $x_j^{\alpha,\beta} < \delta_n^{\alpha,\beta} < x_{j+1}^{\alpha,\beta}$. Hence, we know the position of at least $n-2$ zeros of the polynomial $P_n^{\alpha+1,\beta+1}$. If $j=0$ or $j=n$ we have nothing to prove, since in that case all n zeros of $P_n^{\alpha+1,\beta+1}$ are interlaced with the n zeros of $P_n^{\alpha,\beta}$, according to the property (3).

Now, assume that $1 \le j \le n$. Using the last identity from Lemma 1, we get

$$\begin{aligned}\operatorname{sign} P_n^{\alpha+1,\beta+1}(x_1^{\alpha,\beta}) &= \operatorname{sign}\left[\left(1-(x_1^{\alpha,\beta})^2\right) P_n^{\alpha+1,\beta+1}(x_1^{\alpha,\beta})\right] \\ &= \operatorname{sign}\left[\left(-x_1^{\alpha,\beta}+\delta_n^{\alpha,\beta}\right) P_{n+1}^{\alpha,\beta}(x_1^{\alpha,\beta})\right] = (-1)^n, \\ \operatorname{sign}\left[P_n^{\alpha+1,\beta+1}(x_n^{\alpha,\beta})\right] &= \operatorname{sign}\left[\left(1-(x_n^{\alpha,\beta})^2\right) P_n^{\alpha+1,\beta+1}(x_n^{\alpha,\beta})\right] \\ &= \operatorname{sign}\left[\left(-x_n^{\alpha,\beta}+\delta_n^{\alpha,\beta}\right) P_{n+1}^{\alpha,\beta}(x_n^{\alpha,\beta})\right] = 1.\end{aligned}$$

Using relations (3) and (2), we get

$$-\operatorname{sign} P_n^{\alpha+1,\beta+1}(x_j^{\alpha,\beta}) = -\operatorname{sign} P_n^{\alpha+1,\beta+1}(x_{j+1}^{\alpha,\beta}) = \operatorname{sign} P_n^{\alpha,\beta}(\delta_n^{\alpha,\beta}) = (-1)^{n-j}.$$

Every interval $(x_k^{\alpha,\beta}, x_{k+1}^{\alpha,\beta})$, $k = 1,\dots,j-1,j+1,\dots,n-1$, has at least one zero of $P_n^{\alpha+1,\beta+1}$, and, according to the last relation, each interval $(x_j^{\alpha,\beta}, \delta_n^{\alpha,\beta})$ and $(\delta_n^{\alpha,\beta}, x_{j+1}^{\alpha,\beta})$ has at least one zero. Thus, there are n intervals and the polynomial $P_n^{\alpha+1,\beta+1}$ has n zeros, so each of the above intervals has exactly one zero and the statement of theorem is proved. □

Acknowledgements The author was supported in part by the Serbian Ministry of Science and Technological Development (Project: Orthogonal Systems and Applications, grant number #144004).

References

1. Chihara, T.S.: An Introduction to Orthogonal Polynomials. Gordon and Breach, New York (1978)
2. Dimitrov, D.K., Rodrigues, R.O.: On the behaviour of zeros of Jacobi polynomials. J. Approx. Theory **116**, 224–239 (2002)
3. Driver, K., Jordaan, K.: Interlacing of zeros of shifted sequences of one-parameter orthogonal polynomials. Numer. Math. **107**, 615–624 (2007)
4. Driver, K., Jordaan, K.: Separation theorems for the zeros of certain hypergeometric polynomials. J. Comput. Appl. Math. **199**, 48–55 (2007)
5. Locher, F.: Stability test for linear difference forms, Numerical integration, IV (Oberwolfach, 1992), pp. 215–223, Internat. Ser. Numer. Math. **112** (1993)
6. Mastroianni, G., Milovanović, G.V: Interpolation Processes – Basic Theory and Applications. Springer, Berlin (2008)
7. Mastroianni, G., Occorsio, D.: Interlacing properties of the zeros of the orthogonal polynomials and approximation of the Hilbert transform. Comput. Math. Appl. **30**(3-6), 155–168 (1995)
8. Szegö, G.: Orthogonal Polynomials. American Mathematical Society, New York (1959)

Trigonometric Orthogonal Systems

Aleksandar S. Cvetković and Marija P. Stanić

Dedicated to Professor Gradimir V. Milovanović
on the occasion of his 60th birthday

1 Introduction

The first results on orthogonal trigonometric polynomials of semi-integer degree were given in 1959 by Abram Haimovich Turetzkii (see [27]). A *trigonometric polynomial of semi-integer degree* $n+1/2$ is a trigonometric function of the following form

$$A_{n+1/2}(x) = \sum_{\nu=0}^{n} \left[c_\nu \cos\left(\nu + \frac{1}{2}\right)x + d_\nu \sin\left(\nu + \frac{1}{2}\right)x \right],$$

where $c_\nu, d_\nu \in \mathbb{R}$, $|c_n| + |d_n| \neq 0$. The coefficients c_n and d_n are called *the leading coefficients*.

Let us denote by $\mathcal{T}_n$, $n \in \mathbb{N}_0$, the linear space of all trigonometric polynomials of degree less than or equal to n, i.e., the linear span of $\{1, \cos x, \sin x, \ldots, \cos nx, \sin nx\}$, by $\mathcal{T}$ the set of all trigonometric polynomials, by $\mathcal{T}_n^{1/2}$, $n \in \mathbb{N}_0$, the linear space of all trigonometric polynomials of semi-integer degree less than or equal to $n+1/2$, i.e., the linear span of $\{\cos(k+1/2)x, \sin(k+1/2)x : k = 0, 1, \ldots, n\}$, and by $\mathcal{T}^{1/2}$ the set of all trigonometric polynomials of semi-integer degree.

Aleksandar S. Cvetković
Department of Mathematics and Informatics, Faculty of Sciences and Mathematics, University of Niš, Višegradska 33, 18000 Niš, Serbia, e-mail: aleksandarcvetkovic@pmf.ni.ac.rs

Marija P. Stanić
Department of Mathematics and Informatics, Faculty of Science, University of Kragujevac, Radoja Domanovića 12, 34000 Kragujevac, Serbia, e-mail: stanicm@kg.ac.rs

W. Gautschi et al. (eds.), *Approximation and Computation: In Honor of Gradimir V. Milovanović*, Springer Optimization and Its Applications 42,
DOI 10.1007/978-1-4419-6594-3_8,

The orthogonal trigonometric polynomials of semi-integer degree are connected with quadrature rules with an even maximal trigonometric degree of exactness (with an odd number of nodes). These quadrature rules have application in the numerical integration of 2π-periodic functions. For an integrable and nonnegative weight function w on the interval $[0,2\pi)$, vanishing there only on a set of measure zero, and a given set x_ν, $\nu = 0,1,\ldots,2n$, of distinct points in $[0,2\pi)$, Turetzkii in [27] considered an interpolatory quadrature rule of the form

$$\int_0^{2\pi} t(x)w(x)\,\mathrm{d}x = \sum_{\nu=0}^{2n} w_\nu t(x_\nu), \quad t \in \mathcal{T}_n. \tag{1}$$

Such a quadrature rule can be obtained from the trigonometric interpolation polynomial (cf. [3, 10, 12]).

Definition 1. A quadrature rule of the form

$$\int_0^{2\pi} f(x)w(x)\,\mathrm{d}x = \sum_{\nu=0}^{n} w_\nu f(x_\nu) + R_n(f),$$

where $0 \le x_0 < x_1 < \cdots < x_n < 2\pi$, has trigonometric degree of exactness equal to d if $R_n(f) = 0$ for all $f \in \mathcal{T}_d$ and there exists $g \in \mathcal{T}_{d+1}$ such that $R_n(g) \neq 0$.

Turetzkii tried to increase the trigonometric degree of exactness of a quadrature rule (1) in such a way that he did not specify in advance the nodes x_ν, $\nu = 0,1,\ldots,2n$. His approach is a simulation of the development of Gaussian quadrature rules for algebraic polynomials. He proved that the trigonometric degree of exactness of the quadrature rule (1) is $2n$ if and only if the nodes x_ν $(\in [0,2\pi))$, $\nu = 0,1,\ldots,2n$, are the zeros of a trigonometric polynomial of semi-integer degree $n+1/2$ which is orthogonal on $[0,2\pi)$ with respect to the weight function w to every trigonometric polynomial of semi-integer degree less than or equal to $n-1/2$. It is said that such a quadrature rule is of *Gaussian type*, because it has the maximal trigonometric degree of exactness.

A simple generalization dealing with the translations of the interval $[0,2\pi)$ was given in [18]. Thus, instead of the interval $[0,2\pi)$, one can consider any interval $[L, L+2\pi)$, $L \in \mathbb{R}$. In the sequel of this paper, we use orthogonality with respect to a weight function w on $[-\pi,\pi)$.

It is well known that in the case of a quadrature rule with maximal algebraic degree of exactness, the nodes are the zeros of the corresponding orthogonal algebraic polynomial. Also, in the case of a quadrature rule with an odd maximal trigonometric degree of exactness, the nodes are zeros of orthogonal trigonometric polynomials. Therefore, to obtain a quadrature rule with maximal degree in a subspace of algebraic polynomials one must consider orthogonality in the same subspace. A similar situation exists with a quadrature rule with odd maximal trigonometric degree of exactness, but in the case of even maximal trigonometric degree of exactness one must consider the orthogonality in the subspace of trigonometric polynomials of semi-integer degree. In the general case, the nodes of a quadrature rule with maximal degree of exactness in some space of functions are not zeros of the corresponding orthogonal system of functions (cf. [14]).

Every trigonometric polynomial of semi-integer degree $n+1/2$ can be represented using an algebraic polynomial of degree $2n+1$. An analogous statement for trigonometric polynomials from $\mathcal{T}_n$ was proved in [22, pp. 19–20], and the statement both for $\mathcal{T}_n$ and $\mathcal{T}_n^{1/2}$ was proved in a uniform terminology in [2, Theorem 1.1.1].

Lemma 1. *Let*

$$A_{n+1/2}(x) = \sum_{k=0}^{n} \left[c_k \cos\left(k+\frac{1}{2}\right)x + d_k \sin\left(k+\frac{1}{2}\right)x \right] \in \mathcal{T}_n^{1/2},$$

and $a_k = c_k - \mathrm{i} d_k$, $k = 0,1,\ldots,n$. *Then* $A_{n+1/2}(x)$ *can be represented in the following form:*

$$A_{n+1/2}(x) = \frac{1}{2}\mathrm{e}^{-\mathrm{i}(n+1/2)x} Q_{2n+1}(\mathrm{e}^{\mathrm{i}x}), \tag{2}$$

where $Q_{2n+1}(z)$ *is an algebraic polynomial of degree* $2n+1$, *given by*

$$Q_{2n+1}(z) = \overline{a}_n + \overline{a}_{n-1}z + \cdots + \overline{a}_1 z^{n-1} + \overline{a}_0 z^n + a_0 z^{n+1} + \cdots + a_{n-1}z^{2n} + a_n z^{2n+1}.$$

It is easy to see that $Q_{2n+1}(z) = z^{2n+1}\overline{Q_{2n+1}(1/\overline{z})}$, which means that the polynomial $Q_{2n+1}(z)$ is self-inversive (see [24] and [22, p. 16]). If z is a zero of the polynomial $Q_{2n+1}(z)$, then $x = \arg(z)$ is a zero of $A_{n+1/2}(x)$. That is very important for our numerical method for the construction of trigonometric polynomials of semi-integer degree.

One can consider more general quadrature rules with multiple nodes and maximal trigonometric degree of exactness, i.e., quadrature rules of the following form:

$$\int_{-\pi}^{\pi} f(x)w(x)\,\mathrm{d}x = \sum_{\nu=0}^{2n}\sum_{j=0}^{2s_\nu} A_{j,\nu} f^{(j)}(x_\nu) + R(f), \tag{3}$$

where s_ν are nonnegative integers, $\nu = 0,1,\ldots,2n$, such that $R(f) = 0$ when f is a trigonometric polynomial of degree less than or equal to $d = \sum_{\nu=0}^{2n}(s_\nu+1)-1$. This d is the maximal trigonometric degree of exactness for the quadrature formula (3). Obviously, for $s_0 = s_1 = \cdots = s_{2n} = s$, the maximal trigonometric degree of exactness reduces to $(2n+1)(s+1)-1$. In the case $s_0 = s_1 = \cdots = s_{2n} = 0$, we obtain a quadrature formula with simple nodes.

The first result on quadrature rules with multiple nodes and maximal trigonometric degree of exactness were given for quadrature rules with the same multiplicities of the nodes, i.e., $s_0 = s_1 = \cdots = s_{2n} = s$ by Ghizzetti and Ossicini [7] for the constant weight function $w(x) = 1$. The analogous problem for an arbitrary weight function was solved in [17]. For that purpose, so-called *s-orthogonal trigonometric polynomials of semi-integer degree* with respect to a weight function w on $[-\pi,\pi)$ were used. Such orthogonal trigonometric functions were introduced and studied in [16].

The problem of the existence and uniqueness of a quadrature formula of maximal trigonometric degree of exactness with a given number of free nodes of different multiplicities was considered by Dryanov [4], but only for the weight function $w(x) = 1$. The general case was solved in [21]. In that paper, the so-called *σ-orthogonal trigonometric polynomials of semi-integer degree* with respect to a weight function w on $[-\pi, \pi)$ were introduced and considered.

This paper is organized as follows. Section 2 is devoted to some general properties of orthogonal trigonometric polynomials of semi-integer degree. In Sect. 3 some five-term recurrence relations are presented. Special attention is paid to some classes of weight functions for which recurrence relations have simpler forms. In Sect. 4, orthonormal trigonometric polynomials of semi-integer degree are considered and the Christoffel–Darboux formula is presented. A connection between orthogonal trigonometric polynomials and Szegö polynomials for a certain class of weight functions is given in Sect. 5. Finally, in Sect. 6 a numerical method for constructing orthogonal trigonometric polynomials of semi-integer degree is presented.

2 Orthogonal Trigonometric Polynomials of Semi-Integer Degree

For a given weight function w on $[-\pi, \pi)$, we introduce an inner product of f and g by

$$(f, g) = \int_{-\pi}^{\pi} f(x)g(x)w(x)\,\mathrm{d}x. \tag{4}$$

Turetzkii's approach for developing quadrature rules with maximal trigonometric degree of exactness requires that the trigonometric polynomial of semi-integer degree $A_{n+1/2}$ be orthogonal to every element of $\mathcal{T}_{n-1}^{1/2}$ with respect to the inner product (4), i.e., there must hold

$$\int_{-\pi}^{\pi} A_{n+1/2}(x)t(x)w(x)\,\mathrm{d}x = 0, \quad t \in \mathcal{T}_{n-1}^{1/2}.$$

Since the dimension of $\mathcal{T}_{n-1}^{1/2}$ is $2n$ and $A_{n+1/2}$ has $2n+2$ coefficients, it can be seen that $A_{n+1/2}$ is not determined uniquely up to a multiplicative constant, as in the case of algebraic polynomials. The trigonometric polynomial $A_{n+1/2}$, which is orthogonal with respect to the inner product (4) to every trigonometric polynomial of semi-integer degree less than or equal to $n-1/2$, with given leading coefficients c_n and d_n, however, is uniquely determined (see [27, Sect. 3] and [18]). Obviously, we cannot choose $c_n = d_n = 0$, since in that case we do not have a polynomial of degree $n+1/2$, but $n-1/2$. If we directly compute the nodes, we can fix one of them in advance since for the $2n+1$ nodes we have $2n$ orthogonality conditions.

It is obvious that on any basis of $\mathcal{T}_n^{1/2}$, for all $k = 0, 1, \ldots, n$, we have two different trigonometric polynomials $A_{k+1/2}^{(1)}$ and $A_{k+1/2}^{(2)}$ of semi-integer degree $k+1/2$. In [18] the special choices $c_n = 1$, $d_n = 0$ and $c_n = 0$, $d_n = 1$ were considered and the corresponding orthogonal polynomials of semi-integer degree were denoted by $A_{n+1/2}^{C}$ and $A_{n+1/2}^{S}$, respectively.

Trigonometric polynomials of semi-integer degree $A^C_{n+1/2}$ and $A^S_{n+1/2}$ can be represented in the following expanded forms,

$$A^C_{n+1/2}(x) = \cos\left(n+\frac{1}{2}\right)x + \sum_{\nu=0}^{n-1}\left[c_\nu^{(n)}\cos\left(\nu+\frac{1}{2}\right)x + d_\nu^{(n)}\sin\left(\nu+\frac{1}{2}\right)x\right], \quad (5)$$

$$A^S_{n+1/2}(x) = \sin\left(n+\frac{1}{2}\right)x + \sum_{\nu=0}^{n-1}\left[f_\nu^{(n)}\cos\left(\nu+\frac{1}{2}\right)x + g_\nu^{(n)}\sin\left(\nu+\frac{1}{2}\right)x\right]. \quad (6)$$

Every trigonometric polynomial $A_{n+1/2} \in \mathfrak{T}_n^{1/2}$ is a linear combination of $A^C_{n+1/2}$ and $A^S_{n+1/2}$. Let us now define the linear space

$$\mathfrak{T}^{a,b}_{n+1/2} = \left\{a\cos(n+1/2)x + b\sin(n+1/2)x + t_{n-1/2}(x) : t_{n-1/2} \in \mathfrak{T}^{1/2}_{n-1}\right\},$$

where $a,b \in \mathbb{R}$ are fixed such that $|a|+|b|>0$. Therefore, there exists a unique $A_{n+1/2} \in \mathfrak{T}^{a,b}_{n+1/2}$, such that

$$\int_{-\pi}^{\pi} A_{n+1/2}(x)t(x)w(x)\,\mathrm{d}x = 0, \quad t \in \mathfrak{T}^{1/2}_{n-1}.$$

When we want to emphasize the dependence on a and b, we write $A^{a,b}_{n+1/2}$ for $A_{n+1/2} \in \mathfrak{T}^{a,b}_{n+1/2}$. It is clear that $A^{a,b}_{n+1/2}(x) = aA^C_{n+1/2}(x) + bA^S_{n+1/2}(x)$.

The set $\{A^{a,b}_{k+1/2}, A^{-b,a}_{k+1/2} : k = 0,1,\ldots,n-1,\ |a|+|b|>0, a,b \in \mathbb{R}\} \cup \{A^{a,b}_{n+1/2}\}$ is a basis of $\mathfrak{T}^{a,b}_{n+1/2}$, which is obvious because

$$\cos(k+1/2)x + t(x) = \frac{a}{a^2+b^2}A^{a,b}_{k+1/2}(x) - \frac{b}{a^2+b^2}A_k^{-b,a}(x),$$

$$\sin(k+1/2)x + t(x) = \frac{b}{a^2+b^2}A^{a,b}_{k+1/2}(x) + \frac{a}{a^2+b^2}A^{-b,a}_{k+1/2}(x),$$

where $t \in \mathfrak{T}^{1/2}_{k-1}$, $k = 0,1,\ldots,n$.

The trigonometric polynomial of semi-integer degree $A_{n+1/2}$, which is orthogonal with respect to the inner product (4) to every trigonometric polynomial of semi-integer degree less than or equal to $n-1/2$ has in $[-\pi,\pi)$ exactly $2n+1$ distinct simple zeros (see [27, Theorem 3] and [18, Corollary 2.1]).

We, in particular, consider the case of π-periodic weight function. For such a weight function, orthogonal trigonometric polynomials have the following simple properties.

Theorem 1. *If the weight function is periodic with period π, then the following equalities*

$$A^S_{k+1/2}(x) = (-1)^{k+1}A^C_{k+1/2}(x+\pi), \quad A^C_{k+1/2}(x) = (-1)^k A^S_{k+1/2}(x+\pi), \quad k \in \mathbb{N}_0,$$

hold.

Proof. For $k=0$ this is obvious. Let us introduce the notation

$$B_{k+1/2}(x) = (-1)^{k+1} A^C_{k+1/2}(x+\pi), \quad k \in \mathbb{N}.$$

Since

$$\cos\frac{2\ell+1}{2}(x+\pi) = (-1)^{\ell+1}\sin\frac{2\ell+1}{2}x, \quad \sin\frac{2\ell+1}{2}(x+\pi) = (-1)^{\ell}\cos\frac{2\ell+1}{2}x,$$

it is easy to see that $B_{k+1/2}(x) \in \mathcal{T}^{1/2}_k$ has the leading sine function, i.e., it has the leading coefficients $0, 1$. Starting with the orthogonality conditions for $A^C_{k+1/2}$

$$\int_{-\pi}^{\pi} A^C_{k+1/2}(x) t(x) w(x)\,\mathrm{d}x = 0, \quad t \in \mathcal{T}^{1/2}_{k-1},$$

with the substitution $x := x+\pi$, we obtain

$$\int_{-2\pi}^{0} B_{k+1/2}(x)\widetilde{t}(x) w(x+\pi)\,\mathrm{d}x = 0, \quad \widetilde{t}(x) = t(x+\pi) \in \mathcal{T}^{1/2}_{k-1}.$$

The product of two trigonometric polynomials of semi-integer degree is a trigonometric polynomial, so $B_{k+1/2}(x)\widetilde{t}(x)$ is a trigonometric polynomial, i.e., a periodic function with period 2π. Therefore,

$$\int_{-\pi}^{\pi} B_{k+1/2}(x) t(x) w(x)\,\mathrm{d}x = 0, \quad t \in \mathcal{T}^{1/2}_{k-1}.$$

According to the uniqueness of orthogonal trigonometric polynomials of semi-integer degree with given leading coefficients we have

$$A^S_{k+1/2}(x) = B_{k+1/2}(x) = (-1)^{k+1} A^C_{k+1/2}(x+\pi), \quad k \in \mathbb{N}.$$

The second equality can be proved in a similar way. □

3 Recurrence Relations

It is well known that the orthogonal algebraic polynomials satisfy a three-term recurrence relation (see [1, 6, 10]). Such a recurrence relation is important for the constructive and computational uses of orthogonal polynomials.

In [18] it was proved that the orthogonal trigonometric polynomials of semi-integer degree $A^C_{k+1/2}$ and $A^S_{k+1/2}$, $k \in \mathbb{N}$, satisfy the following five-term recurrence relations:

$$A^C_{k+1/2} = \left(2\cos x - \alpha^{(1)}_k\right) A^C_{k-1/2} - \beta^{(1)}_k A^S_{k-1/2} - \alpha^{(2)}_k A^C_{k-3/2} - \beta^{(2)}_k A^S_{k-3/2}, \quad (7)$$

and

$$A_{k+1/2}^S = \left(2\cos x - \delta_k^{(1)}\right) A_{k-1/2}^S - \gamma_k^{(1)} A_{k-1/2}^C - \gamma_k^{(2)} A_{k-3/2}^C - \delta_k^{(2)} A_{k-3/2}^S, \quad (8)$$

where the recurrence coefficients are given by $\alpha_1^{(2)} = \beta_1^{(2)} = \gamma_1^{(2)} = \delta_1^{(2)} = 0$ and

$$\begin{aligned}
\alpha_k^{(1)} &= \frac{I_{k-1}^S J_{k-1}^C - I_{k-1} J_{k-1}}{D_{k-1}}, & \alpha_k^{(2)} &= \frac{I_{k-1}^C I_{k-2}^S - I_{k-1} I_{k-2}}{D_{k-2}}, \\
\beta_k^{(1)} &= \frac{I_{k-1}^C J_{k-1} - I_{k-1} J_{k-1}^C}{D_{k-1}}, & \beta_k^{(2)} &= \frac{I_{k-1} I_{k-2}^C - I_{k-1}^C I_{k-2}}{D_{k-2}}, \\
\gamma_k^{(1)} &= \frac{I_{k-1}^S J_{k-1} - I_{k-1} J_{k-1}^S}{D_{k-1}}, & \gamma_k^{(2)} &= \frac{I_{k-1} I_{k-2}^S - I_{k-1}^S I_{k-2}}{D_{k-2}}, \\
\delta_k^{(1)} &= \frac{I_{k-1}^C J_{k-1}^S - I_{k-1} J_{k-1}}{D_{k-1}}, & \delta_k^{(2)} &= \frac{I_{k-1}^S I_{k-2}^C - I_{k-1} I_{k-2}}{D_{k-2}},
\end{aligned} \quad (9)$$

where $D_{k-j} = I_{k-j}^C I_{k-j}^S - I_{k-j}^2$, $j = 1,2$, and

$$\begin{aligned}
I_\nu^C &= (A_{\nu+1/2}^C, A_{\nu+1/2}^C), & J_\nu^C &= (2\cos x A_{\nu+1/2}^C, A_{\nu+1/2}^C), \\
I_\nu^S &= (A_{\nu+1/2}^S, A_{\nu+1/2}^S), & J_\nu^S &= (2\cos x A_{\nu+1/2}^S, A_{\nu+1/2}^S), \\
I_\nu &= (A_{\nu+1/2}^C, A_{\nu+1/2}^S), & J_\nu &= (2\cos x A_{\nu+1/2}^C, A_{\nu+1/2}^S).
\end{aligned}$$

Knowing these recurrence coefficients in the five-term recurrence relations (7) and (8), we can obtain the coefficients of the expanded forms (5) and (6) for $A_{n+1/2}^C$ and $A_{n+1/2}^S$ (see [19, Theorem 1]).

In [18], special attention was paid to the case of symmetric weight functions on $(-\pi,\pi)$, since in that case the recurrence relations (7) and (8) have simpler forms. It was proved that when $w(-x) = w(x)$, $x \in (-\pi,\pi)$, the recurrence coefficients $\beta_k^{(j)}$ and $\gamma_k^{(j)}$, $j = 1,2$, $k \in \mathbb{N}$, are equal to zero, and because of that, the five-term recurrence relations reduce to three-term recurrence relations (see [18, Sect. 4]). In that case the trigonometric polynomials of semi-integer degree (5) and (6) reduce to

$$A_{n+1/2}^C(x) = \sum_{\nu=0}^{n} c_\nu^{(n)} \cos\left(\nu + \frac{1}{2}\right)x, \quad c_n^{(n)} = 1$$

and

$$A_{n+1/2}^S(x) = \sum_{\nu=0}^{n} g_\nu^{(n)} \sin\left(\nu + \frac{1}{2}\right)x, \quad g_n^{(n)} = 1,$$

respectively. These orthogonal trigonometric functions can be reduced to algebraic polynomials by using Chebyshev polynomials of the first and the second kind (see [18, Sect. 4] for details).

Simpler forms for the recurrence relations exist also in the case of π-periodic weight functions. By means of Theorem 1, it is easy to prove the following result.

Lemma 2. *In the case of π-periodic weight functions, for all $k \in \mathbb{N}_0$ the equalities $I_k^C = I_k^S$, $I_k = 0$, $J_k^C = -J_k^S$ hold.*

By using Lemma 2, from (9) there follows

$$\alpha_k^{(1)} = -\delta_k^{(1)} = \frac{J_{k-1}^C}{I_{k-1}^C},\ \beta_k^{(1)} = \gamma_k^{(1)} = \frac{J_{k-1}}{I_{k-1}^C},\ \alpha_k^{(2)} = \delta_k^{(2)} = \frac{I_{k-1}^C}{I_{k-2}^C},\ \beta_k^{(2)} = \gamma_k^{(2)} = 0,$$

and because of that, the recurrence relations (7) and (8) reduce to the following four-term recurrence relations

$$A_{k+1/2}^C = \left(2\cos x - \alpha_k^{(1)}\right) A_{k-1/2}^C - \beta_k^{(1)} A_{k-1/2}^S - \alpha_k^{(2)} A_{k-3/2}^C,$$
$$A_{k+1/2}^S = \left(2\cos x + \alpha_k^{(1)}\right) A_{k-1/2}^S - \beta_k^{(1)} A_{k-1/2}^C - \alpha_k^{(2)} A_{k-3/2}^S.$$

Finally, we consider the recurrence relations for the trigonometric polynomials of semi-integer degree $A_{n+1/2}^{a,b}$ and $A_{n+1/2}^{-b,a}$. Starting with the recurrence relations (7) and (8) for $A_{n+1/2}^C$ and $A_{n+1/2}^S$, by using the relations

$$A_{n+1/2}^C = \frac{aA_{n+1/2}^{a,b} - bA_{n+1/2}^{-b,a}}{a^2+b^2}, \quad A_{n+1/2}^S = \frac{bA_{n+1/2}^{a,b} + aA_{n+1/2}^{-b,a}}{a^2+b^2},$$

it is easy to obtain the recurrence relations for $A_{n+1/2}^{a,b}$ and $A_{n+1/2}^{-b,a}$ (see [15, Lemma 2]).

Lemma 3. *The trigonometric polynomials of semi-integer degree $A_{n+1/2}^{a,b}$ and $A_{n+1/2}^{-b,a}$ satisfy the following five-term recurrence relations:*

$$A_{n+1/2}^{a,b} = \left(2\cos x - \widetilde{\alpha}_n^{(1)}\right) A_{n-1/2}^{a,b} - \widetilde{\alpha}_n^{(2)} A_{n-3/2}^{a,b} - \widetilde{\beta}_n^{(1)} A_{n-1/2}^{-b,a} - \widetilde{\beta}_n^{(2)} A_{n-3/2}^{-b,a},$$
$$A_{n+1/2}^{-b,a} = \left(2\cos x - \widetilde{\delta}_n^{(1)}\right) A_{n-1/2}^{-b,a} - \widetilde{\delta}_n^{(2)} A_{n-3/2}^{-b,a} - \widetilde{\gamma}_n^{(1)} A_{n-1/2}^{a,b} - \widetilde{\gamma}_n^{(2)} A_{n-3/2}^{a,b},$$

where

$$\widetilde{\alpha}_n^{(1)} = \frac{a^2\alpha_n^{(1)} + ab(\beta_n^{(1)} + \gamma_n^{(1)}) + b^2\delta_n^{(1)}}{a^2+b^2},\ \widetilde{\alpha}_n^{(2)} = \frac{a^2\alpha_n^{(2)} + ab(\beta_n^{(2)} + \gamma_n^{(2)}) + b^2\delta_n^{(2)}}{a^2+b^2},$$
$$\widetilde{\beta}_n^{(1)} = \frac{a^2\beta_n^{(1)} - ab(\alpha_n^{(1)} - \delta_n^{(1)}) - b^2\gamma_n^{(1)}}{a^2+b^2},\ \widetilde{\beta}_n^{(2)} = \frac{a^2\beta_n^{(2)} - ab(\alpha_n^{(2)} - \delta_n^{(2)}) - b^2\gamma_n^{(2)}}{a^2+b^2},$$
$$\widetilde{\delta}_n^{(1)} = \frac{b^2\alpha_n^{(1)} - ab(\beta_n^{(1)} + \gamma_n^{(1)}) + a^2\delta_n^{(1)}}{a^2+b^2},\ \widetilde{\delta}_n^{(2)} = \frac{b^2\alpha_n^{(2)} - ab(\beta_n^{(2)} + \gamma_n^{(2)}) + a^2\delta_n^{(2)}}{a^2+b^2},$$
$$\widetilde{\gamma}_n^{(1)} = \frac{a^2\gamma_n^{(1)} - ab(\alpha_n^{(1)} - \delta_n^{(1)}) - b^2\beta_n^{(1)}}{a^2+b^2},\ \widetilde{\gamma}_n^{(2)} = \frac{a^2\gamma_n^{(2)} - ab(\alpha_n^{(2)} - \delta_n^{(2)}) - b^2\beta_n^{(2)}}{a^2+b^2},$$

and $\alpha_k^{(i)}$, $\beta_k^{(i)}$, $\gamma_k^{(i)}$, $\delta_k^{(i)}$, $i = 1,2$, are the recurrence coefficients given by (9).

4 Christoffel–Darboux Formula

Let us denote

$$m_n = \begin{bmatrix} I_n^C & I_n \\ I_n & I_n^S \end{bmatrix},$$

and by $\widetilde{m}_n$, $n \in \mathbb{N}_0$, the positive definite square root of m_n (it is proved in [20] that the matrix m_n, $n \in \mathbb{N}_0$, is positive definite), i.e., the unique positive definite matrix such that $m_n = \widetilde{m}_n \widetilde{m}_n$ (see [28]). Since m_n is a symmetric matrix, the matrix $\widetilde{m}_n$ is also symmetric, i.e., it has the following form

$$\widetilde{m}_n = \begin{bmatrix} a_n & b_n \\ b_n & c_n \end{bmatrix},$$

where $a_n^2 + b_n^2 = I_n^C$, $a_n b_n + b_n c_n = I_n$ and $b_n^2 + c_n^2 = I_n^S$.

The following trigonometric polynomials of semi-integer degree $n + 1/2$,

$$\widetilde{A}^C_{n+1/2}(x) = \frac{c_n}{a_n c_n - b_n^2} A^C_{n+1/2}(x) - \frac{b_n}{a_n c_n - b_n^2} A^S_{n+1/2}(x),$$

$$\widetilde{A}^S_{n+1/2}(x) = \frac{a_n}{a_n c_n - b_n^2} A^S_{n+1/2}(x) - \frac{b_n}{a_n c_n - b_n^2} A^C_{n+1/2}(x),$$

satisfy the following equalities (see [20, Theorem 2.1]):

$$\left(\widetilde{A}^C_{n+1/2}, \widetilde{A}^C_{n+1/2}\right) = 1, \quad \left(\widetilde{A}^C_{n+1/2}, \widetilde{A}^S_{n+1/2}\right) = 0 \quad \text{and} \quad \left(\widetilde{A}^S_{n+1/2}, \widetilde{A}^S_{n+1/2}\right) = 1.$$

Because of that, $\widetilde{A}^C_{n+1/2}$ and $\widetilde{A}^S_{n+1/2}$, $n \in \mathbb{N}_0$, are called *orthonormal trigonometric polynomials of semi-integer degree.*

In [20], the following theorem was proved.

Theorem 2 (Christoffel–Darboux formula). *For the orthonormal trigonometric polynomials of semi-integer degree the following formula*

$$\begin{aligned}
&2(\cos x - \cos y) \sum_{k=0}^{n} \left(\widetilde{A}^C_{k+1/2}(x)\widetilde{A}^C_{k+1/2}(y) + \widetilde{A}^S_{k+1/2}(x)\widetilde{A}^S_{k+1/2}(y)\right) \\
&= \frac{(a_{n+1}c_n - b_n b_{n+1})\left(\widetilde{A}^C_{n+3/2}(x)\widetilde{A}^C_{n+1/2}(y) - \widetilde{A}^C_{n+3/2}(y)\widetilde{A}^C_{n+1/2}(x)\right)}{(a_n c_n - b_n^2)} \\
&+ \frac{(b_{n+1}c_n - b_n c_{n+1})\left(\widetilde{A}^S_{n+3/2}(x)\widetilde{A}^C_{n+1/2}(y) - \widetilde{A}^S_{n+3/2}(y)\widetilde{A}^C_{n+1/2}(x)\right)}{(a_n c_n - b_n^2)} \\
&+ \frac{(a_n b_{n+1} - a_{n+1}b_n)\left(\widetilde{A}^C_{n+3/2}(x)\widetilde{A}^S_{n+1/2}(y) - \widetilde{A}^C_{n+3/2}(y)\widetilde{A}^S_{n+1/2}(x)\right)}{(a_n c_n - b_n^2)} \\
&+ \frac{(a_n c_{n+1} - b_n b_{n+1})\left(\widetilde{A}^S_{n+3/2}(x)\widetilde{A}^S_{n+1/2}(y) - \widetilde{A}^S_{n+3/2}(y)\widetilde{A}^S_{n+1/2}(x)\right)}{(a_n c_n - b_n^2)}
\end{aligned}$$

holds.

5 Connection with Szegö Polynomials

Let us denote by $\mathcal{P}_n$ the set of all algebraic polynomials of degree less than or equal to n and by $\mathcal{P}$ the set of all algebraic polynomials.

Szegö polynomials (see [25, p. 287]) are defined to be orthogonal on the unit circle with respect to the following inner product

$$(p,q) = \frac{1}{2\pi}\int_{-\pi}^{\pi} p(\mathrm{e}^{\mathrm{i}x})\overline{q(\mathrm{e}^{\mathrm{i}x})}w(x)\,\mathrm{d}x, \quad p,q \in \mathcal{P}.$$

In [25, p. 289], it was proved that for the special type of weights $w(x) = 1/t_\ell(x)$, where $t_\ell \in \mathcal{T}_\ell$ is strictly positive on $[-\pi,\pi)$, the monic Szegö polynomials can be expressed as

$$\phi_n(\mathrm{e}^{\mathrm{i}x}) = \mathrm{e}^{\mathrm{i}(n-\ell)x}h_\ell(\mathrm{e}^{\mathrm{i}x}), \quad n \geq \ell. \tag{10}$$

To give a connection between orthogonal trigonometric polynomials of semi-integer degree and orthogonal Szegö polynomials on the unit circle, we need the following lemma, which gives a factorization of trigonometric polynomials (see [25, p. 4] and [15]).

Lemma 4. *Let $t_n \in \mathcal{T}_n$, $n \in \mathbb{N}$, be a trigonometric polynomial which is strictly positive on the interval $[-\pi,\pi)$, with exact degree n. Then there exist a unique algebraic polynomial, up to a multiplicative constant of modulus one, $H_n \in \mathcal{P}_n$, with exact degree n, such that*

$$t_n(x) = \mathrm{e}^{-\mathrm{i}nx}H_n(\mathrm{e}^{\mathrm{i}x})H_n^*(\mathrm{e}^{\mathrm{i}x}), \tag{11}$$

where $H_n^*(z) = z^n\overline{H_n(1/\overline{z})}$ and all zeros of H_n have modulus smaller than one.

By using the representation (11), the following theorem was proved in [15].

Theorem 3. *Let $t_\ell \in \mathcal{T}_\ell$, $\ell \in \mathbb{N}_0$, be a strictly positive trigonometric polynomial on $[-\pi,\pi)$, with exact degree ℓ. Then the trigonometric polynomial of semi-integer degree $A_{n+1/2}^{a,b} \in \mathcal{T}_{n+1/2}^{a,b}$, $n \geq \ell$, orthogonal with respect to the weight function $w(x) = 1/t_\ell(x)$ on $[-\pi,\pi)$, is given by*

$$A_{n+1/2}^{a,b}(x) = \frac{a-\mathrm{i}b}{2}\mathrm{e}^{\mathrm{i}(n-\ell+1/2)x}h_\ell(\mathrm{e}^{\mathrm{i}x}) + \frac{a+\mathrm{i}b}{2}\mathrm{e}^{-\mathrm{i}(n-\ell+1/2)x}\overline{h_\ell(\mathrm{e}^{\mathrm{i}x})},$$

where h_ℓ is the monic version of the polynomial H_ℓ from Lemma 4. The five-term recurrence coefficients are given by

$$\widetilde{\alpha}_n^{(1)} = \widetilde{\beta}_n^{(1)} = \widetilde{\gamma}_n^{(1)} = \widetilde{\delta}_n^{(1)} = 0, \quad n \geq \ell+1,$$
$$\widetilde{\beta}_n^{(2)} = \widetilde{\gamma}_n^{(2)} = 0, \quad \widetilde{\alpha}_n^{(2)} = \widetilde{\delta}_n^{(2)} = 1, \quad n \geq \ell+2.$$

Now, from Theorem 3 and (10) it is easy to see that the trigonometric polynomial of semi-integer degree $A^{a,b}_{n+1/2} \in \mathfrak{T}^{a,b}_{n+1/2}$, $n \geq \ell$, orthogonal with respect to the strictly positive weight $w(x) = 1/t_\ell(x)$, $t_\ell \in \mathfrak{T}_n$, can be represented as

$$A^{a,b}_{n+1/2}(x) = \frac{a-\mathrm{i}b}{2}\mathrm{e}^{\mathrm{i}x/2}\phi_n(\mathrm{e}^{\mathrm{i}x}) + \frac{a+\mathrm{i}b}{2}\mathrm{e}^{-\mathrm{i}x/2}\overline{\phi_n(\mathrm{e}^{\mathrm{i}x})}, \quad n \geq \ell,$$

where ϕ_n is the corresponding Szegö polynomial orthogonal on the unit circle. The norm of $A^{a,b}_{n+1/2}$, $n \geq \ell$, is given by

$$\|A^{a,b}_{n+1/2}\|^2 = \pi(a^2+b^2)\exp\left(-\frac{1}{2\pi}\int_{-\pi}^{\pi}\log t_\ell(x)\,\mathrm{d}x\right)$$

(see [15] for details).

6 Numerical Construction

A numerical method for constructing the orthogonal trigonometric polynomials of semi-integer degree, as well as of the corresponding quadratures with maximal trigonometric degree of exactness

$$\int_{-\pi}^{\pi} t(x)w(x)\,\mathrm{d}x = \sum_{\nu=0}^{2n} w_\nu t(\tau_\nu), \quad t \in \mathfrak{T}_{2n}, \tag{12}$$

was presented in [18]. That method was based on the five-term recurrence relations for the orthogonal trigonometric polynomials of semi-integer degree $A^C_{n+1/2}$ and $A^S_{n+1/2}$. The main problem of the procedure presented in [18] is the calculation of the five-term recurrence coefficients. For some special weight functions, explicit formulas for the five-term recurrence coefficients were given in [19].

For calculating the zeros x_ν, $\nu = 0,1,\ldots,2n$, of the orthogonal trigonometric polynomial of semi-integer degree $A_{n+1/2}$ we use the algebraic polynomial $Q_{2n+1}(z)$ introduced in Lemma 1. From the representation (2), we conclude that the zeros of $Q_{2n+1}(z)$ on the unit circle correspond to the zeros of $A_{n+1/2}(x)$ on the interval $[-\pi,\pi)$. Since $A_{n+1/2}(x)$ has $2n+1$ distinct zeros in $[-\pi,\pi)$, the algebraic polynomial $Q_{2n+1}(z)$ has $2n+1$ simple zeros on the unit circle. Using $A^C_{n+1/2}(x)$, the algebraic polynomial $Q_{2n+1}(z)$ has the form

$$Q_{2n+1}(z) = 1 + \overline{a}^{(n)}_{n-1}z + \cdots + \overline{a}^{(n)}_1 z^{n-1} + \overline{a}^{(n)}_0 z^n + a^{(n)}_0 z^{n+1} + \cdots + z^{2n+1},$$

where $a^{(n)}_\nu = c^{(n)}_\nu - \mathrm{i}d^{(n)}_\nu$, $\nu = 0,1,\ldots,n-1$ (see Lemma 1).

In our procedure, we first determine the zeros z_ν, $\nu = 0,1,\dots,2n$, of the algebraic polynomial $Q_{2n+1}(z)$ by the following simultaneous iterative process

$$z_\nu^{(k+1)} = z_\nu^{(k)} - \frac{Q_{2n+1}(z_\nu^{(k)})}{P_k'(z_\nu^{(k)})}, \quad \nu = 0,1,\dots,2n;\ k = 0,1,\dots, \tag{13}$$

where $P_k(z) = \prod_{\nu=0}^{2n}(z - z_\nu^{(k)})$. The starting values must be mutually different, i.e., $z_i^{(0)} \neq z_j^{(0)}$, $i \neq j$. This iterative process converges quadratically, because it is equivalent to the Newton–Kantorovich method applied to the system of Viète formulas (cf. [26]).

For the iterative process (13), we need to calculate the values $Q_{2n+1}(z_\nu^{(k)})$, where $z_\nu^{(k)} \in \mathbb{C}$ in general. Since in (2) all functions are analytic, the relation holds true for all $x \in \mathbb{C}$ by the principle of analytic continuation (see [9]). The values of $A_{n+1/2}(x)$ can be computed accurately using the five-term recurrence relations, so, it holds also for the values of the polynomial Q_{2n+1}. Therefore, we have to compute Q_{2n+1} at $z_\nu^{(k)} \in \mathbb{C}$, i.e., we have to compute $A_{n+1/2}$ at the point $-\mathrm{i}\,\mathrm{Log}(z_\nu^{(k)})$. In [18], it was proved that for different branches of the Log function we get the same results for $Q_{2n+1}(z_\nu^{(k)})$.

Knowing the zeros z_ν, $\nu = 0,1,\dots,2n$, of the algebraic polynomial Q_{2n+1}, the zeros x_ν, $\nu = 0,1,\dots,2n$, of $A_{n+1/2}^C$ can be obtained as $x_\nu = \arg z_\nu \in [-\pi,\pi)$, $\nu = 0,1,\dots,2n$.

As usual, the main problem of an iterative process is the choice of the initial values. That problem was analyzed in detail in [18], where several numerical examples are presented, too.

For the construction of the quadrature formula (12) we also need to calculate weights. When nodes are given, it is possible to calculate weights, and the algorithm for that was described in [18], too.

It is well known that the zeros of orthogonal algebraic polynomials can be computed using the QR-algorithm (see [5,6,8,11]). As it was said in Sect. 3, in a case of a symmetric weight function on $(-\pi,\pi)$, the problem of trigonometric orthogonality reduces to the problem of algebraic orthogonality. Therefore, the zeros of orthogonal trigonometric polynomials can be computed using the QR-algorithm. Using [23] and [13], in [18] it was proved how the weights w_ν, $\nu = 0,1,\dots,2n$, of the quadrature rule (12) can also be constructed using the QR-algorithm. The results obtained are presented in the following two lemmas.

Lemma 5. *Let w be an even weight function on $(-\pi,\pi)$. Let x_ν and ω_ν, $\nu = 1,\dots,n$, be nodes and weights of the n-point Gaussian quadrature rule for the weight function $w(\arccos x)\sqrt{(1+x)/(1-x)}$, $x \in (-1,1)$, constructed for algebraic polynomials. Then, for the quadrature rule of Gaussian type (12) with respect to the weight function w on $(-\pi,\pi)$, we have*

$$w_{2n-\nu-1} = w_\nu = \frac{\omega_{\nu+1}}{1+x_{\nu+1}}, \quad \nu = 0,1,\dots,n-1, \quad w_{2n} = \int_{-\pi}^{\pi} w(x)\,\mathrm{d}x - \sum_{\nu=0}^{2n-1} w_\nu,$$

$$\tau_{2n-\nu-1} = -\tau_\nu = \arccos x_{\nu+1}, \quad \nu = 0,1,\dots,n-1, \quad \tau_{2n} = \pi.$$

Lemma 6. *Let w be an even weight function on $(-\pi,\pi)$. Let x_ν and ω_ν, $\nu=1,\ldots,n$, be nodes and weights for the n-point Gaussian quadrature rule for the weight function $w(\arccos x)\sqrt{(1-x)/(1+x)}$, $x\in(-1,1)$, constructed for algebraic polynomials. Then, for the quadrature rule of Gaussian type (12) with respect to the weight function w on $(-\pi,\pi)$, we have*

$$w_{2n-\nu}=w_\nu=\frac{\omega_{\nu+1}}{1-x_{\nu+1}},\quad \nu=0,1,\ldots,n-1,\quad w_n=\int_{-\pi}^{\pi}w(x)\,\mathrm{d}x-\sum_{\nu=0}^{2n-1}{}' w_\nu,$$

$$\tau_{2n-\nu}=-\tau_\nu=\arccos x_{\nu+1},\quad \nu=0,1,\ldots,n-1,\quad \tau_n=0.$$

Acknowledgements The authors were supported in part by the Serbian Ministry of Science and Technological Development (Project: Orthogonal Systems and Applications, grant number #144004).

References

1. Chihara, T.S.: An Introduction to Orthogonal Polynomials. Gordon and Breach, New York (1978)
2. Cruz-Barroso, R., González-Vera, P., Njåstad, O.: On bi-orthogonal systems of trigonometric functiones and quadrature formulas for periodic integrands. Numer. Algor. **44** (4), 309–333 (2007)
3. DeVore, R.A., Lorentz, G.G.: Constructive Approximation. Springer, Berlin (1993)
4. Dryanov, D.P.: Quadrature formulae with free nodes for periodic functions. Numer. Math. **67**, 441–464 (1994)
5. Gautschi, W.: Algorithm 726: ORTHPOL – A package of routines for generating orthogonal polynomials and Gauss-type quadrature rules. ACM Trans. Math. Software **20** (1), 21–62 (1994)
6. Gautschi, W.: Orthogonal Polynomials: Computation and Approximation. Numerical Mathematics and Scientific Computation, Oxford University Press, Oxford (2004)
7. Ghizzetti, A., Ossicini, A.: Quadrature Formulae. Academie–Verlag, Berlin (1970)
8. Golub, G.H., Welsch, J.H.: Calculation of Gauss quadrature rules. Math. Comput. **23**, 221–230 (1969)
9. Heins, M.: Complex Function Theory. Academic Press, New York (1968)
10. Mastroianni, G., Milovanović, G.V.: Interpolation Processes – Basic Theory and Applications. Springer Monographs in Mathematics, Springer, Berlin (2008)
11. Milovanović, G.V.: Numerical Analysis, Part I. Naučna Knjiga, Beograd (1991) (in Serbian)
12. Milovanović, G.V.: Numerical Analysis, Part II. Naučna Knjiga, Beograd (1991) (in Serbian)
13. Milovanović, G.V., Cvetković, A.S.: Note on a construction of weights in Gauss-type quadrature formula. Facta Univ. Ser. Math. Inform. **15**, 69–83 (2000)
14. Milovanović, G.V., Cvetković, A.S.: Construction of Gaussian type quadrature formulas for Müntz systems. SIAM J. Sci. Comput. **27**, 893–913 (2005)
15. Milovanović, G.V., Cvetković, A.S., Marjanović, Z.M.: Connection of semi-integer trigonometric orthogonal polynomials with Szegö polynomials. NMA 2006, Lecture Notes in Comput. Sci. **4310**, 394–401, Springer, Berlin (2007)
16. Milovanović, G.V., Cvetković, A.S., Stanić, M.P.: Trigonometric orthogonal systems and quadrature formulae with maximal trigonometric degree of exactness. NMA 2006, Lecture Notes in Comput. Sci. **4310**, 402–409, Springer, Berlin (2007)
17. Milovanović, G.V., Cvetković, A.S., Stanić, M.P.: Quadrature formulae with multiple nodes and a maximal trigonometric degree of exactness. In: PAMM · Proc. Appl. Math. Mech. **7**, 2020043–2020044 (2007)

18. Milovanović, G.V., Cvetković, A.S., Stanić, M.P.: Trigonometric orthogonal systems and quadrature formulae. Comput. Math. Appl. **56**, no.11, 2915–2931 (2008)
19. Milovanović, G.V., Cvetković, A.S., Stanić, M.P.: Explicit formulas for five-term recurrence coefficients of orthogonal trigonometric polynomials of semi-integer degree. Appl. Math. Comput. **198**, 559–573 (2008)
20. Milovanović, G.V., Cvetković, A.S., Stanić, M.P.: Christoffel-Darboux formula for orthogonal trigonometric polynomials of semi-integer degree. Facta Univ. Ser. Math. Inform. **23**, 29–37 (2008)
21. Milovanović, G.V., Cvetković, A.S., Stanić, M.P.: Quadrature formulae with multiple nodes and a maximal trigonometric degree of exactness. Numer. Math. **112** (3), 425–448 (2009)
22. Milovanović, G.V., Mitrinović, D.S., Rassias, Th.M.: Topics in Polynomials: Extremal Problems, Inequalites, Zeros. World Scientific, River Edge, NJ (1994)
23. Shohat, J.A.: On a certain formula of mechanical quadratures with non-equidistant ordinates. Trans. Amer. Math. Soc. **31**, 448–463 (1929)
24. Szegö, G.: On bi-orthogonal systems of trigonometric polynomials. Magyar Tud. Akad. Kutató Int. Kőzl **8**, 255–273 (1963)
25. Szegö, G.: Orthogonal Polynomials. Amer. Math. Soc. Colloq. Publ. **23**, 4th ed., Amer. Math. Soc., Providence, R. I., (1975)
26. Tošić, D.Dj., Milovanović, G.V.: An application of Newton's method to simultaneous determination of zeros of a polynomial. Univ. Beograd. Publ. Elektrotehn. Fak. Ser. Mat. Fiz. No 412 – No 460, 175–177 (1973)
27. Turetzkii, A.H.: On quadrature formulae that are exact for trigonometric polynomials. East J. Approx. **11**, 337–359 (2005) (English translation of: Belorus. Gos. Univ. Uch. Zap. Ser. Mat. **1959**, no.1, 31–54)
28. Weisstein, E.W.: Positive Definite Matrix. From MathWorld–A Wolfram Web Resource. http://mathworld.wolfram.com/PositiveDefiniteMatrix.html

Experimental Mathematics Involving Orthogonal Polynomials

Walter Gautschi

Dedicated to Gradimir V. Milovanović on his 60th birthday

2000 *Mathematics Subject Classification*: 26D07 33C45 6504 6505 65D32

1 Introduction

In Wikipedia [20], the term "experimental mathematics" is defined as follows: "*Experimental mathematics is an approach to mathematics in which numerical computation is used to investigate mathematical objects and identify properties and patterns.*" The ultimate goal of experimental mathematics is to encourage, and provide direction for, purely mathematical research, in the hope of thereby extending the boundaries of mathematical knowledge. It is in this spirit that we are going to deal here with a few special topics that involve orthogonal polynomials.

A key to experimental mathematics is *numerical computation*, and that presupposes the existence of a body of reliable computational tools that allows us to generate numerically all entities of interest. In the realm of orthogonal polynomials, we are in the fortunate position of having at disposal a number of well-tested computational techniques for this purpose, supported by a package of software in Matlab, `OPQ` (Orthogonal Polynomials and Quadrature), in the public domain (`http://www.cs.purdue.edu/archives/2002/wxg/codes`). This not only enables but also encourages experimentation in this area of mathematics.

The *mathematical objects* we want to investigate are, on the one hand, Jacobi polynomials and, on the other hand, quadrature formulae. The *properties* and

Walter Gautschi
Department of Computer Sciences, Purdue University, West Lafayette, IN 47907-2066, USA
e-mail: wxg@cs.purdue.edu

W. Gautschi et al. (eds.), *Approximation and Computation: In Honor of Gradimir V. Milovanović*, Springer Optimization and Its Applications 42,
DOI 10.1007/978-1-4419-6594-3_9, © Springer Science+Business Media, LLC 2011

patterns to be identified are, in the former case, inequalities and respective domains of validity–inequalities for zeros of Jacobi polynomials and Bernstein's inequality for Jacobi polynomials–and positivity in the latter case–positivity of Newton–Cotes formulae on zeros of Jacobi polynomials and positivity of generalized Gauss–Radau and Gauss–Lobatto formulae. We will also report on experiments with Gaussian quadrature formulae corresponding to exotic weight functions, for example weight functions exhibiting super-exponential decay at infinity or densely oscillatory behavior at zero. Both are of interest in computing integral transforms that involve modified Bessel functions K_ν of complex order $\nu = \alpha + \mathrm{i}\beta$.

2 Inequalities for Zeros of Jacobi Polynomials

We denote the zeros of the Jacobi polynomial $P_n^{(\alpha,\beta)}(x)$, $\alpha > -1$, $\beta > -1$, by

$$x_{n,r}^{(\alpha,\beta)} = \cos\theta_{n,r}^{(\alpha,\beta)}, \quad r = 1,2,\ldots,n, \tag{1}$$

and assume them, as is customary, in decreasing order,

$$0 < \theta_{n,1}^{(\alpha,\beta)} < \theta_{n,2}^{(\alpha,\beta)} < \cdots < \theta_{n,n}^{(\alpha,\beta)} < \pi. \tag{2}$$

We write $x_n^{(\alpha,\beta)} = \cos\theta_n^{(\alpha,\beta)}$ for the largest zero $x_n^{(\alpha,\beta)} = x_{n,1}^{(\alpha,\beta)}$.

It is well known (cf., e.g., [24, Theorem 8.1.2]) that for r fixed, $r \geq 1$, there holds

$$\lim_{n\to\infty} n\theta_{n,r}^{(\alpha,\beta)} = j_{\alpha,r}, \tag{3}$$

where $j_{\alpha,r}$ is the rth positive zero of the Bessel function J_α. (The speed of convergence is $O(n^{-4})$, as follows from a result of Gatteschi [6]; cf. also [18, Sect. 5.6].) The experiments in this section have to do with the pattern of convergence, specifically with monotonicity.

2.1 Inequalities for the Largest Zero

In [19], we considered the case $r = 1$ of the largest zero $x_{n,1}^{(\alpha,\beta)}$ of the Jacobi polynomial and tried to computationally determine for which values of α and β convergence in (3) is monotone increasing,

$$n\theta_n^{(\alpha,\beta)} < (n+1)\theta_{n+1}^{(\alpha,\beta)}, \quad n = 1,2,3,\ldots. \tag{4}$$

This requires an accurate computation of the quantities $\theta_n^{(\alpha,\beta)} = \cos^{-1} x_n^{(\alpha,\beta)}$, in particular of the largest zero $x_n^{(\alpha,\beta)}$ of $P_n^{(\alpha,\beta)}$, for arbitrary values of the parameters

$\alpha > -1$, $\beta > -1$. Since $x_{n,r}^{(\alpha,\beta)}$ are the nodes of the n-point Gauss–Jacobi quadrature formula, this can be readily accomplished by the OPQ routine

$$\texttt{xw} = \texttt{gauss(n,ab)}, \tag{5}$$

which in the first column of the $(n \times 2)$-array xw furnishes the n nodes $x_{n,r}^{(\alpha,\beta)}$ of the quadrature formula (though in increasing order). The second column contains the corresponding quadrature weights, which here can be ignored. The routine (5) is applicable to Gauss quadrature for any weight function whose recurrence coefficients α_k, β_k in the three-term recurrence relation

$$\pi_{k+1}(x) = (x-\alpha_k)\pi_k(x) - \beta_k\pi_{k-1}(x), \quad k = 0,1,\ldots,n-1, \tag{6}$$

for the respective monic orthogonal polynomials π_k are known (π_{-1} in (6) is to be taken as zero, and β_0, though arbitrary, is assumed to be the zero-order moment of the weight function, i.e., its integral over the interval of orthogonality). In (5), ab is the $(n \times 2)$-array containing in the first column the coefficients $\alpha_0, \alpha_1, \ldots, \alpha_{n-1}$, and in the second column the coefficients $\beta_0, \beta_1, \ldots, \beta_{n-1}$. In the case of Jacobi polynomials, these of course are explicitly known (cf. [8, p. 29]), and are furnished by the OPQ routine

$$\texttt{ab} = \texttt{r\_jacobi(n,a,b)}, \tag{7}$$

where a and b play the role of the Jacobi parameters α and β, respectively.

Extensive experimentation with the routines (5) and (7) revealed (cf. also [13] for a small revision) that the domain of validity $\mathcal{D}$ for (4) in the (α,β)-plane is almost the full domain $\{(\alpha,\beta) : \alpha > -1, \beta > -1\}$, except for a small part near the lower left-hand corner, which has to be deleted. In fact, for $-1 < \alpha < 1$, the lower boundary of $\mathcal{D}$ near this corner is made up of two parts, the straight-line segment $\beta = -\alpha - 1$ $(-1 < \alpha < -1/2)$, on which (4) actually holds for all $n \geq 1$, and the curve $\beta = \beta(\alpha)$ $(-1/2 < \alpha < 1)$, where $\beta(\alpha)$ is the solution of the equation (in β)

$$\begin{aligned} 2\arccos\left(\frac{1}{\alpha+\beta+4}\left[-(\alpha-\beta)+2\sqrt{2+\frac{\alpha\beta-2}{\alpha+\beta+3}}\right]\right) \\ +\arccos\frac{\alpha-\beta}{\alpha+\beta+2} - \pi = 0. \end{aligned} \tag{8}$$

This equation expresses equality in (4) for $n = 1$; see Fig. 1. The point $\alpha = \beta = -1/2$ must be deleted, since for this point one has equality in (4) for all $n \geq 1$.

It is true that these are all experimental results obtained by computation, but the numerical evidence in support of them seems compelling. It may be interesting to note that the inequalities (4) can be *proved* to hold for all n sufficiently large if $\alpha + \beta + 1 > 0$, and to be false for all n sufficiently large if $\alpha + \beta + 1 < 0$ (cf. [13, Theorem in Sect. 2]).

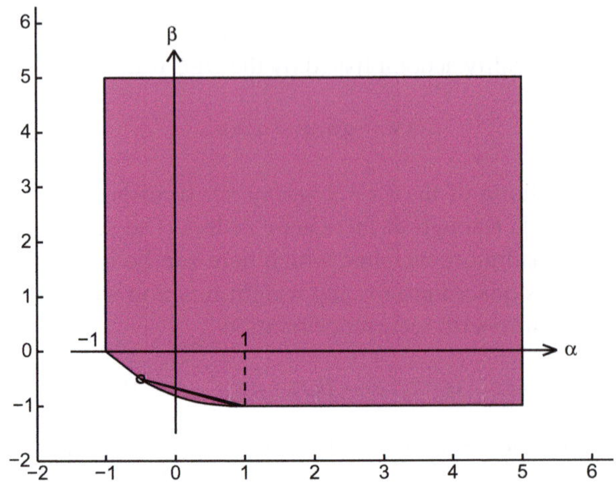

Fig. 1 Conjectured domain of validity for (4)

2.2 Inequalities for All Zeros

It is natural to investigate in a similar manner the case of *all* zeros $x_{n,r}^{(\alpha,\beta)} = \cos\theta_{n,r}^{(\alpha,\beta)}$, $r = 1,2,\dots,n$, and to see to what extent the inequalities (4) continue to hold, that is, for what values of α, β one has

$$n\theta_{n,r}^{(\alpha,\beta)} < (n+1)\theta_{n+1,r}^{(\alpha,\beta)}, \quad r = 1,2,\dots,n;\ n = 1,2,3\dots. \tag{9}$$

While it is again true that for *fixed* r the inequalities (9) are valid for all n sufficiently large if $\alpha+\beta+1>0$ (convergence in (3), therefore, being ultimately monotone), this no longer holds if we allow $r = r(n)$ to grow with n. This can be seen already in the special case of ultraspherical polynomials ($\alpha = \beta$) and $r = n$, in which case the inequalities are found to be false for $n \geq n_0$, where $n_0 = n_0(\alpha)$ depends on α. For example, $n_0(2.2009\dots) = 50$, $n_0(1.0605\dots) = 100$ (cf. [14, Sect. 2]).

Restricting n in (9) to $n = 1,2,\dots,N$, one finds [14] that the domain of validity for the inequalities, $\mathcal{D}_N$, depends on N and is bounded above by an ascending, slightly concave, curve $\mathtt{B}_N$, on the left by the vertical segment at $\alpha = -1$ between $\mathtt{B}_N$ and the α-axis, and below by the diagonally descending line segment $\mathtt{K} = \{(\alpha,\beta) : \beta = -\alpha-1, -1<\alpha<-1/2\}$ followed by a descending, slightly convex, curve $\mathtt{C}_N$ (see Fig. 2).

It is suggestive to conclude from Fig. 2 that as $N \to \infty$, the domain of validity $\mathcal{D} = \mathcal{D}_\infty$ of *all* inequalities in (9) is the horizontal strip $\{(\alpha,\beta) :\ \alpha > -1, |\beta| \leq 1/2\}$ with the lower left-hand corner cut off by the diagonal segment $\mathtt{K}$ (see Fig. 3). This in fact was reinforced by the validity of (9) for $n \leq N = 500$, and for 100 points selected randomly in the strip $\{(\alpha,\beta) :\ -1/2 \leq \alpha \leq 20, |\beta| \leq 1/2\}$.

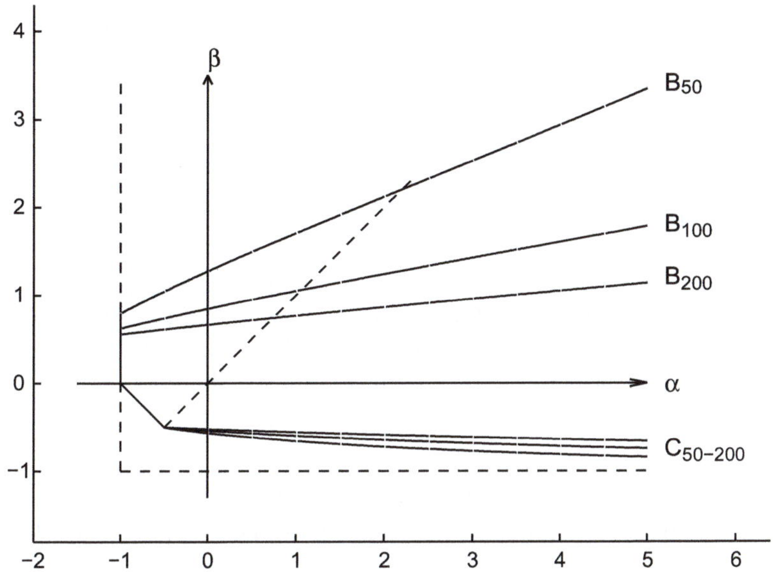

Fig. 2 Domains of validity of the inequalities in (9) when $n = 1, 2, \ldots, N$

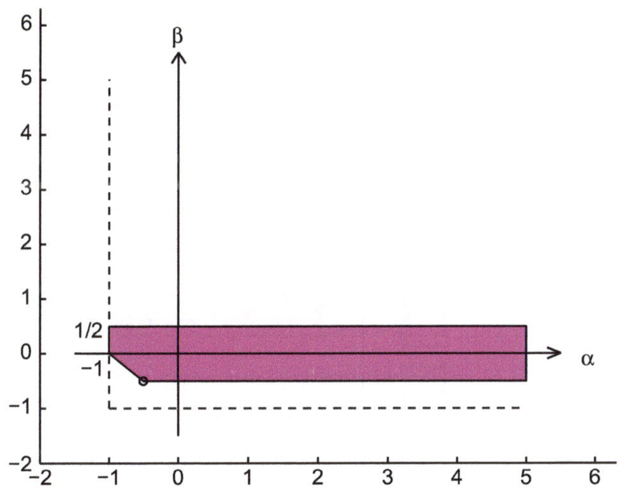

Fig. 3 Domain of validity of (9)

2.3 Modified Inequalities for All Zeros

In [1], Richard Askey suggested to the author to try replacing the factor n in (9) by the factor $n + (\alpha + \beta + 1)/2$, which is often more natural. This led us to consider the modified inequalities

$$(n+(\alpha+\beta+1)/2)\,\theta_{n,r}^{(\alpha,\beta)} > (n+(\alpha+\beta+3)/2)\,\theta_{n+1,r}^{(\alpha,\beta)},\qquad r=1,2,\ldots,n;\; n=1,2,3,\ldots. \tag{10}$$

Clearly, for any *fixed* r, we have

$$\lim_{n\to\infty}((n+(\alpha+\beta+1)/2)\,\theta_{n,r}^{(\alpha,\beta)} = \lim_{n\to\infty}\frac{n+(\alpha+\beta+1)/2}{n}\,n\theta_{n,r}^{(\alpha,\beta)} = j_{\alpha,r}, \tag{11}$$

so that, if (10) holds, convergence in (11) is monotone decreasing.

It is shown in [15] that (10), for fixed r, is true for n sufficiently large if $\alpha^2+3\beta^2>1$, and false for n sufficiently large if $\alpha^2+3\beta^2<1$. Thus, if (10) is to hold for all n and r, then the point (α,β) must lie outside the ellipse

$$\mathcal{E}:\quad \alpha^2+3\beta^2=1$$

or possibly on it. Note, however, that the four points with $|\alpha|=|\beta|=1/2$, which are all on the ellipse, must be excluded, since for them equality holds in (10) for all n and r.

Using the same software as in the preceding sections, we determined by extensive numerical computation that (10) holds in the four rectangular regions (shown in Fig. 4 and extending to infinity in both the α- and β- directions) with corners at

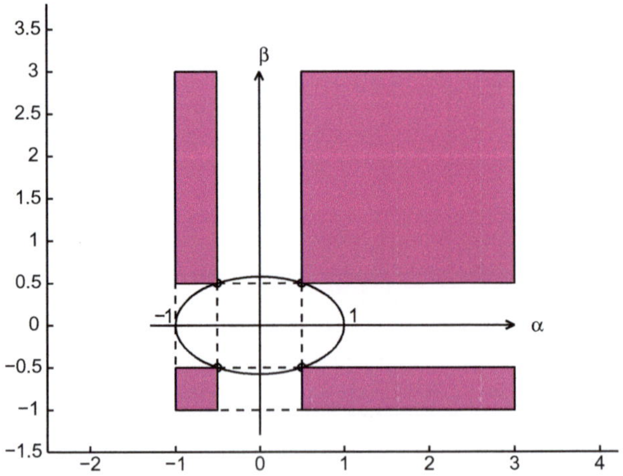

Fig. 4 (Partial) domains of validity of (10)

the four points $|\alpha|=|\beta|=1/2$ on $\mathcal{E}$. The same inequalities also hold in the vertical strips on top and bottom of the ellipse (with base interval $-1/2<\alpha<1/2$), possibly with equality near the ellipse. Inequalities in either direction are observed in the horizontal strip to the right of the ellipse and in the small remaining pieces to the left thereof.

With regard to the region inside the ellipse $\mathcal{E}$, it was found that the inequalities (10) hold with $>$ replaced by $<$ (or by $\leq$ in the two caps on the left and right, where $|\alpha| > 1/2$), and the inequality may occur in both directions in the small caps on top and bottom (where $|\beta| > 1/2$). The inequality on the upward diagonal of the square $|\alpha| < 1/2$, $|\beta| < 1/2$, called "remarkable" by Szegö, has been proved by the Sturm comparison theorem ([24, Sect. 6.3(5)]).

3 Bernstein's Inequality for Jacobi Polynomials

Originally (in 1931, see [2]), Bernstein formulated his inequality for Legendre polynomials,

$$(\sin\theta)^{1/2}|P_n(\cos\theta)| < \sqrt{2/\pi}\, n^{-1/2}, \quad 0 \leq \theta \leq \pi,$$

and showed that the constant $\sqrt{2/\pi}$ is best possible. In the 1980s, the inequality has been slightly sharpened (for example, by replacing n on the right by $n+1/2$) and generalized by various authors to ultraspherical and, eventually, to Jacobi polynomials. A definitive form for the latter was given by Chow, Gatteschi, and Wong in [4], and reads

$$(\sin\tfrac{1}{2}\theta)^{\alpha+1/2}(\cos\tfrac{1}{2}\theta)^{\beta+1/2}|P_n^{(\alpha,\beta)}(\cos\theta)| \leq \frac{\Gamma(q+1)}{\Gamma(1/2)}\binom{n+q}{n}N^{-q-1/2}, \qquad N = n+(\alpha+\beta+1)/2,\; 0 \leq \theta \leq \pi, \tag{12}$$

where $|\alpha| \leq 1/2$, $|\beta| \leq 1/2$, and $q = \max(\alpha,\beta)$. Here also, the constant $\Gamma(q+1)/\Gamma(1/2)$ is best possible [16, Sect. 2].

A matter of some interest is to measure, and compute, the degree of sharpness of the inequality (12) on some domain $\mathcal{D}(n,\alpha,\beta,q)$ (where q may depend on α, β, but such that $q(\alpha,\beta) = q(\beta,\alpha)$, or may be an independent parameter). Another objective is to extend the inequality to more general regions $\mathcal{R}_s = \{(\alpha,\beta) : -1/2 \leq \alpha \leq s, -1/2 \leq \beta \leq s\}$ in the (α,β)-plane, the original region being $\mathcal{R}_{1/2}$.

3.1 Sharpness of Bernstein's Inequality

Upon dividing both sides of (12) by the expression on the right, and letting $x = \cos\theta$, the inequality (12) may be given the form

$$f_n(x;\alpha,\beta,q) \leq 1, \quad -1 \leq x \leq 1. \tag{13}$$

Following [16], given a domain $\mathcal{D}(n,\alpha,\beta,q)$, we define the "local magnitude" of f_n by

$$\rho_n = \rho_n(\alpha,\beta,q) = \|f_n(\,\cdot\,;\alpha,\beta,q)\|_\infty, \tag{14}$$

where the infinity norm is taken over the interval $[-1,1]$; the globally largest and smallest magnitude are then defined, respectively, by

$$\rho_{\mathcal{D}}^{+} = \max_{\mathcal{D}} \rho_n, \quad \text{and} \quad \rho_{\mathcal{D}}^{-} = \min_{\mathcal{D}} \rho_n. \tag{15}$$

If, on the one hand, $\rho_{\mathcal{D}}^{+} \leq 1$, i.e., the inequality (13) holds on $\mathcal{D}$, the quantities in (15) may be interpreted as follows: on a scale from 0 to 1, the *best and worst degree of sharpness on* $\mathcal{D}$ is $\rho_{\mathcal{D}}^{+}$ and $\rho_{\mathcal{D}}^{-}$. If, on the other hand, $\rho_{\mathcal{D}}^{+} > 1$, i.e., the inequality does not hold on $\mathcal{D}$, we can "correct" it by dividing f_n by $\rho_{\mathcal{D}}^{+}$ and by considering the *modified inequality* on $\mathcal{D}$,

$$\hat{f}_n(x;\alpha,\beta,q) \leq 1, \quad -1 \leq x \leq 1, \tag{16}$$

where

$$\hat{f}_n(x;\alpha,\beta,q) = \frac{1}{\rho_{\mathcal{D}}^{+}} f_n(x;,\alpha,\beta,q). \tag{17}$$

By construction, the best degree of sharpness on $\mathcal{D}$ of the modified inequality is $\hat{\rho}_{\mathcal{D}}^{+} = 1$, and the worst degree of sharpness

$$\hat{\rho}_{\mathcal{D}}^{-} = \frac{\rho_{\mathcal{D}}^{-}}{\rho_{\mathcal{D}}^{+}}. \tag{18}$$

To compute the quantities in (15), one takes the maximum and minimum (if necessay) on a sufficiently fine grid of $\mathcal{D}$. This then leaves us with the problem of computing the infinity norm in (14). In the case of Bernstein's inequality this can conveniently be done by computing local extrema of $f_n(x;\alpha,\beta,q)$ in $-1 < x < 1$ and comparing them with the boundary values at $x = \pm 1$. The two routines in (7) and (5) are again heavily engaged in this endeavor, together with a routine for computing the relative extrema; cf. Sects. 4 and 5 of [16] and the Appendix therein for a Matlab script. Because of the reflection formula for Jacobi polynomials, it suffices to consider $\beta \geq \alpha \geq -1/2$, the last inequality by virtue of the fact that $\|f_n\|_\infty = \infty$ if $\alpha < -1/2$.

Table 1 Degree of sharpness of Bernstein's inequality on $\mathcal{R}_{1/2}$

$q \rightarrow$	q^+	q^-	1	0.5	0	−0.5
$\rho_{\mathcal{D}}^{+}$	1.000	1.000	0.999	1.000	1.023	1.000
$\rho_{\mathcal{D}}^{-}$	0.998	0.998	0.813	0.909	0.975	0.917

On the original domain $\mathcal{R}_{1/2} = \{(\alpha,\beta) : |\alpha| \leq 1/2, |\beta| \leq 1/2\}$, taking $\mathcal{D} = \{[5\ 10\ 20\ 50\ 100], \mathcal{R}_{1/2}, q\}$, one obtains the results in Table 1 for selected values of q. Evidently, the choices $q^+ = \max(\alpha,\beta)$, $q^- = \min(\alpha,\beta)$ are by far the best with regard to overall sharpness.

3.2 Bernstein's Inequality on Larger Domains

We now proceed to the larger domains

$$\mathcal{R}_s = \{(\alpha,\beta) : -1/2 \le \alpha \le s, -1/2 \le \beta \le s\}, \quad s > 1/2,$$

and let $\mathcal{D}_s = \{[5\ 10\ 20\ 50\ 100], \mathcal{R}_s, q\}$. Experimentation (with $q = q^+$) revealed that

$$\rho^-_{\mathcal{D}_s} = \rho^-_{\mathcal{D}_{1/2}} \quad \text{for all } s > 1/2, \tag{19}$$

i.e., the minimum of ρ_n on $\mathcal{D}_s$ is always attained in $\mathcal{D}_{1/2}$. Also, the maximum of ρ_n on $\mathcal{D}_s$ is found to be always attained in the upper right-hand corner of $\mathcal{D}_s$,

$$\rho^+_{\mathcal{D}_s} = \max_{\mathcal{D}_s} \rho_n(\alpha,\beta,q^+) = \max_n \rho_n(s,s,s). \tag{20}$$

Using this property, we were able to compute for many values of s both $\rho^+_{\mathcal{D}_s}$ and $\hat{\rho}^-_{\mathcal{D}_s}$ for the modified (in the sense of (16) and (17)) Bernstein's inequality. An extract of the results is given in Table 2. As can be seen from Table 2, the degrees of sharpness on $\mathcal{D}_s$ of the modified Bernstein's inequality, even for s as large as 10, are well within one decimal order of magnitude. Also remarkable is the experimentally observed fact that exactly the same results are obtained if we let n go from 5 up to 200, so that the results in Table 2 are likely to be valid for all $n \ge 5$.

Table 2 Degree of sharpness on $\mathcal{D}_s$ of the modified Bernstein's inequality

s	$\rho^+_{\mathcal{D}_s}$	$\hat{\rho}^-_{\mathcal{D}_s}$
1	1.039	0.961
2	1.120	0.891
5	1.495	0.667
10	3.047	0.327

4 Quadrature Formulae

Our interest in this section is in the positivity of quadrature formulae. Classical Newton–Cotes formulae of moderate to large order are notorious not only for their lack of positivity, but also for their wildly oscillating weights. Newton–Cotes formulae with nonuniformly distributed nodes, however, may well exhibit positivity. A well-known example is Fejér's quadrature formula using Chebyshev nodes of the first and second kind. A Newton–Cotes formula using both kinds of nodes simultaneously, even in a more general setting, has been proposed by Milovanović and conjectured to be positive. This will be further elaborated on in Sect. 4.1. In

Sects. 4.2–4.4, we look at generalized Gauss–Radau and Gauss–Lobatto formulae, for which positivity has been conjectured by us in the past, and has been proved very recently by Joulak and Beckermann.

4.1 Positivity of Weighted Newton–Cotes Formulae

Let w be a (positive) weight function on the interval I, and X_n a set of n distinct points x_k in I. A *weighted Newton–Cotes formula* is an interpolatory quadrature formula of the form

$$\int_I w(x)f(x)\mathrm{d}x = \sum_{x_k\in X_n} w_k f(x_k), \quad f\in\mathbb{P}_{n-1}. \tag{21}$$

Definition We write $(w,X_n)\in \mathrm{NC}_+$ if and only if in (21) there holds

$$w_k>0, \quad \text{all } k. \tag{22}$$

The quadrature rule (21) is then called a *positive* weighted Newton–Cotes formula.

A well-known example is Fejér's quadrature rule [5], for which $w(x)=1$ on $I=[-1,1]$, and $X_n=X_n^T$ or X_n^U, the zeros of the Chebyshev polynomial T_n of the first kind resp. U_n of the second kind. In particular, there holds

$$(w\equiv 1, X_{2n-1}^U)\in \mathrm{NC}_+. \tag{23}$$

Noting that $U_{2n-1}=2T_nU_{n-1}$, we can write

$$X_{2n-1}^U = X_n^T\cup X_{n-1}^U. \tag{24}$$

This is the motivation for the following conjecture.

Conjecture of Milovanović ([23], [22, Sect. 5.1.2]) There holds

$$(w,X_{2n-1})\in \mathrm{NC}_+, \tag{25}$$

where

$$w(x)=(1-x)^{\alpha+1/2}(1+x)^{\beta+1/2} \quad \text{on } [-1,1], \tag{26}$$

and

$$X_{2n-1}=X_n^{P(\alpha,\beta)}\cup X_{n-1}^{P(\alpha+1,\beta+1)}. \tag{27}$$

Here, α and β are arbitrary real numbers greater than -1, and $X_n^{P(\alpha,\beta)}$ the zeros of the Jacobi polynomial $P_n^{(\alpha,\beta)}$, and $X_{n-1}^{P(\alpha+1,\beta+1)}$ the zeros of the Jacobi polynomial $P_{n-1}^{(\alpha+1,\beta+1)}$.

Note that (23), (24) are the special case $\alpha=\beta=-1/2$ of the conjecture.

Testing the conjecture requires a reliable and stable procedure for generating the weighted Newton–Cotes formula (21), that is, the weights w_k, given n and the nodes $x_k \in X_n$. Clearly,

$$w_k = \int_I \ell_k^{(n)}(x)w(x)\mathrm{d}x, \quad k = 1,2,\ldots,n, \tag{28}$$

where $\ell_k^{(n)}$ are the elementary Lagrange interpolation polynomials belonging to the nodes x_k,

$$\ell_k^{(n)}(x) = \prod_{\substack{\ell=1 \\ \ell\neq k}}^{n} \frac{x - x_\ell}{x_k - x_\ell}. \tag{29}$$

As already suggested in [7], the integral in (28) can be computed by $\lfloor \frac{n+1}{2} \rfloor$-point Gaussian quadrature relative to the weight function w, and the Lagrange polynomials $\ell_k^{(n)}$ by means of the barycentric formula

$$\ell_k^{(n)}(x) = \frac{\lambda_k^{(n)}/(x - x_k)}{\sum_{\ell=1}^{n} \lambda_\ell^{(n)}/(x - x_\ell)}, \quad x \neq x_k. \tag{30}$$

Here, $\lambda_k^{(n)}$ are auxiliary quantities that are readily calculated by the recursive scheme (written in pseudoMatlab)

$\lambda_1^{(1)} = 1;$
for k = 2 : n
  for l = 1 : k − 1
    $\lambda_\ell^{(k)} = \lambda_\ell^{(k-1)}/(x_\ell - x_k);$
  end
  $\lambda_k^{(k)} = \prod_{\ell=1}^{k-1} 1/(x_k - x_\ell);$
end

In contrast to [7, Sect. 2.1], where $\lambda_k^{(k)}$ was computed by a sum, causing potentially severe cancellation problems, here, following [3, Sect. 3], we compute it more stably by a product. This is implemented in the `OPQ` routine `NewtCotes.m`, which calls on the routine `gauss.m` to do the integration.

We tested Milovanović's conjecture for $\alpha = -.75 : .25 : 5$, $\beta = \alpha : .25 : 5$, and $n = 2 : 100$ (by symmetry, it suffices to take $\beta \geq \alpha$). We also examined, in part already in [7], the ranges $\alpha = -.9 : .1 : 1.0$, $\beta = \alpha : .1 : 1.0$, and $n = 2 : 100$. In all these tests, the conjecture, without exception, was confirmed.

4.2 Positivity of Generalized Gauss–Radau Formulae

Generalized Gauss–Radau formulae are quadrature formulae of Gauss type, i.e., of maximum algebraic degree of exactness, that involve a boundary point of arbitrary multiplicity $r \geq 2$ (those with $r = 1$ being the ordinary Gauss–Radau formulae). They are thus of the form

$$\int_a^\infty f(x)\mathrm{d}\lambda(x) = \sum_{\rho=0}^{r-1} \lambda_0^{(\rho)} f^{(\rho)}(a) + \sum_{\nu=1}^{n} \lambda_\nu^R f(\tau_\nu^R), \quad f \in \mathbb{P}_{2n-1+r}, \tag{31}$$

where $\mathrm{d}\lambda$ is a positive measure whose support may be bounded or unbounded. (In the former case, the upper limit of the integral could be some b with $a < b < \infty$.) The interior nodes τ_ν^R and weights λ_ν^R are easily obtained (and computed by our routine `gauss.m`) from the n-point Gaussian quadrature formula relative to the modified measure

$$\mathrm{d}\lambda^{[r]}(x) = (x-a)^r \mathrm{d}\lambda(x). \tag{32}$$

The major difficulty in computing generalized Gauss–Radau formulae lies in the boundary weights $\lambda_0^{(\rho)}$, and has been addressed only recently in [9] (see also [17] for additional difficulties when n is very large, of the order of magnitude $n \approx 400$). The method proposed in [9], and implemented in the `OPQ` routine `gradau.m`, is based on the solution of an $(r \times r)$ upper triangular system of linear algebraic equations,

$$Ax = b, \tag{33}$$

where the matrix A is expressible in a somewhat complicated manner in terms of the monic nth-degree polynomial $\pi_{n,r}$ orthogonal with respect to the measure (32), but the diagonal elements, and also the element in position $(r-1, r)$, are more simply expressible as

$$a_{ii} = (i-1)!\pi_{n,r}^2(a), \; i = 1,2,\ldots,r; \quad a_{r-1,r} = -2a_{rr} \sum_{\nu=1}^{n} (\tau_\nu^R - a)^{-1}. \tag{34}$$

The vector $x = [x_j]$ of unknowns in (33), and the right-hand vector $b = [b_i]$, are given by

$$x_j = \lambda_0^{(j-1)}, \; b_i = \int_a^\infty (x-a)^{i-1} \pi_{n,r}^2(x)\mathrm{d}\lambda(x), \quad i,j = 1,2,\ldots,r. \tag{35}$$

With regard to the positivity of (31), it is clear that all interior weights λ_ν^R are positive by definition. When $r = 2$, the same is true for the boundary weights. This follows from the positivity of a_{ii} and b_i, and from $x_2 = b_2/a_{22} > 0$, $x_1 = (b_1 - a_{12}x_2)/a_{11} > 0$, since $a_{12} < 0$ by (34). In the general case $r > 2$, however, positivity of x was left open in [9], but was conjectured to hold, based on extensive computation. Today we know that positivity in fact is a proven theorem; cf. Sect. 4.4.

4.3 Positivity of Generalized Gauss–Lobatto Formulae

Generalized Gauss–Lobatto formulae are similar to generalized Gauss–Radau formulae, except that the interval of integration is necessarily bounded, and there are boundary points of multiplicity $r \geq 2$ at either end of the interval. Thus,

$$\begin{aligned}\int_a^b f(x)\mathrm{d}\lambda(x) = \sum_{\rho=0}^{r-1} \lambda_0^{(\rho)} f^{(\rho)}(a) + \sum_{\nu=1}^{n} \lambda_\nu^L f(\tau_\nu^L) \\ + \sum_{\rho=0}^{r-1} (-1)^\rho \lambda_{n+1}^{(\rho)} f^{(\rho)}(b), \quad f \in \mathbb{P}_{2n-1+2r}.\end{aligned} \tag{36}$$

The signs $(-1)^\rho$ in the last summation are included in anticipation of the fact that $\lambda_0^{(\rho)} = \lambda_{n+1}^{(\rho)}$ in case of symmetry (i.e., $a+b=0$ and $\mathrm{d}\lambda(-x) = \mathrm{d}\lambda(x)$).

For computational methods, which are similar to those for generalized Gauss–Radau formulae indicated in Sect. 4.2, we refer to [9] and [17]. They are implemented in the `OPQ` routine `globatto.m`.

Positivity of (36) is here understood to mean

$$\lambda_0^{(\rho)} > 0,\ \lambda_{n+1}^{(\rho)} > 0, \ \ \rho = 0,1,\ldots,r-1; \quad \lambda_\nu^L > 0, \ \ \nu = 1,2,\ldots,n. \tag{37}$$

The interior weights λ_ν^L, for reasons similar as in Sect. 4.2, are all positive, and so are $\lambda_0^{(\rho)}$, $\lambda_{n+1}^{(\rho)}$ if $r=2$. For general $r>2$, positivity of (36) has been conjectured in [9], again on the basis of extensive computation. In the meantime, it has been proven; cf. Sect. 4.4.

4.4 Positivity of Most General Gauss–Radau/Lobatto Formulae

The question of positivity regarding generalized Gauss–Radau and Gauss–Lobatto formulae has been settled very recently by H. Joulak and B. Beckermann, even for more general formulae of the form

$$\begin{aligned}\int_a^b f(x)\mathrm{d}\lambda(x) = \sum_{\rho=0}^{r-1} \lambda_0^{(\rho)} f^{(\rho)}(a) + \sum_{\nu=1}^{n} \lambda_\nu f(\tau_\nu) \\ + \sum_{\sigma=0}^{s-1} (-1)^\sigma \lambda_{n+1}^{(\sigma)} f^{(\sigma)}(b), \quad f \in \mathbb{P}_{2n-1+r+s},\end{aligned} \tag{38}$$

where possibly $b=\infty$ if $s=0$ or $a=-\infty$ if $r=0$. In fact, we have the following theorem.

Theorem ([21]) *For any positive* $\mathrm{d}\lambda$, *and* $r \geq 0$, $s \geq 0$, *there holds*

$$\begin{aligned} &\lambda_\nu > 0 \;\; (1 \leq \nu \leq n); \\ &\lambda_0^{(\rho)} > 0 \;\; (0 \leq \rho \leq r-1); \;\; \lambda_{n+1}^{(\sigma)} > 0 \;\; (0 \leq \sigma \leq s-1). \end{aligned} \tag{39}$$

The proof of the theorem given in [21] is based on the fact that certain elementary Hermite interpolation polynomials associated with the points a, τ_ν, b of multiplicities $r, 2, s$, respectively, are nonnegative on (a,b).

5 Gauss Quadrature with Exotic Weight Functions

In certain integral transforms involving modified Bessel functions of complex order, the integrand exhibits behavior at infinity, and at zero, that is highly unusual. To properly account for this behavior, it is necessary to develop Gaussian quadrature formulae having weight functions that mimic this behavior. This in turn requires generating the necessary orthogonal polynomials, specifically the coefficients in their three-term recurrence relation. The behavior at infinity is characterized by super-exponential decay, the one at zero by dense oscillation. In the former case, we use a discretized Stieltjes procedure to generate the necessary recurrence coefficients, in the latter case the classical Chebyshev algorithm, executed in symbolic variable-precision computation to counteract the underlying severe ill-conditioning.

5.1 Weight Function Decaying Super-Exponentially at Infinity

The real and imaginary parts of the Macdonald function (or modified Bessel function) $K_\nu(s)$ with complex order $\nu = \alpha + \mathrm{i}\beta$ and $s > 0$ are known to be representable by integral transforms,

$$\begin{aligned} \operatorname{Re} K_{\alpha+\mathrm{i}\beta}(s) &= \int_0^\infty \mathrm{e}^{-s\cosh x} \cosh \alpha x \cos \beta x \, \mathrm{d}x, \\ \operatorname{Im} K_{\alpha+\mathrm{i}\beta}(s) &= \int_0^\infty \mathrm{e}^{-s\cosh x} \sinh \alpha x \sin \beta x \, \mathrm{d}x. \end{aligned} \tag{40}$$

In both, the integrand decays extremely rapidly at infinity, owing to the factor $\cosh x$ in the exponent. The essence of this behavior is captured by the weight function

$$w(x) = \exp(-\mathrm{e}^x), \quad 0 \leq x < \infty, \tag{41}$$

not depending on s. It becomes relevant after a suitable change of variables (cf. [11, Sect. 3]).

The recurrence coefficients for the polynomials orthogonal with respect to the weight function (41) can be generated by a multiple-component discretization procedure, decomposing the interval $[0,\infty]$ into four subintervals and using Fejér quadrature rules as a general-purpose device of discretization (cf. [11, Sect. 2]). The relevant `OPQ` routine is `mcdis.m`, and the first 100 recurrence coefficients generated by it are listed in the file `abmacdonald`. It can be downloaded from the web site cited above in Sect. 1 by clicking on `MCD`. The Matlab commands

```
load -ascii abmacdonald;
xw=gauss(n,abmacdonald);
```

then produce the n-point Gauss quadrature rule relative to the weight function (41), with the first column of `xw` containing the nodes x_ν and the second column the weights w_ν.

To give an example, consider the integral

$$I = \int_0^\infty \frac{\exp(-e^x)}{1+x}\,dx \tag{42}$$

and its approximation $I_n^G = \sum_{\nu=1}^n w_\nu/(1+x_\nu)$ by the n-point Gaussian quadrature rule. In Table 3, we compare it with the n-point Gauss–Laguerre approximation I_n^L of

$$I = e^{-1}\int_0^\infty \frac{1}{(1+\ln(1+t))(1+t)}\,e^{-t}dt.$$

As can be seen, our special Gauss quadrature rule, already for $n=13$ yields 14 correct decimal places, whereas Gauss–Laguerre manages to produce only four.

Table 3 Gauss and Gauss–Laguerre quadrature of the integral (42)

n	I_n^G	I_n^L
1	0.15171877142494	0.108637...
2	0.15936463844634	0.140436...
3	0.15987602667503	0.151468...
4	0.15991604862904	0.155900...
⋮	⋮	⋮
12	0.15991988389384	0.159868...
13	0.15991988389391	0.159886...
14	0.15991988389391	0.159897...

5.2 *Weight Functions Densely Oscillating at Zero*

Integral transforms in which K_ν, $\nu=\alpha+i\beta$, acts as a kernel, called Kontorovich–Lebedev transforms (when $\alpha=0$ or 1/2), yield peculiar behavior of the integrand at zero, caused by the behavior of $K_\nu(x)$ near $x=0$. If $\alpha=0$, for example, the transform is

$$F(\beta)=\int_0^\infty K_{\mathrm{i}\beta}(x)f(x)\mathrm{d}x,$$

which is real-valued for real-valued f. The behavior of $K_{\mathrm{i}\beta}$ near zero is (cf. [12, Sect. 2])

$$K_{\mathrm{i}\beta}(x)\sim\sqrt{\frac{\pi}{\beta\sinh(\pi\beta)}}\,\sin(\beta\ln(2/x)+\gamma),\quad \gamma=\arg\Gamma(1+\mathrm{i}\beta),\ x\downarrow 0,$$

showing that $K_{\mathrm{i}\beta}(x)$ is densely oscillating near $x=0$. This prompted us in [12] to consider Gauss quadrature on $[0,1]$ with the nonnegative weight function

$$w_\beta(x)=1+\sin(\beta\ln(1/x)+\gamma),\quad 0<x\le 1.$$

But how do we find the necessary orthogonal polynomials?

The only way we knew how to do this is by applying the classical Chebyshev algorithm that allows us to generate the required recurrence coefficients directly from the moments of the weight function. The problem is that this approach via moments is quite ill-conditioned. We therefore used a symbolic version `schebyshev.m` of the `OPQ` routine `chebyshev.m` to generate the recurrence coefficients in variable-precision arithmetic (cf. [12, Example 3.5]). This symbolic routine can also be downloaded from the web site mentioned in Sect. 1, by clicking on `SOPQ`[1].

For illustration, we show here the simpler (but not less challenging!) example of the weight function

$$w(x)=1+\sin(1/x)\quad \text{on } [0,1], \tag{43}$$

taken from [10, Sect. 2.1]. Here the moments $m_k=\int_0^1 x^k w(x)\mathrm{d}x$ are computed by

$$m_k=\frac{1}{k+1}+m_k^0,\quad k=0,1,2,\ldots, \tag{44}$$

where m_k^0 are the "core moments" $m_k^0=\int_0^1 x^k\sin(1/x)\mathrm{d}x$ that can be generated recursively by

$$m_{-1}^0=\int_1^\infty\frac{\sin t}{t}\mathrm{d}t=\tfrac{\pi}{2}-\mathrm{Si}(1),$$

$$m_0^0=\int_1^\infty\frac{\sin t}{t^2}\mathrm{d}t=\sin 1-\mathrm{Ci}(1),$$

and

$$m_{k+1}^0=\frac{1}{k+2}\left[\frac{1}{k+1}(\cos 1-m_{k-1}^0)+\sin 1\right],\quad k=0,1,2,\ldots.$$

[1] `SOPQ` is a symbolic counterpart to the package `OPQ`, but far from complete. A worthwhile project for anyone familiar with the symbolic toolbox of Matlab would be to transcribe the entire package `OPQ` into symbolic Matlab.

To give a numerical example, suppose we want to compute the integral

$$I = \int_0^1 f(x)\sin(1/x)\mathrm{d}x,\ f(x) = \tan((\tfrac{1}{2}\pi - \delta)x), \quad 0 < \delta < \tfrac{1}{2}\pi. \tag{45}$$

Table 4 Numerical results for the integral (45) for $\delta = 0.1$

n	I_n
4	1.2716655036125
8	1.2957389942560
12	1.2961790099686
⋮	⋮
32	1.2961861708636
36	1.2961861708636

We write this in the form

$$I = \int_0^1 f(x)[1 + \sin(1/x)]\mathrm{d}x - \int_0^1 f(x)\mathrm{d}x,$$

and compute the first integral by the special Gauss formula for the weight function (43), and the second integral by Gauss–Legendre quadrature on $[0,1]$. The results I_n for $\delta = 0.1$, using n-point Gauss formulae, are shown in Table 4.

Even the special Gauss formula here has some trouble converging, the reason being a pole close to the upper limit of the integral when δ is small.

Other densely oscillating integrals, and also integrals of rapidly decaying functions like $e^{-1/x}$ on $[0,1]$, or $\exp(-1/x - x)$ on $[0,\infty]$, are treated in [10] similarly and with similar success.

References

1. Askey, Richard. Email of May 13, 2008.
2. Bernstein, Serge. Sur les polynomes orthogonaux relatifs a un segment fini, J. Math. 10 (1931), 219–286.
3. Berrut, Jean-Paul and Lloyd N. Trefethen. Barycentric Lagrange interpolation, SIAM Rev. 46 (2004)(3), 501–517.
4. Chow, Yunshyong, L. Gatteschi, and R. Wong. A Bernstein-type inequality for the Jacobi polynomial, Proc. Am. Math. Soc. 121 (1994)(3), 703–709.
5. Fejér, L. Mechanische Quadraturen mit positiven Cotesschen Zahlen, Math. Z. 37 (1933), 287–309.
6. Gatteschi, Luigi. On the zeros of Jacobi polynomials and Bessel functions. In: International conference on special functions: theory and computation (Turin, 1984). Rend. Sem. Mat. Univ. Politec. Torino (Special Issue), pp. 149–177 (1985)
7. Gautschi, Walter. Moments in quadrature problems. Approximation theory and applications, Comput. Math. Appl. 33 (1997)(1–2), 105–118.

8. GAUTSCHI, WALTER. *Orthogonal polynomials: computation and approximation*, Numerical Mathematics and Scientific Computation, Oxford University Press, Oxford, 2004.
9. GAUTSCHI, WALTER. Generalized Gauss–Radau and Gauss–Lobatto formulae, BIT Numer. Math. 44 (2004)(4), 711–720.
10. GAUTSCHI, WALTER. Computing polynomials orthogonal with respect to densely oscillating and exponentially decaying weight fuinctions and related integrals, J. Comput. Appl. Math. 184 (2005)(2), 493–504.
11. GAUTSCHI, WALTER. Numerical quadrature computation of the Macdonald function for complex orders, BIT 45 (2005)(3), 593–603.
12. GAUTSCHI, WALTER. Computing the Kontorovich–Lebedev integral transforms and their inverses, BIT 46 (2006)(1), 21–40.
13. GAUTSCHI, WALTER. On a conjectured inequality for the largest zero of Jacobi polynomials, Numer. Algorithm 49 (2008)(1–4), 195–198.
14. GAUTSCHI, WALTER. On conjectured inequalities for zeros of Jacobi polynomials, Numer. Algorithm 50 (2009)(1), 93–96.
15. GAUTSCHI, WALTER. New conjectured inequalities for zeros of Jacobi polynomials, Numer. Algorithm 50 (2009)(3), 293–296.
16. GAUTSCHI, WALTER. How sharp is Bernstein's inequality for Jacobi polynomials?, Electr. Trans. Numer. Anal. 36 (2009), 1–8.
17. GAUTSCHI, WALTER. High-order generalized Gauss–Radau and Gauss–Lobatto formulae for Jacobi and Laguerre weight functions, Numer. Algorithm 51 (2009)(2), 143–149.
18. GAUTSCHI, WALTER AND CARLA GIORDANO. Luigi Gatteschi's work on aymptotics of special functions and their zeros, Numer. Algorithm 49 (2008)(1–4), 11–31.
19. GAUTSCHI, WALTER AND PAUL LEOPARDI. Conjectured inequalities for Jacobi polynomials and their largest zeros, Numer. Algorithm 45(1–4)(2007), 217–230.
20. `http://en.Wikipedia.org/wiki/Experimental_mathematics`
21. JOULAK, HÉDI AND BERNHARD BECKERMANN. On Gautschi's conjecture for generalized Gauss–Radau and Gauss–Lobatto formulae, J. Comput. Appl. Math. 233 (2009)(3), 768–774.
22. MASTROIANNI, G. AND G.V. MILOVANOVIĆ. *Interpolation processes: basic theory and applications*, Springer Monographs in Mathematics, Springer, Berlin, 2009.
23. MILOVANOVIĆ, GRADIMIR V. Personal communication, December 1993.
24. SZEGÖ, GABOR. *Orthogonal polynomials*, 4th ed., American Mathematical Society, Colloquium Publications, Vol. 23, Amer. Math. Soc., Providence, RI, 1975.

Compatibility of Continued Fraction Convergents with Padé Approximants

Jacek Gilewicz and Radosław Jedynak

Dedicated to Gradimir V. Milovanović on his 60th birthday

1 Introduction

Let f be a complex function having a power expansion $f(z) = \sum_{i=0}^{\infty} c_i z^i$ at the origin. Denote by P_m and Q_n the polynomials of degrees at most m and n, respectively. Assume that $f(0) \neq 0$. If there exist polynomials P_m and Q_n, $Q_n(0) \neq 0$, fulfilling the following condition:

$$f(z) - \frac{P_m(z)}{Q_n(z)} = O(z^{m+n+1}),$$

then the rational function P_m/Q_n is called a *Padé approximant* (PA) to f and it is denoted $[m/n]$. A rational function which is not a Padé approximant (NPA) will be denoted $\{m/n\}$. The two following properties [1] are more relevant to the present study:

1. If PA exists, it is unique.
2. If the degrees of the numerator and the denominator in $[m/n]$ are exactly m and n for all PA, then each PA is different from the other one. In this case, the Padé table of f, that is the table of all PA, is termed *normal*.

The problem of *compatible transformation* of PA was first raised in [1, p. 208–224], where the transformations $\mathcal{T}$ of variables or of functions applied to PA were

Jacek Gilewicz
Centre de Physique Théorique, CNRS, Luminy Case 907, 13288 Marseille Cedex 09, France,
gilewicz@cpt.univ-mrs.fr

Radosław Jedynak
Politechnika Radomska im. K. Pułaskiego, ul. Malczewskiego 20a, 26-600 Radom, Poland,
jedynakr@pr.radom.pl

W. Gautschi et al. (eds.), *Approximation and Computation: In Honor of Gradimir V. Milovanović*, Springer Optimization and Its Applications 42,
DOI 10.1007/978-1-4419-6594-3_10,

considered. First, it was required that $\mathcal{T}$ transforms P_m/Q_n into a rational function, say P_M/Q_N. This reduces considerably the admissible class of compatible transformations, essentially to certain rational and homographic transformations. The actual goal consisted also of establishing the rules guaranteeing that the transformed PA remains PA, that is, the difference

$$\mathcal{T}(f(z)) - \mathcal{T}\left(\frac{P_m(z)}{Q_n(z)}\right) = F - \frac{P_M(z)}{Q_N(z)}$$

must be of order $M+N+1$. If not, P_M/Q_N is NPA. Following the same idea, we analyse in this work the convergents of the continued fractions (CF), which are rational functions and we try to answer the question when they are also PA. In the following, we restrict our analysis to the normal cases of Padé tables.

2 Classical Continued Fractions and Their Relations with Padé Approximants

Let a_i and b_i denote polynomials, then

$$\frac{A_k}{B_k} = a_0 + \cfrac{a_1}{b_1 + \cfrac{a_2}{b_2 + \cfrac{a_3}{b_3 + \ddots \cfrac{a_k}{b_k}}}}$$

is a rational function. Let us recall the usual compact notation:

$$\frac{A_k}{B_k} = a_0 + \frac{a_1}{b_1+} \cdots \frac{a_k}{b_k} = a_0 + \mathop{\mathrm{K}}_{i=1}^{k} \frac{a_i}{b_i}. \tag{1}$$

The sequence (A_k/B_k) is called a *continued fraction* (CF) and is denoted

$$a_0 + \frac{a_1}{b_1+} \frac{a_2}{b_2+} \frac{a_3}{b_3+} \cdots = a_0 + \mathop{\mathrm{K}}_{i=1}^{\infty} \frac{a_i}{b_i}. \tag{2}$$

The polynomials a_i and b_i are called, respectively, the numerators and denominators of CF. Sometimes we say also that a_i/b_i is the ith floor of CF. The rational function A_k/B_k is called the kth *convergent* of CF, $k = 0, 1, 2, \ldots$. The numerators and denominators of each convergent can be calculated by means of the following three-term recurrence relations

$$\begin{pmatrix} A_k \\ B_k \end{pmatrix} = b_k \begin{pmatrix} A_{k-1} \\ B_{k-1} \end{pmatrix} + a_k \begin{pmatrix} A_{k-2} \\ B_{k-2} \end{pmatrix}; \quad \begin{pmatrix} A_{-1} \\ B_{-1} \end{pmatrix} = \begin{pmatrix} 1 \\ 0 \end{pmatrix}; \quad \begin{pmatrix} A_0 \\ B_0 \end{pmatrix} = \begin{pmatrix} a_0 \\ 1 \end{pmatrix}.$$

If (2) represents an expansion in CF of a function, then a_i and b_i are, in all classical cases, polynomials of degree at most two. In other cases, we say that CF has a general form. In this work, we analyze the effects of increasing the degree of certain denominators of CF. In fact, starting from the modified floor, not all convergents remain PA. Our goal consists of identifying PA and NPA.

Suppose we know the Maclaurin series of a function, $f(z) = c_0 + c_1 z + \cdots$. To transform this series into a CF one inverts successively the power series, for instance as follows:

$$\begin{aligned} f(z) &= c_0 + c_1 z + \cdots = c_0 + c_1 z(1 + c_2{}' z + c_3{}' z^2 + \cdots) \\ &= c_0 + \frac{c_1 z}{1 + c_2{}'' z + c_3{}'' z^2 + \cdots} = c_0 + \frac{c_1 z}{1 + c_2{}'' z(1 + c_3{}''' z + \cdots)} \\ &= c_0 + \cfrac{c_1 z}{1 + \cfrac{c_2{}'' z}{1 + \cfrac{c_3{}''' z}{1 + \cdots}}}. \end{aligned}$$

Here we started successively with the first powers. In general, we can invert the series at an arbitrary moment k if $c_k \neq 0$: $f(z) = c_0 + c_1 z + \cdots + c_k z^k (1 + \cdots)$ and we invert the series in the brackets. It is clear that the denominators can be of arbitrary degrees. However, the usual CF have a simple form with small degrees of numerators and denominators and in this case their convergents are always PA as shown in the following classical theorems. Recall that we shall analyze only the normal cases of the Padé table. The two first theorems concern the Stieltjes CF, while Theorem 3 concerns the Jacobi CF.

Theorem 1 ([1]). *Let*

$$f(z) = \sum_{i=0}^{\infty} c_i z^i = c_0 + c_1 + \cdots + c_{k-1} z^{k-1} + z^k \frac{a_0}{1-} \frac{a_1 z}{1-} \frac{a_2 z}{1-} \cdots, \quad k = 1, 2, \ldots, \tag{3}$$

then the successive convergents $n = 0, 1, \ldots, 2m, 2m+1, \ldots$ *of this Stieltjes CF are the following PA*

$$[k-1/0],\ [k/0],\ \ldots,\ [k+m-1/m],\ [k+m/m],\ \ldots. \tag{4}$$

Theorem 2 ([1]). *Let*

$$f(z) = \frac{1}{\sum_{i=0}^{\infty} d_i z^i} = \frac{1}{d_0 + d_1 z + \cdots + d_{k-1} z^{k-1} + z^k \frac{b_0}{1-} \frac{b_1 z}{1-} \frac{b_2 z}{1-} \cdots}, \quad k = 1, 2, \ldots, \tag{5}$$

then the successive convergents $n = 0, 1, \ldots, 2m, 2m+1, \ldots$ *of this Stieltjes CF are the following PA*

$$[0/k-1],\ [0/k],\ \ldots,\ [m/k+m-1],\ [n/k+m],\ \ldots. \tag{6}$$

Theorem 3 ([1]). *Let for* $k = 1, 2, \ldots$

$$f(z) = \sum_{i=0}^{\infty} c_i z^i = c_0 + c_1 + \cdots + c_{k-1} z^{k-1} + z^k \frac{a_0}{1+b_0 z-} \frac{a_1 z^2}{1+b_1 z-} \frac{a_2 z^2}{1+b_2 z-} \cdots, \quad (7)$$

then the nth, $n \geq 0$*, convergents of this Jacobi CF are the* $[n+k-1/n]$ *PA.*

3 Method of Analysis and Simplified Notation

In the following we denote by m/n a rational function the numerator of which is of degree m and the denominator is of degree n. This rational function can be a Padé approximant and in this case it is denoted $[m/n]$, otherwise it is not a Padé approximant (NPA) and it is denoted $\{m/n\}$. We say that each coefficient of a power series of a function represents one piece of information. It represents its value or the value of a certain derivative. Then, if the convergent of CF reduced to a rational function of type m/n was calculated from $(m+n+1)$ pieces of information, then it is identified with PA $[m/n]$. Else, i.e., if it is calculated from less than $(m+n+1)$ pieces of information, it is not a Padé approximant: NPA or $\{m/n\}$. Consequently, we introduce a new notation where the indices indicate the rank of information and the letters are without importance. For instance, the CF (3) in this notation will be written as follows:

$$k = 0: \quad \frac{a_1}{1+} \frac{a_2 z}{1+} \frac{a_3 z}{1+} \frac{a_4 z}{1+} \cdots, \quad (8)$$

$$k = 1: \quad a_1 + \frac{a_2 z}{1+} \frac{a_3 z}{1+} \frac{a_4 z}{1+} \cdots,$$

$$k = 2: \quad a_1 + a_2 z + \frac{a_3 z}{1+} \frac{a_4 z}{1+} \cdots, \qquad \text{etc.}$$

In each case the last coefficient of the 4th convergent is denoted a_4, but it signifies only that each convergent in question is built with four pieces of information, each a_k, however, being different from the other one.

Now we turn our attention to the calculation of the degrees of the numerator and the denominator of a convergent (1). Our interest being directed only to the degrees, this leads to simplification of the notation of (1) following the rule

$$\frac{\deg A_k}{\deg B_k} = \frac{\max(\deg(b_k A_{k-1}), \deg(a_k A_{k-2}))}{\max(\deg(b_k B_{k-1}), \deg(a_k B_{k-2}))}. \quad (9)$$

The following example gives the "info-notation" and the "degree-notation" of the same CF, where the second floor is modified with respect to the classical CF:

$$a_1 + \frac{a_2 z}{1 + a_3 z + a_4 z^2 +} \frac{a_5 z^3}{1+} \frac{a_6 z}{1+} \frac{a_7 z}{1+} \frac{a_8 z}{1+} \cdots \equiv 0 + \frac{1}{2+} \frac{3}{0+} \frac{1}{0+} \frac{1}{0+} \frac{1}{0+} \cdots. \quad (10)$$

On the left-hand side, we can read the number of pieces of information indicated by the indices, while on the right-hand side we read the degrees of the numerators and the denominators. We say that the modification of the second floor in this example is a "two-degree modification," because we jumped two powers before inverting the series factorized by z^3. Our goal consists of analyzing the convergents following the first modification and, essentially, of selecting those of them, which remain PA. In the following we observe, that only the one-degree modifications of CF produce interesting effects.

4 Main Results

The following theorem shows that it suffices to restrict our investigations to the case of the one-degree modifications.

Theorem 4. *Let the kth floor of the Stieltjes CF (4) be subjected to a two-degree modification, then all of its convergents, starting from the kth, become NPA.*

Proof. Without loss of generality, we can consider the case of the modification of the first floor:

Convergent n	0	1	2	3	4	5	6	7	...
CF	a_1+	$\dfrac{a_2z}{1+a_3z+a_4z^2+}$	$\dfrac{a_5z^3}{1+}$	$\dfrac{a_6z}{1+}$	$\dfrac{a_7z}{1+}$	$\dfrac{a_8z}{1+}$	$\dfrac{a_9z}{1+}$	$\dfrac{a_{10}z}{1+}$	...
# information	1	4	5	6	7	8	9	10	...
PA or NPA	[0/0]	{2/2}	{3/3}	{3/3}	{4/4}	{4/4}	{5/5}	{5/5}	...

It is clear that all convergents starting from the first are built with an insufficient number of informations. □

The main general theorems are the following.

Theorem 5. *Let the Stieltjes CF (3) be subjected only to one-degree modifications and let its kth floor be the last modified floor leading to NPA convergent, then, alternately, the $(k+2n)$th, $n \geq 0$, convergents and the $(k+2n+1)$th convergents of this CF are NPA and PA, respectively.*

Proof. The *last modified floor leading to the NPA convergent* means that all eventual interruptions of a sequence of modifications contain an even number of floors. It guarantees that the convergent containing the next floor is NPA. This will become clear after the following proof. The proof of Theorem 5 is separated in a number of different cases presented in the following examples.

Example 1. Successive convergents of the following CF

$$a_1 + \frac{a_2z}{1+a_3z+} \frac{a_4z^2}{1+} \frac{a_5z}{1+} \frac{a_6z}{1+} \frac{a_7z}{1+} \cdots$$

are alternately PA and NPA:

$$\frac{A_0}{B_0} = a_1 = [0/0], \quad \frac{A_1}{B_1} = a_1 + \frac{a_2 z}{1 + a_3 z} = [1/1], \tag{11}$$
$$k \geq 1: \ \frac{A_{2k}}{B_{2k}} = \{k+1/k+1\}; \quad \frac{A_{2k+1}}{B_{2k+1}} = [k+1/k+1].$$

Proof. $A_1/B_1 = [1/1]$ is PA because it contains three pieces of information. The next convergent following the modified floor is: $A_2/B_2 = a_1 + a_2 z/(1 + a_3 z + a_4 z^2) = \{2/2\}$. It is NPA because it is built from only four pieces of information, while PA $[2/2]$ needs five. Here the floor 2 is the "last modified floor" of Theorem 5. The above perturbation is "repaired" by the next classical floor, but the following convergent is again NPA and so one. To prove this, we use the degree-notation (10):

$$\frac{\deg A_0}{\deg B_0} = \frac{0}{0}; \quad \frac{\deg A_1}{\deg B_1} = \frac{\max(1+0, 1+0)}{\max(1+0, 1*0)} = \frac{1}{1}; \quad \frac{\deg A_2}{\deg B_2} = \frac{\max(0+1, 2+0)}{\max(0+1, 2+0)} = \frac{2}{2}.$$

For instance, $\frac{2}{2}$ denotes a rational function of type $2/2$. The observed results are presented in the following table:

Convergent n	0	1	2	3	4	5	...	$2k$	$2k+1$
# information	1	3	4	5	6	7	...	$2k+2$	$2k+3$
$\deg A_n = \deg B_n$	0	1	2	2	3	3	...	$k+1$	$k+1$
[ ]:PA or {}:NPA	[ ]	[ ]	{}	[ ]	{}	[ ]	...	{}	[ ]

Beginning from the 2nd convergent the regularity expressed in (11) occurs. □

Example 2. The following modification of a Stieltjes fraction

$$a_1 + a_2 z + \cdots + a_k z^{k-1} + \frac{a_{k+1} z^k}{1 + a_{k+2} z +} \frac{a_{k+3} z^2}{1+} \frac{a_{k+4} z}{1+} \frac{a_{k+5} z}{1+} \frac{a_{k+6} z}{1+} \frac{a_{k+7} z}{1+} \cdots$$

leads to an effect similar to that in Example 1 shifted to the position k:

$$\frac{A_0}{B_0} = [k-1/0]; \quad \frac{A_1}{B_1} = [k/1];$$
$$m \geq 1: \ \frac{A_{2m}}{B_{2m}} = \{k+m/m+1\}; \quad \frac{A_{2m+1}}{B_{2m+1}} = [k+m/m+1]$$

Proof.

$$\underset{[k-1/0]}{a_1 + a_2 z + \cdots + a_k z^{k-1}} + \underset{[k/1]}{\frac{a_{k+1} z^k}{1 + a_{k+2} z +}} \ \underset{\{k+1/2\}}{\frac{a_{k+3} z^2}{1+}} \ \underset{[k+1/2]}{\frac{a_{k+4} z}{1+}} \ \underset{\{k+2/3\}}{\frac{a_{k+5} z}{1+}} \ \underset{[k+2/3]}{\frac{a_{k+6} z}{1+}} \ \cdots$$

To verify this result, it suffices to read the index of the last coefficient in each floor of CF which indicates the number of pieces of information used and next to add the degree of the corresponding numerator to the degree of the denominator of the rational functions indicated in the bottom line. □

Example 3. The following interior modification of the Stieltjes fraction

$$a_1 + \frac{a_2 z}{1+} \frac{a_3 z}{1+} \cdots \frac{a_k z}{1+} \frac{a_{k+1} z}{1 + a_{k+2} z +} \frac{a_{k+3} z^2}{1+} \frac{a_{k+4} z}{1+} \frac{a_{k+5} z}{1+} \cdots$$

leads to an effect similar to that in Example 1 shifted to the position k. All convergents up to

$$\frac{A_k}{B_k} = a_1 + \cdots + \frac{a_{k+1} z}{1 + a_{k+2} z}$$

are PA and:

$$\begin{aligned}
k = 2m: \quad & \frac{A_0}{B_0} = [0/0],\ \frac{A_1}{B_1} = [1/0], \ldots, \frac{A_{2m-2}}{B_{2m-2}} = [m-1/m-1],\\
& \frac{A_{2m-1}}{B_{2m-1}} = [m/m-1],\ \frac{A_{2m}}{B_{2m}} = [m+1/m];\\
n \geq 0: \quad & \frac{A_{2m+2n}}{B_{2m+2n}} = [m+n+1/m+n];\ \frac{A_{2m+2n+1}}{B_{2m+2n+1}} = \{m+n+2/m+n+1\};\\
k = 2m+1: \quad & \frac{A_0}{B_0} = [0/0], \ldots, \frac{A_{2m+1}}{B_{2m+1}} = [m+1/m+1];\\
n \geq 0: \quad & \frac{A_{2m+2n+1}}{B_{2m+2n+1}} = [m+n+1/m+n+1];\\
& \frac{A_{2m+2n+2}}{B_{2m+2n+2}} = \{m+n+2/m+n+2\}.
\end{aligned}$$

Proof. The degrees of the numerators and the denominators of successive convergents can be easily calculated by means of (9) and next compared to the number of pieces of information used. For instance for

$$\begin{aligned}
k = 2m: \quad & \frac{\deg A_{2m}}{\deg B_{2m}} = \frac{\max(1+m, 1+m-1)}{\max(1+m-1, 1+m-1)} = \frac{m+1}{m};\\
& \frac{\deg A_{2m+1}}{\deg B_{2m+1}} = \frac{\max(1+m, 2+m)}{\max(m, 2+m-1)} = \frac{m+2}{m+1}; \ldots
\end{aligned}$$

where the first is the last of successive PA in the beginning and the second is the first NPA. The result could have been anticipated by looking over the previous examples. □

Comment. If each floor of the CF of the last Example is modified, then it becomes a Jacobi CF (7).

Example 4. Let us start from the modified CF of Example 1. If we proceed to increase the degrees of the successive denominators to degree 1 up to the kth convergent:

CF	a_1+	$\frac{a_2 z}{1+a_3 z+}$	$\frac{a_4 z^2}{1+a_5 z+}$	$\cdots$	$\frac{a_{2k} z^2}{1+a_{2k+1} z+}$	$\frac{a_{2k+2} z^2}{1+}$	$\frac{a_{2k+3} z}{1+}$	$\frac{a_{2k+4} z}{1+}$	$\cdots$
$\frac{A_i}{B_i}$	$[0/0]$	$[1/1]$	$[2/2]$	$\cdots$	$[k/k]$	$\{k+1/k+1\}$	$[k+1/k+1]$	$\{k+2/k+2\}$	$\cdots$

then all convergents up to kth are PA, the $(k+1)$th is NPA, the $(k+2)$th is again PA and so on.

Example 5. Let us start from the modified CF of Example 1. If one of the convergents which is PA (here the 3rd) is modified

$$\begin{array}{c|cccccccccc}
\text{CF} & a_1+ & \dfrac{a_2 z}{1+a_3 z+} & \dfrac{a_4 z^2}{1+} & \dfrac{a_5 z}{1+a_6 z+} & \dfrac{a_7 z^2}{1+} & \dfrac{a_8 z}{1+} & \dfrac{a_9 z}{1+} & \dfrac{a_{10} z}{1+} & \dots \\
\hline
\dfrac{A_i}{B_i} & [0/0] & [1/1] & \{2/2\} & \{3/3\} & \{4/4\} & \{4/4\} & \{5/5\} & \{5/5\} & \dots
\end{array}$$

then all successive convergents become NPA. On the contrary, if we modify one of the NPA convergents, for instance the $(k+3)$th convergent

$$\begin{array}{cccccccccc}
a_1+ & \dfrac{a_2 z}{1+a_3 z+} & \dfrac{a_4 z^2}{1+} & \dfrac{a_5 z}{1+} & \dfrac{a_6 z}{1+} & \dots & \dfrac{a_{2k+2} z}{1+a_{2k+3} z+} & \dfrac{a_{2k+4} z^2}{1+} & \dfrac{a_{2k+5} z}{1+} & \dots \\
\hline
[0/0] & [1/1] & \{2/2\} & [2/2] & \{3/3\} & \dots & [k+1/k+1] & \{k+2/k+2\} & [k+2/k+2] & \dots
\end{array}$$

then we obtain the usual situation where the $(k+3)$th convergent

$$\frac{A_{k+3}}{B_{k+3}} = a_1 + \dots + \frac{a_{2k+2} z}{1 + a_{2k+3} z} = [k+1/k+1]$$

becomes PA, the next is NPA, the following is again PA, the next NPA, and so on.

The proofs of Examples 4 and 5 are similar to the previous proofs: they consist of comparing the degrees of convergents with the number of pieces of information used. Example 4 is a particular case of Example 5: all modifications which follow the first one concern NPA convergents. This completes the proof of Theorem 5. □

Theorem 5 characterizes the modification of the CF (3). The following theorem characterizes the CF (5).

Theorem 6. *Let the Stieltjes CF (5) be subjected only to one-degree modifications and let its kth floor be the last modified floor leading to NPA convergent, then, alternately, the $(k+2n)$th, $n \geq 0$, convergents and the $(k+2n+1)$th convergents of this CF are NPA and PA respectively.*

It is an exact copy of Theorem 5, except that in all examples one must interchange the degrees of the numerators and denominators. For instance (cf. Example 2), the convergents of

$$f(z) = \frac{1}{\sum_{i=0}^{\infty} d_i z^i} = \cfrac{1}{d_1 + d_2 z + \dots + d_k z^{k-1} + z^k \dfrac{d_{k+1}}{1+d_{k+2} z+} \dfrac{d_{k+3} z^2}{1+} \dfrac{d_{k+4} z}{1+} \dfrac{d_{k+5} z}{1+} \dots}$$

are

$$\frac{A_0}{B_0} = [0/k-1], \quad [1/k], \quad \{2/k+1\}, \quad [2/k+1], \quad \{3/k+2\}, \quad [3/k+2], \dots .$$

We can also consider the effects of different transformations in PA of the denominators of CF, for instance:

$$f(z) = \frac{a_1}{1+zg(z)} = \frac{a_1}{1+z([m/n]_g(z)+O(z^{m+n+1}))}.$$

Here, for all $n \leq m+1$ the convergent $a_1/(1+z[m/n]_g(z))$ is PA $[n/m+1]$ of f.

A more general example is given by the following theorem.

Theorem 7. *Let*

$$\begin{aligned} f(z) &= a_1 + a_2 z + \cdots + a_k z^{k-1} \\ &\quad + \frac{a_{k+1}z^k}{1+} \frac{a_{k+2}z}{1+} \cdots \frac{a_{k+i-1}z}{1+} \frac{a_{k+i}z}{1+a_{k+i+1}z+\ldots+a_{k+i+m+n}z^{m+n}+\ldots} \\ &= a_1 + a_2 z + \cdots + a_k z^{k-1} + \frac{a_{k+1}z^k}{1+} \frac{a_{k+2}z}{1+} \cdots \frac{a_{k+i-1}z}{1+} \frac{a_{k+i}z}{\dfrac{P_m(z)}{Q_n(z)}+O(z^{m+n+1})} \\ &= a_1 + a_2 z + \cdots + a_k z^{k-1} + \frac{a_{k+1}z^k}{1+} \frac{a_{k+2}z}{1+} \cdots \frac{a_{k+i-1}z}{1+} \frac{a_{k+i}zQ_n(z)}{P_m(z)+} \frac{O(z^{m+n+1})}{1}, \end{aligned}$$

then the ith convergent becomes PA only for $m=n$ ($[n+k+l-1/n+l]$ if $i=2l$ and $[n+k+l/n+l]$ if $i=2l+1$) and for $m=n+1$ ($[n+k+l/n+l]$ if $i=2l$ and $[n+k+l/n+l+1]$ if $i=2l+1$).

Proof. The proof is based on (6), (9) and the condition of compatibility

$$\deg A_i + \deg B_i + 1 \leq k+i+m+n, \tag{12}$$

where the right-hand side represents the number of pieces of information.

For $i=2l$: $\deg A_i = \max(m+k+l-1, n+k+l-1)$ and $\deg B_i = \max(m+l-1, n+l)$. Now, if $m \leq n$ then $A_i/B_i = n+k+l-1/n+l$ and (12) gives $m=n$. If $m>n$ then $A_i/B_i = m+k+l-1/m+l-1$ and (12) gives $m=n+1$.

For $i=2l+1$: $\deg A_i = \max(m+k+l-1, n+k+l)$ and $\deg B_i = \max(m+l-1, n+l)$. Now, if $m \leq n$ then $A_i/B_i = n+k+l/n+l$ and (12) gives $m=n$. If $m>n$ then $A_i/B_i = m+k+l-1/m+l$ and (12) gives $m=n+1$.

This completes the proof of Theorem 7. □

Remark 1. Note that all PA from Theorems 5, 6 and from the Examples 1–5 have also different CF representation given in the classical Theorems 1–3.

Conclusion. All presented results were proved by means of simple algebraic verifications. As trivial as they may be, to the best of our knowledge they were not published before.

Reference

1. Gilewicz, J.: Approximants de Padé. Lecture Notes in Mathematics, vol. 667, Springer, Berlin (1978)

Orthogonal Decomposition of Fractal Sets

Ljubiša M. Kocić, Sonja Gegovska - Zajkova, and Elena Babače

Dedicated to Professor Gradimir V. Milovanović
on the occasion of his 60th birthday

1 Introduction

Let $\{w_i \,|\, i = 1,2,\ldots,n\}$, $n > 1$, be a set of contractive affine mappings defined on the complete Euclidean metric space $(\mathbb{R}^m, d_E)$, $m > 1$, with

$$w_i(\mathbf{x}) = \mathbf{A}_i\mathbf{x} + \mathbf{b}_i, \quad \mathbf{x} \in \mathbb{R}^m, \ i = 1,2,\ldots,n,$$

where $\mathbf{A}_i$ is an $m \times m$ real matrix and $\mathbf{b}_i$ is an m-dimensional real vector.

Supposing that the Lipschitz factors, $s_i = \mathrm{Lip}\{w_i\}$, satisfy $|s_i| < 1$, $i = 1,2,\ldots,n$, the system $S = \{\mathbb{R}^m; w_i, i = 1,2,\ldots,n\}$ is called a *(hyperbolic) Iterated Function System* (IFS). Associated with a hyperbolic IFS, the so-called *Hutchinson operator* $W_S : \mathcal{H}(\mathbb{R}^m) \to \mathcal{H}(\mathbb{R}^m)$, defined by

$$W_S(B) = \bigcup_{i=1}^{n} w_i(B), \quad \forall B \in \mathcal{H}(\mathbb{R}^m)$$

Ljubiša M. Kocić
University of Niš, Faculty of Electronic Engineering, POBox 73, 18 000 Niš, Serbia,
e-mail: kocic@elfak.ni.ac.rs

Sonja Gegovska - Zajkova, and Elena Babače
Ss Cyril and Methodius University, Faculty of Electrical Engineering and Information Technologies, P.O. Box 574, Skopje, Macedonia,
e-mail: szajkova@feit.ukim.edu.mk, elena.babace@feit.ukim.edu.mk

W. Gautschi et al. (eds.), *Approximation and Computation: In Honor of Gradimir V. Milovanović*, Springer Optimization and Its Applications 42,
DOI 10.1007/978-1-4419-6594-3_11,

performs a contractive mapping on the complete metric space $(\mathcal{H}(\mathbb{R}^m), h)$ with contractivity factor $s = \max_i\{s_i\}$ [1]. Here, $\mathcal{H}(\mathbb{R}^m)$ is the space of nonempty compact subsets of $\mathbb{R}^m$ and h stands for the Hausdorff metric induced by d_E, i.e.,

$$h(A,B) = \max\left\{\max_{a\in A}\min_{b\in B} d_E(a,b),\ \max_{b\in B}\min_{a\in A} d_E(b,a)\right\}, \quad A,B \in \mathcal{H}(\mathbb{R}^m).$$

According to the contraction mapping theorem, W_S has a unique fixed point, $A \in \mathcal{H}(\mathbb{R}^m)$, called the *attractor* of the IFS, satisfying

$$A = W_S(A) = \bigcup_{i=1}^{n} w_i(A).$$

Definition 1. A (nondegenerate) *m-dimensional simplex* (or just *simplex*) is the convex hull of a set of $m+1$ affinely independent points $\mathbf{p}_1, \mathbf{p}_2, \dots, \mathbf{p}_{m+1}$ in Euclidean space of dimension m or higher, $\widehat{\mathbf{P}}_m = \mathrm{conv}\{\mathbf{p}_1, \mathbf{p}_2, \dots, \mathbf{p}_{m+1}\}$. The vertices of $\widehat{\mathbf{P}}_m$ will be denoted by $\mathbf{P}_m$ and represented by the vector $\mathbf{P}_m = [\mathbf{p}_1^{\mathrm{T}}\ \mathbf{p}_2^{\mathrm{T}}\ \cdots\ \mathbf{p}_{m+1}^{\mathrm{T}}]^{\mathrm{T}}$.

Let $\mathbf{S} = [s_{ij}]_{i,j=1}^{m+1}$ be an $(m+1)\times(m+1)$ row-stochastic real matrix (its rows sum up to 1).

Definition 2. We refer to the linear mapping $\mathcal{L} : \mathbb{R}^{m+1} \to \mathbb{R}^{m+1}$, such that $\mathcal{L}(\mathbf{x}) = \mathbf{S}^{\mathrm{T}}\mathbf{x}$ as the linear mapping *associated with* $\mathbf{S}$.

Definition 3. Let $\widehat{\mathbf{P}}_m$ be a nondegenerate simplex and let $\{\mathbf{S}_i\}_{i=1}^{n}$ be a set of real square nonsingular row-stochastic matrices of order $m+1$. The system $\Omega(\widehat{\mathbf{P}}_m) = \{\widehat{\mathbf{P}}_m; \mathbf{S}_1, \mathbf{S}_2, \dots, \mathbf{S}_n\}$ is called a *(hyperbolic) Affine invariant IFS (AIFS)*, provided that the linear mappings associated with $\mathbf{S}_i$ are contractions in $(\mathbb{R}^{m+1}, d_E)$ [3–5].

2 Hyperplane of Areal Coordinates

Let $\mathbf{r} = (\xi_1, \xi_2, \dots, \xi_{m+1})$ be a point in $(m+1)$-dimensional real space $\mathbb{R}^{m+1}$, which can be seen as a column vector as well, $\mathbf{r} = [\xi_1\ \xi_2\ \cdots\ \xi_{m+1}]^{\mathrm{T}}$. Furthermore, let $\mathcal{E}_{m+1} = \left\{\mathbf{e}_i = [\delta_{ij}]_{j=1}^{m+1}, i = 1, \dots, m+1\right\}$, be the set of unit vectors constituting the standard orthonormal basis in $\mathbb{R}^{m+1}$. A simple $(m+1)$-order identity matrix gives a matrix representation of this basis,

$$\mathbf{E}_{m+1} = \begin{bmatrix} 1 & 0 & & 0 \\ 0 & 1 & & 0 \\ & & \ddots & \\ 0 & 0 & & 1 \end{bmatrix} = \begin{bmatrix} \mathbf{e}_1^{\mathrm{T}} \\ \mathbf{e}_2^{\mathrm{T}} \\ \vdots \\ \mathbf{e}_{m+1}^{\mathrm{T}} \end{bmatrix} = [\mathbf{e}_1^{\mathrm{T}}\ \mathbf{e}_2^{\mathrm{T}}\ \cdots\ \mathbf{e}_{m+1}^{\mathrm{T}}]^{\mathrm{T}}.$$

Moreover, the matrix $\mathbf{E}_{m+1}$ also represents the so-called *standard simplex*, $\widehat{\mathbf{T}}_m = \mathrm{conv}\{\mathbf{e}_1, \mathbf{e}_2, \dots, \mathbf{e}_{m+1}\}$. Being of dimensionality m, the simplex $\widehat{\mathbf{T}}_m$ has $m+1$

vertices and $m(m+1)/2$ edges or the same number of linearly dependent edge vectors. Choosing the vertex $\mathbf{e}_{m+1}$ as the common origin and setting $\mathbf{u}_j = \mathbf{e}_j - \mathbf{e}_{m+1}$, $j = 1,2,\ldots,m$, one gets the set of linearly independent vectors, $\mathcal{U}_m = \{\mathbf{u}_1, \mathbf{u}_2, \ldots, \mathbf{u}_m\}$, which is called the *basis generated by the simplex* $\widehat{\mathbf{T}}_m$. Note that the vectors $\mathbf{u}_i$ are rows of an $m \times (m+1)$ matrix $\mathbf{U}_m$ obtained from the identity matrix $\mathbf{E}_{m+1}$ by dropping the last row and then replacing the last 0 – column with the (-1) - column (or by appending the (-1) - column to the m-identity matrix $\mathbf{E}_m$)

$$\mathbf{U}_m = \left[\begin{array}{ccc|c} 1 & 0 & 0 & -1 \\ 0 & 1 & 0 & -1 \\ & & \ddots & \\ 0 & 0 & 1 & -1 \end{array}\right] = [\mathbf{E}_m \mid -1] = \begin{bmatrix} \mathbf{u}_1^{\mathrm{T}} \\ \mathbf{u}_2^{\mathrm{T}} \\ \vdots \\ \mathbf{u}_m^{\mathrm{T}} \end{bmatrix}.$$

The span of the vector set $\mathcal{U}_m$ forms an m-dimensional vector space, the hyperplane $\mathbf{V}^m = \operatorname{span}\{\mathbf{u}_1, \mathbf{u}_2, \ldots, \mathbf{u}_m\} \subseteq \mathbb{R}^{m+1}$. If we denote $V^m = \mathbf{e}_{m+1} + \mathbf{V}^m$, then V^m is an affine space that can easily be transformed into a vector space if we take $\mathbf{e}_{m+1}$ as origin. So, from now on, it will be considered that V^m is a vector space, unless it is differently specified.

Note that the following lemma holds:

Lemma 1. *V^m is identical to the m-dimensional hyperplane* affi $\{\mathbf{e}_1, \mathbf{e}_2, \ldots, \mathbf{e}_{m+1}\} \subseteq \mathbb{R}^{m+1}$, *i.e., for any $\mathbf{v} \in V^m$, there exists a set of $m+1$ constants $\{a_1, \ldots, a_{m+1}\}$, $\sum_{i=1}^{m+1} a_i = 1$, such that $\mathbf{v} = \sum_{i=1}^{m+1} a_i \mathbf{e}_i$.*

Proof. By definition, any vector $\mathbf{v} \in \mathbf{V}^m$ is a linear combination of the vectors of $\mathcal{U}_m$, i.e., $\mathbf{v} = \sum_{i=1}^{m} \alpha_i \mathbf{u}_i = \sum_{i=1}^{m} \alpha_i (\mathbf{e}_i - \mathbf{e}_{m+1})$, where $\alpha_i \in \mathbb{R}$. Then, for every $\mathbf{r} \in V^m = \mathbf{e}_{m+1} + \mathbf{V}^m$,

$$\mathbf{r} = \mathbf{e}_{m+1} + \sum_{i=1}^{m} \alpha_i (\mathbf{e}_i - \mathbf{e}_{m+1}) = \sum_{i=1}^{m} \alpha_i \mathbf{e}_i + \mathbf{e}_{m+1} \left(1 - \sum_{i=1}^{m} \alpha_i\right).$$

Setting $a_i = \alpha_i$, $i = 1, \ldots, m$ and $a_{m+1} = 1 - \sum_{i=1}^{m} \alpha_i$, clearly implies that $\sum_{i=1}^{m+1} a_i = 1$. □

The basis $\mathcal{U}_m = \{\mathbf{u}_1, \mathbf{u}_2, \ldots, \mathbf{u}_m\}$ is neither orthogonal nor normalized. Namely,

$$\langle \mathbf{u}_i, \mathbf{u}_j \rangle = \begin{cases} 1, & i \neq j, \\ 2, & i = j. \end{cases}$$

To transform $\mathcal{U}_m$ into an orthonormal basis $\mathcal{V}_m = \{\mathbf{v}_1, \mathbf{v}_2, \ldots, \mathbf{v}_m\}$ in V^m, the Gram–Schmidt procedure is applied,

$$\mathbf{v}_k^* = \begin{cases} \mathbf{u}_1, & k = 1, \\ \mathbf{u}_k - \sum_{i=1}^{k-1} \dfrac{\langle \mathbf{v}_i^*, \mathbf{u}_k \rangle}{\langle \mathbf{v}_i^*, \mathbf{v}_i^* \rangle} \mathbf{v}_i^*, & k = 2, \ldots, m, \end{cases} \tag{1}$$

and

$$\mathbf{v}_k = \frac{\mathbf{v}_k^*}{\sqrt{\langle \mathbf{v}_k^*, \mathbf{v}_k^* \rangle}}, \quad k = 1, 2, \ldots, m.$$

This new orthonormal basis is represented by the $m \times (m+1)$ matrix

$$\mathbf{V}_m = \begin{bmatrix} v_{11} & v_{12} & & v_{1,m+1} \\ v_{21} & v_{22} & & v_{2,m+1} \\ & & \ddots & \\ v_{m,1} & v_{m,2} & & v_{m,m+1} \end{bmatrix} = \begin{bmatrix} \mathbf{v}_1^{\mathrm{T}} \\ \mathbf{v}_2^{\mathrm{T}} \\ \vdots \\ \mathbf{v}_m^{\mathrm{T}} \end{bmatrix},$$

whose row-space is the hyperplane V^m.

Example 1. Take $m = 2$. The standard simplex is an equilateral triangle $\widehat{\mathbf{T}}_2 = \mathrm{conv}\{\mathbf{e}_1, \mathbf{e}_2, \mathbf{e}_3\}$, Fig. 1 (left). Then, $\mathbf{u}_1 = [1\ 0\ -1]^{\mathrm{T}}$, $\mathbf{u}_2 = [0\ 1\ -1]^{\mathrm{T}}$. By the Gram–Schmidt procedure of orthogonalization, the basis $\mathcal{U}_2 = \{\mathbf{u}_1, \mathbf{u}_2\}$ transforms into $\mathcal{V}_2 = \{\mathbf{v}_1, \mathbf{v}_2\}$, where one gets

$$\mathbf{v}_1 = \left[\frac{\sqrt{2}}{2}\ 0 - \frac{\sqrt{2}}{2}\right]^{\mathrm{T}}, \quad \mathbf{v}_2 = \left[-\frac{\sqrt{6}}{6}\ \frac{\sqrt{6}}{3} - \frac{\sqrt{6}}{6}\right]^{\mathrm{T}}.$$

All coordinates are given with respect to the basis $\{\mathbf{e}_1, \mathbf{e}_2, \mathbf{e}_3\}$.

The next step is establishing a relation between the coordinates of the point $\mathbf{r} \in V^m$ w.r.t. the basis $\mathcal{E}_{m+1}$, $\mathbf{r} = [\xi_1\ \xi_2\ \cdots\ \xi_{m+1}]^{\mathrm{T}}$, and the coordinates w.r.t. the orthonormal basis $\mathcal{V}_m$, $\mathbf{x} = [x_1\ x_2\ \cdots\ x_m]^{\mathrm{T}}$ of the same point. It is enough to give the

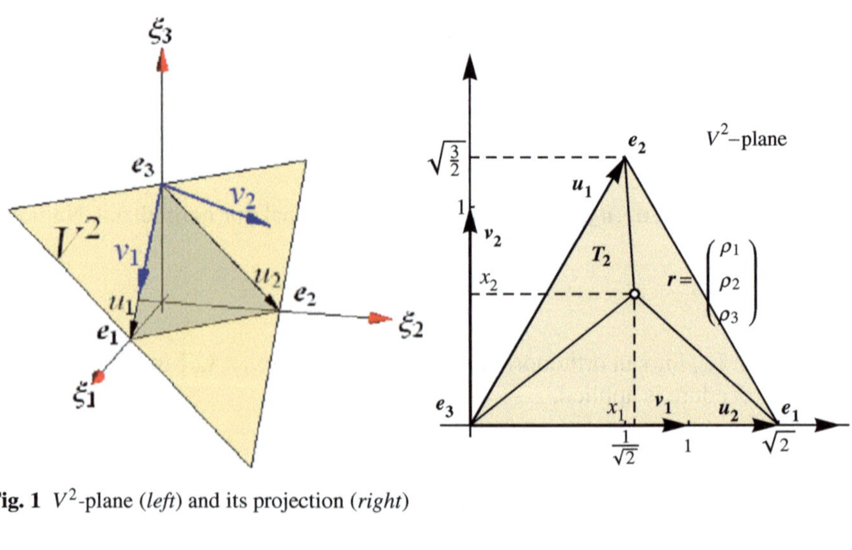

Fig. 1 V^2-plane (*left*) and its projection (*right*)

reduced relation between **r** and **x**, since (because of Lemma 1) the last coordinate of **r** is dispensable. The reduced relation takes the matrix form

$$\overline{\mathbf{r}} = \begin{bmatrix} v_{11} & v_{21} & & v_{m,1} \\ v_{12} & v_{22} & & v_{m,2} \\ & & \ddots & \\ v_{1,m} & v_{2,m} & & v_{m,m} \end{bmatrix} \begin{bmatrix} x_1 \\ x_2 \\ \\ x_m \end{bmatrix} = \mathbf{Q}_m \mathbf{x}, \tag{2}$$

where $\overline{\mathbf{r}} = [\xi_1\ \xi_2\ \cdots\ \xi_m]^{\mathrm{T}}$ is the truncated vector **r**, $\mathbf{Q}_m = \left(\overline{\mathbf{V}}_m\right)^{\mathrm{T}}$, $\overline{\mathbf{V}}_m$ is the truncated matrix $\mathbf{V}_m$ obtained by dropping the last column of $\mathbf{V}_m$. Since the matrix $\mathbf{Q}_m$ plays the key role in relating $[\xi_1\ \xi_2\ \cdots\ \xi_{m+1}]$ coordinates with $[x_1\ x_2\ \cdots\ x_m]$, it is important to have an explicit formula for $\mathbf{Q}_m$. It is given by the following lemma:

Lemma 2. *The matrix* $\mathbf{Q}_m = [q_{ij}]_{i,j=1}^{m}$ *is given by*

$$q_{ij} = \begin{cases} \dfrac{-1}{\sqrt{j(j+1)}}, & i < j, \\ \sqrt{\dfrac{i}{i+1}}, & i = j, \\ 0, & i > j. \end{cases}$$

Proof. The combination of (1) and the induction principle gives

$$\mathbf{v}_i^* = \Big[-\frac{1}{i}\ \cdots\ -\frac{1}{i}\ \underbrace{1}_{i\text{–th place}}\ 0\ \cdots\ 0\ -\frac{1}{i}\Big]^{\mathrm{T}}, \quad i = 1,2,\ldots,m,$$

$\mathbf{v}_i^*$ is an $(m+1)$-dimensional vector. Since $\|\mathbf{v}_i^*\| = \sqrt{(i+1)/i}$, the vector $\mathbf{v}_i$ has the form

$$\mathbf{v}_i = \Big[-\frac{1}{\sqrt{i(i+1)}}\ \cdots\ -\frac{1}{\sqrt{i(i+1)}}\ \underbrace{\sqrt{\frac{i}{i+1}}}_{i\text{–th place}}\ 0\ \cdots\ 0\ -\frac{1}{\sqrt{i(i+1)}}\Big]^{\mathrm{T}},$$

$i = 1,2,\ldots,m$. Taking into account that $\mathbf{v}_i^{\mathrm{T}}$ are the rows in the matrix $\mathbf{V}_m$, and $\mathbf{Q}_m = \left(\overline{\mathbf{V}}_m\right)^{\mathrm{T}}$, one gets

$$\mathbf{Q}_m = \begin{bmatrix} 1/\sqrt{2} & -1/\sqrt{2\cdot 3} & -1/\sqrt{3\cdot 4} & \cdots & -1/\sqrt{m(m+1)} \\ 0 & \sqrt{2/3} & -1/\sqrt{3\cdot 4} & \cdots & -1/\sqrt{m(m+1)} \\ 0 & 0 & \sqrt{3/4} & \cdots & -1/\sqrt{m(m+1)} \\ & & & \ddots & \\ 0 & 0 & 0 & \cdots & \sqrt{m/(m+1)} \end{bmatrix}. \quad \square$$

Lemma 3. *The inverse transformation of (2) is*

$$\mathbf{x} = \mathbf{Q}_m^{-1}\overline{\mathbf{r}}.$$

Explicitly, $\mathbf{Q}_m^{-1} = [q_{ij}^*]_{i,j=1}^m$ *is given by*

$$q_{ij}^* = \begin{cases} \dfrac{1}{\sqrt{i(i+1)}}, & i < j, \\ \sqrt{\dfrac{i+1}{i}}, & i = j, \\ 0, & i > j. \end{cases}$$

Proof. The existence of the inverse matrix of $\mathbf{Q}_m$ is ensured by the fact that det $\mathbf{Q}_m = 1/\sqrt{m+1}$. The reader easily verifies that

$$\mathbf{Q}_m^{-1} = \begin{bmatrix} \sqrt{2/1} & 1/\sqrt{1\cdot 2} & 1/\sqrt{1\cdot 2} & \cdots & 1/\sqrt{1\cdot 2} \\ 0 & \sqrt{3/2} & 1/\sqrt{2\cdot 3} & \cdots & 1/\sqrt{2\cdot 3} \\ 0 & 0 & \sqrt{4/3} & \cdots & 1/\sqrt{3\cdot 4} \\ & & & \ddots & \\ 0 & 0 & 0 & \cdots & \sqrt{(m+1)/m} \end{bmatrix}. \tag{3}$$

□

Remark 1. As a consequence of the last lemma, it can be concluded that the $\mathcal{V}_m$-coordinates of the vertices of the standard simplex $\widehat{\mathbf{T}}_m$, $\mathbf{t}_1, \mathbf{t}_2, \ldots, \mathbf{t}_m$, $\mathbf{t}_i = (t_{ij})_{j=1}^m$, $1 \le i \le m+1$, are given by the matrix $\mathbf{T}_m^{\mathrm{T}} = [\mathbf{t}_1 \;\; \mathbf{t}_2 \;\; \cdots \;\; \mathbf{t}_m] = [\mathbf{Q}_m^{-1} \,|\, \mathbf{0}]$. More precisely,

$$t_{ij} = \begin{cases} \dfrac{1}{\sqrt{i(i+1)}}, & j < i \le m, \\ \sqrt{\dfrac{i+1}{i}}, & i = j \le m, \\ 0, & i < j \text{ and } i = m+1,\ 1 \le j \le m. \end{cases}$$

This can be easily proven starting from the relation

$$\mathbf{t}_i = \mathbf{Q}_m^{-1}\overline{\mathbf{e}}_i, \quad i = 1, 2, \ldots, m+1.$$

Example 2. For some low-dimensional cases, one has (Fig. 2):

$$m=2: \mathbf{t}_1 = [\sqrt{2} \quad 0]^{\mathrm{T}}, \qquad \mathbf{t}_2 = \begin{bmatrix} \frac{\sqrt{2}}{2} & \frac{\sqrt{6}}{2} \end{bmatrix}^{\mathrm{T}},$$
$$\mathbf{t}_3 = [0 \quad 0]^{\mathrm{T}};$$
$$m=3: \mathbf{t}_1 = [\sqrt{2} \quad 0 \quad 0]^{\mathrm{T}}, \qquad \mathbf{t}_2 = \begin{bmatrix} \frac{\sqrt{2}}{2} & \frac{\sqrt{6}}{2} & 0 \end{bmatrix}^{\mathrm{T}},$$
$$\mathbf{t}_3 = \begin{bmatrix} \frac{\sqrt{2}}{2} & \frac{\sqrt{6}}{6} & \frac{2\sqrt{3}}{3} \end{bmatrix}^{\mathrm{T}}, \qquad \mathbf{t}_4 = [0 \quad 0 \quad 0]^{\mathrm{T}};$$
$$m=4: \mathbf{t}_1 = [\sqrt{2} \quad 0 \quad 0 \quad 0]^{\mathrm{T}}, \qquad \mathbf{t}_2 = \begin{bmatrix} \frac{\sqrt{2}}{2} & \frac{\sqrt{6}}{2} & 0 & 0 \end{bmatrix}^{\mathrm{T}},$$
$$\mathbf{t}_3 = \begin{bmatrix} \frac{\sqrt{2}}{2} & \frac{\sqrt{6}}{6} & \frac{2\sqrt{3}}{3} & 0 \end{bmatrix}^{\mathrm{T}}, \ \mathbf{t}_4 = \begin{bmatrix} \frac{\sqrt{2}}{2} & \frac{\sqrt{6}}{6} & \frac{\sqrt{3}}{6} & \frac{\sqrt{5}}{2} \end{bmatrix}^{\mathrm{T}},$$
$$\mathbf{t}_5 = [0 \quad 0 \quad 0 \quad 0]^{\mathrm{T}}.$$

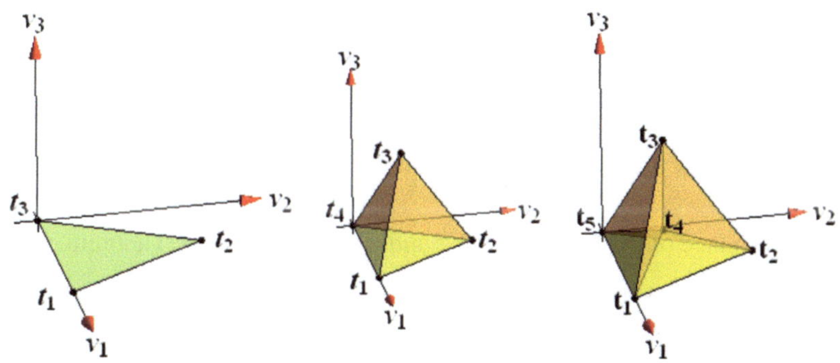

Fig. 2 2-, 3- and 4-dimensional standard simplices

Now, every point $\mathbf{r} \in V^m$ may have two different coordinate identifications, the rectangular coordinates w.r.t. the basis $\mathcal{V}_m$, $\mathbf{x} = (x_1, x_2, \ldots, x_m)$ and the so-called *areal coordinates* (sometimes called *generalized barycentric* coordinates) generated by the simplex $\widehat{\mathbf{T}}_m$ and denoted by $\mathbf{r} = (\rho_1, \rho_2, \ldots, \rho_{m+1})$. The name comes from the definition of these coordinates in $\mathbb{R}^2$, as ratio of triangular areas.

Definition 4. The areal coordinates of the point $\mathbf{r} \in V^m$ w.r.t. a simplex $\widehat{\mathbf{P}}_m$ are defined as

$$\rho_i = \text{area}\, \widehat{\mathbf{P}}^i_m / \text{area}\ \widehat{\mathbf{P}}_m, \quad i = 1, 2, \ldots, m+1,$$

where area $\widehat{\mathbf{P}}_m$ is the signed m-dimensional volume of the simplex $\widehat{\mathbf{P}}_m$ and $\widehat{\mathbf{P}}^i_m$ is the simplex derived from $\widehat{\mathbf{P}}_m$ by substituting the i-th vertex $\mathbf{p}_i$ by the point $\mathbf{r}$.

It follows from the definition that areal coordinates have the unity partition property

$$\sum_{i=1}^{m+1} \rho_i = 1. \tag{4}$$

This means that one areal coordinate is dispensable and the full information of the position of the point $\mathbf{r}$ is contained in the truncated areal vector $\bar{\mathbf{r}} = [\rho_1\ \rho_2\ \cdots\ \rho_m]^{\mathrm{T}}$.

Lemma 4. *The signed area of the standard simplex $\widehat{\mathbf{T}}_m \subset \mathbb{R}^{m+1}$ is*

$$\operatorname{area} \widehat{\mathbf{T}}_m = \frac{(-1)^m}{m!}\sqrt{m+1}, \quad m = 1,2,3,\ldots. \tag{5}$$

Proof. By the known formula for the signed area of the simplex [2],

$$\operatorname{area} \widehat{\mathbf{T}}_m = \frac{1}{m!}\begin{vmatrix} 1 & \sqrt{2/1} & 0 & 0 & \cdots & 0 \\ 1 & 1/\sqrt{1\cdot 2} & \sqrt{3/2} & 0 & & 0 \\ 1 & 1/\sqrt{1\cdot 2} & 1/\sqrt{2\cdot 3} & \sqrt{4/3} & & 0 \\ \vdots & & & & \ddots & \\ 1 & 1/\sqrt{1\cdot 2} & 1/\sqrt{2\cdot 3} & 1/\sqrt{3\cdot 4} & \cdots & \sqrt{(m+1)/m} \\ 1 & 0 & 0 & 0 & & 0 \end{vmatrix}, \tag{6}$$

where the determinant is of order $m+1$. Expanding along the elements of the last row gives

$$\operatorname{area} \widehat{\mathbf{T}}_m = \frac{(-1)^m}{m!}\prod_{k=1}^{m}\sqrt{\frac{k+1}{k}} = \frac{(-1)^m}{m!}\sqrt{m+1}. \quad \square$$

The importance of the hyperplane V^m hides in the following Theorem.

Theorem 1. *Let $\mathbf{r} \in \mathbb{R}^{m+1}$ and let $[\rho_1\ \rho_2\ \cdots\ \rho_{m+1}]^{\mathrm{T}}$ be the vector of areal coordinates of $\mathbf{r}$ w.r.t. the canonical simplex $\widehat{\mathbf{T}}_m \subset \mathbb{R}^{m+1}$. Let $[\xi_1\ \xi_2\ \cdots\ \xi_{m+1}]^{\mathrm{T}}$ be the coordinates of $\mathbf{r}$ w.r.t. the orthonormal basis $\mathcal{E}_{m+1} = \{\mathbf{e}_1, \mathbf{e}_2, \ldots, \mathbf{e}_{m+1}\}$. Then, the areal coordinates of $\mathbf{r}$ coincide with its orthonormal coordinates if and only if $\mathbf{r} \in V^m$. In other words*

$$[\rho_1\ \rho_2\ \cdots\ \rho_{m+1}] = [\xi_1\ \xi_2\ \cdots\ \xi_{m+1}] \Leftrightarrow \mathbf{r} \in V^m.$$

Proof. Let $\mathbf{r} \in \mathrm{V}^m$ and let $\mathbf{T}_m = [\mathbf{t}_1\ \mathbf{t}_2\ \cdots\ \mathbf{t}_{m+1}]^{\mathrm{T}}$, where $\mathbf{t}_i$ are the vertices of $\widehat{\mathbf{T}}_m$ in the $\mathcal{V}_m$-basis. Let $\mathbf{r} = \sum_{i=1}^{m+1} \rho_i \mathbf{t}_i$ and $\mathbf{r} = \sum_{i=1}^{m+1} \xi_i \mathbf{e}_i$. On the basis of Lemma 1, and because of the property (4), it is enough to show that $\rho_i = \xi_i$, $i = 1,2,\ldots,m$. To apply formula (5), one needs to evaluate $\operatorname{area} \widehat{\mathbf{T}}_m^i$, which one gets from (6) by replacing its i-th row,

$$\left[1\ \frac{1}{\sqrt{1\cdot 2}}\ \frac{1}{\sqrt{2\cdot 3}}\ \frac{1}{\sqrt{3\cdot 4}}\ \cdots\ \frac{1}{\sqrt{(i-1)i}}\ \sqrt{\frac{i+1}{i}}\ 0\ \cdots\ 0\right],$$

with $[1\ x_1\ \cdots\ x_{i-1}\ x_i\ x_{i+1}\ \cdots\ x_m]$, where $[x_1\ x_2\ \cdots\ x_m]^{\mathrm{T}}$ are the coordinates of the point $\mathbf{r}$ in the $\mathcal{V}_m$-basis. In both cases, the $(i+1)$-st place is on the upper side diagonal. Expansion along the last row results in

$$\operatorname{area}\widehat{\mathbf{T}}^i_m = \frac{(-1)^m}{m!}\begin{vmatrix} \sqrt{2/1} & 0 & 0 & \cdots & 0 \\ 1/\sqrt{1\cdot 2} & \sqrt{3/2} & 0 & & 0 \\ \vdots & & & \ddots & \\ x_1 & \cdots & x_i & \cdots & x_m \\ & & & \ddots & \\ 1/\sqrt{1\cdot 2} & 1/\sqrt{2\cdot 3} & 1/\sqrt{3\cdot 4} & \cdots & \sqrt{(m+1)/m} \end{vmatrix}. \tag{7}$$

Since $[x_1\ x_2\ \cdots\ x_m]^{\mathrm{T}} = \mathbf{Q}_m^{-1}\,[\xi_1\ \xi_2\ \cdots\ \xi_m]^{\mathrm{T}}$, one gets

$$x_i = \sqrt{\frac{i+1}{i}}\xi_i + \frac{\xi_{i+1} + \ldots + \xi_m}{\sqrt{i(i+1)}}.$$

The value of the determinant in (7) is $\sqrt{m+1}\,\xi_i$, i.e.,

$$\operatorname{area}\widehat{\mathbf{T}}^i_m = \frac{(-1)^m}{m!}\sqrt{m+1}\,\xi_i.$$

Therefore

$$\rho_i = \left(\operatorname{area}\widehat{\mathbf{T}}^i_m\right) / \left(\operatorname{area}\widehat{\mathbf{T}}_m\right) = \xi_i, \quad i = 1,2,\ldots,m.$$

Now, let $\mathbf{r} = [\rho_1\ \rho_2\ \cdots\ \rho_{m+1}]^{\mathrm{T}} = [\xi_1\ \xi_2\ \cdots\ \xi_{m+1}]^{\mathrm{T}}$. Since $1 = \sum_{i=1}^{m+1}\rho_i = \sum_{i=1}^{m+1}\xi_i$, the point $\mathbf{r} = \sum_{i=1}^{m+1}\xi_i\mathbf{e}_i$ belongs to $\operatorname{affi}\{\mathbf{e}_1,\mathbf{e}_2,\ldots,\mathbf{e}_{m+1}\}$. □

So it can be seen that the matrices $\mathbf{Q}_m$ and $\mathbf{Q}_m^{-1}$ effect the transformation between the areal coordinate system defined by the standard simplex $\widehat{\mathbf{T}}_m$ and the $\mathcal{V}_m$-basis in V^m.

3 Changing AIFS to IFS and Back

We note that an affine transformation of the hyperplane V^m (isomorphic to $\mathbb{R}^m$) is uniquely determined by a pair of nondegenerate simplices; it is convenient to choose the standard simplex, represented by its vertices $\mathbf{T}_m$, as the original, and some arbitrary simplex, represented by the vertices $\mathbf{T}'_m$, as its image. Note that this transformation is affine in the rectangular basis $\mathcal{V}_m$, while it is linear in the areal basis.

– *Orthonormal coordinates*: The simplex $\widehat{\mathbf{T}}_m$ is given by its vertices in the $\mathcal{V}_m$-basis, $\{\mathbf{t}_1,\mathbf{t}_2,\ldots,\mathbf{t}_m\}$, or by the $m\times(m+1)$ matrix $\mathbf{T}_m = [\mathbf{t}_1\ \mathbf{t}_2\ \cdots\ \mathbf{t}_m]^{\mathrm{T}}$, and the affine mapping of the hyperplane V^m is given by any nondegenerate simplex $\widehat{\mathbf{T}}'_m$ defined by the new $m \times (m+1)$ matrix of vertices, $\mathbf{T}'_m = [\mathbf{t}'_1\ \mathbf{t}'_2\ \cdots\ \mathbf{t}'_m]^{\mathrm{T}}$. The affine mapping $\mathcal{A} : \mathbf{T}_m \to \mathbf{T}'_m$ then is defined by the pair $(\mathbf{A}_m, \mathbf{b}_m)$, where $\mathbf{A}_m$ is a regular square $m \times m$ matrix and $\mathbf{b}_m$ is an m-dimensional vector of translation. So, the number of parameters involved in $\mathcal{A}$ is $m^2 + m = m(m+1)$.

– *Areal coordinates*: Since $\mathbf{T}_m$ is represented by the $(m+1)$-unit matrix $\mathbf{E}_{m+1}$, and $\mathbf{T}'_m$ by a row-stochastic square matrix $\mathbf{S}_{m+1}$ of order $(m+1)$, an affine transformation of the hyperplane V^m, $\mathcal{L} : \mathbf{T}_m \to \mathbf{T}'_m$, is given by $\mathcal{L}(\mathbf{e}_i) = \mathbf{S}^T_{m+1}\mathbf{e}_i = \mathbf{e}'_i$, $i = 1, 2, \ldots, m+1$, where $\mathbf{e}_i$ and $\mathbf{e}'_i$ are the areal coordinates of the vertices of $\mathbf{T}_m$ and $\mathbf{T}'_m$ w.r.t. $\mathbf{T}_m$. This simply means that any regular row-stochastic square matrix $\mathbf{S}_{m+1} = [s_{ij}]^{m+1}_{i,j=1}$ defines an affine transformation of V^m. Since the last column of $\mathbf{S}_{m+1}$ is a linear combination of the previous m columns, the number of parameters that characterize the affine mapping $\mathcal{L}$ is $m(m+1)$. By the same reason, instead of $\mathbf{S}_{m+1}$, its truncated version $\overline{\mathbf{S}}_{m+1}$ (i.e., $\mathbf{S}_{m+1}$ with the last column dropped) will be used.

Now, there are two problems: First, given $\mathbf{S}_{m+1}$ (which means $\mathcal{L}$), find the pair $(\mathbf{A}_m, \mathbf{b}_m)$ or the affine mapping $\mathcal{A}$. The second is the inverse, given $\mathcal{A}$ by $(\mathbf{A}_m, \mathbf{b}_m)$, find $\mathcal{L}$, i.e. $\mathbf{S}_{m+1}$.

Theorem 2. *Let the affine transformation $\mathcal{L} : \mathbf{T}_m \to \mathbf{T}'_m$ be given by a row-stochastic matrix $\mathbf{S}_{m+1}$. Then, the pair $(\mathbf{A}_m = [a_{ij}]_{m\times m}, \mathbf{b}_m = [b_i]_m)$ performing the same affine transformation in $\mathcal{V}_m$-coordinates is given by*

$$\begin{bmatrix} \mathbf{A}_m^{\mathrm{T}} \\ \hline \mathbf{b}_m^{\mathrm{T}} \end{bmatrix} = \left[\overline{\mathbf{E}}_{m+1} \left(\mathbf{Q}_m^{-1}\right)^{\mathrm{T}} \middle| \mathbf{1}\right]^{-1} \overline{\mathbf{S}}_{m+1} \left(\mathbf{Q}_m^{-1}\right)^{\mathrm{T}}, \tag{8}$$

where $\overline{\mathbf{E}}_{m+1}$ and $\overline{\mathbf{S}}_{m+1}$ are the unit matrix $\mathbf{E}_{m+1}$ and the matrix $\mathbf{S}_{m+1}$ without their last columns, respectively, and $\mathbf{Q}_m^{-1}$ is given by (3). *The inverse transform reads*

$$\overline{\mathbf{S}}_{m+1} = \left[\overline{\mathbf{E}}_{m+1} \left(\mathbf{Q}_m^{-1}\right)^{\mathrm{T}} \middle| \mathbf{1}\right] \begin{bmatrix} \mathbf{A}_m^{\mathrm{T}} \\ \hline \mathbf{b}_m^{\mathrm{T}} \end{bmatrix} \mathbf{Q}_m^{\mathrm{T}}. \tag{9}$$

Proof. Let the square row-stochastic matrix $\mathbf{S}_{m+1}$ of order m+1 be given. As stated above, it defines an affine transformation of the hyperplane V^m, and maps $\mathbf{T}_m \to \mathbf{T}'_m$. The same mapping, acting in $\mathcal{V}_m$-coordinates, maps the vertices $\{\mathbf{t}_1, \mathbf{t}_2, \ldots, \mathbf{t}_m\}$ of $\widehat{\mathbf{T}}_m$ into the vertices $\{\mathbf{t}'_1, \mathbf{t}'_2, \ldots, \mathbf{t}'_m\}$ of $\widehat{\mathbf{T}}'_m$ by $\mathbf{t}'_i = A_m\mathbf{t}_i + \mathbf{b}_m$, $1 \le i \le m+1$. This leads to the system

$$\begin{bmatrix} t_{11} & \cdots & t_{1m} & 1 \\ \vdots & & \vdots & \vdots \\ t_{m+1,1} & \cdots & t_{m+1,m} & 1 \end{bmatrix} \begin{bmatrix} a_{11} & \cdots & a_{1m} \\ \vdots & & \vdots \\ a_{1,m} & \cdots & a_{m,m} \\ b_1 & \cdots & b_m \end{bmatrix} = \begin{bmatrix} t'_{11} & \cdots & t'_{1m} \\ \vdots & & \vdots \\ t'_{m+1,1} & \cdots & t'_{m+1,m} \end{bmatrix}, \tag{10}$$

where the sizes of the matrices are $(m+1)\times(m+1)$, $(m{+}1){\times}m$ and $(m{+}1){\times}m$, respectively. More concisely, the system (10) can be written as

$$[\mathbf{T}_m \,|\, \mathbf{1}] \cdot \left[\begin{array}{c}\mathbf{A}_m^{\mathrm{T}} \\ \hline \mathbf{b}_m^{\mathrm{T}}\end{array}\right] = \mathbf{T}'_m. \tag{11}$$

The matrix $[\mathbf{T}_m \,|\, \mathbf{1}]$ is invertible and $\mathbf{T}_m = \left(\mathbf{Q}_m^{-1}\right)^{\mathrm{T}} \overline{\mathbf{E}}_{m+1}$ (see the remark after Lemma 3); analogously, $\mathbf{T}'_m = \overline{\mathbf{S}}_{m+1}\left(\mathbf{Q}_m^{-1}\right)^{\mathrm{T}}$. Now, (11) gives

$$\left[\begin{array}{c}\mathbf{A}_m^{\mathrm{T}} \\ \hline \mathbf{b}_m^{\mathrm{T}}\end{array}\right] = [\mathbf{T}_m \,|\, \mathbf{1}]^{-1}\,\mathbf{T}'_m = \left[\overline{\mathbf{E}}_{m+1}\left(\mathbf{Q}_m^{-1}\right)^{\mathrm{T}} \,|\, \mathbf{1}\right]^{-1} \overline{\mathbf{S}}_{m+1}\left(\mathbf{Q}_m^{-1}\right)^{\mathrm{T}},$$

which is (8), the formula that calculates $\mathbf{A}_m$ and $\mathbf{b}_m$, when $\mathbf{S}_{m+1}$ is given. The converse (9) immediately follows. □

Remark 2. The equivalent forms of (8) and (9) are

$$\left[\begin{array}{c|c}\mathbf{A}_m^{\mathrm{T}} & \mathbf{0} \\ \hline \mathbf{b}_m^{\mathrm{T}} & 1\end{array}\right] = \left[\begin{array}{c|c}\mathbf{Q}_m^{\mathrm{T}} & \mathbf{Q}_m^{\mathrm{T}}(-\mathbf{1}) \\ \hline \mathbf{0}^{\mathrm{T}} & 1\end{array}\right] \left[\overline{\mathbf{S}}_{m+1} \,|\, \mathbf{1}\right] \left[\begin{array}{c|c}\left(\mathbf{Q}_m^{-1}\right)^{\mathrm{T}} & \mathbf{0} \\ \hline \mathbf{0}^{\mathrm{T}} & 1\end{array}\right], \tag{12}$$

and

$$\left[\overline{\mathbf{S}}_{m+1} \,|\, \mathbf{1}\right] = \left[\begin{array}{c|c}\left(\mathbf{Q}_m^{-1}\right)^{\mathrm{T}} & \mathbf{1} \\ \hline \mathbf{0}^{\mathrm{T}} & 1\end{array}\right] \left[\begin{array}{c|c}\mathbf{A}_m^{\mathrm{T}} & \mathbf{0} \\ \hline \mathbf{b}_m^{\mathrm{T}} & 1\end{array}\right] \left[\begin{array}{c|c}\mathbf{Q}_m^{\mathrm{T}} & \mathbf{0} \\ \hline \mathbf{0}^{\mathrm{T}} & 1\end{array}\right].$$

Theorem 3. *One eigenvalue of the matrix* $\mathbf{S}_{m+1}$ *is* 1*, the other* m *eigenvalues coincide with the eigenvalues of* $\mathbf{A}_m$*. In other words,* $\operatorname{sp}\mathbf{S}_{m+1} = \operatorname{sp}\mathbf{A}_m \cup \{1\}$.

Proof. Using the fact $\det(\lambda\mathbf{E}_m - \mathbf{BCD}) = \det\mathbf{B}\cdot\det(\lambda\mathbf{B}^{-1}\mathbf{D}^{-1} - \mathbf{C})\cdot\det\mathbf{D}$, (where $\mathbf{B},\mathbf{C},\mathbf{D}$ are arbitrary square matrices of order $m \in N$, such that $\mathbf{B}$ and $\mathbf{D}$ are nonsingular), and using (12), we can obtain the characteristic polynomial of the matrix

$$\left[\begin{array}{c|c}\mathbf{A}_m^{\mathrm{T}} & \mathbf{0} \\ \hline \mathbf{b}_m^{\mathrm{T}} & 1\end{array}\right],$$

as follows:

$$\begin{aligned}
&\det\left(\lambda\mathbf{E}_{m+1} - \left[\begin{array}{c|c}\mathbf{A}_m^{\mathrm{T}} & \mathbf{0} \\ \hline \mathbf{b}_m^{\mathrm{T}} & 1\end{array}\right]\right) \\
&= \det\left(\lambda\mathbf{E}_{m+1} - \left[\begin{array}{c|c}\mathbf{Q}_m^{\mathrm{T}} & \mathbf{Q}_m^{\mathrm{T}}(-\mathbf{1}) \\ \hline \mathbf{0}^{\mathrm{T}} & 1\end{array}\right] \left[\overline{\mathbf{S}}_{m+1} \,|\, \mathbf{1}\right] \left[\begin{array}{c|c}\left(\mathbf{Q}_m^{-1}\right)^{\mathrm{T}} & \mathbf{0} \\ \hline \mathbf{0}^{\mathrm{T}} & 1\end{array}\right]\right) \\
&= \det\mathbf{Q}_m \cdot \det\left(\lambda \left[\begin{array}{c|c}(\mathbf{Q}_m^{-1})^{\mathrm{T}} & \mathbf{1} \\ \hline \mathbf{0}^{\mathrm{T}} & 1\end{array}\right] \left[\begin{array}{c|c}\mathbf{Q}_m^{\mathrm{T}} & \mathbf{0} \\ \hline \mathbf{0}^{\mathrm{T}} & 1\end{array}\right] - \left[\overline{\mathbf{S}}_{m+1} \,|\, \mathbf{1}\right]\right) \cdot \det\mathbf{Q}_m^{-1}
\end{aligned}$$

$$
\begin{aligned}
&= \det\left(\lambda \left[\begin{array}{c|c} \mathbf{E}_m & \mathbf{1} \\ \hline \mathbf{0}^{\mathrm{T}} & 1 \end{array}\right] - \left[\overline{\mathbf{S}}_{m+1} \,|\, \mathbf{1}\right]\right) \\
&= \begin{vmatrix} \lambda - s_{11} & -s_{12} & \dots & -s_{1m} & \lambda - 1 \\ -s_{21} & \lambda - s_{22} & \dots & -s_{2m} & \lambda - 1 \\ & & \ddots & & \\ -s_{m+1,1} & -s_{m+1,2} & \dots & -s_{m+1,m} & \lambda - 1 \end{vmatrix} && (13) \\
&= \det\left(\lambda \mathbf{E}_{m+1} - \mathbf{S}_{m+1}\right). && (14)
\end{aligned}
$$

If we multiply the first m columns of the determinant (13) by (-1) and add to the last column, we obtain (14). On the contrary,

$$
\det\left(\lambda \mathbf{E}_{m+1} - \left[\begin{array}{c|c} \mathbf{A}_m^{\mathrm{T}} & \mathbf{0} \\ \hline \mathbf{b}_m^{\mathrm{T}} & 1 \end{array}\right]\right) = (\lambda - 1)\det\left(\lambda \mathbf{E}_m - \mathbf{A}_m\right). \quad \square
$$

Theorem 4. *The determinants of* $\mathbf{A}_m$ *and* $\mathbf{S}_{m+1}$ *have the same value,*

$$
\det \mathbf{A}_m = \det \mathbf{S}_{m+1}. \tag{15}
$$

Proof. By the equality (12), one gets

$$
\det\left[\begin{array}{c|c} \mathbf{A}_m^{\mathrm{T}} & \mathbf{0} \\ \hline \mathbf{b}_m^{\mathrm{T}} & 1 \end{array}\right] = \det\left[\begin{array}{c|c} \mathbf{Q}_m^{\mathrm{T}} & \mathbf{Q}_m^{\mathrm{T}}(-\mathbf{1}) \\ \hline \mathbf{0}^{\mathrm{T}} & 1 \end{array}\right] \cdot \det\left[\overline{\mathbf{S}}_{m+1} \,|\, \mathbf{1}\right] \cdot \det\left[\begin{array}{c|c} \left(\mathbf{Q}_m^{-1}\right)^{\mathrm{T}} & \mathbf{0} \\ \hline \mathbf{0}^{\mathrm{T}} & 1 \end{array}\right]
$$

$$
= \det \mathbf{Q}_m \cdot \begin{vmatrix} s_{11} & s_{12} & \dots & s_{1m} & 1 \\ s_{21} & s_{22} & \dots & s_{2m} & 1 \\ & & \ddots & & \\ s_{m+1,1} & s_{m+1,2} & \dots & s_{m+1,m} & 1 \end{vmatrix} \cdot \det \mathbf{Q}_m^{-1} = \begin{vmatrix} s_{11} & s_{12} & \dots & s_{1,m+1} \\ s_{21} & s_{22} & \dots & s_{2,m+1} \\ & & \ddots & \\ s_{m+1,1} & s_{m+1,2} & \dots & s_{m+1,m+1} \end{vmatrix}.
$$

The last determinant, $\det \mathbf{S}_{m+1}$ can be obtained from $\det\left[\overline{\mathbf{S}}_{m+1} \,|\, \mathbf{1}\right]$ by multiplying the first m columns of $\det\left[\overline{\mathbf{S}}_{m+1} \,|\, \mathbf{1}\right]$ by (-1) and adding to its last column. Therefore, (15) is valid. □

References

1. Barnsley, M.F.: Fractals Everywhere. Academic, San Diego (1993)
2. Iwata, Sh.: On the geometry of the n-dimensional simplex. Math. Mag., Vol. **35**, No 5, 273–277, Mathematical Association of America (1962)
3. Kocić, Lj.M., Simoncelli, A.C.: Shape predictable IFS representations. In: Emergent Nature, M.M. Novak (ed), 435–436, World Scientific, Singapore (2002)
4. Kocić, Lj.M., Simoncelli, A.C.: Cantor Dust by AIFS. FILOMAT (Nis) **15** 265–276 (2001)
5. Kocić, Lj.M., Simoncelli, A.C.: Stochastic approach to affine invariant IFS. In: Prague Stochastics'98 (Proc. 6th Prague Symp., Aug. 23–28, 1998, M. Hruskova, P. Lachout and J.A. Visek eds), Vol II, Charles Univ. and Academy of Sciences of Czech Republic, Union of Czech Mathematicians and Physicists, Prague (1998)

Positive Trigonometric Sums and Starlike Functions

Stamatis Koumandos

Dedicated to Gradimir V. Milovanović on the occasion of his 60-th anniversary

1 Introduction

Positive trigonometric sums appear in various branches of Mathematics and they have numerous and surprising applications. Over the years, there has been a persistent interest in the positivity of certain special trigonometric sums, partly due to their own beauty, but perhaps largely due to their applications in other areas and the fact that they turn out to be precursors of other families of inequalities for finite sums of classical orthogonal polynomials.

There is a close connection between positivity results for trigonometric sums and complex function theory. This interrelation has stimulated many results in both fields. The extent and variety of this interplay has been demonstrated in several instances in the past century.

Although these topics are considered classical, there has been a revived interest over the last few years, and some interesting new results have been obtained.

In this paper, we give a systematic account of some recent results on positive trigonometric sums motivated by, and applied to, specific problems of geometric function theory. Most of these results sharpen and extend some classical ones, seen from a different point of view. This work aims to present some recent and current investigations concerning generalizations and extensions of the celebrated Vietoris' inequalities for trigonometric sums. The new inequalities are used to solve some problems of geometric function theory concerning the partial sums of starlike functions. Furthermore, some new positivity results for sums of Gegenbauer polynomials are obtained as applications.

Stamatis Koumandos
Department of Mathematics and Statistics, University of Cyprus, P.O. Box 20537, 1678 Nicosia, CYPRUS, e-mail: skoumand@ucy.ac.cy

W. Gautschi et al. (eds.), *Approximation and Computation: In Honor of Gradimir V. Milovanović*, Springer Optimization and Its Applications 42,
DOI 10.1007/978-1-4419-6594-3_12,

In this paper, we survey some recent results, give new and simpler proofs, and provide some additional results and comments. We also present some current research and developments and give further generalizations and extensions of some previously established theorems. We also pose some demanding conjectures and discuss some open problems.

2 Positive Trigonometric Sums

From the beginning of the twentieth century, there has been interest in finding positive trigonometric sums and finding uses for them. In 1910, Fejér conjectured that

$$\sum_{k=1}^{n} \frac{\sin k\theta}{k} > 0 \quad \text{for all } n \in \mathbb{N} \text{ and } 0 < \theta < \pi. \tag{1}$$

This conjecture was proved a little later by D. Jackson [13] and independently by T. H. Gronwall [12]. After the publication of these proofs, inequality (1) attracted the attention of several mathematicians, who offered new and shorter proofs. It is worth mentioning that E. Landau [20] found a ten-line proof of (1) and this is the one appearing in Zygmund's book [33, p. 62] and also in the book [21, p. 306].

W. H. Young obtained in [32] an analog of (1) for cosine sums. He proved that

$$1+\sum_{k=1}^{n} \frac{\cos k\theta}{k} > 0 \quad \text{for all } n \in \mathbb{N} \text{ and } 0 < \theta < \pi. \tag{2}$$

Inequalities (1), (2) have led to several generalizations, extensions and applications of various kinds. We refer the reader to the monographs [3] and [21] for a historical account of related results and background information on the subject, and also to the research article [1] and the references given therein for some recent results on refinements of (1) and (2).

There is another positive cosine sum obtained by Turán [30] and used to prove that the positivity of the partial sums of a trigonometric series does not imply that it is a Fourier series of a function in $L^2([-\pi,\pi])$. The result is also given on pp. 248–249 of N. K. Bary's book [7] and it states that

$$1+\sum_{k=1}^{n} \beta_k \cos k\theta > 0 \quad \text{for all } n \in \mathbb{N} \text{ and } 0 < \theta < \pi, \tag{3}$$

where the coefficients β_k are defined by

$$\frac{1}{(1-z)^{1/2}} = 1+\sum_{k=1}^{\infty} \beta_k z^k, \quad |z|<1.$$

Of course, we have

$$\beta_k = \frac{(2k-1)!!}{(2k)!!} = \frac{1\cdot 3\cdot 5\cdots(2k-1)}{2\cdot 4\cdot 6\cdots 2k} = 2^{-2k}\binom{2k}{k}, \quad k=1,2,\ldots,$$

and β_k has order of magnitude $k^{-1/2}$ as opposed to the order of magnitude k^{-1} for the coefficients in (1) and (2). However, inequality (3) is not strong enough to imply (2), as the case $n=1$ reveals.

In 1958, L. Vietoris [31] proved a surprising and deep result, which provides a substantial improvement of (1), (2) and (3).

As this paper aims to highlight some recent extensions of Vietoris' Theorem, we present it in more detail in the next section.

2.1 Vietoris' Inequalities

Vietoris gave sufficient conditions on the coefficients of a general class of sine and cosine sums that ensure their simultaneous positivity in $(0,\pi)$. His result is the following.

Theorem 1. *Suppose that $a_0 \geq a_1 \geq \cdots \geq a_n \geq \cdots > 0$ and $2ka_{2k} \leq (2k-1)a_{2k-1}$, for all $k \geq 1$. Then for all positive integers n, we have*

$$\sum_{k=0}^{n} a_k \cos k\theta > 0, \quad 0<\theta<\pi, \tag{4}$$

$$\sum_{k=1}^{n} a_k \sin k\theta > 0, \quad 0<\theta<\pi. \tag{5}$$

Vietoris observed that (4) and (5) are equivalent to the corresponding inequalities for the specific case in which $a_k = \gamma_k$, where the sequence γ_k is defined as follows:

$$\gamma_0=\gamma_1=1 \quad \text{and} \quad \gamma_{2k}=\gamma_{2k+1}=\beta_k=\frac{1\cdot 3\cdot 5\cdots(2k-1)}{2\cdot 4\cdot 6\cdots 2k}, \; k=1,2,\ldots,$$

where β_k is as in (3), and γ_k is the extreme case of equality in the defining inequalities for the numbers a_k.

Taking $a_0=1$ and $a_k=1/k$, $k\geq 1$ in Theorem 1, we immediately obtain (1) and (2). It is readily seen that the sequence β_k does not satisfy the condition of the Theorem above. It is, however, the case that (3) is obtained by this Theorem. We recall that another consequence of Theorem 1 is the following.

Corollary 1. *Let $\beta_0=1$ and let the sequence β_k be as in* (3) *for $k\geq 1$. If $0\leq\rho\leq 1/2$ then for all positive integers n, we have*

$$\sum_{k=0}^{n} \beta_k \cos(2k+\rho)\theta > 0, \quad 0<\theta<\pi. \tag{6}$$

(See [6, p. 299]). Now, take $\rho=0$ in Corollary 1 to see that Vietoris' Theorem, in fact, implies (3).

Vietoris' Theorem is perfectly natural for the case of positive sine sums. Suppose that $a_0 \geq a_1 \geq \cdots \geq a_n \geq \cdots > 0$. The condition

$$\sum_{k=1}^{n}(-1)^{k-1}ka_k \geq 0 \quad \text{for all } n \geq 2,\ a_1 > 0, \tag{7}$$

is necessary for the validity of (5). Indeed, divide (5) by $\sin\theta$ and take the limit as $\theta \to \pi$ to obtain (7). This explains why there is no analogue of (3) for sine sums.

Observe next that for a positive sequence (a_k) the condition (7) is equivalent to

$$\sum_{k=1}^{n}((2k-1)a_{2k-1} - 2ka_{2k}) \geq 0 \quad \text{for all } n \geq 1. \tag{8}$$

Thus, if the sequence a_k is chosen so that for each n the above sum vanishes and $a_1 = 1$, $a_{2k+1} = a_{2k}$, $k = 0, 1, \ldots$, we obtain the extremal sequence

$$a_0 = a_1 = 1 \quad \text{and} \quad a_{2k} = a_{2k+1} = \frac{1\cdot 3\cdot 5\cdots(2k-1)}{2\cdot 4\cdot 6\cdots 2k}, \quad k = 1, 2, \ldots.$$

However, for any sequence a_k satisfying the condition of Theorem 1, inequality (8) clearly holds. All of this gives an explanation of where the coefficient condition in Vietoris' Theorem comes from.

R. Askey and J. Steinig [6] have given an alternate version of Vietoris' proof and have also performed a valuable service in drawing attention to this theorem and indicating several applications of it. More specifically, it can be used to obtain sharp estimates for the location of zeros of a class of trigonometric polynomials whose coefficients satisfy certain growth conditions. Theorem 1 has also some remarkable applications in problems dealing with positive quadrature methods. It is worth mentioning that Vietoris' result suggested some more general inequalities for sums of Jacobi polynomials as well as various new summation and transformation formulas for hypergeometric series. Details of these and some historical comments regarding Vietoris' Theorem are in [3, 4] and [2, p. 371].

Applying Theorem 1, S. Ruscheweyh obtained in [26] some coefficient conditions which ensure that certain analytic functions are starlike univalent in the unit disk. Recently, S. Ruscheweyh and L. Salinas gave in [28] a beautiful interpretation of Vietoris' Theorem in geometric function theory.

In 1995, A. S. Belov proved in [8] that the condition (7) is also sufficient for the validity of (5), and this result is the last word on positivity of all partial sums of sine series with positive and decreasing coefficients. Accordingly, inequality (5) cannot hold for all n and $\theta \in (0, \pi)$ under a condition weaker than (7). Belov proved that the condition (7) implies also the positivity of the corresponding cosine sums (4) and therefore his result is stronger than Theorem 1. Condition (7), however, is not natural for cosine sums and this suggests that there should be a further sharpening of Vietoris' cosine inequality. This problem has been dealt with by several authors and a complete account and detailed references to relevant work is given in the recent paper [15].

The generalization of Vietoris' cosine inequality given in the next section is very natural and it is suggested by specific problems of geometric function theory dealing with partial sums of starlike functions. These will be presented in Sect. 4. Applications to the theory of starlike functions also suggest that there should be a natural extension of (6) which will lead to a generalization of (3) as well. This is given in Theorem 3 of Sect. 3.2.

3 Extensions of Vietoris' Cosine Inequality

The problem of sharpening Vietoris' cosine inequality can be formulated as follows: Suppose that (a_k), $k=0,1,\dots$ is a (decreasing) sequence of positive numbers. Determine the smallest $\alpha \in (0,1)$ such that the conditions

$$\frac{a_{2k+1}}{a_{2k}} \le 1,\ k=0,1,2\dots \quad \text{and} \quad \frac{a_{2k}}{a_{2k-1}} \le 1-\frac{\alpha}{k},\ k=1,2,\dots \tag{9}$$

imply that

$$\sum_{k=0}^{n} a_k \cos k\theta > 0 \quad \text{for all } n \in \mathbb{N} \text{ and } 0<\theta<\pi. \tag{10}$$

Let c_k be the extremal sequence for which we have equality in both relations in (9) for all k and $c_0 = 1$. We find that

$$c_{2k} = c_{2k+1} = \frac{(1-\alpha)_k}{k!}, \quad k=0,1,\dots,$$

where $(a)_k$ is the Pochhammer notation defined by

$$(a)_0 = 1, \quad (a)_k = a(a+1)\cdots(a+k-1) = \frac{\Gamma(k+a)}{\Gamma(a)},\ k\ge 1.$$

It is sufficient to determine the smallest $\alpha \in (0,1)$ for which (10) holds when $a_k = c_k$, $k=0,1,2,\dots$. Then for any sequence a_k satisfying (9), inequality (10) follows by a summation by parts, because the sequence a_k/c_k, $k=0,1,2\dots$ is clearly decreasing.

The following Theorem gives a complete solution to the problem above and it has been first established in [15]. Note that Vietoris' result corresponds to the case $\alpha = 1/2$, because

$$2^{-2k}\binom{2k}{k} = \frac{(1/2)_k}{k!}, \quad k=1,2,\dots.$$

Theorem 2. *Let* $0<\alpha<1$ *and*

$$c_0 = c_1 = 1, \quad c_{2k} = c_{2k+1} = \frac{(1-\alpha)_k}{k!},\ k=1,2,\dots.$$

For all positive integers n and $0<\theta<\pi$, we have $\sum_{k=0}^{n} c_k \cos k\theta > 0$, when $\alpha \geq \alpha_0$, where α_0 is the unique solution in $(0,1)$ of the equation

$$\int_0^{3\pi/2} \frac{\cos t}{t^\alpha}\, \mathrm{d}t = 0.$$

Furthermore, α_0 is best possible in the sense that for any $\alpha < \alpha_0$ we have

$$\lim_{n\to\infty} \min \left\{ \sum_{k=0}^{n} c_k \cos k\theta \,:\, \theta \in (0,\pi) \right\} = -\infty. \tag{11}$$

The numerical value of α_0 is $\alpha_0 = 0.3084437\ldots$ and this is the Littlewood–Salem–Izumi number. See the discussion in [33, Chap. V, Theorem 2.29] and see also [33, p. 379] for more information regarding the origin of α_0.

The case of Theorem 2 where n is odd is particularly interesting and important to applications, as will be demonstrated in Sect. 5. We give a new proof of this case in Sect. 3.2, which is shorter than the one given in [15]. This will be incorporated into a more general sharp result, which is presented in Sect. 3.2 (see Theorem 3). It is necessary to insert first some additional remarks concerning this case.

3.1 Remarks

1. We observe that for $a_{2k} = a_{2k+1}$, we have the following symmetric formula

$$\sin\frac{\theta}{2} \sum_{k=0}^{2n+1} a_k \cos k\theta = \cos\frac{\theta}{2} \sum_{k=1}^{2n+1} a_k \sin k(\pi - \theta).$$

This implies that the inequalities

$$\sum_{k=0}^{2n+1} a_k \cos k\theta > 0,\ \ 0<\theta<\pi \quad \text{and} \quad \sum_{k=1}^{2n+1} a_k \sin k\theta > 0,\ \ 0<\theta<\pi$$

are equivalent.

Consequently, if the sequence c_k is as in Theorem 2, then for all positive integers n and $0<\theta<\pi$, we have

$$\sum_{k=1}^{2n+1} c_k \sin k\theta > 0$$

if and only if $\alpha \geq \alpha_0$, while

$$\sum_{k=1}^{2n} c_k \sin k\theta > 0,$$

precisely when $\alpha \geq 1/2$, in view of the condition (8) and Vietoris' result (5).

2. Let the sequence c_k be as in Theorem 2. Relation (11) says that the corresponding cosine sums are unbounded below when $\alpha < \alpha_0$. Here we give another proof of this result which simplifies the argument given in [15, p.10].

We observe that

$$\sum_{k=0}^{2n+1} c_k \cos k\theta = 2\cos\frac{\theta}{2} \sum_{k=0}^{n} \frac{(1-\alpha)_k}{k!} \cos\left(2k+\frac{1}{2}\right)\theta. \tag{12}$$

Using the limit formula

$$\lim_{n\to\infty} \left(\frac{\theta}{n}\right)^{1-\alpha} \sum_{k=0}^{n} \frac{(1-\alpha)_k}{k!} \cos\left[\left(2k+\frac{1}{2}\right)\frac{\theta}{2n}\right] = \frac{1}{\Gamma(1-\alpha)} \int_0^\theta \frac{\cos t}{t^\alpha}\,\mathrm{d}t \tag{13}$$

(cf. [14, p. 85]), we see that a necessary condition for the positivity of the sums in (12) is

$$\int_0^\theta \frac{\cos t}{t^\alpha}\,\mathrm{d}t \geq 0 \quad \text{for all} \quad \theta > 0.$$

It is known that (cf. [33, Chap. V, Theorem 2.29]) the above integral is negative for $\theta = 3\pi/2$ and $\alpha < \alpha_0$. By (13) we obtain

$$\lim_{n\to\infty} \sum_{k=0}^{n} \frac{(1-\alpha)_k}{k!} \cos\left[\left(2k+\frac{1}{2}\right)\frac{3\pi}{4n}\right] = -\infty$$

for all $\alpha < \alpha_0$, which in combination with (12) completes the proof of (11).

3. It follows from (12) and the discussion above that the case of Theorem 2 where n is odd is equivalent to the statement: For all positive integers n

$$\sum_{k=0}^{n} \frac{(1-\alpha)_k}{k!} \cos\left(2k+\frac{1}{2}\right)\theta > 0, \quad 0 < \theta < \pi, \tag{14}$$

precisely when $1 > \alpha \geq \alpha_0$. Making the transformation $\theta \to \pi - \theta$, we see that (14) is equivalent to

$$\sum_{k=0}^{n} \frac{(1-\alpha)_k}{k!} \sin\left(2k+\frac{1}{2}\right)\theta > 0, \quad 0 < \theta < \pi, \tag{15}$$

and this inequality holds if and only if $1 > \alpha \geq \alpha_0$.

In the next section, we give generalizations and extensions of both (14) and (15) which contain a generalization of Corollary 1 as well.

3.2 Further Generalizations and Related Results

We first prove the following Theorem.

Theorem 3. *Let $d_k := (1-\alpha)_k/k!$ and $\rho \in [0,1]$. For all $0 \leq \rho \leq 1/2$, we have*

$$\sum_{k=0}^{n} d_k \cos(2k+\rho)\theta > 0 \quad \textit{for all } n \in \mathbb{N} \textit{ and } 0 < \theta < \pi \tag{16}$$

if and only if $1 > \alpha \geq \alpha_0$. *When* $1/2 < \rho \leq 1$, *inequality* (16) *fails to hold for appropriate* n *and* θ, *for all* $\alpha \in [0,1)$.

The case $\rho = 1/2$ of this Theorem is inequality (14). Here we give a new proof of this result which is simpler than the one given in [15]. The proof will be broken down into a sequence of lemmas which will demonstrate the relative difficulty in estimating trigonometric sums of this type.

Lemma 1. *Let* (a_k), $k = 0,1,\ldots$ *be any decreasing sequence of positive real numbers such that* $a_0 > a_N$. *Then, we have*

$$\sum_{k=0}^{N} a_k \cos k\theta > 0 \quad \textit{for} \quad 0 \leq \theta \leq \frac{\pi}{N}.$$

Proof. This is given in [31]. See also [6]. □

Lemma 2. *Let* (b_k), $k = 0,1,\ldots$ *be any decreasing sequence of positive real numbers. Then for all positive integers* n, m *such that* $n \geq m \geq 1$, *and all real numbers* θ, *we have*

$$2\sin\theta \sum_{k=0}^{n} b_k \cos\left(2k+\frac{1}{2}\right)\theta \geq 2\sin\theta \sum_{k=0}^{m-1} b_k \cos\left(2k+\frac{1}{2}\right)\theta - b_m - b_m \sin\left(2m-\frac{1}{2}\right)\theta.$$

Proof. This follows by a summation by parts. □

Let the sequence d_k, $k = 0,1,2\ldots$, be as in Theorem 3. Then, for trigonometric sums with coefficients d_k, we have the following general expression (cf. [19, (3.8)]).

$$\begin{aligned} &2^{1-\alpha}\theta^{-\alpha}\Gamma(1-\alpha)\sum_{k=0}^{n} d_k \,\mathrm{e}^{2\mathrm{i}k\theta} \qquad (17)\\ &= 2^{1-\alpha}\theta^{-\alpha}\Gamma(1-\alpha)F(\theta) + g_n(\theta) - \sigma_n(\theta) - \tau_n(\theta)\\ &+ 2^{1-\alpha}\theta^{-\alpha}\Gamma(1-\alpha)\sum_{k=n+1}^{\infty} \Delta_k \,\mathrm{e}^{2\mathrm{i}k\theta}, \end{aligned}$$

where

$$\begin{aligned} F(\theta) &:= \sum_{k=0}^{\infty} d_k \mathrm{e}^{2\mathrm{i}k\theta} - \frac{\theta}{\sin\theta}\frac{\mathrm{e}^{\mathrm{i}(1-\alpha)\pi/2}}{(2\theta)^{1-\alpha}},\\ g_n(\theta) &:= \frac{1}{\sin\theta}\int_0^{(2n+1)\theta} \frac{\mathrm{e}^{\mathrm{i}t}}{t^{\alpha}}\,\mathrm{d}t,\\ \sigma_n(\theta) &:= 2^{1-\alpha}\theta^{-\alpha}\frac{\theta}{\sin\theta}\sum_{k=n+1}^{\infty} A_k(\theta),\\ \tau_n(\theta) &:= 2^{1-\alpha}\theta^{-\alpha}\frac{\theta}{\sin\theta}\sum_{k=n+1}^{\infty} B_k(\theta), \end{aligned}$$

$$A_k(\theta) := \int_0^{1/2} (L(k,t) - M(k,t))\mathrm{e}^{2\mathrm{i}\theta(k-t)}\,\mathrm{d}t,$$
$$B_k(\theta) := \int_0^{1/2} 2\mathrm{i}\sin(2\theta t)L(k,t)\mathrm{e}^{2\mathrm{i}k\theta}\,\mathrm{d}t,$$
$$L(k,t) := \int_0^t \frac{\alpha}{(k+s)^{\alpha+1}}\,\mathrm{d}s,$$
$$M(k,t) := \int_0^t \frac{\alpha}{(k-t+s)^{\alpha+1}}\,\mathrm{d}s,$$
$$\Delta_k := \frac{1}{\Gamma(1-\alpha)k^{\alpha}} - \frac{(1-\alpha)_k}{k!}, \quad k = 1,2,\dots.$$

Next we give some estimates of the terms on the right-hand side of (17) involving the functions defined above.

Lemma 3. *For* $0 < \theta \le \theta_0 < \pi$*, we have*

$$2^{1-\alpha}\theta^{-\alpha}\,\mathrm{Re}\left[F(\theta)\mathrm{e}^{\mathrm{i}\theta/2}\right] \ge (1-\alpha)\cos\left(\frac{\alpha\pi}{2}\right) - \Lambda(\theta_0), \tag{18}$$

where

$$\Lambda(\theta) := \frac{1}{\sin\theta}\left[1 - \left(\frac{\sin\theta}{\theta}\right)^{\alpha}\right].$$

Proof. Similar to the proof of Proposition 2 of [19]. □

Lemma 4. *Let* $\mu_n(\theta) := \mathrm{Re}\left[g_n(\theta)\mathrm{e}^{\mathrm{i}\theta/2}\right]$ *and* $\alpha = \alpha_0$*. For* $0 < \theta \le \theta_0 < \pi$*, we have the estimate*

$$\mu_n(\theta) \ge \frac{1}{\sin\theta_0}\int_0^{(3\pi-\theta_0)/2} \frac{\cos(t+\theta_0/2)}{t^{\alpha_0}}\,\mathrm{d}t. \tag{19}$$

Proof. Clearly

$$\mu_n(\theta) = \frac{1}{2\sin(\theta/2)}\int_0^{(2n+1)\theta} \frac{\cos t}{t^{\alpha}}\,\mathrm{d}t - \frac{1}{2\cos(\theta/2)}\int_0^{(2n+1)\theta} \frac{\sin t}{t^{\alpha}}\,\mathrm{d}t.$$

It is known that for $\alpha \ge \alpha_0$, we have

$$\int_0^x \frac{\cos t}{t^{\alpha}}\,\mathrm{d}t \ge 0 \quad \text{for all } x > 0,$$

while for all $0 < \alpha < 1$, we have

$$\int_0^x \frac{\sin t}{t^{\alpha}}\,\mathrm{d}t > 0 \quad \text{for all } x > 0.$$

Thus, when $\alpha = \alpha_0$, it follows from the above that

$$\mu_n(\theta) \ge \frac{1}{\sin\theta_0}\int_0^{(2n+1)\theta} \frac{\cos(t+\theta_0/2)}{t^{\alpha_0}}\,\mathrm{d}t \ge \frac{1}{\sin\theta_0}\int_0^{(3\pi-\theta_0)/2} \frac{\cos(t+\theta_0/2)}{t^{\alpha_0}}\,\mathrm{d}t,$$

which is the desired result. □

Remark 1. Note that the right-hand side of (19) is negative for all $0 < \theta_0 < \pi$. Since $\alpha_0 = 0.3084437\ldots$, we obtain from (19)

$$\mu_n(\theta) \geq -0.61137\ldots, \quad 0 < \theta < \pi/22.$$

This slightly improves the lower bound -0.62 obtained in [15, Lemma 4] in a different way. Taking $\theta_0 = \pi/18$ in (19) we find that

$$\mu_n(\theta) \geq -0.614626\ldots, \quad 0 < \theta < \pi/18. \tag{20}$$

The numerical values above are obtained using Maple 12.

Lemma 5. *For* $0 < \theta < \pi$ *and* $0 < \alpha < 1$, *we have*

$$\left| \sum_{k=n+1}^{\infty} A_k(\theta) \right| < \frac{\alpha}{8} \frac{1}{n^{\alpha+1}}, \qquad \left| \sum_{k=n+1}^{\infty} B_k(\theta) \right| < \frac{\theta}{\sin\theta} \frac{\alpha}{6} \frac{1}{n^{\alpha+1}}.$$

Proof. Similar to the proof of Lemma 1 of [18]. See also [15, Lemma 1] and [19, Proposition 1]. □

Using the estimates of Lemma 5, we easily obtain the following.

Lemma 6. *For* $\pi/(2n+1) \leq \theta < \pi$ *and* $0 < \alpha < 1$, *we have*

$$\mathrm{Re}\left[\sigma_n(\theta)e^{i\theta/2}\right] \leq \frac{\alpha}{4} \frac{\theta}{\sin\theta} \frac{1}{n}, \quad \mathrm{Re}\left[\tau_n(\theta)e^{i\theta/2}\right] \leq \frac{\alpha}{3} \left(\frac{\theta}{\sin\theta}\right)^2 \frac{1}{n}.$$

Lemma 7. *For* $\pi/(2n+1) \leq \theta < \pi/2$ *and* $0 < \alpha \leq 2/3$, *we have*

$$\left| \sum_{k=n+1}^{\infty} \Delta_k e^{2ik\theta} \right| \leq \frac{\alpha(1-\alpha)}{2} \frac{1}{\Gamma(1-\alpha)} \frac{1}{(n+1)^{\alpha}}. \tag{21}$$

Proof. This estimate has been obtained in various steps of increasing generality in [19, Proposition 4], [16, Proposition 1] and [17, Theorem 5]. □

We are now in a position to give a proof of Theorem 3.

Proof. We first prove (16) for $\rho = 1/2$. It is sufficient to prove this inequality for $\alpha = \alpha_0$. The case $\alpha > \alpha_0$ follows by a summation by parts. Let

$$U_n(\theta) := \sum_{k=0}^{n} d_k \cos\left(2k + \frac{1}{2}\right)\theta.$$

It follows from (12) and Lemma 1 that for all positive integers n, we have

$$U_n(\theta) > 0 \quad \text{for } 0 \leq \theta \leq \frac{\pi}{2n+1}. \tag{22}$$

Then we define

$$V_m(\theta) := 2\sin\theta \sum_{k=0}^{m-1} d_k \cos\left(2k+\frac{1}{2}\right)\theta - d_m - d_m \sin\left(2m-\frac{1}{2}\right)\theta.$$

It is easy to see that $V_m(\theta)$ is a polynomial in $t=\sin(\theta/2)$. By elementary computations we find that for $m=1,2,\ldots 9$ the polynomial $V_m(\theta)$ has a unique root θ_m in $(0,\pi)$ and that

$$V_m(\theta) > 0, \quad \theta_m < \theta < \pi. \tag{23}$$

We also find that

$$0 < \theta_m < \frac{\pi}{2m+1}, \quad m=1,2,\ldots,8; \quad \frac{\pi}{19} < \theta_9 = 0.1668\ldots < \frac{\pi}{18}, \tag{24}$$

see also [15, Proposition 4]. Note that in these computations we have taken $\alpha=\alpha_0$. By Lemma 2, we also have that for all $n \geq m \geq 1$

$$2\sin\theta U_n(\theta) \geq V_m(\theta), \quad 0<\theta<\pi.$$

Combining this with (22), (23), and (24), we infer that

$$U_n(\theta)>0,\ 0<\theta<\pi,\ n=1,2,\ldots,8; \quad U_n(\theta)>0,\ \frac{\pi}{18}\leq\theta<\pi,\ n\geq 9.$$

To prove that $U_n(\theta)$ is positive in the remaining cases $n\geq 9$ and $\pi/(2n+1)<\theta<\pi/18$, we write

$$U_n(\theta) = \mathrm{Re}\left[\mathrm{e}^{\mathrm{i}\theta/2}\sum_{k=0}^{n} d_k \mathrm{e}^{2ik\theta}\right],$$

use the expression (17) and the estimates obtained in Lemmas 3–7. In particular, using (21), we find that for $\pi/(2n+1)\leq\theta<\pi/2$ we have

$$2^{1-\alpha}\theta^{-\alpha}\Gamma(1-\alpha)\,\mathrm{Re}\left[\mathrm{e}^{\mathrm{i}\theta/2}\sum_{k=n+1}^{\infty}\Delta_k \mathrm{e}^{2ik\theta}\right] > -\frac{\alpha(1-\alpha)}{\pi^{\alpha}}.$$

Finally, using this together with (18), (20), and Lemma 6, we conclude from (17) that, for $n\geq 9$ and $\pi/(2n+1)<\theta<\pi/18$,

$$\begin{aligned}2^{1-\alpha_0}\theta^{-\alpha_0}\Gamma(1-\alpha_0)U_n(\theta) &> \Gamma(1-\alpha_0)\left\{(1-\alpha_0)\cos\left(\frac{\alpha_0\pi}{2}\right)-\Lambda\left(\frac{\pi}{18}\right)\right\}\\ &-0.615-\frac{\alpha_0\pi}{648\sin(\pi/18)}-\frac{\alpha_0}{8748}\left(\frac{\pi}{\sin(\pi/18)}\right)^2-\frac{\alpha_0(1-\alpha_0)}{\pi^{\alpha_0}}>0.005,\end{aligned}$$

and this completes the proof of (16) for $\rho=1/2$.

Consider now $0\leq\rho<1/2$. Taking into account Remark 3 of Sect. 3.1 and the elementary identity

$$\cos(2k+\rho)\theta = \cos\left(2k+\frac{1}{2}\right)\theta\cos\left(\frac{1}{2}-\rho\right)\theta+\sin\left(2k+\frac{1}{2}\right)\theta\sin\left(\frac{1}{2}-\rho\right)\theta,$$

we deduce that inequality (16) is valid for $0 < \theta < \pi$ when $\alpha_0 \le \alpha < 1$. To prove that this is the best possible range of α for the validity of this inequality, we need an extended version of (13), namely

$$\lim_{n\to\infty}\left(\frac{\theta}{n}\right)^{1-\alpha}\sum_{k=0}^{n}\frac{(1-\alpha)_k}{k!}\cos\left[(2k+\rho)\frac{\theta}{2n}\right]=\frac{1}{\Gamma(1-\alpha)}\int_0^{\theta}\frac{\cos t}{t^{\alpha}}\,\mathrm{d}t, \tag{25}$$

which is obtained by a method similar to the one described in [14]. Then, as in Remark 2 of Sect. 3.1, we conclude that the sums in (16) are unbounded below for $\alpha < \alpha_0$.

Next consider the case $1/2 < \rho < 1$. We use a limit formula similar to the above, which is

$$\lim_{n\to\infty}\left(\frac{\theta}{n}\right)^{1-\alpha}\sum_{k=0}^{n}\frac{(1-\alpha)_k}{k!}\cos\left[(2k+\rho)\left(\pi-\frac{\theta}{2n}\right)\right]=\frac{1}{\Gamma(1-\alpha)}\int_0^{\theta}\frac{\cos(t-\rho\pi)}{t^{\alpha}}\mathrm{d}t.$$

Taking $\theta = (\rho - 1/2)\pi$, we see that the integral of the preceding line is negative for all $0 \le \alpha < 1$, therefore inequality (16) cannot hold for $1/2 < \rho < 1$, appropriate θ, n sufficiently large and any value of α in $[0,1)$.

Finally, when $\rho = 1, 0 \le \alpha < 1$, it is evident that all the sums in (16) are negative for $\theta = \pi$, therefore these sums must assume also negative values near π.

This completes the proof of Theorem 3. □

The result of Theorem 3 is best possible and it provides a substantial generalization of Corollary 1. It contains also the following remarkable special case.

Corollary 2. *For all positive integers n and all real θ, we have*

$$\sum_{k=0}^{n}\frac{(1-\alpha)_k}{k!}\cos k\theta > 0 \tag{26}$$

if and only if $\alpha_0 \le \alpha < 1$. The sums in (26) *are unbounded below when $\alpha < \alpha_0$.*

This result has been first established in [18] and it plays an important role in the context of positive sums of Gegenbauer polynomials and starlike functions. Details on this will be presented in the next sections. Notice also that the case $\alpha = 1/2$ of (26) is Turán's inequality (3).

Inequality (26) has been a precursor of several sharp results and inequalities obtained in our recent papers [14–19]. It has been also the starting point of the development of new techniques to obtain sharp results of this type and also the motivation for improvement and refinement of some older ideas.

It is of interest to note that inequality (26) can be obtained from Theorem 2 by simply observing that

$$\sum_{k=0}^{n}d_k\cos 2k\theta=\frac{1}{2}\left[\sum_{k=0}^{2n+1}c_k\cos k\theta+\sum_{k=0}^{2n+1}c_k\cos k(\pi-\theta)\right],$$

where

$$c_{2k} = c_{2k+1} = d_k = \frac{(1-\alpha)_k}{k!}, \quad k = 0,1,2,\ldots,$$

and the sharpness of the result with respect α can be proven using formula (25) for $\rho = 0$.

However, the method described in this section can be applied to give a direct proof of (26), which is shorter than the one given in [18]. Notice that this inequality admits a simple proof for $1 \le n \le 3$ and also for $n \ge 4$ and $\pi/2 \le \theta < \pi$ (cf. [18, p. 200]). The case $0 < \theta < \pi/n$ is settled by a direct application of Lemma 1. For the remaining cases, let $T_n(\theta) := \sum_{k=0}^{n} d_k \cos 2k\theta$, with d_k as above. We take the real part in representation (17) and keep in mind that we need to establish the positivity of $T_n(\theta)$ only for $\alpha = \alpha_0$ and $n \ge 4$, $\pi/(2n) \le \theta \le \pi/4$. We observe that in this case we have $\operatorname{Re} g_n(\theta) \ge 0$. The estimate obtained using Lemma 7 is unaffected. Then, an obvious modification of Lemmas 3 and 6 yields

$$2^{1-\alpha_0}\theta^{-\alpha_0}\Gamma(1-\alpha_0)T_n(\theta) > \Gamma(1-\alpha_0)\left\{(1-\alpha_0)\cos\left(\frac{(3\alpha_0+1)\pi}{8}\right) - \Lambda\left(\frac{\pi}{4}\right)\right\}$$
$$-\frac{\alpha_0\pi\sqrt{2}}{64} - \frac{\alpha_0\pi^2}{96} - \frac{\alpha_0(1-\alpha_0)}{\pi^{\alpha_0}} > 0.3,$$

and this completes the proof.

Given the equivalence of (14) with (15), it is natural to look for an analog of Theorem 3 for sine sums. It follows from (14), (15) and the elementary identity

$$\sin(2k+\rho)\theta = \sin\left(2k+\frac{1}{2}\right)\theta\cos\left(\rho-\frac{1}{2}\right)\theta + \cos\left(2k+\frac{1}{2}\right)\theta\sin\left(\rho-\frac{1}{2}\right)\theta,$$

that

$$\sum_{k=0}^{n} d_k \sin(2k+\rho)\theta > 0, \quad 0 < \theta < \pi, \tag{27}$$

for $1/2 \le \rho \le 1$, when $1 > \alpha \ge \alpha_0$, but this result is best possible with respect to α only when $\rho = 1/2$. For instance, when $\rho = 1$ inequality (27) holds for all $\alpha \in [0,1)$ and this follows from the nonnegativity of the classical Fejér kernel. Another sharp case of (27) has been recently obtained in [17]. This is the following.

Theorem 4. *Let $d_k := (1-\alpha)_k/k!$. For all positive integers n, we have*

$$\sum_{k=0}^{n} d_k \sin\left(2k+\frac{1}{4}\right)\theta > 0, \quad 0 < \theta < \pi, \tag{28}$$

precisely when $\alpha \ge \alpha_1$, where α_1 is the unique solution in $(0,1)$ of the equation

$$\int_0^{5\pi/4} \frac{\cos\left(t+\frac{\pi}{4}\right)}{t^\alpha}\,\mathrm{d}t = 0.$$

In this case, we have $\alpha_1 = 0.6144334\ldots$, *and the sums in* (28) *are unbounded below when* $\alpha < \alpha_1$.

In the light of the above results we have been led to the following conjecture which appears to be supported by numerical experimentation.

Conjecture 1. Let $\mu = 1-\alpha$. Write $d_k := (\mu)_k/k!$, $\mu \in (0,1]$. For $\rho \in (0,1]$ the inequality

$$\sum_{k=0}^{n} d_k \sin(2k+\rho)\theta > 0 \tag{29}$$

holds for all $n = 1,2,\ldots$ and $0 < \theta < \pi$, precisely when $0 < \mu \le \mu^*(\rho)$, where $\mu^*(\rho)$ is the unique solution in $(0,1]$ of the equation

$$\int_0^{(\rho+1)\pi} \sin(t-\rho\pi)t^{\mu-1}\,dt = 0.$$

A limiting case of the sums in (29) can be obtained using the asymptotic formula

$$\lim_{n\to\infty}\left(\frac{\theta}{n}\right)^{\mu}\sum_{k=0}^{n}\frac{(\mu)_k}{k!}\sin\left[(k+\rho)\left(\pi-\frac{\theta}{2n}\right)\right] = -\frac{1}{\Gamma(\mu)}\int_0^{\theta}\sin(t-\rho\pi)t^{\mu-1}\,dt. \tag{30}$$

Hence, a necessary condition for the validity of (29) is the nonpositivity of the integral in (30) for all θ, and in particular for $\theta = (\rho+1)\pi$:

$$I(\mu) := \int_0^{(\rho+1)\pi} \sin(t-\rho\pi)t^{\mu-1}\,dt \le 0.$$

It can be shown that the function $I(\mu)$ is strictly increasing on $(0,1)$ and since $I(0) = -\infty$ and $I(1) > 0$, we conclude that the equation $I(\mu) = 0$ has a unique solution in $(0,1)$, which is the function $\mu^*(\rho)$. Therefore, inequality (29) cannot hold for $\mu > \mu^*(\rho)$, $0 < \rho < 1$, appropriate θ and n sufficiently large.

Our methods of proving Theorems 3 and 4 are quite powerful and they can be adapted, in principle, to settle the above conjecture for other particular values of $\rho \in (0,1)$. The proof of this conjecture as a whole is, nevertheless, a very hard problem. To be able to make some additional progress with it, it seems necessary to obtain more information concerning the behavior of the implicitly defined function $\mu^*(\rho)$.

It is perhaps interesting to note that $|\mu^*(\rho) - \sin(\rho\pi/2)| < 0.02$, $\rho \in (0,1)$. Also, computation using Maple 12 yields $\mu^*(1/4) = \mu_1 = 1-\alpha_1 = 0.385566\ldots$, $\mu^*(1/2) = \mu_0 = 1-\alpha_0 = 0.691556\ldots$, $\mu^*(3/4) = 0.907689\ldots$. It is easy to verify that $\mu^*(1) = 1$.

Figure 1 shows a graph of the function $\mu^*(\rho)$, which suggests also that this function is strictly increasing and concave on $(0,1)$.

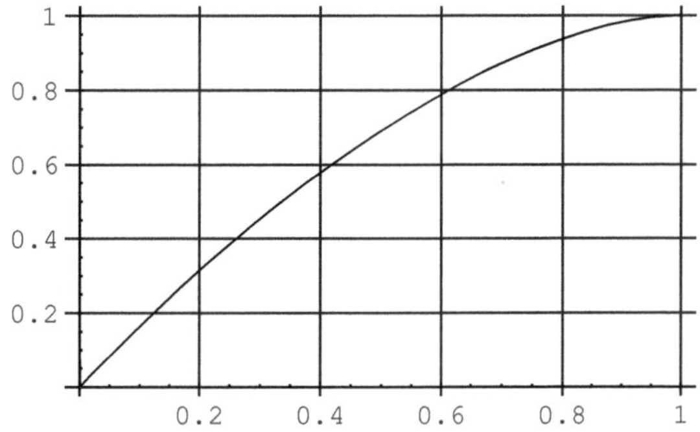

Fig. 1 The graph of $\mu^*(\rho)$

4 Starlike Functions

The sharp inequalities for trigonometric sums presented in the preceding section, as already mentioned, are motivated by specific problems dealing with properties of the partial sums of certain classes of analytic functions. We recall some definitions and basic facts about such functions.

Let $\mathbb{D} = \{z \in \mathbb{C} : |z| < 1\}$ be the unit disk in the complex plane $\mathbb{C}$ and $A(\mathbb{D})$ the space of analytic functions in $\mathbb{D}$. For $\lambda < 1$, let $\mathcal{S}_\lambda$ be the family of functions f starlike of order λ, that is, the family consisting of those functions $f \in A(\mathbb{D})$ satisfying $f(0) = f'(0) - 1 = 0$ and $\mathrm{Re}\,(zf'(z)/f(z)) > \lambda$ for all $z \in \mathbb{D}$. The class $\mathcal{S}_\lambda$ has been introduced by M. S. Robertson in [23] and since then it became the subject of systematic study by many researchers. Here we suppose also that $\lambda \geq 0$ and this implies that $\mathcal{S}_\lambda \subset \mathcal{S}$, where $\mathcal{S}$ is the usual class of normalized univalent functions in the unit disk.

It should be noted that the functions

$$g_\lambda(z) := \frac{z}{(1 - \mathrm{e}^{-it}z)^{2-2\lambda}}$$

belong to $\mathcal{S}_\lambda$ for all $t \in \mathbb{R}$ and they are, in fact, the most important members of this family. It is well known that $\mathcal{S}_\lambda$ is a compact set of the locally convex topological vector space $A(\mathbb{D})$ with respect to the topology given by uniform convergence on compact subsets of $\mathbb{D}$. The functions $g_\lambda(z)$ are the extreme points of the closed convex hull of the set $\mathcal{S}_\lambda$, see [9].

We denote by $s_n(f,z) = \sum_{k=0}^{n} a_k z^k$ the n-th partial sum of $f(z) = \sum_{k=0}^{\infty} a_k z^k$. In [24], the following theorem has been shown.

Theorem 5. *Let $f \in \mathcal{S}_\lambda$, $\lambda \geq 1/2$. Then we have $s_n(f/z,z) \neq 0$ for all $z \in \mathbb{D}$, $n \in \mathbb{N}$.*

Note that $s_n(f/z,z)$ are polynomials taking the value 1 in the origin. It is natural to seek information about the image domain $s_n(f/z,z)(\mathbb{D})$. Starting from the result of Theorem 5, we study the following problems.

Problems. Let $f \in \mathcal{S}_\lambda$, $1 > \lambda \geq 1/2$.

1. Determine the best possible range of λ for which $\mathrm{Re}\, s_n(f/z,z) > 0$ for all $z \in \overline{\mathbb{D}}$, $n \in \mathbb{N}$.
2. Given $\rho \in (0,1]$. Consider the more general question of determining the best possible range of λ such that $|\arg s_n(f/z,z)| \leq \rho\pi$ for all $z \in \mathbb{D}$, $n \in \mathbb{N}$.
3. Find generalizations and extensions of the results of 1 and 2 by considering suitable sequences of nonvanishing analytic functions in $\mathbb{D}$.

The next Theorem gives a complete solution to the first problem and has been established in [18].

Theorem 6. *For $f \in \mathcal{S}_\lambda$, we have $\mathrm{Re}\, s_n(f/z,z) > 0$ for all $n \in \mathbb{N}$ and $z \in \overline{\mathbb{D}}$ if and only if $\lambda_0 \leq \lambda < 1$, where λ_0 is the unique solution in $(1/2,1)$ of the equation*

$$\int_0^{3\pi/2} t^{1-2\lambda} \cos t \, \mathrm{d}t = 0.$$

The numerical value of λ_0 is $\lambda_0 = (1+\alpha_0)/2 = 0.654222\ldots$, where α_0 is as in Theorem 2.

The proof of this result relies on the following observations. Let $f \in \mathcal{S}_\lambda$. Then there exists a probability measure ω on $[0,2\pi)$ such that

$$f(z) = \int_0^{2\pi} \frac{z}{(1-\mathrm{e}^{-\mathrm{i}t}z)^{2-2\lambda}} \, \mathrm{d}\omega(t),$$

see [9, Theorem 3] and [25, p.51]. Suppose that $f(z) = z\sum_{k=0}^{\infty} a_k z^k \in \mathcal{S}_\lambda$. Then we have

$$a_k = \widehat{\omega}(k) \frac{(2-2\lambda)_k}{k!}, \quad k = 0,1,\ldots, \tag{31}$$

where $\widehat{\omega}(k)$ are the Fourier coefficients of the measure ω.
Since $s_n(f/z,z) = \sum_{k=0}^{n} a_k z^k$, we obtain

$$\mathrm{Re}\, s_n(f/z,\mathrm{e}^{\mathrm{i}\theta}) = \int_0^{2\pi} \sum_{k=0}^{n} \frac{(2-2\lambda)_k}{k!} \cos k(\theta - t) \, \mathrm{d}\omega(t),$$

and by the minimum principle for harmonic functions it is sufficient to prove the positivity of the integral above. But, the trigonometric sum in the integrand is positive precisely when $\lambda \geq \lambda_0$ and this is an immediate consequence of Corollary 2. It should be noted that Theorem 6 is sharp with respect to λ. Consider the extremal function $z/(1-z)^{2-2\lambda}$ of $\mathcal{S}_\lambda$ and observe that the conclusion of this Theorem fails to hold for this function when $\lambda < \lambda_0$, by invoking again Corollary 2.

Theorem 6 enables us to establish sharp positivity results for sums of Gegenbauer (ultraspherical) polynomials. These are given in the next section.

4.1 Positive Sums of Gegenbauer Polynomials

Let $C_k^\nu(x)$ be the Gegenbauer polynomial of degree k and order $\nu > 0$, defined by the generating function

$$G_\nu(z,x) := (1-2xz+z^2)^{-\nu} = \sum_{k=0}^{\infty} C_k^\nu(x) z^k, \quad x \in [-1,1].$$

It is easy to see that $zG_\nu(z,x) \in \mathcal{S}_{1-\nu}$.

An important special case of Theorem 6 is the following.

Corollary 3. *For* $0 < \nu \le \nu_0 = 1 - \lambda_0 = 0.345778\ldots$ *we have*

$$\operatorname{Re} \sum_{k=0}^{n} C_k^\nu(x) z^k > 0 \quad \textit{for all } z \in \overline{\mathbb{D}},\ n \in \mathbb{N},\ x \in [-1,1]$$

and, in particular,

$$\sum_{k=0}^{n} C_k^\nu(x) \cos k\theta > 0 \quad \textit{for all } \theta \in \mathbb{R},\ n \in \mathbb{N},\ x \in [-1,1]. \tag{32}$$

This result is best possible with respect to ν.

Recalling that

$$C_k^\nu(1) = \frac{(2\nu)_k}{k!}, \quad k = 0,1,2,\ldots$$

and taking into consideration inequality (26), we see that, by Theorem 6, the case $x=1$ of (32) implies its validity for all $x \in [-1,1]$.

There is another way of obtaining (32) from (26). It follows from (31) that

$$\frac{C_k^\nu(x)}{C_k^\nu(1)} = \int_0^{2\pi} \cos kt \, \mathrm{d}\omega_{x,\nu}(t), \tag{33}$$

where $\omega_{x,\nu}(t)$ is a probability measure on $[0,2\pi)$ which depends on x, ν but is independent of k. This result was originally established in [29], then generalized in [5]. Hence

$$\sum_{k=0}^{n} C_k^\nu(x) \cos k\theta = \frac{1}{2} \int_0^{2\pi} \sum_{k=0}^{n} \frac{(2\nu)_k}{k!} [\cos k(\theta + t) + \cos k(\theta - t)] \, \mathrm{d}\omega_{x,\nu}(t),$$

which is to say that the statement (26) implies the conclusion (32).

Using (33), we obtain a different generalization of (26). This is given next.

Corollary 4. *For all positive integers n and* $-1 < x < 1$ *we have*

$$\sum_{k=0}^{n} \frac{(1-\alpha)_k}{k!} \frac{C_k^\nu(x)}{C_k^\nu(1)} > 0 \quad \textit{for all } 1 > \alpha \ge \alpha_0 \textit{ and } \nu \ge 0. \tag{34}$$

This is proved in a different way in [15, Corollary 2]. Note that $C_k^0(\cos\theta)/C_k^0(1) = \cos k\theta$.

The condition $\alpha \geq \alpha_0$ in (34) can be replaced by $\alpha \geq 0$ when ν is sufficiently large. The best possible range of ν so that (34) holds for all $\alpha \in [0,1)$ can be determined. Indeed, it is proved in [10] that for all positive integers n and $-1 < x < 1$, we have

$$\sum_{k=0}^{n} \frac{C_k^\nu(x)}{C_k^\nu(1)} > 0 \quad \text{for } \nu \geq \nu' = \alpha' + \frac{1}{2}, \tag{35}$$

where α' is the unique solution in $(-1/2, 0)$ of the equation

$$\int_0^{j_{\alpha,2}} t^{-\alpha} J_\alpha(t)\,\mathrm{d}t = 0,$$

and J_α is the Bessel function of the first kind of order α with $j_{\alpha,2}$ being its second positive root. Moreover, for every $\nu < \nu'$ these sums are unbounded below in $(-1,1)$. The numerical value of ν' is $\nu' = 0.23061297\ldots$.

A summation by parts shows that inequality (34) holds for all $\alpha \in [0,1)$ provided that $\nu \geq \nu'$, and the result is sharp because of (35).

4.2 Subordination and Convolution of Analytic Functions

The Problem 2 in this section can be reformulated in terms of the notion of subordination of analytic functions. We recall the definition.

Definition 1. Let $f(z), g(z) \in A(\mathbb{D})$. We say that $f(z)$ is subordinate to $g(z)$ if there exists a function $\varphi(z) \in A(\mathbb{D})$ satisfying $\varphi(0) = 0$ and $|\varphi(z)| < 1$ such that

$$f(z) = g(\varphi(z)), \quad \forall z \in \mathbb{D}.$$

Subordination is denoted by $f(z) \prec g(z)$.

If $f(z) \prec g(z)$, then $f(0) = g(0)$ and $f(\mathbb{D}) \subset g(\mathbb{D})$. Conversely, if $g(z)$ is univalent and $f(0) = g(0)$ and $f(\mathbb{D}) \subset g(\mathbb{D})$ then $f(z) \prec g(z)$. We refer the reader to [22, Chap. 2] for proofs of several properties related to subordination of analytic functions.

We recall also that, for $0 < p < 2$, the function

$$u_p(z) := \left(\frac{1+z}{1-z}\right)^p, \quad z \in \mathbb{D},$$

is univalent in $\mathbb{D}$ and maps $\mathbb{D}$ onto the sector $\{\zeta \in \mathbb{C} : |\arg \zeta| < p\pi/2\}$. Thus, the statement of Problem 2 is equivalent to:

Given $\rho \in (0,1]$. Determine the best possible range of λ such that

$$s_n(f/z, z) \prec \left(\frac{1+z}{1-z}\right)^{2\rho} \quad \text{for all } n \in \mathbb{N}, \ z \in \mathbb{D}. \tag{36}$$

It is immediately obvious that the result of Theorem 6 corresponds to the case $\rho = 1/2$. The method of the proof of this Theorem reveals that it is sufficient to establish the required result for the extremal function $z/(1-z)^{2-2\lambda}$ of $\mathcal{S}_\lambda$, and we seek to deduce the result (36) as a whole in similar fashion. To simplify expressions, we introduce the following classes of analytic functions. Let $\mathcal{A}_0$ be the set of analytic functions f in $\mathbb{D}$ normalized by $f(0) = 1$. For $\mu > 0$ we set $f_\mu(z) := 1/(1-z)^\mu$ and

$$\mathcal{F}_\mu = \left\{ f \in \mathcal{A}_0 : \mathrm{Re}\,\frac{zf'(z)}{f(z)} > -\frac{\mu}{2}, z \in \mathbb{D} \right\}.$$

It is clear that $f_\mu \in \mathcal{F}_\mu$ and that $f \in \mathcal{F}_\mu$ if and only if $zf \in \mathcal{S}_{1-\mu/2}$. For all $f \in \mathcal{F}_\mu$, we have $f \prec f_\mu$ (cf. [22, p. 50]). Thus, the result of Theorem 6 can be reformulated as follows:

Let $f \in \mathcal{F}_\mu$. We have $\mathrm{Re}\, s_n(f,z) > 0$ *for all $n \in \mathbb{N}$ and $z \in \overline{\mathbb{D}}$ if and only if* $0 < \mu \le \mu_0 = 1 - \alpha_0 = 0.691556\ldots$.

In particular, let

$$s_n^\mu(z) := \sum_{k=0}^{n} \frac{(\mu)_k}{k!} z^k$$

be the n-th partial sum of the Taylor series expansion around the origin of the extremal element $f_\mu(z) = 1/(1-z)^\mu$. It follows from the above that

$$\mathrm{Re}\, s_n^\mu(z) > 0 \quad \text{for all } \; n \in \mathbb{N}, \; z \in \overline{\mathbb{D}} \tag{37}$$

if and only if $0 < \mu \le \mu_0 = 1 - \alpha_0 = 0.691556\ldots$, and this is a very natural way of generalizing (3).

There is a standard way of obtaining (36) from the corresponding result for the special case in which f is the extremal element zf_μ, when $0 < \rho < 1/2$. We recall that for $f(z) = \sum_{k=0}^\infty a_k z^k$ and $g(z) = \sum_{k=0}^\infty b_k z^k$ in $A(\mathbb{D})$, the Hadamard product (or convolution) $f * g$ is defined as $(f * g)(z) := \sum_{k=0}^\infty a_k b_k z^k$. We also define $\widetilde{f_\mu} \in \mathcal{A}_0$ to be the unique solution of the equation $f_\mu * \widetilde{f_\mu} = 1/(1-z)$. For our purpose, we need the following proposition, which is a special case of a more general result given in [25, p.55].

Proposition 1. *For $\mu > 0$, $f \in \mathcal{F}_\mu$ and h an arbitrary analytic function in $\mathbb{D}$, define*

$$T_h(z) := \left(f * \widetilde{f_\mu} * h \right)(z), \quad z \in \mathbb{D}.$$

Then we have $T_h(\mathbb{D}) \subset \mathrm{conv}(h(\mathbb{D}))$, where $\mathrm{conv}(A)$ stands for the convex hull of the set A.

For $f \in \mathcal{F}_\mu$, we observe that $s_n(f,z) = \left(f * \widetilde{f_\mu} * s_n^\mu \right)(z)$. Therefore, by Proposition 1, we deduce that

$$s_n(f,z)(\mathbb{D}) \subset \mathrm{conv}(s_n^\mu(\mathbb{D})). \tag{38}$$

Since the sector $\{\zeta \in \mathbb{C} : |\arg \zeta| < \rho\pi\}$ is a convex set when $0 < \rho < 1/2$, in view of (38), the statement that

$$s_n^{\mu}(z) \prec \left(\frac{1+z}{1-z}\right)^{2\rho} \quad \text{for all } n \in \mathbb{N},\ z \in \mathbb{D}, \tag{39}$$

implies the conclusion (36) for $f \in \mathcal{S}_{1-\mu/2}$, which is the same as

$$s_n(f,z) \prec \left(\frac{1+z}{1-z}\right)^{2\rho} \quad \text{for all } n \in \mathbb{N},\ z \in \mathbb{D}, \tag{40}$$

for $f \in \mathcal{F}_{\mu}$.

Suppose that $\rho \in (0,1)$. The determination of the best possible range of μ such that the relation (39) is valid turns out to be a quite hard problem. An interesting special case is the following.

Theorem 7. *For $\rho = 1/4$, the relation* (40) *holds true precisely when $0 < \mu \le \mu_1 = 1 - \alpha_1 = 0.385566\ldots$, with α_1 as in Theorem 4.*

Proof. It follows from the discussion above that we need only to establish (39) for $\rho = 1/4$. This is equivalent to $\operatorname{Re}\left\{\left[s_n^{\mu}(z)\right]^2\right\} > 0$. We shall prove that this inequality holds for all $z \in \overline{\mathbb{D}}$. By the minimum principle of harmonic functions, it is sufficient to prove the above inequality only for the boundary points $z = e^{2i\theta}$, $0 < \theta < \pi$. We observe that in this case we have

$$\operatorname{Re}\left\{\left[s_n^{\mu}\left(e^{2i\theta}\right)\right]^2\right\} = 2v_n(\theta)v_n(\pi-\theta), \tag{41}$$

where

$$v_n(\theta) := \sum_{k=0}^{n} \frac{(\mu)_k}{k!} \cos\left(2k\theta + \frac{\pi}{4}\right).$$

The positivity of the left-hand side of (41) will follow from the positivity of the trigonometric sum $v_n(\theta)$ for $0 < \theta < \pi$. To prove the latter, we write

$$\cos\left(2k\theta + \frac{\pi}{4}\right) = \sin\frac{\theta}{4}\cos\left[\left(2k+\frac{1}{4}\right)(\pi-\theta)\right] + \cos\frac{\theta}{4}\sin\left[\left(2k+\frac{1}{4}\right)(\pi-\theta)\right].$$

Using Theorem 3, Theorem 4 together with the observation that $0.385566\ldots = \mu_1 = 1-\alpha_1 < 1-\alpha_0 = \mu_0 = 0.691556\ldots$, we infer that $v_n(\theta) > 0$ for all n and $0 < \theta < \pi$, when $0 < \mu \le \mu_1 = 1-\alpha_1$. Then by (41), we deduce that for all n

$$\operatorname{Re}\left\{\left[s_n^{\mu}\left(e^{2i\theta}\right)\right]^2\right\} > 0, \quad 0 < \theta < \pi, \tag{42}$$

when $0 < \mu \le \mu_1 = 1-\alpha_1$. An asymptotic analysis similar to the one in the proof of Theorem 3 reveals that inequality $v_n(\theta) > 0$ is sharp with respect to μ, that is, the sums $v_n(\theta)$ are unbounded below, when $\mu > \mu_1$.

Observe also that inequality (42) is equivalent to $\left|\arg\left[s_n^{\mu}(\mathrm{e}^{2\mathrm{i}\theta})\right]\right| < \pi/4$ and, in turn, $0 < \arg\left[\mathrm{e}^{\mathrm{i}\pi/4}s_n^{\mu}(\mathrm{e}^{2\mathrm{i}\theta})\right] < \pi/2$, that is, $\mathrm{Re}\left[\mathrm{e}^{\mathrm{i}\pi/4}s_n^{\mu}\left(\mathrm{e}^{2\mathrm{i}\theta}\right)\right] = v_n(\theta) > 0$ and

$$\mathrm{Im}\left[\mathrm{e}^{\mathrm{i}\pi/4}s_n^{\mu}\left(\mathrm{e}^{2\mathrm{i}\theta}\right)\right] := w_n(\theta) = \sum_{k=0}^{n}\frac{(\mu)_k}{k!}\sin\left(2k\theta+\frac{\pi}{4}\right) > 0.$$

Consequently, inequality (42) cannot hold when $\mu > \mu_1$ for appropriate n and θ, thus this result is also sharp with respect to μ. We note, in passing, that inequalities $v_n(\theta) > 0$ and $w_n(\theta) > 0$ are equivalent for $0 < \theta < \pi$.

The proof of the theorem is complete. □

5 Generalizations and Extensions

The passage from (39) to (40) is not clear in the case where $1/2 < \rho < 1$, as the convexity argument that uses (38) is no longer effective. On the contrary, there should be an extension of (37) which is similar to the extension of (26) given by Theorem 2. It is the aim of this section to tackle these questions that correspond essentially to Problem 3 of Sect. 4. We give a statement that strengthens (39) and leads to a result that implies (40) for all $\rho \in (0,1]$.

Definition 2. For $\rho \in (0,1]$ define $\mu(\rho)$ as the maximal number such that for all $z \in \mathbb{D}$, $n \in \mathbb{N}$,

$$(1-z)^{\rho}s_n^{\mu}(z) \prec \left(\frac{1+z}{1-z}\right)^{\rho}, \tag{43}$$

holds for all $0 < \mu \leq \mu(\rho)$.

It follows from [27, Theorem 1.1], see also [19], that (43) holds true for $0 < \mu \leq \rho$, therefore we obtain the estimate $\rho \leq \mu(\rho)$, for all $\rho \in (0,1]$. In particular, we seek to show that the equality sign in this inequality holds only when $\rho = 1$.

It is easy to see that (43) implies (39) for all $\rho \in (0,1]$. We observe that (43) also implies

$$\mathrm{Re}\left[(1-z)^{2\rho-1}s_n^{\mu}(z)\right] > 0 \tag{44}$$

for $\rho \in (0,1)$, and that (43) and (44) are equivalent when $\rho = 1$.

When $\rho = 1/2$, (39) and (44) are equivalent and become the statement (37). It is not hard to see that when $0 < \rho < 1/2$ the relation (39) implies (44), whereas relation (44) implies (39) for $1/2 < \rho \leq 1$.

The value $\mu(1/2)$ has been determined in [19] and it is given in the Theorem below.

Theorem 8. *When $\rho = 1/2$, relation* (43) *holds if and only if $0 < \mu \leq \mu_0 = 1-\alpha_0 = 0.691556\ldots$, where α_0 is as in Theorem 2.*

Some basic facts in the proof of this Theorem are, first, the observation that for $\rho = 1/2$, (43) is equivalent to $\mathrm{Re}\left\{(1-z)\left[s_n^{\mu}(z)\right]^2\right\} > 0$, and second, that the minimum principle for harmonic functions reduces the verification of this inequality only for $z = \mathrm{e}^{2\mathrm{i}\theta}$, $0 < \theta < \pi$. We set

$$P_n(\theta) := \left(1 - \mathrm{e}^{2\mathrm{i}\theta}\right)\left\{\sum_{k=0}^{n} \frac{(\mu)_k}{k!}\mathrm{e}^{2\mathrm{i}k\theta}\right\}^2$$

and try to prove that $\mathrm{Re}\, P_n(\theta) > 0$ for all $n \in \mathbb{N}$, $0 < \theta < \pi$. We find that

$$\begin{aligned} \mathrm{Re}\, P_n(\theta) = & \left(\sum_{k=0}^{2n+1} c_k \cos k\theta\right)\left(\sum_{k=0}^{2n+1} c_k \cos k(\pi-\theta)\right) \\ & + \left(\sum_{k=1}^{2n+1} c_k \sin k\theta\right)\left(\sum_{k=1}^{2n+1} c_k \sin k(\pi-\theta)\right), \end{aligned}$$

where $c_0 = c_1 = 1$, $c_{2k} = c_{2k+1} = (\mu)_k/k!$, $k = 1, 2, \ldots$, and the desired result follows from Theorem 2 and Remark 1 of Sect. 3.1.

As mentioned earlier, Theorem 8 implies inequality (37). The result of Theorem 3 helps to explain why the best possible range of μ for the validity of both results is exactly the same.

Observe also that $\mu_0 = 1 - \alpha_0 = \mu(1/2) = \mu^*(1/2)$, where the function μ^* is defined in Conjecture 1 of Sect. 3. It is natural to seek other values of $\rho \in (0,1]$ for which the functions $\mu(\rho)$ and $\mu^*(\rho)$ are equal. □

For $z = \mathrm{e}^{\mathrm{i}\theta}$, (44) is equivalent to the trigonometric inequality

$$\sum_{k=0}^{n} \frac{(\mu)_k}{k!} \sin\left[\left(k + \rho - \frac{1}{2}\right)\theta - \rho\pi\right] < 0, \quad 0 < \theta < 2\pi. \tag{45}$$

A limiting case of this inequality can be obtained using the asymptotic formula

$$\lim_{n\to\infty}\left(\frac{\theta}{n}\right)^{\mu}\sum_{k=0}^{n}\frac{(\mu)_k}{k!}\sin\left[\left(k+\rho-\frac{1}{2}\right)\frac{\theta}{n}-\rho\pi\right] = \frac{1}{\Gamma(\mu)}\int_0^{\theta}\sin\left(t-\rho\pi\right)t^{\mu-1}\,\mathrm{d}t.$$

Then, as in (30) we deduce that inequality (45) cannot hold for $\mu > \mu^*(\rho)$, $0 < \rho < 1$, and appropriate n and θ, where $\mu^*(\rho)$ is the function appearing in Conjecture 1. On the contrary, since (43) implies (44), it follows from the above that for all $\rho \in (0,1]$ we have the estimate $\mu(\rho) \le \mu^*(\rho)$. Recall that for $\rho = 1$, (43), (44) are equivalent. In this case, inequality (45) holds for all $0 < \mu \le 1$, and this follows from the nonnegativity of the classical Fejér kernel. Therefore, $\mu(1) = \mu^*(1) = 1$. These observations in combination with the result of Theorem 8 led us to the following challenging conjectures, originally posed and discussed in [19].

Conjecture 2. For all $\rho \in (0,1]$, we have $\mu(\rho) = \mu^*(\rho)$.

Conjecture 3. Let $\rho \in (0,1]$. Then inequality (44) holds precisely when $0 < \mu \leq \mu^*(\rho)$.

It follows from the discussion above that for $\rho = 1$ the two conjectures coincide and their assertion is true. When $0 < \rho < 1$ Conjecture 2 implies Conjecture 3.

For $z = \mathrm{e}^{2\mathrm{i}\theta}$, $0 < \theta < \pi$, relation (43) is equivalent to

$$\mathrm{Re}\left\{\left(1-\mathrm{e}^{2\mathrm{i}\theta}\right)\left[s_n^{\mu}\left(\mathrm{e}^{2\mathrm{i}\theta}\right)\right]^{1/\rho}\right\} > 0,$$

and this, in turn, is equivalent to

$$0 < \arg\left[\mathrm{e}^{\mathrm{i}\rho\theta} s_n^{\mu}\left(\mathrm{e}^{2\mathrm{i}\theta}\right)\right] < \rho\pi. \tag{46}$$

The last inequality implies that

$$\sum_{k=0}^{n} \frac{(\mu)_k}{k!} \sin(2k+\rho)\theta > 0 \tag{47}$$

when $0 < \rho \leq 1$. Conversely, suppose that inequality (47) is valid for $\rho \in (0,1]$ and all $0 < \theta < \pi$. Then we also have

$$\sum_{k=0}^{n} \frac{(\mu)_k}{k!} \sin\left[(2k+\rho)(\pi-\theta)\right] > 0,$$

which is equivalent to $\mathrm{Im}\left[\mathrm{e}^{\mathrm{i}\rho(\theta-\pi)} s_n^{\mu}\left(\mathrm{e}^{2\mathrm{i}\theta}\right)\right] < 0$, and, in turn,

$$-(1-\rho)\pi < \arg\left[\mathrm{e}^{\mathrm{i}\rho\theta} s_n^{\mu}\left(\mathrm{e}^{2\mathrm{i}\theta}\right)\right] < \rho\pi. \tag{48}$$

Similarly, inequality (47) is equivalent to $0 < \arg\left[\mathrm{e}^{\mathrm{i}\rho\theta} s_n^{\mu}\left(\mathrm{e}^{2\mathrm{i}\theta}\right)\right] < \pi$. A combination of this with (48) yields (46) and therefore the relation (43). We conclude from the above that Conjectures 1 and 2 are equivalent for all $\rho \in (0,1]$. We have also given the geometric interpretation of the inequality (47). By a similar argument, it can be shown that inequality (47) implies (45).

Conjectures 1–2 hold for $\rho = 1/2$ by Theorem 8. Moreover, Theorem 4 ensures that the statement of these conjectures is true for $\rho = 1/4$. In this case, relation (43) implies inequality (42). Nevertheless, inequality (42) admits the direct proof given in Sect. 4.2.

In addition, Conjecture 3 has been verified in [19] for $\rho = 3/4$, using the positivity of a trigonometric sum, which is a remarkable special case of (45). This is

$$\sum_{k=0}^{n} \frac{(\mu)_k}{k!} \cos\left[\left(2k+\frac{1}{2}\right)\theta - \frac{\pi}{4}\right] > 0, \quad 0 < \theta < \pi,$$

and the inequality holds precisely when $0<\mu\leq\mu^*(\frac{3}{4})=0.907689\ldots$. Its proof uses a refined technique which develops the ideas of the proof of Theorem 3.

There is another reason for which inequality (45) is of interest. Suppose that $1/2\leq\rho<1$. Let $\nu=\rho-1/2$ and rewrite (45) as

$$\sum_{k=0}^{n}\frac{(\mu)_k}{k!}\sin\left[(k+\nu)t+\left(\frac{1}{2}-\nu\right)\pi\right]>0,\quad 0<t<2\pi. \tag{49}$$

For the Gegenbauer polynomials $C_k^\nu(x)$, we have the Dirichlet–Mehler formula

$$\frac{C_k^\nu(\cos\theta)}{C_k^\nu(1)}=\frac{2^\nu(\sin\theta)^{1-2\nu}}{B(\nu,1/2)}\int_0^\theta\frac{\cos(k+\nu)t}{(\cos t-\cos\theta)^{1-\nu}}\,dt,$$

see [11]. Replacing θ by $\pi-\theta$ and making the change of variables $t=\pi-u$ in this integral we easily get

$$\frac{C_k^\nu(\cos\theta)}{C_k^\nu(1)}=\frac{2^\nu(\sin\theta)^{1-2\nu}}{B(\nu,1/2)}\int_\theta^\pi\frac{\sin[(k+\nu)t+(1/2-\nu)\pi]}{(\cos\theta-\cos t)^{1-\nu}}\,dt.$$

This in combination with (49) entails

$$\sum_{k=0}^{n}\frac{(\mu)_k}{k!}\frac{C_k^\nu(x)}{C_k^\nu(1)}>0,\quad -1<x<1. \tag{50}$$

It follows from the discussion in Sect. 4.1 that this inequality is valid for all $0<\mu\leq1$ when $\nu\geq\nu'$. For the range $0\leq\nu<\nu'$, Corollary 4 confirms that (50) holds for $0<\mu\leq\mu^*(1/2)$, the case $\nu=0$ being sharp. Inequality (49), once established, would extend the range of μ to $0<\mu\leq\mu^*(\nu+1/2)$ for all $0<\nu<\nu'$ and this because the function $\mu^*(\rho)$ is strictly increasing on $(0,1)$. As mentioned above, inequality (49) is a consequence of (47), that is, of Conjecture 1. Thus, the validity of this Conjecture (especially for $\rho\in(1/2,1)$) would provide a powerful tool to attack the problem of extending the range of μ for which (50) is valid. □

Finally, we describe how the relation (43) can be generalized to a form involving the partial sums of any function f in the class $\mathfrak{F}_\mu$, $0<\mu\leq1$. Recall that $s_n^\mu(z)$ is the partial sum of the extremal element $f_\mu(z)=1/(1-z)^\mu$ of $\mathfrak{F}_\mu$. Then we write

$$(1-z)^\rho s_n^\mu(z)=\frac{s_n^\mu(z)}{(\Phi_{\rho,\mu}*f_\mu)(z)},$$

where $\Phi_{\rho,\mu}(z):={}_2F_1(\rho,1;\mu;z)$ is the Gaussian hypergeometric function.

A generalization of Proposition 1, see [25, p. 55] and compare also [19, Lemma 3], shows that when $\rho\leq\mu\leq1$ and $f\in\mathfrak{F}_\mu$, we have

$$\frac{s_n(f,z)}{\Phi_{\rho,\mu}*f}(\mathbb{D})\subset\operatorname{conv}\frac{s_n^\mu}{\Phi_{\rho,\mu}*f_\mu}(\mathbb{D}).$$

On the contrary, when $0 < \mu \le \rho \le 1$ and $f \in \mathfrak{F}_\mu$, it is shown in [19, Theorem 5] that

$$\frac{s_n(f,z)}{\Phi_{\rho,\mu} * f} \prec (1-z)^\rho.$$

Since for all $\rho \in (0,1]$, we have $(1-z)^\rho \prec ((1+z)/(1-z))^\rho$ and the function $((1+z)/(1-z))^\rho$ is convex univalent in $\mathbb{D}$, combining the above with (43), we arrive at the following.

Theorem 9. *Let $\rho \in (0,1]$ and $f \in \mathfrak{F}_\mu$. Then for all $n \in \mathbb{N}$, we have*

$$\frac{s_n(f,z)}{\Phi_{\rho,\mu} * f} \prec \left(\frac{1+z}{1-z}\right)^\rho, \quad z \in \mathbb{D},$$

for $0 < \mu \le \mu(\rho)$.

This theorem enables us to derive relation (40) for the same range of μ. We have, in fact, the following.

Corollary 5. *Let $\rho \in (0,1]$ and $f \in \mathfrak{F}_\mu$. Then for all $n \in \mathbb{N}$ and $z \in \mathbb{D}$, we have*

$$s_n(f,z) \prec \left(\frac{1+z}{1-z}\right)^{2\rho} \quad \text{for } 0 < \mu \le \mu(\rho). \tag{51}$$

Proof. For $0 < \mu \le 1$ and $f \in \mathfrak{F}_\mu$ we have $|\arg s_n(f,z)| \le \mu\pi$ for all $z \in \mathbb{D}$, $n \in \mathbb{N}$, see [27, Corollary 1.2] and compare [18, Remark 1]. When $0 < \mu \le \rho$ the above relation clearly implies (51). When $\rho < \mu \le 1$ and $f \in \mathfrak{F}_\mu$ we have $\Phi_{\rho,\mu} * f \in \mathfrak{F}_\rho$, (cf. [25, p. 49]), hence $\Phi_{\rho,\mu} * f \prec f_\rho$, that is, $|\arg(\Phi_{\rho,\mu} * f)| \le \rho\pi/2$. Combine this with Theorem 9 to obtain (51). Recall that $\rho \le \mu(\rho)$ for all $\rho \in (0,1]$.

The proof is complete. □

Acknowledgements This research was supported by a grant from the Leventis Foundation.

References

1. Alzer, H., Koumandos, S.: On the partial sums of a Fourier series. Constr. Approx. **27**, 253–268 (2008)
2. Andrews, G.E, Askey, R., Roy, R.: Special Functions. Cambridge University Press, Cambridge (1999)
3. Askey, R.: Orthogonal polynomials and special functions. Regional Conf. Lect. Appl. Math. **21**. SIAM, Philadelphia (1975)
4. Askey, R.: Vietoris's inequalities and hypergeometric series. In: Milovanović, G. V. (ed.) Recent progress in inequalities, pp. 63–76. Kluwer, Dordrecht (1998)
5. Askey, R., Fitch, J.: Integral representations for Jacobi polynomials and some applications. J. Math. Anal. Appl. **26**, 411–437 (1969)
6. Askey, R., Steinig, J.: Some positive trigonometric sums. Trans. Am. Math. Soc. **187**, 295–307 (1974)

7. Bary, N.K.: A Treatise on Trigonometric Series. Vol I, Pergamon, New York (1964)
8. Belov, A.S.: Examples of trigonometric series with nonnegative partial sums. (Russian) Mat. Sb. **186**, 21–46 (1995); (English translation) **186**, 485–510 (1995)
9. Brickman, L., Hallenbeck, D.J., MacGregor, T.H., Wilken, D.R.: Convex hulls and extreme points of families of starlike and convex functions. Trans. Am. Math. Soc. **185**, 413–428 (1973)
10. Brown, G., Koumandos, S., Wang, K.: Positivity of basic sums of ultraspherical polynomials. Analysis **18/4**, 313–331 (1998)
11. Erdélyi, A., Magnus, W., Oberhettinger, F., Tricomi, F.G.: Higher Transcendental Functions. Vol. **2**. McGraw-Hill, New York (1953)
12. Gronwall, T.H.: Über die Gibbssche Erscheinung und die trigonometrischen Summen $\sin x + \frac{1}{2}\sin 2x + \ldots + \frac{1}{n}\sin nx$. Math. Ann. **72**, 228–243 (1912)
13. Jackson, D.: Über eine trigonometrische Summe. Rend. Circ. Mat. Palermo **32**, 257–262 (1911)
14. Koumandos, S.: Positive trigonometric sums and applications. Ann. Math. Inform. **33**, 77–91 (2006)
15. Koumandos, S.: An extension of Vietoris's inequalities. Ramanujan J. **14**, 1–38 (2007)
16. Koumandos, S.: Monotonicity of some functions involving the gamma and psi functions. Math. Comp. **77**, 2261–2275 (2008)
17. Koumandos, S., Lamprecht, M.: On a conjecture for trigonometric sums and starlike functions II. J. Approx. Theory **162**, 1068–1084 (2010)
18. Koumandos, S., Ruscheweyh, S.: Positive Gegenbauer polynomial sums and applications to starlike functions. Constr. Approx. **23**, 197–210 (2006)
19. Koumandos, S., Ruscheweyh, S.: On a conjecture for trigonometric sums and starlike functions. J. Approx. Theory **149**, 42–58 (2007)
20. Landau, E.: Über eine trigonometrische Ungleichung. Math. Z. **37**, 36 (1933)
21. Milovanović, G.V, Mitrinović, D.S., Rassias, Th.M.: Topics in Polynomials: Extremal Problems, Inequalities, Zeros. World Sci. Publ., Singapore (1994)
22. Pommerenke, C.: Univalent Functions. Vandenhoeck and Ruprecht, Göttingen (1975)
23. Robertson, M.S.: On the theory of univalent functions. Ann. Math. (2) **37**, 374–408 (1936)
24. Ruscheweyh, S.: On the Kakeya-Eneström theorem and Gegenbauer polynomial sums. SIAM J. Math. Anal. **9**, 682–686 (1978)
25. Ruscheweyh, S.: Convolutions in geometric function theory. Sem. Math. Sup. **83**, Les Presses de l'Université de Montréal (1982)
26. Ruscheweyh, S.: Coefficient conditions for starlike functions. Glasgow Math. J. **29**, 141–142 (1987)
27. Ruscheweyh, S., Salinas, L.: On starlike functions of order $\lambda \in [1/2, 1)$. Ann. Univ. Mariae Curie-Sklodowska **54**, 117–123 (2000)
28. Ruscheweyh, S., Salinas, L.: Stable functions and Vietoris' theorem. J. Math. Anal. Appl. **291**, 596–604 (2004)
29. Seidel, W., Szász, O.: On positive harmonic functions and ultraspherical polynomials. J. Lond. Math. Soc. **26**, 36–41 (1951)
30. Turán, P.: Egy Steinhausfele problémáról. Mat. Lapok **4**, 263–275 (1953)
31. Vietoris, L.: Über das Vorzeichen gewisser trigonometrischer Summen. S.-B. Öster. Akad. Wiss. **167**, 125–135 (1958); Teil II: Anzeiger Öster. Akad. Wiss. **167**, 192–193 (1959)
32. Young, W.H.: On a certain series of Fourier. Proc. Lond. Math. Soc. (2) **11**, 357–366 (1913)
33. Zygmund, A.: Trigonometric Series. 3rd ed. Cambridge University Press, Cambridge (2002)

Part III
Quadrature Formulae

Quadrature Rules for Unbounded Intervals and Their Application to Integral Equations

G. Monegato and L. Scuderi

Dedicated to Professor Gradimir V. Milovanović on the occasion of his 60th birthday

1 Introduction

Compared with the case of integrals defined on bounded intervals, only a few quadrature rules have been proposed for the numerical integration of functions defined on the real (positive) semiaxis or on the real axis. The Gauss–Laguerre and Gauss–Hermite rules are the classical and best known ones. However, when the integrand function, though smooth and nonoscillatory, does not have the proper decay at infinity, for example, if it is only of algebraic type, it is well known that their performance is rather poor. In this case some alternatives are being proposed.

When the domain of integration is the real semiaxis, and the integrand function is smooth and nonoscillatory, in [4] and [5] quadrature rules with some kind of (weighted) rational degree of exactness have been constructed. When the main goal is the computation of an integral, a very efficient rule is also the one based on the double exponential formula described in [10], although this requires an ad hoc selection of a parameter h, which depends upon the function and the number of quadrature nodes one wants to use.

For the real axis the double exponential rule presented in [10] (see also [9]) generally gives quite accurate results if one chooses properly the value of the parameter the rule depends upon. The key ingredients of the above-mentioned double exponential rules are: (double) exponential type change of variable and the subsequent application of a truncated trapezoidal rule.

G. Monegato and L. Scuderi
Dipartimento di Matematica, Politecnico di Torino, Corso Duca degli Abruzzi 24, 10129 Torino, Italia, giovanni.monegato@polito.it, letizia.scuderi@polito.it

W. Gautschi et al. (eds.), *Approximation and Computation: In Honor of Gradimir V. Milovanović*, Springer Optimization and Its Applications 42,
DOI 10.1007/978-1-4419-6594-3_13, © Springer Science+Business Media, LLC 2011

In these last years, the first author and Mastroianni [8] have considered integrals of smooth (nonoscillatory) functions defined on the positive semiaxis. In particular, they have proposed to use a properly truncated version of the Gauss–Laguerre formula, after having introduced, when needed, an exponential type change of variable. These rules have then been used to construct Nyström type interpolants for second-kind Fredholm integral equations. Further results along this line have been obtained by Mastroianni and Milovanović in [6] and by De Bonis and Mastroianni in [2].

A survey of the above quadrature rules, including some numerical testing and new remarks, will be presented in Sect. 2.

In Sect. 3 we will examine similar approaches for integrals of nonoscillatory smooth functions defined on the whole real axis and having only algebraic decay at ∞. In particular, we propose four new alternative methods. One of them, after having introduced a proper change of variable of exponential type, requires the use of a Gaussian rule defined by the weight function $w(x) = [\cosh(x)]^{-1}$. The other two, in some sense, extend the idea applied by Gautschi in [5] for the $(0,\infty)$ case. Several numerical tests will be presented.

In Sect. 4 we will use some of the rules we have examined in the earlier two sections to solve some second-kind Fredholm integral equations.

2 Quadrature Formulae on Half-Infinite Interval $(0,\infty)$

In this section, we examine the computation of integrals of the form

$$\int_0^{\infty} g(x)\,\mathrm{d}x \tag{1}$$

with nonoscillatory integrand functions $g(x) = O(x^{-\rho})$, $\rho > 1$, decaying only algebraically to zero for $x \to \infty$. We consider essentially three quadrature formulae; they are based on preliminary changes of variables followed by the application of Gaussian formulae or the trapezoidal rule.

The first quadrature formula is defined by the change of variable

$$x = \sinh(pt) \tag{2}$$

with $p > 0$, after which the n-point Gauss–Laguerre quadrature formula, possibly truncated to its first $\lfloor n/l \rfloor$ points with $l = 2,3,4$, is applied:

$$\begin{aligned} \int_0^{\infty} g(x)\,\mathrm{d}x &= p\int_0^{\infty} g(\sinh(pt))\cosh(pt)\,\mathrm{d}t \\ &= \int_0^{\infty} f(t)\exp(-t)\,\mathrm{d}t \approx \sum_{i=1}^{\lfloor n/l \rfloor} w_i^L f\left(t_i^L\right), \end{aligned} \tag{3}$$

where we have set $f(t) := pg(\sinh(pt))\cosh(pt)\exp(t)$ and denoted by $\{w_i^L, t_i^L\}$, the weights and the nodes of the n-point Gauss–Laguerre formula, respectively. In the sequel, we refer to (3) by the acronym SGL(p,l), where S stands for the change of variable sinh and GL for the Gauss–Laguerre quadrature rule used. A similar quadrature formula has been proposed in [8], but with the change of variable $x = \exp(pt) - 1$ instead of (2); even if the two transformations have a similar behavior, numerically the hyperbolic sine function sometimes seems to be just slightly more effective than the exponential function.

The parameter p in (2) has to be chosen in a proper way; if one wants to apply a truncated Gauss–Laguerre formula and, hence, to use a reduced number of quadrature points with respect to n, then as in [8] we propose to choose p in terms of the decay rate ρ of g by

$$p = p_t := \frac{1}{\rho - 1},$$

which implies $f(t) = O(1)$, $t \to \infty$. In this way, by taking into account that the weights w_i^L decay exponentially as i increases, the terms of the quadrature sum from a certain value of i up to n are negligible. The above value p_t is not necessarily the optimal value, but it seems to be a reasonable and robust one. Indeed, for larger values than p_t the transformed integrand function decays to zero exponentially and the interval where the transformed integrand function is significant becomes too small. Therefore, to have enough quadrature nodes within that interval, we have to take n sufficiently large. This is also the reason why SGL(p_t,l) for small n gives poor accuracy.

If, instead of the above truncated Gauss–Laguerre formula, one wants to apply the complete one, then we suggest to choose a value of p smaller than p_t, in such a way that $f(t) \to \infty$ as $t \to \infty$ and all the terms $w_i^L f(t_i^L)$ in (3) take significant values. However, since the Gauss–Laguerre weights decay to zero exponentially while the integrand function increases exponentially, to avoid machine underflow and/or overflow and consequently mathematically undefined operations like $0 \cdot \infty$, we stop the quadrature sum as soon as we encounter the first term $w_i^L f(t_i^L)$ that is smaller than tol times the partial sum obtained so far, tol being the accuracy tolerance with which we desire to compute the original integral. In our case, tol will coincide with the double precision machine accuracy eps. In practice, also in this case, we consider a truncated Gauss–Laguerre quadrature rule; however, in this case the cutoff point is not fixed a priori.

As regards the parameter p, on the basis of numerous and various numerical tests, we suggest to choose

$$p = p_c := \frac{1}{5(\rho - 1)}.$$

The second quadrature formula we consider has been proposed in [5]. It is defined for integrand functions $g(x) = x^{\alpha} G(x)$, where $G(x) = (1+x)^{-\beta} \overline{G}(x)$ with a "nice" function $\overline{G}(x)$ and $\alpha > -1, \beta - \alpha > 1$. To construct it, introduce first the Möbius transformation

$$x = \frac{1-t}{1+t}, \tag{4}$$

which maps $(0,\infty)$ into $(-1,1)$, and then apply the n-point Gaussian formula with respect to the Jacobi weight $(1-t)^{\alpha}(1+t)^{\beta-\alpha-2}$ with properly chosen β. We obtain:

$$\begin{aligned}\int_0^{\infty} x^{\alpha} G(x)\,\mathrm{d}x &= \int_{-1}^{1}\left(\frac{1-t}{1+t}\right)^{\alpha} G\left(\frac{1-t}{1+t}\right)\frac{2}{(1+t)^2}\,\mathrm{d}t \\ &= \int_{-1}^{1}\frac{2}{(1+t)^{\beta}} G\left(\frac{1-t}{1+t}\right)(1-t)^{\alpha}(1+t)^{\beta-\alpha-2}\,\mathrm{d}t \\ &\approx \sum_{i=1}^{n}\frac{2w_i^{(\alpha,\beta-\alpha-2)}}{\left(1+t_i^{(\alpha,\beta-\alpha-2)}\right)^{\beta}} G\left(\frac{1-t_i^{(\alpha,\beta-\alpha-2)}}{1+t_i^{(\alpha,\beta-\alpha-2)}}\right),\end{aligned} \tag{5}$$

where we have denoted by $\{w_i^{(\alpha,\beta-\alpha-2)}, t_i^{(\alpha,\beta-\alpha-2)}\}$ the weights and the nodes of the n-point Gaussian rule associated with the Jacobi weight $(1-t)^{\alpha}(1+t)^{\beta-\alpha-2}$, respectively. In the following, we refer to (5) by MGJ(α,β), since (5) is characterized by the Möbius transformation and the Gauss–Jacobi quadrature rule.

In [5, p. 440] it has been shown that MGJ(α,β) is exact for the rational functions

$$G(x) = \frac{1}{(1+x)^{\beta+k}}, \quad k = 0,1,\dots,2n-1.$$

Moreover, if $\overline{G}((1-t)/(1+t))$ is analytic in the disc $|t| \le r$, $r > 1$, which implies that $\overline{G}(x)$ must be necessarily analytic outside the disc with center at $-(r^2+1)/(r^2-1)$ and radius $2r/(r^2-1)$, $r > 1$, then convergence of MGJ(α,β) is exponentially fast.

When we apply this rule to an integral of type (1), we will take $\beta = \rho$ (and $\alpha = 0$). Indeed, for this choice the transformed integrand function behaves like a constant in the neighborhood of -1, while for values of β larger or smaller than ρ it is singular at $t = -1$ or flat close to -1, respectively. Therefore, we reject these latter values since the resulting endpoint behavior of the integrand function could adversely affect the accuracy of the Gaussian rule. The numerical tests below have confirmed that the choice $\beta = \rho$ is the best one.

The third quadrature formula we consider has been proposed in [10]. It is based on the double exponential transformation

$$x = \exp(\pi/2\sinh(t)) \tag{6}$$

and on the application of the trapezoidal rule with a mesh size h,

$$\begin{aligned}\int_0^{\infty} g(x)\,\mathrm{d}x &= \frac{\pi}{2}\int_{-\infty}^{\infty} g\left(\exp\left(\frac{\pi}{2}\sinh(t)\right)\right)\exp\left(\frac{\pi}{2}\sinh(t)\right)\cosh(t)\,\mathrm{d}t \\ &\approx \frac{\pi}{2}h\sum_{i=-M}^{M} g\left(\exp\left(\frac{\pi}{2}\sinh(ih)\right)\right)\exp\left(\frac{\pi}{2}\sinh(ih)\right)\cosh(ih)\end{aligned} \tag{7}$$

with M sufficiently large. We refer to (7) by $\mathrm{DET}(h)$, since (7) is characterized by the double exponential transformation and the trapezoidal rule. As we will see in the numerical tests, the accuracy of the approximations depends on a proper choice of h and M. As regards M, similarly to $\mathrm{SGL}(p,l)$, starting from $i=0$, the quadrature sum is truncated to $-M_1 \le i \le M_2$ as soon as the first ignored quadrature terms are smaller than eps times the partial sum obtained so far. How to choose h properly is still an open problem; indeed, in [10] the authors treat the above problem only in the case of integrals over the finite interval $(-1,1)$. The value of h we have chosen to obtain the numerical results reported in the tables below seems to be a good one.

In Table 1, we compare the above three quadrature formulae by applying them to the following integral:

$$\int_0^\infty \frac{\mathrm{d}x}{((x-a)^2+b^2)^{3/2}} = \frac{1}{b^2} + \frac{a}{b^2\sqrt{a^2+b^2}}. \tag{8}$$

In this table, as well as in the following ones, we denote by n_t the number of quadrature nodes actually used by each truncated quadrature formula described above; moreover, the sign "–" means that the full relative accuracy (i.e., 14 significant digits in our case) has been achieved. When the exact value of the integral is not analytically known, we have computed a reference value by using the most accurate formula among those proposed, with n_t sufficiently large.

Table 1 Relative errors for integral (8) with $a=1$ and $b=1$

n	n_t	SGL(1/2,2)	n_t	SGL(1/10,1)	n_t	MGJ(0,3)	n_t	DET(1/n)
4	2	4.56 – 02	4	2.09 – 01	4	1.19 – 01	30	2.84 – 05
8	4	4.44 – 02	8	3.97 – 03	8	1.56 – 03	58	1.18 – 10
16	8	5.83 – 03	16	5.45 – 06	16	5.76 – 06	112	–
32	16	1.64 – 04	32	5.60 – 09	32	4.14 – 11	222	–
64	32	8.68 – 06	61	6.94 – 13	64	–	439	–
128	64	1.79 – 08	92	–	128	–	872	–
256	128	3.80 – 12	133	–	256	–	1,730	–
512	256	–	189	–	512	–	3,435	–

We note that small values of h in $\mathrm{DET}(h)$ require a large number of nodes before the quadrature sum can be truncated. Further, we remark that if we consider $l=4$ in $\mathrm{SGL}(1/2,l)$, we obtain the same order of accuracy reached with $l=2$ and $n_t = 64,128,256$, but by using only the half of the quadrature nodes, i.e., $n_t = 32,64,128$. As a consequence, we have that all the quadrature rules reach the double precision by using a number of quadrature nodes less than $n_t = 128$. However, as Fig. 1 shows, if we choose increasing values of a and we fix, for example, $b=1$ and $n_t = 128$, the corresponding relative accuracy worsens.

In Fig. 2, we show that a similar phenomenon occurs also when we choose decreasing values of b and we fix, for example, $a=1$ and $n_t = 128$.

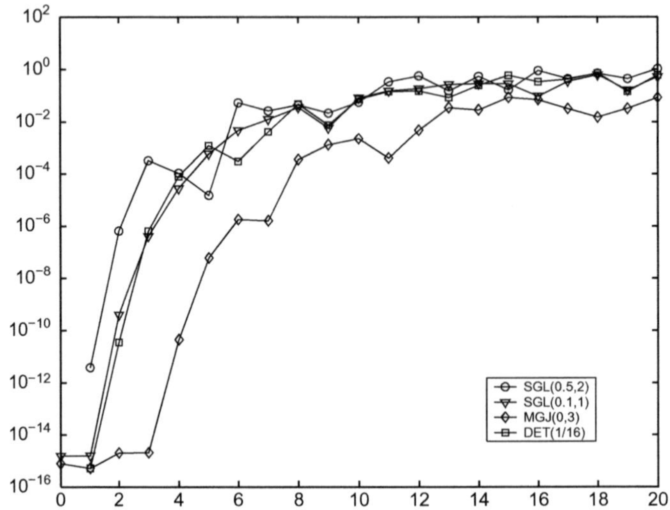

Fig. 1 Relative errors for integral (8) with $a = 0:1:20$, $b = 1$, by using $n_t = 128$ quadrature nodes

The reason of these drawbacks is the closeness to the interval of integration of the complex poles $a \pm ib$; indeed, all three changes of variable map the original poles into poles whose imaginary part becomes increasingly smaller than the original one as a increases or b decreases. In the particular case considered, or when the integrand function is given by that in (8) multiplied by a smooth function having no poles very close to the interval of integration, to preserve full relative accuracy for any given value of a and b, it is convenient to use in $\text{SGL}(p,l)$ the change of variable

$$x = a + b\sinh(pt) \tag{9}$$

instead of (2), before applying the (truncated) Gauss–Laguerre quadrature rule. This is a situation that one may encounter, for example, in boundary integral equation methods, where the integrand function in (8) is nothing but the kernel r^{-3} of the integral equation.

In the sequel, this new numerical method will be named $\text{SGLm}(p,l)$, since it is obtained from $\text{SGL}(p,l)$ by modifying only the change of variable.

Since for $n = 128$ we obtain full relative accuracy by means of $\text{SGLm}(p,l)$ for all the values of a and b considered in Figs. 1 and 2, in Figs. 3 and 4 below we report the relative accuracy achieved with $n = 32$ and by varying a and b, respectively.

What regards the choice of the parameters p in $\text{SGL}(p,l)$, β in $\text{MGJ}(0,\beta)$, and h in $\text{DET}(h)$, in Fig. 5 we show how the relative accuracy given by the corresponding rules depends upon them. In particular, in Fig. 5 we have fixed $n_t = 64$ (in $\text{DET}(h)$: $-32 \le i \le 31$) and chosen first the values of $p = h$ and then those of β equally spaced between 0.01 and 0.8 and 1.1 and 5, respectively.

Notice that the empirical optimal parameters are about 0.5 for $\text{SGL}(p,2)$, 0.07 for $\text{SGL}(p,1)$, and 0.1 for $\text{DET}(h)$ and exactly $\beta = 2$ or $\beta = 3$ in $\text{MGJ}(0,\beta)$. Moreover,

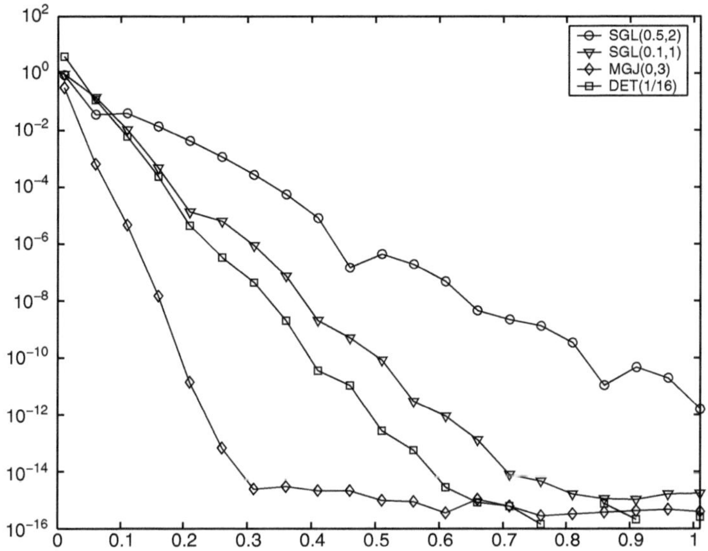

Fig. 2 Relative errors for integral (8) with $b = 0.01 : 0.05 : 1.01$, $a = 1$, by using $n_t = 128$ quadrature nodes

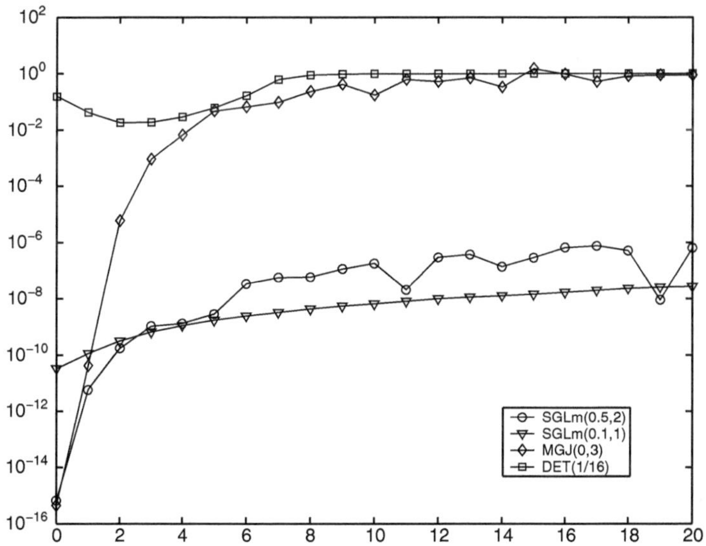

Fig. 3 Relative errors for integral (8) with $a = 0 : 1 : 20$, $b = 1$, by using $n_t = 32$ quadrature nodes

for $\mathrm{MGJ}(0,\beta)$ as well as for $\mathrm{DET}(h)$ and $\mathrm{SGL}(p,1)$ a small variation of the corresponding parameter implies a large variation in the relative error; finally, values of h larger than 0.2 yield undefined mathematical operations.

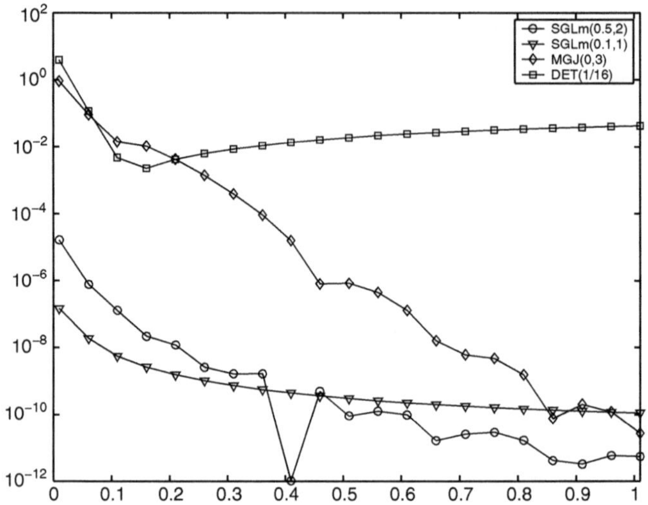

Fig. 4 Relative errors for integral (8) with $b = 0.01 : 0.05 : 1.01$, $a = 1$, by using $n_t = 32$ quadrature nodes

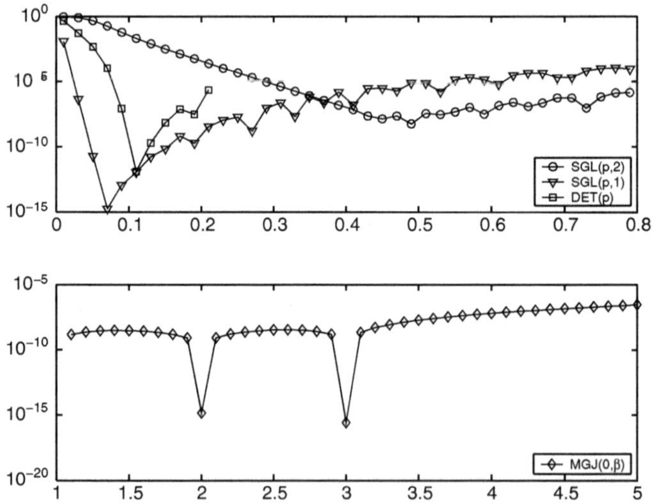

Fig. 5 Relative errors for integral (8) with $a = 1$, $b = 1$, by choosing $p = h = 0.01 : 0.02 : 0.8$ and $\beta = 1.1 : 0.1 : 5$ and using $n_t = 64$ quadrature nodes

In Table 2, we compare the above-mentioned quadrature formulae in the case of the following integral:

$$\int_0^{\infty} \frac{dx}{x^4 - 3x^3 + 2x^2 + 1} \approx 2.9974180827565. \qquad (10)$$

Table 2 Relative errors for integral (10)

n	n_t	SGL(1/3,4)	n_t	SGL(1/15,1)	n_t	MGJ(0,4)	n_t	DET(1/n)
4	1	9.09 − 01	4	7.16 − 01	4	3.49 − 01	28	4.31 − 02
8	2	8.43 − 01	8	1.85 − 01	8	1.09 − 01	54	2.48 − 03
16	4	6.94 − 01	16	3.55 − 02	16	6.37 − 02	107	1.87 − 05
32	8	8.48 − 02	32	4.39 − 03	32	1.60 − 03	209	2.04 − 10
64	16	1.54 − 02	61	3.00 − 04	64	1.22 − 05	415	−
128	32	6.84 − 03	92	9.21 − 06	128	1.27 − 10	823	−
256	64	6.21 − 04	133	1.89 − 08	256	−	1,632	−
512	128	7.97 − 06	190	3.49 − 11	512	−	3,240	−
1,024	256	1.88 − 07	268	−	1,024	−	6,432	−

In Table 3, we further compare the above-mentioned quadrature formulae by applying them to the integral:

$$\int_0^{\infty} \frac{\sqrt{x+2}}{(x-1)^2+1}\,\mathrm{d}x \approx 5.0112918536702. \tag{11}$$

Table 3 Relative errors for integral (11)

n	n_t	SGL(2,4)	n_t	SGL(2/5,1)	n_t	MGJ(0,3/2)	n_t	DET(1/n)
4	1	3.95 − 01	4	8.27 − 01	4	4.53 − 02	36	6.59 − 06
8	2	2.10 − 01	8	9.07 − 01	8	2.44 − 03	69	1.97 − 11
16	4	3.00 − 02	16	5.58 − 02	16	3.67 − 06	136	−
32	8	1.22 − 02	32	3.21 − 05	32	6.66 − 12	268	−
64	16	1.03 − 03	61	3.45 − 07	64	−	532	−
128	32	8.19 − 06	92	5.22 − 10	128	−	1,057	−
256	64	1.78 − 06	133	4.27 − 14	256	−	2,101	−
512	128	4.56 − 10	190	−	512	−	4,178	−
1,024	256	2.27 − 13	268	−	1,024	−	8,309	−

By taking different values of the corresponding parameters for SGL(p,l) and MGJ$(0,\beta)$, in particular values that exceed those in Table 3 by $1/2$, the relative accuracy of SGL(p,l) with $l=1$ and $l=4$ remains the same as those of Table 3, while that of MGJ$(0,\beta)$ worsens hugely. Indeed, the relative accuracy of MGJ$(0,\beta)$ with $\beta=2$ for (11) decreases very slowly and for $n_t = 512$ the corresponding relative error is only $3.39-04$.

As last example, we consider

$$\int_0^{\infty} \frac{\log(x+2)}{(x+2)^2+1}\,\mathrm{d}x \approx 0.80859839124679. \tag{12}$$

This integral has been already considered in [8]; there, it was shown that for integrand functions with a log-factor, SGL(p,l) with $l=1$ or $l=4$ outperforms MGJ$(0,\beta)$. In Table 4, we report the relative errors obtained for integral (12).

Table 4 Relative errors for integral (12)

n	n_t	SGL(1,2)	n_t	SGL(1/5,1)	n_t	MGJ(0,2)	n_t	DET(1/n)
4	2	3.66 − 01	4	4.44 − 01	4	4.06 − 02	33	5.10 − 13
8	4	4.60 − 02	8	4.73 − 02	8	1.10 − 02	64	−
16	8	4.57 − 04	16	1.80 − 04	16	2.88 − 03	125	−
32	16	2.14 − 09	32	1.07 − 09	32	7.40 − 04	248	−
64	32	1.32 − 12	61	−	64	1.88 − 04	490	−
128	64	−	92	−	128	4.73 − 05	972	−
256	128	−	133	−	256	1.19 − 05	1,932	−
512	256	−	190	−	512	2.97 − 06	3,839	−

However, we will show in Table 5 that if we introduce in (12) first the change of variable

$$x = y^q, \quad q > 1 \tag{13}$$

and then apply MGJ$(0,\beta)$ with an appropriate β depending on q (in this case $\beta = q+1$), we obtain for MGJ$(0,\beta)$ numerical approximations of (12) better than those in Table 4. Indeed, in Table 5 we report the relative errors given by the application of the above formulae to the integral

$$\int_0^\infty \frac{qy^{q-1}\log(y^q+2)}{(y^q+2)^2+1}\,\mathrm{d}y \approx 0.80859839124679, \quad q = 4, \tag{14}$$

which comes out from (12) after having introduced (13). Notice that the performance of SGL$(p,1)$ is almost analogous to that of Table 4, while the accuracy of SGL(p,l) with $l > 1$ has worsened.

By examining the function that results by applying the transformation $x = (1-\mathrm{t})/(1+t)$ to the integrand of (12), one may observe that this function is not smooth, because of the log factor. Indeed, after having introduced the Möbius transformation, this factor gives rise to a corresponding one of the form $\log((t+3)/(t+1))$, which is (weakly) singular at the interval endpoint $t = -1$. The (smoothing) change of variable (13) has the effect of transforming this factor into a new one having a weak singularity of the form $(1+t)^{q-1}\log(1+t)$, which can be made as weak as we like by taking q large enough.

Table 5 Relative errors for integral (14)

n	n_t	SGL(1/4,2)	n_t	SGL(1/20,1)	n_t	MGJ(0,5)	n_t	DET(1/n)
4	2	9.65 − 01	4	9.70 − 01	4	1.98 − 01	22	6.43 − 05
8	4	2.55 − 01	8	2.85 − 01	8	2.67 − 03	42	2.67 − 09
16	8	4.34 − 03	16	1.22 − 03	16	5.96 − 05	81	−
32	16	3.59 − 05	32	8.10 − 09	32	1.06 − 09	158	−
64	32	3.53 − 07	61	−	64	1.08 − 12	313	−
128	64	4.63 − 10	92	−	128	−	620	−
256	128	−	133	−	256	−	1,226	−
512	256	−	190	−	512	−	2,429	−

3 Quadrature Formulae on Infinite Interval $(-\infty, \infty)$

In this section, we consider the numerical evaluation or discretization of integrals of the form

$$\int_{-\infty}^{\infty} g(x)\,\mathrm{d}x \tag{15}$$

with nonoscillatory integrand functions $g(x) = O(|x|^{-\rho})$, $\rho > 1$, decaying algebraically to zero for $x \to \pm\infty$. Obviously, to compute (15), we may proceed by rewriting it as follows:

$$\int_{-\infty}^{\infty} g(x)\,\mathrm{d}x = \int_{0}^{\infty} g(x)\,\mathrm{d}x + \int_{0}^{\infty} g(-x)\,\mathrm{d}x \tag{16}$$

and then applying one of the quadrature formulae of Sect. 2 to each integral of the right-hand side in (16). However, we will also consider different approaches based on five quadrature formulae: four of them new and one already proposed in [8]. This latter is the analog of (7), except for the double exponential transformation, which in this case is

$$x = \sinh\left(\frac{\pi}{2}\sinh(t)\right). \tag{17}$$

Therefore, it is defined as follows:

$$\begin{aligned} \int_{-\infty}^{\infty} g(x)\,\mathrm{d}x &= \frac{\pi}{2}\int_{-\infty}^{\infty} g\Big(\sinh\Big(\frac{\pi}{2}\sinh(t)\Big)\Big)\cosh\Big(\frac{\pi}{2}\sinh(t)\Big)\cosh(t)\,\mathrm{d}t \\ &\approx \frac{\pi}{2}h\sum_{i=-M}^{M} g\Big(\sinh\Big(\frac{\pi}{2}\sinh(ih)\Big)\Big)\cosh\Big(\frac{\pi}{2}\sinh(ih)\Big)\cosh(ih) \end{aligned} \tag{18}$$

with a mesh size h and M sufficiently large. We refer to (18) by DE2T(h), since (18) is characterized by the double exponential transformation (17), which is different from (6), and the trapezoidal rule.

Like for DET(h), the accuracy of the approximations given by DE2T(h) depends on a proper choice of h and M. In our numerical tests, we will choose h and M as we did in DET(h).

The second quadrature formula we consider consists in the introduction of the change of variable (2) and in the application of the n-point Gauss–Hermite quadrature formula, possibly truncated to its first $\lfloor n/l \rfloor$ points with $l = 2,3,4$,

$$\begin{aligned} \int_{-\infty}^{\infty} g(x)\,\mathrm{d}x &= p\int_{-\infty}^{\infty} g(\sinh(pt))\cosh(pt)\,\mathrm{d}t \\ &= \int_{-\infty}^{\infty} f(t)\exp(-t^2)\,\mathrm{d}t \approx \sum_{i=i_0}^{\lfloor n-i_0+1\rfloor} w_i^H f\left(t_i^H\right), \end{aligned} \tag{19}$$

where we have set $i_0 := \lfloor n(l-1)/(2l)+1 \rfloor$ and

$$f(t) := pg(\sinh(pt))\cosh(pt)\exp(t^2)$$

and denoted by $\{w_i^H, t_i^H\}$ the weights and the nodes of the n-point Gauss–Hermite quadrature formula, respectively. In the sequel, we refer to (19) by SGH(p,l), since (19) is characterized by the sinh transformation and the Gauss–Hermite quadrature rule. In the numerical tests below we will take $p = 1$, $l = 1$.

Notice that (19) can be considered as the analog of SGL(p,l) for the interval $(-\infty,\infty)$, although the behavior of the transformed integrand function $f(t)$ in (19), when $t \to \pm\infty$, is quite different.

To have a transformed function $f(t)$ with the same exponential behavior of that in (3) for $t \to \infty$, instead of the Gauss–Hermite weight function we may consider $w(x) = 1/\cosh(x) = 2/(\exp(x)+\exp(-x))$ and then apply the corresponding Gaussian quadrature rule. Incidentally, we notice that this weight function apparently has been first considered by Lindelöf [7, Sect. 2.4.6].

The weights and the nodes of this nonclassical (symmetric) Gaussian quadrature rule can be computed by means of the classical eigenvalue method [3, p. 118] (see also [1]). This requires the preliminary computation of the three-term recurrence coefficients for the associated orthonormal polynomials $p_n^*(x)$. Since it is known that the corresponding orthogonal polynomials $p_n(x)$, $n = 1,2,\ldots$, with leading coefficient equal to 1, satisfy the following recurrence relation:

$$p_{-1}(x) = 0, \quad p_0(x) = 1, \quad p_{n+1}(x) = xp_n(x) - \frac{\pi^2 n^2}{4} p_{n-1}(x),$$

the coefficients of the recurrence relation satisfied by $p_n^*(x)$ can be easily derived from those in the preceding recurrence relation by using the normalization constant (see [7])

$$h_n = \int_{-\infty}^{\infty} p_n^2(x) \frac{1}{\cosh(x)} \, dx = \pi \left(\frac{\pi}{2}\right)^{2n} (n!)^2.$$

Thus, denoting by $\{w_i, t_i\}$ the weights and the nodes of the n-point Gauss formula with respect to the weight $w(x)$ defined above, we have the following quadrature rule:

$$\int_{-\infty}^{\infty} g(x)\, dx = p \int_{-\infty}^{\infty} g(\sinh(pt)) \cosh(pt)\, dt = \int_{-\infty}^{\infty} \frac{f(t)}{\cosh(t)}\, dt \approx \sum_{i=i_0}^{\lfloor n-i_0+1 \rfloor} w_i f(t_i), \tag{20}$$

where we have set $i_0 := \lfloor n(l-1)/(2l)+1 \rfloor$ and

$$f(t) := pg(\sinh(pt)) \cosh(pt) \cosh(t).$$

In the sequel, we refer to (20) by SG(p,l), since (20) is characterized by the sinh transformation and the Gaussian quadrature rule with respect to the weight function $w(x) = 1/\cosh(x)$.

Similarly to SGL(p,l), one might choose the parameter $p = p_t := 1/(\rho - 1)$ for $l \neq 1$, which implies $f(t) = O(1)$, $t \to \pm\infty$. In this way, by taking into account that the weights w_i become smaller and smaller for large values of i, the terms of the quadrature sum from a certain value of i up to n are negligible. Unfortunately, in practice, the truncation of the quadrature sum defined by $\mathrm{SG}(p,l)$ does not work as well as for $\mathrm{SGL}(p,l)$. Therefore, in the sequel we have always considered $l=1$ in $\mathrm{SG}(p,l)$ and we have chosen $p=p_c:=1/(5(\rho-1))$, this last value being a good choice in the various experiments performed.

Further, we propose a new numerical quadrature. It is based on the introduction of the change of variable

$$x=\sinh\left(\frac{1}{2}\log\frac{1-t}{1+t}\right)=\sinh(\operatorname{atanh}(-t))=\frac{-t}{\sqrt{1-t^2}}, \tag{21}$$

which maps $(-\infty,\infty)$ into $(-1,1)$ and on the application of a proper Gauss–Jacobi quadrature rule. More precisely,

$$\begin{aligned}\int_{-\infty}^{\infty} g(x)\,\mathrm{d}x &= \int_{-1}^{1} g\left(-\frac{t}{\sqrt{1-t^2}}\right)\frac{1}{(1-t^2)^{3/2}}\,\mathrm{d}t \\ &= \int_{-1}^{1}\left(g\left(-\frac{t}{\sqrt{1-t^2}}\right)\frac{1}{(1-t^2)^{\gamma}}\right)\left(1-t^2\right)^{\gamma-3/2}\mathrm{d}t \\ &\approx \sum_{i=1}^{n}\frac{w_i^{(\gamma-3/2)}}{\left(1-\left(t_i^{(\gamma-3/2)}\right)^2\right)^{\gamma}} g\left(-\frac{t_i^{(\gamma-3/2)}}{\sqrt{1-\left(t_i^{(\gamma-3/2)}\right)^2}}\right),\end{aligned} \tag{22}$$

where we have denoted by $\{w_i^{(\gamma-3/2)},t_i^{(\gamma-3/2)}\}$ the weights and the nodes of the n-point Gaussian rule with respect to the Jacobi weight $(1-t^2)^{\gamma-3/2}$ respectively, with $\gamma>1/2$. In the sequel, we refer to (22) by $\mathrm{SLGJ}(\gamma)$, since (22) is characterized in particular by the sinh and log transformations and a Gauss–Jacobi quadrature rule.

By following a reasoning similar to that made for $\mathrm{MGJ}(\alpha,\beta)$, when we apply $\mathrm{SLGJ}(\gamma)$ to an integral of type (15), we will take $\gamma=\rho/2$.

It is immediate to verify that (22) is exact (i.e., it yields an error equal to zero) whenever

$$g(x)=\frac{1}{(1+x^2)^{\gamma+k}},\quad k=0,1,\ldots,n-1.$$

As a consequence, by modifying (21) as follows: $x=a-bt/\sqrt{1-t^2}$, we can use (22) to compute exactly all the integrals of the type

$$\int_{-\infty}^{\infty}\frac{\mathrm{d}x}{((x-a)^2+b^2)^{\gamma+k}},\quad k=0,1,\ldots,n-1.$$

As regards the convergence of (22), if we denote by R_n^J the remainder term in the Gauss–Jacobi formula,

$$\int_{-1}^{1} h(t)\left(1-t^2\right)^{\gamma-3/2}\mathrm{d}x=\sum_{i=1}^{n} w_i^{(\gamma-3/2)} h\left(t_i^{(\gamma-3/2)}\right)+R_n^J(h)$$

and by R_n the remainder term of the quadrature formula

$$\int_{-\infty}^{\infty} g(x)\,\mathrm{d}x = \sum_{i=1}^{n} W_i g(T_i) + R_n(g)$$

with

$$W_i = \frac{w_i^{(\gamma-3/2)}}{\left(1-\left(t_i^{(\gamma-3/2)}\right)^2\right)^{\gamma}}, \quad T_i = -\frac{t_i^{(\gamma-3/2)}}{\sqrt{1-\left(t_i^{(\gamma-3/2)}\right)^2}},$$

we have $R_n(g) = R_n^J(h)$, $h(t) = (1-t^2)^{-\gamma} g\left(-t/\sqrt{1-t^2}\right)$. Therefore, for integrand functions of the type

$$g(x) = (1+x^2)^{-\gamma} G(x), \tag{23}$$

we have

$$h(t) = G\left(-\frac{t}{\sqrt{1-t^2}}\right), \quad -1 < t < 1,$$

and, by taking into account that the inverse function of $x = -t/\sqrt{1-t^2}$ is $t = -x/\sqrt{1+x^2}$,

$$G(x) = h\left(-\frac{x}{\sqrt{1+x^2}}\right), \quad -\infty < x < \infty.$$

Since the convergence $R_n^J(h) \to 0$ is guaranteed for any continuous function h (in fact, bounded Riemann-integrable) in the closed interval $[-1,1]$, we have that $R_n(f) \to 0$ for any (integrable) G which is continuous in $(-\infty,\infty)$. Moreover, if $h(t)$ is analytic in a disc $|t| < r$, with $r > 1$, which implies that G must necessarily be analytic outside the regions defined by $|1/z^2+1| \geq 1/r^2$ and represented in Fig. 6 for $r = 2,\ldots,5$, then the convergence is exponentially fast, the rate of convergence being larger the larger $r > 1$. Notice that as r increases, the "circular" regions in Fig. 6 become smaller and degenerate to the points $z = \pm \mathrm{i}$ when $r \to \infty$.

Incidentally, note that (22) can be obtained in a different way. Indeed, if we first introduce in (15) the change of variable $x = \tan(t)$,

$$\begin{aligned} \int_{-\infty}^{\infty} g(x)\,\mathrm{d}x &= \int_{-\pi/2}^{\pi/2} g(\tan(t)) \frac{1}{\cos^2(t)}\,\mathrm{d}t \\ &= \int_{-\pi/2}^{\pi/2} g\left(\frac{\sin(t)}{\sqrt{1-\sin^2(t)}}\right) \frac{1}{1-\sin^2(t)}\,\mathrm{d}t, \end{aligned} \tag{24}$$

and then set $t = -\mathrm{asinh}(y)$, which maps $(-\pi/2,\pi/2)$ into $(-1,1)$, we obtain the same transformed integral as in (22), which can be computed by the n-point Gauss–Jacobi quadrature rule with respect to the Jacobi weight $(1-t^2)^{\gamma-3/2}$. We have to remark that in the literature (see [3, p. 200], [10, p. 737]) the substitution $x = \tan(t)$ has already been proposed for integrand function $g(x)$ regular at $\pm\infty$, and combined

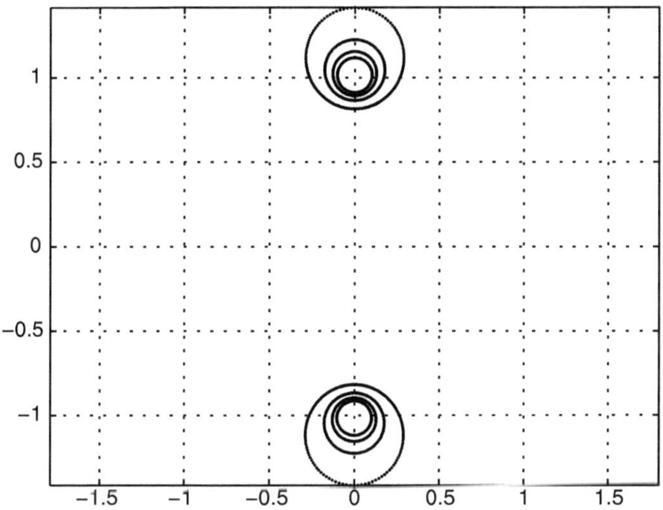

Fig. 6 Regions defined by $|1/z^2+1| \geq 1/r^2$ with $r = 2,\ldots,5$

with the trapezoidal rule since the integrand is a periodic function in t. However, as it is well known, the trapezoidal rule excels only for regular integrand functions; otherwise, it is not sufficiently efficient. In contrast, as we will see in the numerical tests, our method is efficient for nonregular functions also.

The first example we consider is chosen to illustrate how the rate of convergence of (22) depends on the analyticity properties of the function $h(t)$. To this aim, we choose

$$G_r(x) = \frac{1}{x^2 + r/(r+1)}$$

and we compare the quadrature formulae SLGJ(γ), MGJ$(0,\beta)$ and the trapezoidal rule applied to (24) to which we refer by TT (acronym of the tan transformation and the trapezoidal rule), for the computation of the following integral:

$$\int_{-\infty}^{\infty} \frac{G_r(x)}{(x^2+1)^{12.5}}\,\mathrm{d}x. \tag{25}$$

In Table 6 we report the relative errors for integral (25) and $r = 1.1, 2, 5, 10$. Notice that the complex poles $\pm \mathrm{i}\sqrt{(r+1)/r}$ represent the minimum and maximum values of the top and bottom curves, respectively, drawn in Fig. 6.

To use the same number of quadrature nodes, MGJ(α,β) has been applied to each of the two integrals on the right-hand side in (16) with half as many quadrature nodes compared to the other approaches.

Finally, we propose another new quadrature formula. It is based on the change of variable

$$x = \frac{2t}{1-t^2}, \tag{26}$$

Table 6 Relative errors for integral (25) and $r = 1.1, 2, 5, 10$

	SLGJ(13.5)	MGJ(0,27)	TT	SLGJ(13.5)	MGJ(0,27)	TT
n		$r = 1.1$			$r = 2$	
4	1.03 − 05	1.70 − 01	6.33 − 01	1.15 − 06	1.77 − 01	6.11 − 01
8	1.23 − 09	3.47 − 03	1.57 − 02	1.80 − 11	3.68 − 03	1.30 − 02
16	–	3.31 − 06	1.91 − 08	–	4.02 − 06	1.14 − 09
32	–	5.44 − 13	–	–	4.63 − 12	–
64	–	–	–	–	–	–
		$r = 5$			$r = 10$	
4	3.46 − 08	1.83 − 01	5.94 − 01	2.29 − 09	1.85 − 01	5.89 − 01
8	–	3.80 − 03	1.12 − 02	–	3.83 − 03	1.06 − 02
16	–	4.46 − 06	1.89 − 11	–	4.56 − 06	1.54 − 12
32	–	9.01 − 12	–	–	1.05 − 11	–
64	–	–	–	–	–	–

which maps $(-\infty,\infty)$ into $(-1,1)$, and on the application of an appropriate Gauss–Jacobi quadrature rule:

$$\begin{aligned}\int_{-\infty}^{\infty} g(x)\,\mathrm{d}x &= \int_{-1}^{1} g\left(\frac{2t}{1-t^2}\right)\frac{2(1+t^2)}{(1-t^2)^2}\,\mathrm{d}t \\ &= \int_{-1}^{1}\left(g\left(\frac{2t}{1-t^2}\right)\frac{2(1+t^2)}{(1-t^2)^{\gamma}}\right)(1-t^2)^{\gamma-2}\,\mathrm{d}t \quad (\gamma > -1) \\ &\approx \sum_{i=1}^{n}\frac{2w_i^{(\gamma-2)}\left(1+\left(t_i^{(\gamma-2)}\right)^2\right)}{\left(1-\left(t_i^{(\gamma-2)}\right)^2\right)^{\gamma}}\, g\left(\frac{2t_i^{(\gamma-2)}}{1-\left(t_i^{(\gamma-2)}\right)^2}\right),\end{aligned} \tag{27}$$

where we have denoted by $\{w_i^{(\gamma-2)}, t_i^{(\gamma-2)}\}$ the weights and the nodes of the n-point Gaussian rule with respect to the Jacobi weight $(1-t^2)^{\gamma-2}$, respectively.

It is immediate to verify that (27) is exact (i.e., it yields an error equal to zero) whenever $g(x) = 1/\left(\sqrt{1+x^2}+1\right)^{\gamma+k}$, $k = 0,1,\ldots,n-2$.

In the sequel we refer to (27) by M2GJ(γ), since the negative transformation (26) is the mean value of a Möbius transformation and its reciprocal: $x = ((1-t)/(1+t) - (1+t)/(1-t))/2$. By the acronym M2GJ, we point out the fact that (27) is characterized by two Möbius transformations and a Gauss–Jacobi quadrature rule.

For a (smooth) function $g(x)$ decaying at $\pm\infty$ like $|x|^{-\rho}$, $\rho > 1$, we set $\gamma = \rho$.

Next we will compare the numerical approaches SLGJ(γ), MGJ(α,β), and M2GJ(γ), which reduce the original integration domain $(-\infty,\infty)$ to $(-1,1)$, with SGH(p,l), SG(p,l), and DE2T(h), which leave unchanged the original integration domain. In the following, if the total number of quadrature nodes n_t is not explicitly specified, then $n_t = n$.

As first example we compare the above-mentioned quadrature formulae on the computation of the following integral:

$$\int_{-\infty}^{\infty} \frac{\mathrm{d}x}{(x^2+1)^{5/4}} = \mathrm{Beta}\left(\frac{1}{2}, \frac{3}{4}\right), \tag{28}$$

already considered in [10].

Table 7 Relative errors for integral (28)

n	SLGJ(5/4)	MGJ(0,5/2)	M2GJ(5/2)	SGH(1,1)	SG(2/15,1)	n_t	DE2T(1/n)
4	–	4.88 – 02	2.36 – 03	4.08 – 02	2.10 – 01	27	–
8	–	1.66 – 03	2.73 – 06	7.04 – 03	2.00 – 02	55	–
16	–	1.68 – 06	2.81 – 12	5.78 – 04	1.49 – 04	109	–
32	–	1.49 – 12	–	1.68 – 05	6.90 – 09	217	–
64	–	–	–	1.13 – 07	–	433	–
128	– –	–	–	9.64 – 11	–	859	–

We remark that by applying TT, i.e., the trapezoidal rule to (24) for $h = \pi/512$, we obtain only 8.34 – 05 as relative error. In contrast, in this particular case, the approach SLGJ(γ), with $\gamma = 5/4$, is exact, as shown in Table 7.

Next, we consider the following integral:

$$\int_{-\infty}^{\infty} \frac{\mathrm{d}x}{x^4 - 3x^3 + 2x^2 + 1} \approx 3.6275987284684, \tag{29}$$

whose relative errors are reported in Table 8. Notice that in this case SLGJ(γ) gives the best performance; it reaches full relative accuracy with a number of nodes less than that for the other formulae; namely with $n_t = 184$.

Table 8 Relative errors for integral (29)

n	SLGJ(2)	MGJ(0,4)	M2GJ(4)	SGH(1,1)	SG(1/15,1)	n_t	DE2T(1/n)
4	2.49 – 01	9.43 – 02	4.37 – 01	4.02 – 01	7.03 – 01	23	2.20 – 01
8	2.50 – 01	2.88 – 01	2.75 – 01	1.60 – 01	2.55 – 01	43	2.63 – 02
16	6.07 – 02	9.02 – 02	1.27 – 01	8.32 – 02	1.66 – 02	87	3.85 – 04
32	6.42 – 03	5.26 – 02	6.58 – 03	2.33 – 02	3.04 – 02	173	2.71 – 08
64	2.32 – 05	1.32 – 03	1.96 – 04	1.68 – 02	1.29 – 02	343	–
128	1.18 – 09	1.00 – 05	4.23 – 08	5.89 – 03	2.57 – 03	683	–

We now consider the following integral:

$$\int_{-\infty}^{\infty} \frac{\sqrt{x^2+2}}{((x-1)^2+1)^{5/3}} \mathrm{d}x \approx 3.66435178586511; \tag{30}$$

the corresponding relative errors, reported in Table 9, show that SLGJ(7/6) is again the most accurate formula.

Table 9 Relative errors for integral (30)

n	SLGJ(7/6)	MGJ(0,7/3)	M2GJ(7/3)	SGH(1,1)	SG(3/20,1)	n_t	DE2T(1/n)
4	6.34 − 02	1.70 − 01	1.19 − 01	1.68 − 01	2.30 − 01	29	7.55 − 04
8	9.30 − 03	6.61 − 02	3.84 − 02	1.78 − 02	9.51 − 03	57	1.49 − 07
16	1.12 − 05	7.13 − 03	2.62 − 04	5.12 − 03	1.79 − 04	113	–
32	7.15 − 11	1.98 − 05	6.08 − 08	4.77 − 06	4.00 − 05	225	–
64	–	9.73 − 11	–	2.43 − 06	1.13 − 05	445	–
128	–	–	–	6.83 − 09	2.11 − 06	887	–

Finally, we consider

$$\int_{-\infty}^{\infty} \frac{\log(x^2+2)}{((x-1)^2+1)^{5/3}}\,\mathrm{d}x \approx 2.33559889291266. \tag{31}$$

Table 10 Relative errors for integral (31)

n	SLGJ(5/3)	MGJ(0,10/3)	M2GJ(10/3)	SGH(1,1)	SG(3/35,1)	n_t	DE2T(1/n)
4	1.48 − 01	5.09 − 01	2.17 − 01	1.67 − 01	7.75 − 01	25	5.73 − 04
8	8.88 − 02	8.83 − 02	1.31 − 02	3.30 − 02	9.47 − 02	49	2.51 − 07
16	1.53 − 03	4.67 − 03	3.63 − 04	5.33 − 03	1.62 − 03	99	–
32	3.33 − 04	8.05 − 06	5.11 − 07	2.64 − 05	1.82 − 05	195	–
64	6.83 − 05	1.11 − 07	2.29 − 08	3.16 − 06	5.27 − 07	389	–
128	1.37 − 05	4.77 − 09	9.63 − 10	1.02 − 08	9.13 − 10	773	–

For integral (31) M2GJ$(10/3)$ outperforms all the other formulae. Notice that in this case the numerical approximations of SG$(p,1)$ are more accurate than SLGJ(γ) and MGJ(α,β); however, as in Sect. 1, we can improve the relative accuracy of SLGJ(γ) and MGJ(α,β) by introducing the preliminary change of variable $x=y^q$, with an odd integer q, and hence to apply the above-mentioned formulae to the following integral:

$$\int_{-\infty}^{\infty} \frac{qy^{q-1}\log(y^{2q}+2)}{((y^q-1)^2+1)^{5/3}}\,\mathrm{d}y \approx 2.33559889291266. \tag{32}$$

Notice that, although SLGJ(γ) was less accurate than MGJ(α,β) in Table 10 (i.e., for $q=1$), in Table 11 for $q=3$ it reaches a precision which is better than that of MGJ(α,β); contrary to the approach SLGJ(γ), the accuracy of MGJ(α,β) and M2GJ(γ) for $q=3$ is worse than that for $q=1$.

However, there is a class of integrand functions for which SLGJ(γ) gives approximations of the corresponding integral much less accurate than those of MGJ(α,β) or M2GJ(γ). It consists of a product of combinations of powers, integer or not, and

Table 11 Relative errors for integral (32)

n	SLGJ(12/3)	MGJ(0,24/3)	M2GJ(24/3)
4	1.47 − 01	8.83 − 01	2.57 − 01
8	2.38 − 01	3.42 − 01	4.16 − 01
16	4.96 − 02	3.39 − 01	4.95 − 02
32	9.06 − 04	9.50 − 03	8.67 − 03
64	3.65 − 07	8.81 − 04	3.19 − 05
128	4.55 − 12	5.70 − 08	1.66 − 10

polynomials such that it has only complex poles. In Tables 12 and 13, we report the relative errors for the following two integrals:

$$\int_{-\infty}^{\infty} \frac{\sqrt{x^2+1}+x+1}{(x^2+x+2)^2}\,\mathrm{d}x \approx 1.35099223786288, \tag{33}$$

$$\int_{-\infty}^{\infty} \frac{x\sqrt{x^2+1}+1}{(x^2+x+2)^2}\,\mathrm{d}x \approx -3.39053915109880-02, \tag{34}$$

respectively.

Table 12 Relative errors for integral (33)

n	SLGJ(3/2)	MGJ(0,3)	M2GJ(3)	SGH(1,1)	SG(1/10,1)	n_t	DE2T(1/n)
4	1.64 − 03	1.40 − 02	1.09 − 02	1.96 − 02	3.37 − 01	24	5.71 − 06
8	1.32 − 03	7.91 − 03	5.75 − 04	3.32 − 03	3.39 − 02	48	1.48 − 12
16	1.44 − 04	1.19 − 04	1.98 − 07	1.71 − 05	2.75 − 04	95	−
32	1.80 − 05	7.50 − 09	−	6.25 − 07	1.99 − 08	188	−
64	2.28 − 06	−	−	8.46 − 10	2.93 − 10	374	−
128	2.88 − 07	−	−	−	3.46 − 12	743	−

Table 13 Relative errors for integral (34)

n	SLGJ(1)	MGJ(0,2)	M2GJ(2)	SGH(1,1)	SG(1/5,1)	n_t	DE2T(1/n)
4	5.90 − 00	4.82 − 00	4.92 − 00	2.01 − 00	3.03 − 00	31	3.61 − 04
8	7.78 − 01	1.66 − 00	1.24 − 01	3.71 − 02	2.58 − 01	63	1.93 − 10
16	1.92 − 01	4.08 − 04	9.09 − 05	1.03 − 02	3.15 − 02	127	−
32	4.76 − 02	3.01 − 07	1.17 − 11	3.96 − 05	2.61 − 04	251	−
64	1.19 − 02	−	−	5.34 − 08	3.58 − 04	499	−
128	2.96 − 03	−	−	7.04 − 12	4.17 − 05	995	−

Finally, we consider two integrals whose integrand functions have real and weak singularities. For these integrals, all the above quadrature formulae perform very poorly; however, it is possible to improve their accuracy by introducing a proper polynomial change of variable. Indeed, in the case of the following integral:

$$\int_{-\infty}^{\infty} \frac{|x-1|^{1/4}}{(x^2+x+2)^2}\,\mathrm{d}y \approx -6.50814879957545-01 \tag{35}$$

it is enough to introduce the change $x = 1+t^q$, with an odd q. For example, for $q=3$, we have obtained the relative errors of Table 14. Incidentally, we notice that for $q=1$, DE2T gives $8.34-05$ with $n_t = 1423$ and the other formulae attain an accuracy larger or equal to 10^{-4} for $n=256$.

Table 14 Relative errors for integral (35) with the change of variable $x = 1+t^3$

n	SLGJ(37/8)	MGJ(0,37/4)	M2GJ(37/4)	SGH(1,1)	SG(4/165,1)	n_t	DE2T(1/n)
4	7.52 − 01	9.96 − 01	1.27 − 00	9.64 − 01	1.00 − 00	15	5.21 − 01
8	4.94 − 01	1.73 − 00	4.89 − 01	1.34 − 01	9.99 − 01	31	8.36 − 02
16	1.37 − 01	2.15 − 01	2.59 − 01	4.90 − 01	3.75 − 01	61	1.00 − 03
32	3.96 − 03	2.56 − 03	1.27 − 02	3.02 − 01	9.10 − 03	119	8.27 − 08
64	2.04 − 06	3.77 − 03	1.02 − 04	3.10 − 02	2.89 − 04	237	–
128	2.28 − 13	1.43 − 06	7.56 − 09	4.03 − 03	9.00 − 06	469	–

The previous change of variable can also be used to smooth a weak logarithmic singularity, as for example in the following integral,

$$\int_{-\infty}^{\infty} \frac{\log|x-1|}{(x^2+x+2)^2}\,\mathrm{d}y \approx 1.73461779301527-01. \tag{36}$$

In Table 15 we report the relative errors obtained for $q=5$.

In Table 15, we have not considered DE2T$(1/n)$ since the corresponding quadrature sum requires the evaluation of the transformed integrand function at 0 and this latter is not defined there.

Table 15 Relative errors for integral (36) with the change of variable $x = 1+t^5$

n	SLGJ(8)	MGJ(0,16)	M2GJ(16)	SGH(1,1)	SG(1/45,1)
4	1.67 − 00	1.02 − 00	2.42 − 00	2.19 − 00	1.00 − 00
8	3.57 − 00	1.93 − 00	2.47 − 00	6.33 − 01	1.01 − 00
16	1.04 − 00	1.39 − 01	1.33 − 00	2.30 − 00	1.22 − 00
32	1.69 − 01	3.23 − 01	2.47 − 01	3.59 − 00	2.57 − 02
64	3.13 − 04	1.64 − 01	2.46 − 02	1.78 − 00	7.14 − 03
128	1.90 − 07	1.80 − 03	3.89 − 05	7.77 − 01	1.89 − 04

4 An Application to Some Integral Equations

In this section, we apply the most efficient quadrature formulae we have proposed in Sects. 2 and 3, to construct a Nyström method for the solution of integral equations of the following type:

$$f(y) - \mu \int_a^b k(x,y) f(x)\,\mathrm{d}x = g(y), \tag{37}$$

where $k(x,y)$ and $g(y)$ are known functions, μ is a given real number and $(a,b) = (0,\infty)$ or $(a,b) = (-\infty,\infty)$. Since all the quadrature formulae we have presented are based on a preliminary change of variable (that here we generally denote by $x = \varphi(t)$) followed by the application of a Gaussian formula, the corresponding Nyström method is obtained first by setting $x = \varphi(t)$, $y = \varphi(s)$ in (37)

$$f(\varphi(s)) - \mu \int_{\varphi^{-1}(a)}^{\varphi^{-1}(b)} k(\varphi(t),\varphi(s)) f(\varphi(t)) \varphi'(t)\,\mathrm{d}t = g(\varphi(s))$$

and then by using the chosen Gaussian formula, combined with φ and yielding one of the formulae of Sects. 2 or 3,

$$f_n(\varphi(s)) - \mu \sum_{i=1}^{n} w_i k(\varphi(t_i),\varphi(s)) f_n(\varphi(t_i)) \varphi'(t_i) = g(\varphi(s)).$$

After having collocated the new approximate equation at the quadrature nodes t_k, $k = 1,\ldots,n$, and solved the resulting linear system of order n for the unknowns $F_n(t_k) := f_n(\varphi(t_k))$, the approximate solution $F_n(s)$ is defined by the Nyström interpolation formula

$$F_n(s) = \mu \sum_{i=1}^{n} w_i k(\varphi(t_i),\varphi(s)) F_n(t_i) \varphi'(t_i) + g(\varphi(s)). \tag{38}$$

Finally, the approximation of the original solution is given by $f_n(z) = F_n(\varphi^{-1}(z))$.

In the case $(a,b) = (0,\infty)$, we have considered the quadrature schemes $\mathrm{SGL}(p,l)$ and $\mathrm{MGJ}(\alpha,\beta)$, defined in (3) and (5), respectively. As a first example, we have set

$$\mu = \frac{1}{8}, \quad k(x,y) = \frac{1}{1+(1+x^2)+(1+y^2)}, \quad g(y) = \frac{1}{1+(1+y^2)} \tag{39}$$

in (37). This example has been recently considered in [8]. In Table 16 we report the relative errors for the two corresponding Nyström interpolants $f_n^{\mathrm{SGL}}(z)$ and $f_n^{\mathrm{MGJ}}(z)$, evaluated for example at $z = 0.5$.

For both quadrature schemes, the corresponding linear systems are well conditioned, since their condition numbers are, respectively, 1.06 and 1.07, for any value of n.

Table 16 Relative errors for the Nyström interpolants $F_n^{\mathrm{SGL}}(\varphi^{-1}(0.5))$ and $F_n^{\mathrm{MGJ}}(\varphi^{-1}(0.5))$ defined by (38), with the inputs (39)

n	SGL$(1,2)$	SGL$(1/5,1)$	MGJ$(0,2)$
4	5.70 − 05	3.91 − 05	2.29 − 05
8	2.19 − 06	1.67 − 08	1.06 − 08
16	1.32 − 08	1.02 − 11	7.50 − 11
32	2.14 − 11	–	3.62 − 13
64	–	–	–
128	–	–	–

For the second example, we have set

$$\mu = \frac{1}{\pi}, \; k(x,y) = \frac{x\sin(\chi\pi)}{x^2+2xy\cos(\chi\pi)+y^2}, \; \chi = \frac{1}{3}, \; g(y) = \frac{1}{\pi(1+(1+y^2))} \tag{40}$$

in (37). The kernel k of this example appears in the boundary integral equation reformulation of a Dirichlet problem for Laplace's equation on a simple wedge, whose arms are straight line segment of infinite length and the central angle is $(1-\chi)\pi$, with $0<\chi<1$. In Table 17 we report the relative errors for the two corresponding Nyström interpolants $f_n^{\mathrm{SGL}}(z)$ and $f_n^{\mathrm{MGJ}}(z)$, evaluated at $z=0.5$.

Table 17 Relative errors for the Nyström interpolants $F_n^{\mathrm{SGL}}(\varphi^{-1}(0.5))$ and $F_n^{\mathrm{MGJ}}(\varphi^{-1}(0.5))$ defined by (38), with the inputs (40)

n	SGL$(1,2)$	SGL$(1/5,1)$	MGJ$(0,2)$
4	1.95 − 03	2.30 − 03	5.60 − 04
8	9.13 − 04	7.30 − 05	2.63 − 05
16	1.43 − 04	1.31 − 05	7.66 − 07
32	2.49 − 05	2.19 − 06	1.88 − 08
64	4.09 − 06	3.47 − 07	4.18 − 10
128	6.44 − 07	5.32 − 08	8.54 − 12

Contrary to the previous example, the condition numbers of the linear systems for (40) do not remain constant, but they slightly increase with n and are about equal to 45 for SGL(p,l) $(l=1,2)$ and 17 for MGJ(α,β), when $n=128$.

Finally, for the case $(a,b)=(-\infty,\infty)$ we have considered the quadrature schemes SLGJ(γ) and M2GJ(γ), defined in (22) and (27), respectively. As an example for this case, we have considered (37) with the input functions (39), all defined on $(-\infty,\infty)$. In Table 18 we report the relative errors for the Nyström interpolants $F_n^{\mathrm{SLGJ}}(\varphi^{-1}(0.5))$ and $F_n^{\mathrm{M2GJ}}(\varphi^{-1}(0.5))$.

The condition numbers of the corresponding linear systems are 1.39 and 1.32 for the methods SLGJ and M2GJ, respectively.

Table 18 Relative errors for the Nyström interpolants $F_n^{\mathrm{SLGJ}}(\varphi^{-1}(0.5))$ and $F_n^{\mathrm{M2GJ}}(\varphi^{-1}(0.5))$ defined by (38), with the inputs (39)

n	SLGJ(1)	M2GJ(2)
4	8.56 − 04	5.54 − 03
8	2.09 − 03	9.63 − 03
16	6.75 − 06	1.24 − 05
32	7.31 − 07	9.63 − 09
64	4.78 − 08	1.29 − 13
128	2.86 − 09	−

As last example, we have considered (37) with the following input functions:

$$\mu = \frac{1}{\pi}, \quad k(x,y) = \frac{1}{y^2+b^2}\frac{a^3}{a^2+(y-x)^2}, \quad g(y) = 1 \tag{41}$$

with a, b to be assigned. The kernel k is encountered in atomic and nuclear physics.

In Table 19 we report the relative errors for the Nyström interpolants $F_n^{\mathrm{SLGJ}}(\varphi^{-1}(0.5))$ and $F_n^{\mathrm{M2GJ}}(\varphi^{-1}(0.5))$, by setting $a = 0.1, 1$ and $b = 1$ in (41).

Table 19 Relative errors for the Nyström interpolants $F_n^{\mathrm{SLGJ}}(\varphi^{-1}(0.5))$ and $F_n^{\mathrm{M2GJ}}(\varphi^{-1}(0.5))$ defined by (38), with the inputs (41)

	$a = 0.1$		$a = 1$	
n	SLGJ(1)	M2GJ(2)	SLGJ(1)	M2GJ(2)
4	7.53 − 03	2.63 − 03	4.43 − 03	7.03 − 02
8	2.49 − 03	2.24 − 03	1.84 − 03	2.30 − 03
16	8.21 − 04	9.58 − 04	2.65 − 04	1.75 − 04
32	4.86 − 06	9.92 − 04	4.11 − 05	1.26 − 05
64	6.09 − 07	4.76 − 05	6.72 − 06	1.04 − 06
128	3.96 − 09	3.61 − 08	1.14 − 06	9.21 − 08

5 Conclusions

In Sects. 2 and 3 we have proposed several quadrature rules to compute integrals of functions defined on unbounded intervals and having a mild (algebraic) decay at ∞. We point out, however, that for all these rules, except the truncated Gauss–Laguerre and Gauss–Hermite ones, the corresponding (known) error estimates are of pointwise type, that is, they hold for a given function. Therefore, they are not useful to prove the stability property of the associated Nyström method. Thus, although rules like MGJ, SLGJ, and M2GJ in general have shown better performances, even when they are used to solve an integral equation by means of the associated Nyström method, error estimates needed to prove the stability of the latter are not yet available.

Another advantage of the truncated Gauss–Laguerre and Gauss–Hermite rules is that they only require the function to be smooth and the knowledge of an estimate of the integrand function decay at ∞. For the MGJ, SLGJ, and M2GJ rules, if the exact behavior is not known, the accuracy generally decreases significantly. Moreover, as we have seen in example (12), in the case of the rules based on the Gauss–Jacobi formulas the preliminary change of variable can introduce extra (weak) singularities that will then reduce the accuracy of the overall quadrature, unless one introduces a further change of variable (see (14)). This is a severe drawback since in the case of integral equations, we generally do not know the analytic representation of the solution.

References

1. Cvetković, A.S., Milovanović, G.V.: The mathematica package “Orthogonal Polynomials”, Facta Universitatis (Niš). Ser. Math. Inform. **19**, 17–36 (2004)
2. De Bonis, M.C., Mastroianni, G.: Nyström method for systems of integral equations on the real semiaxis. IMA J. Numer. Anal. **29**, 632–650 (2009)
3. Davis, P.J., Rabinowitz, P.: Methods of numerical integration. Academic Press, New York (1984)
4. Evans, G.A.: Some new thoughts on Gauss-Laguerre quadrature. Int. J. Comput. Math. **82**, 721–730 (2005)
5. Gautschi, W.: Quadrature formulae on half-infinite intervals. BIT **31**, 438–446 (1991)
6. Mastroianni, G., Milovanović, G.V.: Some numerical methods for second kind Fredholm integral equation on the real semiaxis. IMA J. Numer. Anal. **29**, 1046–1066 (2009) doi:10.1093/imanum/drn056
7. Mastroianni, G., Milovanović, G.V.: Interpolation processes. Basic theory and applications. Springer, Berlin (2008)
8. Mastroianni, G., Monegato, G.: Some new applications of truncated Gauss-Laguerre quadrature formulas. Numer. Alg. **49**, 283–297 (2008) doi: 10.1007/s11075-008-9191-x
9. Mori, M.: Discovery of the double exponential transformation and its developments. Publ. RIMS Kyoto Univ. **41**, 897–935 (2005)
10. Takahasi, H., Mori, M.: Double exponential formulas for numerical integration. Publ. RIMS Kyoto Univ. **9**, 721–741 (1974)

Gauss-Type Quadrature Formulae for Parabolic Splines with Equidistant Knots

Geno Nikolov and Corina Simian

Dedicated to Professor Gradimir V. Milovanović on the occasion of his 60th birthday

1 Introduction

We consider quadrature formulae of the type

$$Q[f] = \sum_{i=1}^{n} a_i f(\xi_i), \quad 0 \le \xi_1 < \cdots < \xi_n \le 1, \tag{1}$$

that approximate the definite integral

$$I[f] := \int_0^1 f(x)\,\mathrm{d}x.$$

The classical approach for construction of quadrature formulae is based on the concept of *algebraic degree of precision*. The quadrature formula (1) is said to have algebraic degree of precision m (in brief, ADP(Q) $= m$), if the remainder functional $R[Q;f] := I[f] - Q[f]$ vanishes whenever f is an algebraic polynomial of degree at most m, and $R[Q;f] \neq 0$ when f is a polynomial of degree $m+1$. The pursuit of quadrature formula (1) with the highest possible ADP leads to the classical quadrature formulae of Gauss, Radau and Lobatto. They are uniquely determined

Geno Nikolov
Faculty of Mathematics and Informatics, Sofia University St. Kliment Ohridski, 5 James Baurchier Boulevard, 1164 Sofia, Bulgaria, geno@fmi.uni-sofia.bg

Corina Simian
Lucian Blaga University of Sibiu, Ion Raţiu 5-7, 550012 Sibiu, Romania, corinafirst@yahoo.com

W. Gautschi et al. (eds.), *Approximation and Computation: In Honor of Gradimir V. Milovanović*, Springer Optimization and Its Applications 42,
DOI 10.1007/978-1-4419-6594-3_14,

by having ADP equal to $2n-1$, $2n-2$ and $2n-3$, respectively, where, in addition, the left (right) Radau quadrature formula has as a node the left (right) end point of the integration interval, while the Lobatto quadrature formulae have both end points of the integration interval as nodes.

In [8] the first-named author initiated the study of Gauss-type quadrature formulae associated with spaces of polynomial splines. Instead of looking for the highest possible ADP, it is required that a quadrature formula is exact for spaces of polynomial splines of maximal dimension. Two main reasons explain the interest in such quadratures. First, in many cases spline functions of relatively low degree provide much better approximation than spaces of algebraic polynomials of the same dimension. Second, it has been shown in [5] that the Gauss-type quadrature formulae associated with the spaces of splines of given degree and with equidistant knots are asymptotically optimal in certain Sobolev classes of functions.

The concept of optimal quadrature formulae originates from the 40s of the twentieth century, its founders being Kolmogorov, Sard, and Nikolskii. For a given normed linear space X of functions defined in $[0,1]$, we denote by $\mathcal{E}(Q,X)$ the error of a quadrature formula Q of the form (1) in the unit ball of X,

$$\mathcal{E}(Q,X) := \sup_{\|f\|_X \le 1} |R[Q;f]|.$$

Then we look for the best possible choice of the coefficients $\{a_i\}$ and the nodes $\{\xi_i\}$, setting

$$\mathcal{E}_n(X) := \inf_Q \mathcal{E}(Q,X).$$

If the quantity $\mathcal{E}_n(X)$ is attained for a quadrature formula $Q^{\rm opt}$ of the form (1), then $Q^{\rm opt}$ is said to be an optimal quadrature formula of the type (1) in the space X. Of particular interest is the case when X is some of the Sobolev spaces of functions

$$\begin{aligned}\widetilde{W}_p^r &:= \{f \in C^{r-1}[0,1],\ f-\text{one periodic},\ f^{(r-1)}\ \text{abs. cont.},\ \|f\|_p < \infty\},\\ W_p^r &:= \{f \in C^{r-1}[0,1],\ f^{(r-1)}\ \text{abs. cont.}, \|f\|_p < \infty\},\end{aligned}$$

where

$$\|f\|_p := \left(\int_0^1 |f(t)|^p\,\mathrm{d}t\right)^{1/p},\ \text{if}\ 1 \le p < \infty,\ \text{and}\ \|f\|_\infty = \sup_{t\in(0,1)} \text{vrai}\,|f(t)|.$$

In all periodic Sobolev spaces $\widetilde{W}_p^r$, there is a unique (up to translation) optimal quadrature formula, namely, the rectangle rule. This is a result of Zhensykbaev [9], special cases having been obtained earlier by Motornii [7] and Ligun [6]. The existence and uniqueness of optimal quadrature formulae in the nonperiodic Sobolev spaces W_p^r is equivalent to existence and uniqueness of monosplines of degree r of minimal L_q-deviation ($1/p+1/q=1$). This result is due to Zhensykbaev [10] and, in a more general setting, for quadrature formulae involving derivatives of the integrand, to Bojanov [1,2]. Unfortunately, except for some special cases with $r=1$ and $r=2$, the optimal quadrature formulae in the non-periodic Sobolev spaces are unknown. Obviously, $\mathcal{E}_n(\widetilde{W}_p^r) \le \mathcal{E}_n(W_p^r)$, and it is known that (see Brass [3]) for $1 < p \le \infty$,

$$\lim_{n\to\infty} \frac{\mathcal{E}_n(\widetilde{W}_p^r)}{\mathcal{E}_n(W_p^r)} = 1.$$

Denote by π_s the set of all algebraic polynomials of degree at most s. Given $r \in \mathbb{N}$ and a mesh $\Delta = \{x_i\}_{i=1}^{m-1}$, $0 = x_0 < x_1 < \cdots < x_{m-1} < x_m = 1$, let $S_{r-1,\Delta}$ be the linear space of spline functions of degree $r-1$ with simple knots defined by Δ,

$$S_{r-1,\Delta} := \{f \in C^{r-2}[0,1] \,:\, f|_{(x_{i-1},x_i)} \in \pi_{r-1},\ i = 1,\dots,m\}.$$

The dimension of the space $S_{r-1,\Delta}$ is $m+r-1$, and a basis for $S_{r-1,\Delta}$ is given by

$$\{1, x, \dots, x^{r-1}, (x-x_1)_+^{r-1}, \dots, (x-x_{m-1})_+^{r-1}\},$$

where $u_+ := \max\{u, 0\}$.

According to the fundamental theorem of algebra for monosplines satisfying boundary conditions [4, Theorem 0.1], for any given set of knots Δ such that $\dim S_{r-1,\Delta} = 2n$, there exists a unique n-point Gaussian quadrature formula Q_n^G, and a unique $(n+1)$-point Lobatto quadrature formula Q_{n+1}^L with $\xi_1 = 0$ and $\xi_{n+1} = 1$ as nodes, which are exact for all functions from $S_{r-1,\Delta}$. Similarly, in the case $\dim S_{r-1,\Delta} = 2n-1$, there exist unique n-point left Radau quadrature formula $Q_n^{R,l}$ (with $\xi_1 = 0$) and right Radau quadrature formula $Q_n^{R,r}$ (with $\xi_n = 1$), which are exact for all functions from $S_{r-1,\Delta}$.

In [5] it has been shown that the Gauss, Radau and Lobatto quadrature formulae associated with the spaces of spline functions with equidistant knots are asymptotically optimal in W_∞^r for all r, and in W_p^r, $1 \le p < \infty$, when r is odd. In other words, for these values of r and p and with "$*$" standing for "G," "L," "R,l," and "R,r," there holds

$$\lim_{n\to\infty} \frac{\mathcal{E}(Q_n^*, W_p^r)}{\mathcal{E}_n(W_p^r)} = 1.$$

In the present paper, we provide algorithms for the explicit construction of Gauss, Lobatto and Radau quadrature formulae associated with the spaces of parabolic splines with equidistant knots. Sharp estimates for the quantities $\mathcal{E}(Q_n^*, W_\infty^3)$, which improve upon known upper estimates for $\mathcal{E}_n(W_\infty^3)$ are given. The paper is organized as follows. In the next section, we provide some facts about spline functions and the Peano kernel representation of the quadrature remainder. Section 3 is devoted to Gauss quadrature formulae Q_n^G associated with spaces of parabolic splines with equidistant knots. Specifically, the section contains: algorithm for construction of Q_n^G; derivation of an explicit formula for $c_{3,\infty}(Q_n^G)$; estimates for the Peano error constant, the weights, and the nodes of Q_n^G; and application of these estimates for obtaining sharp bounds for $c_{3,\infty}(Q_n^G)$. The same questions for the Lobatto quadratures Q_n^L are studied in Sect. 4. In the last section we briefly consider the left Radau quadratures, and discuss the possibility for replacement of the Gauss and Lobatto quadratures by some modified trapezium rules.

2 Spline Functions and Peano Kernels of Quadratures

If $\mathcal{L}$ is a linear functional defined on $C[0,1]$, which vanishes on the space π_s of algebraic polynomials of degree not exceeding s, then, according to a classical result of Peano, for $r \in \mathbb{N}$, $1 \le r \le s+1$ and $f \in W_1^r[0,1]$, $\mathcal{L}$ admits the integral representation

$$\mathcal{L}[f] = \int_0^1 K_r(t) f^{(r)}(t)\,\mathrm{d}t,$$

where, for $t \in [0,1]$, $K_r(t) = \mathcal{L}\left[(\cdot - t)_+^{r-1}/(r-1)!\right]$. In the case when $\mathcal{L}$ is the remainder $R[Q;\cdot]$ of a quadrature formula Q with algebraic degree of precision s, the function $K_r(t) = K_r(Q;t)$ is referred to as the rth Peano kernel of Q. For Q as in (1), explicit representations for $K_r(Q;t)$, $t \in [0,1]$, are

$$K_r(Q;t) = \frac{(1-t)^r}{r!} - \frac{1}{(r-1)!}\sum_{i=1}^{n} a_i(\xi_i - t)_+^{r-1} \tag{2}$$

and

$$K_r(Q;t) = (-1)^r \left[\frac{t^r}{r!} - \frac{1}{(r-1)!}\sum_{i=1}^{n} a_i(t - \xi_i)_+^{r-1}\right]. \tag{3}$$

If the integrand f belongs to the Sobolev space W_p^r $(1 \le p \le \infty)$, then from

$$R[Q;f] = \int_0^1 K_r(Q;t) f^{(r)}(t)\,\mathrm{d}t$$

and Hölder's inequality one obtains the sharp error estimate

$$|R[Q;f]| \le c_{r,p}(Q)\|f^{(r)}\|_p, \quad \text{where } c_{r,p}(Q) = \|K_r(Q;\cdot)\|_q,\ p^{-1} + q^{-1} = 1.$$

In other words, we have $\mathcal{E}(Q, W_p^r) = c_{r,p}(Q)$.

The function $K_r(Q;t)$ is also called a monospline of degree r with simple knots $\{\xi_i : \xi_i \in (0,1)\}$. It is easily seen that

$$K_r^{(j)}(Q;0) = 0 \quad \text{for } j = 0,1,\ldots,r-1-\alpha, \tag{4}$$
$$K_r^{(j)}(Q;1) = 0 \quad \text{for } j = 0,1,\ldots,r-1-\beta, \tag{5}$$

where

$$\alpha = \begin{cases} 0, & \text{if } \xi_1 > 0, \\ 1, & \text{if } \xi_1 = 0, \end{cases} \quad \text{and } \beta = \begin{cases} 0, & \text{if } \xi_n < 1, \\ 1, & \text{if } \xi_n = 1. \end{cases}$$

From $K_r(Q;x) = R[Q;(\cdot - x)_+^{r-1}/(r-1)!]$ it is seen that $K_r(Q;x) = 0$ for some $x \in (0,1)$ if and only if Q is exact for the spline function $f(t) = (t-x)_+^{r-1}$. Thus, in order that a quadrature formula Q has *maximal spline degree of precision*, i.e., Q

is exact for a space of spline functions of degree $r-1$ with maximal dimension, it is necessary and sufficient that the corresponding monospline $K_r(Q;\cdot)$ has a maximal number of zeros in $(0,1)$. According to the fundamental theorem of algebra for monosplines [4], every monospline of the form (2) and (3), which satisfies (4) and (5), has at most $2n-r-\alpha-\beta$ distinct zeros in $(0,1)$. Conversely, given $2n-r-\alpha-\beta$ points in $(0,1)$, say, $x_1 < x_2 < \cdots < x_{2n-r-\alpha-\beta}$, there exists a unique monospline $K_r(t)$ of the form (3), which vanishes at $\{x_i\}_{i=1}^{2n-r-\alpha-\beta}$ and has zeros at $t=0$ and $t=1$ of multiplicities $r-1-\alpha$ and $r-1-\beta$, respectively. In view of the one-to-one correspondence between monosplines and quadrature formulae, associated with K_r there exists a n-point quadrature formula Q of the form (1), which is exact for the space $S_{r-1,\Delta}$ of dimension $2n-\alpha-\beta$, where $\Delta=\{x_1,\ldots,x_{2n-r-\alpha-\beta}\}$. This property explains why such a quadrature formula is called a Gauss (respectively, Radau, Lobatto) quadrature formula associated with $S_{r-1,\Delta}$. Let us point out that, as in the case of classical Gauss, Radau and Lobatto quadrature formulae, all the nodes of the Gauss-type quadratures associated with spaces of spline functions lie in $[0,1]$, and all their weights are positive [4, Theorem 7.1].

We conclude this section with some facts about B-splines, which will be needed in the sequel. A B-spline of degree $r-1$ with knots $t_0 < \cdots < t_r$ is the function $B(t) := (\cdot - t)_+^{r-1}[t_0,t_1,\ldots,t_r]$, where $f[t_0,\ldots,t_r]$ stands for the rth order divided difference of f with nodes $t_0,\ldots,t_r$. Clearly, $B(t)$ is a spline function of order $r-1$ with knots $t_0,\ldots,t_r$. The following properties of B-splines are well known:

(i) $B(t) > 0$ for $t \in (t_0,t_r)$;
(ii) $B(t) = 0$ for $t \notin [t_0,t_r]$;
(iii) $\int_{-\infty}^{\infty} B(t)\,\mathrm{d}t = 1/r$.

Finally, we recall a way for constructing a basis of B-splines for a space of spline functions. Given $\Delta=\{x_i\}_{i=1}^{n-r}$, $0 < x_1 < \cdots < x_{n-r} < 1$, we have $\dim S_{r-1,\Delta} = n$. We choose arbitrarily $2r$ distinct knots, $x_{-r+1} < \cdots < x_0 \le 0$ and $1 \le x_{n-r+1} < \cdots < x_n$, and set $B_k(t) = (\cdot - t)_+^{r-1}[x_{k-r},x_{k-r+1},\ldots,x_k]$ for $k=1,\ldots,n$. Then $\{B_k(t)\}_{k=1}^{n}$ form a basis for $S_{r-1,\Delta}$ on the interval $[0,1]$.

3 Gaussian Quadrature Formulae for Parabolic Splines

3.1 The Construction

We assume here that $n \ge 6$. Let $m = 2n-2$, $x_k = k/m$ for $k=-2,-1,\ldots,2n$, and $\Delta = \{x_k\}_1^{2n-3}$. The space $S = S_{3,\Delta}$ has dimension $2n$, and a basis for S on $[0,1]$ is formed by $\{B_k(t)\}_{k=1}^{2n}$, where $B_k(t) = (\cdot - t)_+^2[x_{k-3},x_{k-2},x_{k-1},x_k]$ for $k=1,\ldots,2n$. The explicit form of $B_k(t)$ is given by

$$B_k(t) = \begin{cases} 0, & \text{if } t \le x_{k-3}, \\ \dfrac{m^3}{6}\left(t - \dfrac{k-3}{m}\right)^2, & \text{if } t \in (x_{k-3}, x_{k-2}], \\ \dfrac{m^3}{6}\left(t - \dfrac{k-3}{m}\right)^2 - \dfrac{m^3}{2}\left(t - \dfrac{k-2}{m}\right)^2, & \text{if } t \in (x_{k-2}, x_{k-1}], \\ \dfrac{m^3}{6}\left(\dfrac{k}{m} - t\right)^2, & \text{if } t \in (x_{k-1}, x_k], \\ 0, & \text{if } t \ge x_k. \end{cases} \tag{6}$$

We shall construct the n-point Gaussian quadrature formula Q_n^G,

$$Q_n^G[f] = \sum_{k=1}^{n} a_{k,n}^G f(\tau_{k,n}^G), \quad 0 < \tau_{1,n}^G < \cdots < \tau_{n,n}^G < 1,$$

associated with S (for notational simplicity, we suppress the dependence on n and write henceforth a_k^G and τ_k^G instead of $a_{k,n}^G$ and $\tau_{k,n}^G$). In view of the uniqueness of Q_n^G, it must be symmetric, therefore it suffices to find only half of the coefficients and the nodes of Q_n^G, say, those with indices less than or equal to $[(n+1)/2]$. They are determined from the conditions Q_n^G to be exact for the basis functions $B_k(t)$. We have to place the n nodes $\{\tau_k^G\}_{k=1}^n$ into the $2n-2$ intervals $\{[x_i, x_{i+1}]\}_{i=0}^{2n-3}$, so that, roughly, half of those intervals must be free of nodes. Properties (i)–(iii) of B-splines imply that every two consecutive intervals $[x_i, x_{i+1}]$ must contain at least one node of Q_n^G. Moreover, since Q_n^G is exact for $B_1(t)$, at least one node of Q_n^G is located in $[x_0, x_1)$. Therefore, we assume that

$$\tau_k^G \in (x_{2k-2}, x_{2k-1}) \quad \text{for } k = 1, \ldots, \left[\frac{n}{2}\right],$$

and set

$$\tau_k^G := \frac{2k-2+\theta_k}{m}, \quad k = 1, \ldots, \left[\frac{n}{2}\right], \text{ where } \theta_k \in (0,1). \tag{7}$$

We determine a_1^G and θ_1 (and hence τ_1^G) from the condition Q_n^G to be exact for $B_1(t)$ and $B_2(t)$. This condition reads as

$$a_1^G B_1(\tau_1^G) = I[B_1] = \frac{1}{18}, \quad a_1^G B_2(\tau_1^G) = I[B_2] = \frac{5}{18}.$$

In view of (6), we have

$$B_1(\tau_1^G) = \frac{m}{6}(1-\theta_1)^2, \quad B_2(\tau_1^G) = \frac{m}{6}(1 + 2\theta_1 - 2\theta_1^2),$$

and replacement in the above equations yields

$$\theta_1 = \frac{6-2\sqrt{2}}{7}, \quad \tau_1^G = \frac{6-2\sqrt{2}}{7m}, \quad a_1^G = \frac{(2\sqrt{2}-1)^2}{3m}.$$

Next, we determine θ_2 (hence τ_2^G, by (7)) and a_2^G from the condition Q_n^G to be exact for $B_3(t)$ and $B_4(t)$:

$$a_1^G B_3(\tau_1^G) + a_2^G B_3(\tau_2^G) = I[B_3] = \frac{1}{3}, \quad a_2^G B_4(\tau_2^G) = I[B_4] = \frac{1}{3}.$$

From (6) we find

$$B_3(\tau_1^G) = \frac{m}{6}\theta_1^2, \quad B_3(\tau_2^G) = \frac{m}{6}(1-\theta_2)^2, \quad B_4(\tau_2^G) = \frac{m}{6}(1+2\theta_2-2\theta_2^2).$$

By substituting these expressions and the values found for a_1^G and τ_1^G in the above system, we consecutively get

$$\begin{cases} a_2^G \cdot \dfrac{m}{6}(1-\theta_2)^2 & = \dfrac{1}{3} - \dfrac{2(\sqrt{2}-1)^2}{9}, \\ a_2^G \cdot \dfrac{m}{6}(1+2\theta_2-2\theta_2^2) & = \dfrac{1}{3}, \end{cases} \Rightarrow \frac{(1-\theta_2)^2}{1+2\theta_2-2\theta_2^2} = \frac{4\sqrt{2}-3}{3} =: c_2.$$

The resulting quadratic equation in θ_2 has only one solution belonging to the interval $(0,1)$, namely,

$$\theta_2 = \frac{1-c_2}{1+c_2+\sqrt{c_2(1+3c_2)}}.$$

We then determine a_2^G and τ_2^G by

$$a_2^G = \frac{2}{m} \cdot \frac{1}{1+2\theta_2-2\theta_2^2}, \quad \tau_2^G = \frac{2+\theta_2}{m}.$$

Assume that a_{k-1}^G and τ_{k-1}^G are known for some k, $3 \le k \le [(n-1)/2]$. Then a_k^G and τ_k^G are found from the condition $Q_n^G[B_{2k-1}] = Q_n^G[B_{2k}] = I[B_{2k-1}] = I[B_{2k}]$ ($= 1/3$, in view of property (iii) of B-splines). Since

$$B_{2k-1}(\tau_{k-1}^G) = \frac{m}{6}\theta_{k-1}^2, \quad B_{2k-1}(\tau_k^G) = \frac{m}{6}(1-\theta_k)^2, \quad B_{2k}(\tau_k^G) = \frac{m}{6}(1+2\theta_k-2\theta_k^2)$$

and

$$a_{k-1}^G = \frac{2}{m} \cdot \frac{1}{1+2\theta_{k-1}-2\theta_{k-1}^2},$$

the latter condition is equivalent to the system

$$\begin{cases} a_k^G \cdot \dfrac{m}{6}(1-\theta_k)^2 & = \dfrac{1}{3} - \dfrac{1}{3}\dfrac{\theta_{k-1}^2}{1+2\theta_{k-1}-2\theta_{k-1}^2}, \\ a_k^G \cdot \dfrac{m}{6}(1+2\theta_k-2\theta_k^2) & = \dfrac{1}{3}. \end{cases}$$

Hence,

$$\frac{(1-\theta_k)^2}{1+2\theta_k-2\theta_k^2}=1-\frac{\theta_{k-1}^2}{1+2\theta_{k-1}-2\theta_{k-1}^2}=\frac{(1-\theta_{k-1})(1+3\theta_{k-1})}{1+2\theta_{k-1}-2\theta_{k-1}^2}=:c_k.$$

Therefore, for $3\le k\le[(n-1)/2]$, the coefficients a_k^G and the weights τ_k^G are determined through the following procedure:

$$c_k := \frac{(1-\theta_{k-1})(1+3\theta_{k-1})}{1+2\theta_{k-1}-2\theta_{k-1}^2}, \tag{8}$$

$$\theta_k = \frac{1-c_k}{1+c_k+\sqrt{c_k(1+3c_k)}}, \tag{9}$$

$$\tau_k^G = \frac{2k-2+\theta_k}{m}, \quad a_k^G=\frac{2}{m}\cdot\frac{1}{1+2\theta_k-2\theta_k^2}. \tag{10}$$

For determining the nodes of Q_n^G around the middle of the integration interval and the corresponding weights, we consider separately the cases of even and odd n.

Case A. $n=2M+1$. We have so far found a_k^G, τ_k^G for $k=1,\dots,M$, ensuring that Q_n^G calculates exactly the integrals of $\{B_k(t)\}_{k=1}^{2M}$. By the symmetry of Q_n^G, it follows that $\tau_{M+1}^G=x_{2M}=1/2$.

The coefficient a_{M+1}^G is determined so that $Q_n^G[B_{2M+1}]=I[B_{2M+1}]=1/3$, i.e., by

$$a_M^G B_{2M+1}(\tau_M^G)+a_{M+1}^G B_{2M+1}(\tau_{M+1}^G)=\frac{1}{3}.$$

In view of

$$a_M^G=\frac{2}{m}\cdot\frac{1}{1+2\theta_M-2\theta_M^2}, \quad B_{2M+1}(\tau_M^G)=\frac{m}{6}\theta_M^2, \quad \text{and } B_{2M+1}(\tau_{M+1}^G)=\frac{m}{6},$$

the latter condition yields

$$a_{M+1}^G=\frac{2}{m}\cdot\frac{(1-\theta_M)(1+3\theta_M)}{1+2\theta_M-2\theta_M^2}. \tag{11}$$

Since Q_n^G is symmetric, we determine the remaining weights and nodes by

$$a_k^G=a_{n+1-k}^G, \quad \tau_k^G=1-\tau_{n+1-k}^G \quad \text{for } k=M+2,\dots,n. \tag{12}$$

Notice that (12) ensures that Q_n^G is exact for $B_k(t)$, $k=2M+2,\dots,2n$.

Case B. $n=2M$. In this case a_k^G, τ_k^G are found for $k=1,\dots,M-1$, ensuring that Q_n^G calculates exactly the integrals of $\{B_k(t)\}_{k=1}^{2M-2}$. We have $\tau_M^G\in(x_{2M-2},x_{2M-1})$, i.e., $\tau_M^G\in(x_{2M-2},1/2)$, and by symmetry, $\tau_{M+1}^G=1-\tau_M^G$, $\tau_{M+1}^G\in(x_{2M-1},x_{2M})$.

Now a_M^G and τ_M^G are determined by the condition $Q_n^G[B_{2M-1}] = Q_n^G[B_{2M}] = 1/3$, i.e.,

$$\begin{cases} a_{M-1}^G B_{2M-1}(\tau_{M-1}^G) + a_M^G B_{2M-1}(\tau_M^G) = 1/3, \\ a_M^G B_{2M}(\tau_M^G) + a_M^G B_{2M+1}(\tau_M^G) \qquad = 1/3, \end{cases}$$

where in the second equation we have used symmetry to replace a_{M+1}^G by a_M^G and $B_{2M}(\tau_{M+1}^G)$ by $B_{2M+1}(\tau_M^G)$. Since a_{M-1}^G is obtained from (10) with $k = M$, and

$$B_{2M-1}(\tau_{M-1}^G) = \frac{m}{6}\theta_M^2, \quad B_{2M-1}(\tau_M^G) = \frac{m}{6}(1-\theta_M)^2,$$
$$B_{2M}(\tau_M^G) = \frac{m}{6}(1+2\theta_M - 2\theta_M^2), \quad B_{2M+1}(\tau_M^G) = \frac{m}{6}\theta_M^2,$$

the above system becomes

$$\begin{cases} a_M^G \cdot \dfrac{m}{6}(1-\theta_M)^2 \qquad = \dfrac{1}{3} - \dfrac{1}{3}\dfrac{\theta_{M-1}^2}{1+2\theta_{M-1}-2\theta_{M-1}^2}, \\ a_M^G \cdot \dfrac{m}{6}(1+2\theta_M - \theta_M^2) = \dfrac{1}{3}. \end{cases}$$

From here, we get

$$\frac{(1-\theta_M)^2}{1+2\theta_M-\theta_M^2} = 1 - \frac{\theta_{M-1}^2}{1+2\theta_{M-1}-2\theta_{M-1}^2} = \frac{(1-\theta_{M-1})(1+3\theta_{M-1})}{1+2\theta_{M-1}-2\theta_{M-1}^2} =: c_M.$$

The solution of the resulting quadratic equation with respect to θ_M, which belongs to $(0,1)$, is given by

$$\theta_M = \frac{1-c_M}{1+c_M+\sqrt{2c_M(1+c_M)}}. \tag{13}$$

Then τ_M^G and a_M^G are obtained by

$$\tau_M^G = \frac{2M-2+\theta_M}{m}, \quad a_M^G = \frac{2}{m}\cdot\frac{1}{1+2\theta_M-\theta_M^2}. \tag{14}$$

The remaining nodes and weights of Q_n^G are obtained by symmetry, using (12).

3.2 A Formula for $c_{3,\infty}(Q_n^G)$

The error constant $c_{3,\infty}(Q_n^G)$ is calculated as

$$c_{3,\infty}(Q_n^G) = \int_0^1 \left|K_3(Q_n^G)\right| \, dt,$$

and the fact that the zeros of $K_3(Q_n^G;t)$ are equidistant enables us to derive an explicit formula for $c_{3,\infty}(Q_n^G)$, which turns out to be very useful for estimation.

Theorem 1. *For $n \ge 4$, the error constant $c_{3,\infty}(Q_n^G)$ can be represented as*

$$c_{3,\infty}(Q_n^G) = \frac{1}{192(n-1)^3} - \frac{1}{4(n-1)} \sum_{k=1}^{[n/2]} a_k^G (\tau_k^G - x_{2k-2})^2 + \frac{1}{3} \sum_{k=1}^{[n/2]} a_k^G (\tau_k^G - x_{2k-2})^3.$$

Proof. The cases of odd and even n require separate consideration, but as the proof in both cases is almost identical, we present proof only for the case of odd n, say, $n = 2M+1$ (and $m = 4M$).

According to (2) and (3), the third Peano kernel of Q_n^G is given by

$$K_3(Q_n^G;t) = \frac{(1-t)^3}{6} - \frac{1}{2}\sum_{i=1}^{n} a_i^G (\tau_i^G - t)_+^2 = -\frac{t^3}{6} + \frac{1}{2}\sum_{i=1}^{n} a_i^G (t - \tau_i^G)_+^2,$$

and the zeros of $K_3(Q_n^G;t)$ in $(0,1)$ are $\{x_k\}_{k=1}^{2n-3}$, all simple. Thus, $K_3(Q_n^G;t)$ changes its sign at those points only, and it is easy to see from the above representation that

$$\operatorname{sign} K_3(Q_n^G;t) = (-1)^{k+1} \quad \text{for } t \in (x_k, x_{k+1}),\ k = 0, \dots, 2n-3. \tag{15}$$

Since Q_n^G is symmetric, $K_3(Q_n^G;t)$ is an odd function with respect to $t = \frac{1}{2}(= x_{2M})$. In view of (15), we have

$$\begin{aligned}
c_{3,\infty}(Q_n^G) &= 2\int_0^{x_{2M}} |K_3(Q_n^G;t)|\,\mathrm{d}t = 2\sum_{k=0}^{2M-1} (-1)^{k-1} \int_{x_k}^{x_{k+1}} K_3(Q_n^G;t)\,\mathrm{d}t \\
&= 2\left[\sum_{k=1}^{M} \int_{x_{2k-1}}^{x_{2k}} K_3(Q_n^G;t)\,\mathrm{d}t - \sum_{k=1}^{M} \int_{x_{2k-2}}^{x_{2k-1}} K_3(Q_n^G;t)\,\mathrm{d}t\right] \\
&= -\frac{1}{12}\sum_{k=1}^{M} (x_{2k}^4 - x_{2k-1}^4) + \frac{1}{12}\sum_{k=1}^{M} (x_{2k-1}^4 - x_{2k-2}^4) \\
&\quad + \frac{1}{3}\sum_{k=1}^{M}\sum_{i=1}^{n} a_i^G \left[(x_{2k} - \tau_i^G)_+^3 - (x_{2k-1} - \tau_i^G)_+^3\right] \\
&\quad - \frac{1}{3}\sum_{k=1}^{M}\sum_{i=1}^{n} a_i^G \left[(x_{2k-1} - \tau_i^G)_+^3 - (x_{2k-2} - \tau_i^G)_+^3\right] \\
&= -\frac{1}{12}\sum_{k=1}^{M} (x_{2k-2}^4 - 2x_{2k-1}^4 + x_{2k}^4) \\
&\quad + \frac{1}{3}\sum_{k=1}^{M}\sum_{i=1}^{k} a_i^G \left[(x_{2k} - \tau_i^G)^3 - (x_{2k-1} - \tau_i^G)^3\right] \\
&\quad - \frac{1}{3}\sum_{k=1}^{M}\left[\sum_{i=1}^{k} a_i^G \left[(x_{2k-1} - \tau_i^G)^3 - (x_{2k-2} - \tau_i^G)^3\right] + a_{k+1}^G (x_{2k-2} - \tau_k^G)^3\right].
\end{aligned}$$

For the last equality we have used that $\tau_k^G \in (x_{2k-2}, x_{2k-1})$ for $k = 1, \dots, M$ and $\tau_{M+1}^G = x_{2M} = 1/2$. To simplify the double sums above, we make use of the identity

$$(x-\tau)^2 + (x-\tau)(y-\tau) + (y-\tau)^2 = \frac{3}{2}(x-\tau)^2 + \frac{3}{2}(y-\tau)^2 - \frac{1}{2}(x-y)^2,$$

and the fact that Q_n^G is exact for parabolic spline functions with knots $\{x_k\}_{k=1}^{2n-3}$. We have

$$
\begin{aligned}
&\sum_{k=1}^{M}\sum_{i=1}^{k} a_i^G\left[(x_{2k}-\tau_i^G)_+^3-(x_{2k-1}-\tau_i^G)_+^3\right]\\
&=\frac{1}{m}\sum_{k=1}^{M}\sum_{i=1}^{k} a_i^G\left[(x_{2k}-\tau_i^G)^2+(x_{2k}-\tau_i^G)(x_{2k-1}-\tau_i^G)+(x_{2k-1}-\tau_i^G)^2\right]\\
&=\frac{1}{m}\sum_{k=1}^{M}\sum_{i=1}^{k} a_i^G\left[\frac{3}{2}(x_{2k}-\tau_i^G)^2+\frac{3}{2}(x_{2k-1}-\tau_i^G)^2-\frac{1}{2m^2}\right]\\
&=\frac{3}{2m}\sum_{k=1}^{M}\sum_{i=1}^{n} a_i^G\left[(x_{2k}-\tau_i^G)_+^2+(x_{2k-1}-\tau_i^G)_+^2\right]-\frac{1}{2m^3}\sum_{k=1}^{M}\sum_{i=1}^{k} a_i^G\\
&=\frac{3}{2m}\sum_{k=1}^{M}\int_0^1\left[(x_{2k}-t)_+^2+(x_{2k-1}-t)_+^2\right]\mathrm{d}t-\frac{1}{2m^3}\sum_{k=1}^{M}\sum_{i=1}^{k} a_i^G\\
&=\frac{1}{2m}\sum_{k=1}^{M}(x_{2k}^3+x_{2k-1}^3)-\frac{1}{2m^3}\sum_{k=1}^{M}\sum_{i=1}^{k} a_i^G.
\end{aligned}
$$

In a similar manner we get

$$
\begin{aligned}
&\sum_{k=1}^{M}\sum_{i=1}^{k} a_i^G\left[(x_{2k-1}-\tau_i^G)_+^3-(x_{2k-2}-\tau_i^G)_+^3\right]\\
&=\frac{1}{m}\sum_{k=1}^{M}\sum_{i=1}^{k} a_i^G\left[(x_{2k-1}-\tau_i^G)^2+(x_{2k-1}-\tau_i^G)(x_{2k-2}-\tau_i^G)+(x_{2k-2}-\tau_i^G)^2\right]\\
&=\frac{1}{m}\sum_{k=1}^{M}\sum_{i=1}^{k} a_i^G\left[\frac{3}{2}(x_{2k-1}-\tau_i^G)^2+\frac{3}{2}(x_{2k-2}-\tau_i^G)^2-\frac{1}{2m^2}\right]\\
&=\frac{3}{2m}\sum_{k=1}^{M}\left[\sum_{i=1}^{n} a_i^G\left[(x_{2k-1}-\tau_i^G)_+^2+(x_{2k-2}-\tau_i^G)_+^2\right]+a_k^G(x_{2k-2}-\tau_k^G)^2\right]\\
&\quad-\frac{1}{2m^3}\sum_{k=1}^{M}\sum_{i=1}^{k} a_i^G\\
&=\frac{3}{2m}\sum_{k=1}^{M}\int_0^1\left[(x_{2k-1}-t)_+^2+(x_{2k-2}-t)_+^2\right]\mathrm{d}t\\
&\quad-\frac{1}{2m^3}\sum_{k=1}^{M}\sum_{i=1}^{k} a_i^G+\frac{3}{2m}\sum_{k=1}^{M} a_k^G(x_{2k-2}-\tau_k^G)^2\\
&=\frac{1}{2m}\sum_{k=1}^{M}(x_{2k-1}^3+x_{2k-2}^3)-\frac{1}{2m^3}\sum_{k=1}^{M}\sum_{i=1}^{k} a_i^G+\frac{3}{2m}\sum_{k=1}^{M} a_k^G(x_{2k-2}-\tau_k^G)^2.
\end{aligned}
$$

Replacement of the double sums in the last representation of $c_{3,\infty}(Q_n^G)$ yields

$$\begin{aligned} c_{3,\infty}(Q_n^G) &= -\frac{1}{12}\sum_{k=1}^{M}(x_{2k-2}^4 - 2x_{2k-1}^4 + x_{2k}^4) + \frac{1}{6m}(x_{2M}^3 - x_0^3) \\ &\quad -\frac{1}{3}\sum_{k=1}^{M} a_{k+1}^G (x_{2k-2}-\tau_k^G)^3 - \frac{1}{2m}\sum_{k=1}^{M} a_{k+1}^G (x_{2k-2}-\tau_k^G)^2 \\ &= -\frac{1}{12}\sum_{k=1}^{M}(x_{2k-2}^4 - 2x_{2k-1}^4 + x_{2k}^4) + \frac{1}{48m} \\ &\quad +\frac{1}{3}\sum_{k=1}^{[n/2]} a_{k+1}^G (\tau_k^G - x_{2k-2})^3 - \frac{1}{4(n-1)}\sum_{k=1}^{[n/2]} a_{k+1}^G (\tau_k^G - x_{2k-2})^2. \end{aligned}$$

Recall that $n = 2M+1$ and $m = 4M = 2(n-1)$. Now the proof of Theorem 1 simplifies to the verification of the algebraic identity

$$-\frac{1}{12}\sum_{k=1}^{M}(x_{2k-2}^4 - 2x_{2k-1}^4 + x_{2k}^4) + \frac{1}{192M} = \frac{1}{1536M^3}, \quad \text{where } x_k = \frac{k}{4M}.$$

The latter identity is shown to be true by the Euler–MacLaurin summation formula. Theorem 1 is proved. □

3.3 Estimates for $c_{3,\infty}(Q_n^G)$

Theorem 1 enables us to derive sharp lower and upper bounds for $c_{3,\infty}(Q_n^G)$. Let us point out that the calculation of the exact error constants of quadratures is rarely possible (the case of compound quadrature formulae is excluded).

Theorem 2. *For $n \geq 8$, the error constant $c_{3,\infty}(Q_n^G)$ satisfies the inequalities*

$$c_{3,\infty}(Q_n^G) > \frac{1}{192(n-1)^3} - \frac{1}{198.210443039(n-1)^4}$$

and

$$c_{3,\infty}(Q_n^G) < \frac{1}{192(n-1)^3} - \frac{1}{198.21044304(n-1)^4}.$$

Remark 1. As is well known, $\mathcal{E}_n(\widetilde{W}_\infty^3) = 1/(192n^3)$, hence the first inequality in Theorem 2 sharpens the inequality $c_{3,\infty}(Q_n^G) > 1/(192n^3)$. On the other hand, the second inequality in Theorem 2 improves upon the estimate $c_{3,\infty}(Q_n^G) < 1/(192(n-1)^3)$, found in [5].

For the proof of Theorem 2, we need to estimate the weights and nodes of Q_n^G.

Lemma 1. *Assume that $n \geq 6$, and set $\tau_k^G = (2k-2+\theta_k)/m$ for $1 \leq k \leq [(n+1)/2]$. Then for $2 \leq k \leq [n/2]$,*

$$0 < \frac{2}{m} - a_k^G < \frac{4\theta_k}{m} \quad \textit{and } 0 \leq \theta_k < \frac{4}{128^{2^{k-2}}}.$$

Proof. We follow the notations from Sect. 3.1. The first inequality follows easily from the representation of a_k^G in (10) and (14), having in mind that $\theta_k \in [0,1)$. The second inequality in Lemma 1 is directly verified to be true for $k=2$, as $\theta_2 = 0.031\cdots < 1/32$. For $3 \leq k < [n/2]$ or $k = [n/2]$ and n odd, we observe that, according to (9),

$$\theta_k = \frac{1-c_k}{1+c_k+\sqrt{c_k(1+3c_k)}} \leq \frac{1-c_k}{1+c_k+\sqrt{4c_k^2}} = \frac{1-c_k}{1+3c_k} \tag{16}$$

(here we have used that $0 < c_k < 1$). By replacing c_k with the expression in (8) we get

$$\frac{1-c_k}{1+3c_k} = \frac{\theta_{k-1}^2}{4+8\theta_{k-1}-11\theta_{k-1}^2} < \left(\frac{\theta_{k-1}}{2}\right)^2,$$

hence

$$\theta_k < \left(\frac{\theta_{k-1}}{2}\right)^2 \quad \text{for } 3 \leq k < \left[\frac{n}{2}\right]. \tag{17}$$

Inequality (17) is also true when $n = 2M$ and $k = M$, in this case we apply (13) in (16). The proof of the second inequality in Lemma 1 is accomplished by induction, using (17) and $\theta_2 < 1/32$. □

Proof of Theorem 2. With the notation and results from Sect. 3.1, we apply Theorem 1 to obtain

$$\begin{aligned} c_{3,\infty}(Q_n^G) &= \frac{1}{192(n-1)^3} - \frac{1}{16(n-1)^3}\sum_{k=1}^{[n/2]} a_k^G \theta_k^2 + \frac{1}{24(n-1)^3}\sum_{k=1}^{[n/2]} a_k^G \theta_k^3 \\ &= \frac{1}{192(n-1)^3} - \frac{1}{48(n-1)^3}\sum_{k=1}^{[n/2]} a_k^G \theta_k^2 (3-2\theta_k). \end{aligned}$$

The upper estimate for $c_{3,\infty}(Q_n^G)$ is obtained as follows:

$$\begin{aligned} c_{3,\infty}(Q_n^G) &\leq \frac{1}{192(n-1)^3} - \frac{1}{48(n-1)^3}\sum_{k=1}^{3} a_k^G \theta_k^2 (3-2\theta_k) \\ &= \frac{1}{192(n-1)^3} - \frac{(3-\sqrt{2})^2}{504(n-1)^4} - \frac{1}{48(n-1)^4}\sum_{i=2}^{3} \frac{\theta_i^2(3-2\theta_i)}{1+2\theta_i-2\theta_i^2} \\ &< \frac{1}{192(n-1)^3} - \frac{1}{198.21044304(n-1)^4}. \end{aligned}$$

For the final result, we used the exact value of θ_1 and the numerically calculated $\theta_2 = 0.03103799435856347\ldots$ and $\theta_3 = 0.000227316882450987\ldots$

For the lower bound, we have

$$c_{3,\infty}(Q_n^G) \ge \frac{1}{192(n-1)^3} - \frac{(3-\sqrt{2})^2}{504(n-1)^4}$$
$$- \frac{1}{48(n-1)^4}\left[\frac{\theta_2^2(3-2\theta_2)}{1+2\theta_2-2\theta_2^2} + \frac{\theta_3^2(3-2\theta_3)}{1+2\theta_3-2\theta_3^2}\right] - \frac{1}{16(n-1)^4}\sum_{k=4}^{[n/2]}\theta_k^2.$$

The last sum is estimated with the help of Lemma 1,

$$\sum_{k=4}^{[n/2]}\theta_k^2 < \sum_{k=4}^{\infty}\left(\frac{4}{128^{2^{k-2}}}\right)^2 = 16\sum_{k=4}^{\infty}\frac{1}{128^{2^{k-1}}} < 16\sum_{k=1}^{\infty}\frac{1}{128^{8k}} = \frac{16}{128^8-1}.$$

Hence, a lower bound for $c_{3,\infty}(Q_n^G)$ can be obtained by subtracting from the upper one the term $1/((128^8-1)(n-1)^4)$. This yields the second inequality in Theorem 2. □

In [5], the asymptotic optimality of Gaussian quadratures Q_n^G in $W_p^3[0,1]$ and, in particular, the inequality $c_{3,\infty}(Q_n^G) < 1/(192(n-1)^3)$, was proved by pointwise comparison of $K_3(Q_n^G;t)$ with the adjusted Bernoulli monospline $H_{3,n}(t)$, given by

$$H_{3,n}(t) = -\frac{1}{(n-1)^3}B_3^*((n-1)t),$$

where $B_3^*(t)$ is the one-periodic extension of the Bernoulli polynomial

$$B_3(t) = \frac{t^3}{6} - \frac{t^2}{4} + \frac{1}{12}. \qquad (18)$$

It has been shown in [5] that $|K_3(Q_n^G;t)| \le |H_{3,n}(t)|$ for every $t \in [0,1]$. In fact, except for small neighborhoods of the end points of $[0,1]$, the graphs of $K_3(Q_n^G;t)$ and $H_{3,n}(t)$ are indistinguishable (see Fig. 1).

4 Lobatto Quadrature Formulae for Parabolic Splines

4.1 The Construction

For $n \in \mathbb{N}$, $n \ge 6$, we set $m = 2n-4$ and $x_k = k/m$, $k = -2,-1,\ldots,2n-2$. The space $S = S_{3,\Delta}$ with $\Delta = \{x_k\}_{k=1}^{2n-5}$ has dimension $2n-2$, and a basis for S is formed by the B-splines $\{B_k(t)\}_{k=1}^{2n-2}$, defined in Sect. 3.1. We shall construct the n-point Lobatto quadrature formula Q_n^L, associated with S,

$$Q_n^L[f] = \sum_{k=1}^{n} a_k^L f(\tau_k^L), \quad 0 = \tau_1^L < \cdots < \tau_n^L = 1.$$

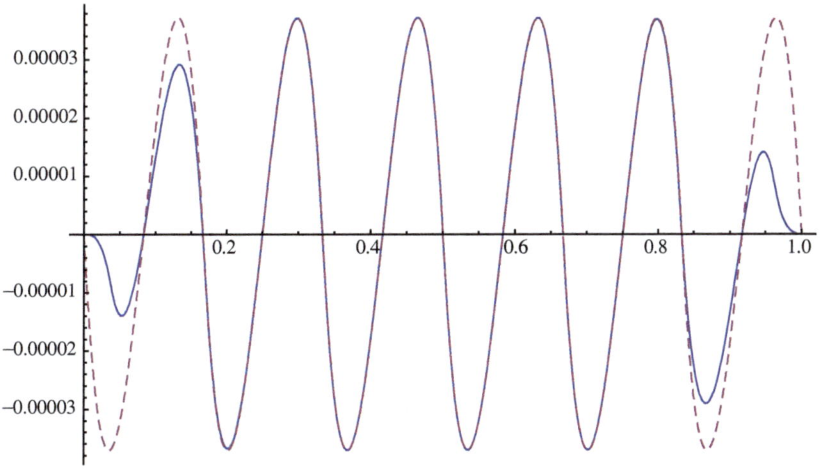

Fig. 1 The graphs of $K_3(Q_7^G;t)$ (*solid*) and $H_{3,7}(t)$ (*dashed*)

Again, Q_n^L is symmetric, therefore we shall determine only the left half of the nodes and the corresponding weights of Q_n^L. We assume that $\tau_k^L \in (x_{2k-3}, x_{2k-2})$ for $k=2$, $\ldots,[n/2]$, and set

$$\tau_k^L = \frac{2k-3+\theta_k}{m}, \quad k=2,\ldots,\left[\frac{n}{2}\right], \quad \theta_k \in (0,1). \tag{19}$$

The coefficient a_1^L is found from $Q_n^L[B_1] = a_1^L B_1(0) = a_1^L \cdot m/6 = I[B_1] = 1/18$,

$$a_1^L = \frac{1}{3m}.$$

Next, a_2^L and τ_2^L are determined from $Q_n^L[B_2] = I[B_2] = 5/18$ and $Q_n^L[B_3] = I[B_3] = 1/3$. Hence,

$$\begin{cases} a_1^L B_2(\tau_1^L) + a_2^L B_2(\tau_2^L) = \dfrac{5}{18}, \\ a_2^L B_3(\tau_2^L) = \dfrac{1}{3}, \end{cases} \implies \begin{cases} a_2^L B_2(\tau_2^L) = \dfrac{2}{9}, \\ a_2^L B_3(\tau_2^L) = \dfrac{1}{3} \end{cases}$$

(since $B_2(\tau_1^L) = B_1(\tau_1^L)$ and $a_1^L B_1(\tau_1^L) = 1/18$). Hence $B_2(\tau_2^L)/B_3(\tau_2^L) = 2/3$, and on using $B_2(\tau_2^L) = m/6 \cdot (1-\theta_2)^2$, $B_3(\tau_2^L) = m/6 \cdot (1+2\theta_2 - 2\theta_2^2)$, we find

$$\theta_2 = \frac{5-3\sqrt{2}}{7}, \quad \tau_2^L = \frac{1+\theta_2}{m}, \quad a_2^L = \frac{2(3-\sqrt{2})^2}{3m} = \frac{(3-\sqrt{2})^2}{3(n-2)}.$$

Now, assuming that a_{k-1}^L and τ_{k-1}^L are known for some k, $3 \le k < [(n+1)/2]$, we determine a_k^L and τ_k^L from the condition Q_n^L to be exact for $B_{2k-2}(t)$ and $B_{2k-1}(t)$.

This leads to the system

$$\begin{cases} a_{k-1}^L B_{2k-2}(\tau_{k-1}^L) + a_k^L B_{2k-2}(\tau_k^L) = \dfrac{1}{3}, \\ \qquad\qquad\qquad\qquad a_k^L B_{2k-1}(\tau_k^L) = \dfrac{1}{3}. \end{cases}$$

From (6) and (19), we have $B_{2k-2}(\tau_{k-1}^L) = m/6 \cdot \theta_{k-1}^2$, $B_{2k-2}(\tau_k^L) = m/6 \cdot (1-\theta_k)^2$, and $B_{2k-1}(\tau_k^L) = m/6 \cdot (1+2\theta_k - 2\theta_k^2)$. Therefore the above system is equivalent to

$$\begin{cases} a_{k-1}^L \theta_{k-1}^2 + a_k^L (1-\theta_k)^2 = \dfrac{2}{m}, \\ \qquad\qquad\qquad a_k^L = \dfrac{2}{m} \cdot \dfrac{1}{1+2\theta_k - 2\theta_k^2}. \end{cases}$$

It leads to the following equation for θ_k:

$$\frac{(1-\theta_k)^2}{1+2\theta_k - 2\theta_k^2} = 1 - a_{k-1}^L \frac{m}{2} \theta_{k-1}^2 =: c_k. \tag{20}$$

Using the second equation of the last system with k replaced $k-1$ we obtain

$$c_k = 1 - \frac{\theta_{k-1}^2}{1+2\theta_{k-1} - 2\theta_{k-1}^2}. \tag{21}$$

It is easily checked that $0 < c_k < 1$ whenever $\theta_{k-1} \in (0,1)$. Equation (20) has only one solution $\theta_k \in (0,1)$, given by

$$\theta_k = \frac{1-c_k}{1+c_k+\sqrt{c_k(1+3c_k)}}. \tag{22}$$

Then for τ_k^L and a_k^L we find

$$\tau_k^L = \frac{2k-3+\theta_k}{m}, \quad a_k^L = \frac{2}{m} \cdot \frac{1}{1+2\theta_k - 2\theta_k^2}. \tag{23}$$

Applying (21)–(23), we calculate a_k^L and τ_k^L recursively for $3 \le k \le [(n-1)/2]$. The weights and the nodes of Q_n^L around the middle of the integration interval are determined by the same arguments as in the Gaussian case. Precisely, if $n = 2M+1$, then $\tau_{M+1}^L = 1/2$ and a_{M+1}^L is calculated by (11) (with "G" replaced by "L"); if $n = 2M$, then θ_M is calculated by (13), and

$$\tau_M^L = \frac{2M-3+\theta_M}{m}, \quad a_M^L = \frac{2}{m} \cdot \frac{1}{1+2\theta_M - \theta_M^2}. \tag{24}$$

The remaining weights and nodes of Q_n^L are determined by symmetry.

4.2 Estimates for $c_{3,\infty}(Q_n^L)$

We start with the analog of Theorem 1.

Theorem 3. *For $n \geq 5$ the error constant $c_{3,\infty}(Q_n^L)$ can be represented as*

$$c_{3,\infty}(Q_n^L) = \frac{1}{192(n-2)^3} - \frac{1}{4(n-2)} \sum_{k=1}^{[n/2]} a_k^L(\tau_k^L - x_{2k-3})^2 + \frac{1}{3} \sum_{k=1}^{[n/2]} a_k^L(\tau_k^L - x_{2k-3})^3.$$

Proof. Again, there are some minor differences in the cases of odd and even n; here we choose to present the proof in the case of even n, say, $n = 2M$ (and $m = 4M-4$). The proof of the case $n = 2M+1$ is similar and is omitted.

For $t \in [0,1]$, the third Peano kernel of Q_n^L is given by

$$K_3(Q_n^L;t) = \frac{(1-t)^3}{6} - \frac{1}{2}\sum_{i=2}^{n} a_i^L(\tau_i^L - t)_+^2 = -\frac{t^3}{6} + \frac{1}{2}\sum_{i=1}^{n} a_i^L(t - \tau_i^L)_+^2.$$

All the zeros of $K_3(Q_n^L;t)$ in $(0,1)$ are $\{x_k\}_{k=1}^{2n-5}$, and they are simple. Hence, $K_3(Q_n^L;t)$ changes its sign at those points only, and

$$\operatorname{sign} K_3(Q_n^L;t) = (-1)^k \quad \text{for } t \in (x_k, x_{k+1}),\ k = 0,\ldots,2n-5.$$

Moreover, $K_3(Q_n^L;t)$ is an odd function with respect to $t = 1/2 (= x_{2M-2})$. Therefore,

$$\begin{aligned}
c_{3,\infty}(Q_n^L) &= 2\int_0^{x_{2M-2}} |K_3(Q_n^L;t)|\,\mathrm{d}t = 2\sum_{k=0}^{2M-3}(-1)^k \int_{x_k}^{x_{k+1}} K_3(Q_n^L;t)\,\mathrm{d}t \\
&= 2\left[\sum_{k=1}^{M-1}\int_{x_{2k-2}}^{x_{2k-1}} K_3(Q_n^L;t)\,\mathrm{d}t - \sum_{k=1}^{M-1}\int_{x_{2k-1}}^{x_{2k}} K_3(Q_n^L;t)\,\mathrm{d}t\right] \\
&= -\frac{1}{12}\sum_{k=1}^{M-1}(x_{2k-1}^4 - x_{2k-2}^4) + \frac{1}{12}\sum_{k=1}^{M-1}(x_{2k}^4 - x_{2k-1}^4) \\
&\quad + \frac{1}{3}\sum_{k=1}^{M-1}\sum_{i=1}^{n} a_i^L\left[(x_{2k-1} - \tau_i^L)_+^3 - (x_{2k-2} - \tau_i^L)_+^3\right] \\
&\quad - \frac{1}{3}\sum_{k=1}^{M-1}\sum_{i=1}^{n} a_i^L\left[(x_{2k} - \tau_i^L)_+^3 - (x_{2k-1} - \tau_i^L)_+^3\right].
\end{aligned}$$

We simplify the double sums, taking into account that $\tau_k^L \in (x_{2k-3}, x_{2k-2}]$ for $k = 1, \ldots, M$, and using the exactness of Q_n^L for the functions from S. For the first double sum we obtain

$$\begin{aligned}
&\sum_{k=1}^{M-1}\sum_{i=1}^{n} a_i^L\left[(x_{2k-1} - \tau_i^L)_+^3 - (x_{2k-2} - \tau_i^L)_+^3\right] \\
&\quad = \sum_{k=1}^{M-1}\sum_{i=1}^{k} a_i^L\left[(x_{2k-1} - \tau_i^L)^3 - (x_{2k-2} - \tau_i^L)^3\right]
\end{aligned}$$

$$
\begin{aligned}
&= \frac{1}{m}\sum_{k=1}^{M-1}\sum_{i=1}^{k} a_i^L \left[\frac{3}{2}(x_{2k-1}-\tau_i^L)^2+\frac{3}{2}(x_{2k-2}-\tau_i^L)^2-\frac{1}{2m^2}\right]\\
&= \frac{3}{2m}\sum_{k=1}^{M-1}\sum_{i=1}^{n} a_i^L \left[(x_{2k-1}-\tau_i^L)_+^2+(x_{2k-2}-\tau_i^L)_+^2\right]-\frac{1}{2m^3}\sum_{k=1}^{M-1}\sum_{i=1}^{k} a_i^L\\
&= \frac{3}{2m}\sum_{k=1}^{M-1}\int_0^1 \left[(x_{2k-1}-t)_+^2+(x_{2k-2}-t)_+^2\right]\mathrm{d}t-\frac{1}{2m^3}\sum_{k=1}^{M-1}\sum_{i=1}^{k} a_i^L\\
&= \frac{1}{2m}\sum_{k=1}^{M-1}(x_{2k-1}^3+x_{2k-2}^3)-\frac{1}{2m^3}\sum_{k=1}^{M-1}\sum_{i=1}^{k} a_i^L.
\end{aligned}
$$

For the second double sum we have

$$
\begin{aligned}
&\sum_{k=1}^{M-1}\sum_{i=1}^{n} a_i^L\left[(x_{2k}-\tau_i^L)_+^3-(x_{2k-1}-\tau_i^L)_+^3\right]\\
&\quad= \sum_{k=1}^{M-1}\sum_{i=1}^{k+1} a_i^L\left[(x_{2k}-\tau_i^L)^3-(x_{2k-1}-\tau_i^L)^3\right]+\sum_{k=1}^{M-1} a_{k+1}^L(x_{2k-1}-\tau_{k+1}^L)^3\\
&\quad= \frac{1}{m}\sum_{k=1}^{M-1}\sum_{i=1}^{k+1} a_i^L\left[\frac{3}{2}(x_{2k}-\tau_i^L)^2+\frac{3}{2}(x_{2k-1}-\tau_i^L)^2-\frac{1}{2m^2}\right]\\
&\qquad+\sum_{k=1}^{M-1} a_{k+1}^L(x_{2k-1}-\tau_{k+1}^L)^3\\
&\quad= \frac{3}{2m}\left[\sum_{k=1}^{M-1}\sum_{i=1}^{n} a_i^L\left[(x_{2k}-\tau_i^L)_+^2+(x_{2k-1}-\tau_i^L)_+^2\right]+\sum_{k=1}^{M-1} a_{k+1}^L(x_{2k-1}-\tau_{k+1}^L)^2\right]\\
&\qquad+\sum_{k=1}^{M-1} a_{k+1}^L(x_{2k-1}-\tau_{k+1}^L)^3-\frac{1}{2m^3}\sum_{k=1}^{M-1}\sum_{i=1}^{k+1} a_i^L\\
&\quad= \frac{3}{2m}\sum_{k=1}^{M-1}\int_0^1\left[(x_{2k}-t)_+^2+(x_{2k-1}-t)_+^2\right]\mathrm{d}t+\frac{3}{2m}\sum_{k=1}^{M-1} a_{k+1}^L(x_{2k-1}-\tau_{k+1}^L)^2\\
&\qquad+\sum_{k=1}^{M-1} a_{k+1}^L(x_{2k-1}-\tau_{k+1}^L)^3-\frac{1}{2m^3}\sum_{k=1}^{M-1}\sum_{i=1}^{k+1} a_i^L\\
&\quad= \frac{1}{2m}\sum_{k=1}^{M-1}(x_{2k}^3+x_{2k-1}^3)+\frac{3}{2m}\sum_{k=2}^{M} a_k^L(\tau_k^L-x_{2k-3})^2\\
&\qquad-\sum_{k=2}^{M} a_k^L(\tau_k^L-x_{2k-3})^3-\frac{1}{2m^3}\sum_{k=1}^{M-1}\sum_{i=1}^{k+1} a_i^L.
\end{aligned}
$$

Replacing the double sums in $c_{3,\infty}(Q_n^L)$ by the expressions found above, we get

$$
\begin{aligned}
c_{3,\infty}(Q_n^L) &= \frac{1}{12}\sum_{k=1}^{M-1}(x_{2k-2}^4-2x_{2k-1}^4+x_{2k}^4)+\frac{1}{6m}(x_0^3-x_{2M-2}^3)+\frac{1}{6m^3}\sum_{k=2}^{M} a_k^L\\
&\quad-\frac{1}{2m}\sum_{k=2}^{M} a_k^L(\tau_k^L-x_{2k-3})^2+\frac{1}{3}\sum_{k=2}^{M} a_k^L(\tau_k^L-x_{2k-3})^3
\end{aligned}
$$

$$= \frac{1}{12}\sum_{k=1}^{M-1}(x_{2k-2}^4 - 2x_{2k-1}^4 + x_{2k}^4) - \frac{1}{48m} + \frac{1}{12m^3} - \frac{1}{6m^3}a_1^L$$
$$- \frac{1}{2m}\sum_{k=2}^{M} a_k^L(\tau_k^L - x_{2k-3})^2 + \frac{1}{3}\sum_{k=2}^{M} a_k^L(\tau_k^L - x_{2k-3})^3$$
$$= \frac{1}{12}\sum_{k=1}^{M-1}(x_{2k-2}^4 - 2x_{2k-1}^4 + x_{2k}^4) - \frac{1}{48m} + \frac{1}{12m^3}$$
$$- \frac{1}{2m}\sum_{k=1}^{[n/2]} a_k^L(\tau_k^L - x_{2k-3})^2 + \frac{1}{3}\sum_{k=1}^{[n/2]} a_k^L(\tau_k^L - x_{2k-3})^3.$$

The proof is completed by verification of the identity

$$\frac{1}{12}\sum_{k=1}^{M-1}(x_{2k-2}^4 - 2x_{2k-1}^4 + x_{2k}^4) - \frac{1}{192(M-1)} + \frac{1}{768(M-1)^3} = \frac{1}{1536(M-1)^3},$$

where $x_k = k/(4(M-1))$. The latter identity is shown to be true again by means of the Euler–MacLaurin formula (it can also be easily reduced to the final identity required for the proof of Theorem 1). □

Lemma 2. *Assume that $n \geq 7$, and set $\tau_k^L = (2k-3+\theta_k)/m$ for $2 \leq k \leq [(n+1)/2]$. Then for $2 \leq k \leq [n/2]$,*

$$0 < \frac{2}{m} - a_k^L < \frac{4\theta_k}{m} \quad \textit{and} \;\; 0 \leq \theta_k < \frac{4}{32^{2^{k-2}}}.$$

Proof. The estimates for a_k^L are verified directly for $k = 2$, and for $k > 2$ follow easily from (23), (24) and $\theta_k \in (0,1)$. As for the estimates for θ_k, (21) and (22) are all the ingredients necessary for reproducing the inequality (17) for $k \geq 3$. Then we can accomplish the proof by induction with respect to k. For $k = 2$, we have $\theta_2 = 0.108194\cdots < 1/8$, showing the validity of the estimates for θ_2. Now the induction step is performed with the help of the inequality $\theta_k < (\theta_{k-1}/2)^2$. Lemma 2 is proved. □

Theorem 3 and Lemma 2 are combined for the derivation of sharp bounds for $c_{3,\infty}(Q_n^L)$.

Theorem 4. *For $n \geq 7$, the error constant $c_{3,\infty}(Q_n^L)$ satisfies the inequalities*

$$c_{3,\infty}(Q_n^L) < \frac{1}{192(n-2)^3} - \frac{1}{247.42420487(n-2)^4}$$

and

$$c_{3,\infty}(Q_n^L) > \frac{1}{192(n-2)^3} - \frac{1}{247.4242048(n-2)^4}.$$

Proof. From Theorem 3 and the representation $\tau_k^L = x_{2k-3} + \theta_k/m$ with $\theta_1 = 1$ and $0 < \theta_k < 1$ for $k \le [n/2]$, we deduce that

$$c_{3,\infty}(Q_n^L) = \frac{1}{192(n-2)^3} - \frac{1}{12(n-2)} \sum_{k=1}^{[n/2]} a_k^L (\tau_k^L - x_{2k-3})^2 (3 - 2\theta_k),$$

and all the terms in the above sum are positive. Therefore, for $n \ge 7$, an upper bound for $c_{3,\infty}(Q_n^L)$ can be obtained by taking only three summands instead of the whole sum:

$$\begin{aligned} c_{3,\infty}(Q_n^L) &\le \frac{1}{192(n-2)^3} - \frac{1}{12(n-2)} \sum_{k=1}^{3} a_k^L (\tau_k^L - x_{2k-3})^2 (3 - 2\theta_k) \\ &\le \frac{1}{192(n-2)^3} - \frac{1}{48(n-2)^4} \left[\frac{1}{6} + \frac{(5-3\sqrt{2})^2}{21} + \frac{\theta_3^2 (3 - 2\theta_3)}{1 + 2\theta_3 - 2\theta_3^2} \right]. \end{aligned}$$

The numerical value of θ_3 is $\theta_3 = 0.00246976636532\ldots$, and substitution in the above estimate yields the upper bound for $c_{3,\infty}(Q_n^L)$ in Theorem 4.

A lower bound for $c_{3,\infty}(Q_n^L)$ is obtained as follows:

$$\begin{aligned} c_{3,\infty}(Q_n^L) &= \frac{1}{192(n-2)^3} - \frac{1}{12(n-2)} \sum_{k=1}^{3} a_k^L (\tau_k^L - x_{2k-3})^2 (3 - 2\theta_k) \\ &\quad - \frac{1}{4(n-2)} \sum_{k=4}^{[n/2]} a_k^k (\tau_k^L - x_{2k-3})^2 \\ &\ge \frac{1}{192(n-2)^3} - \frac{1}{12(n-2)} \sum_{k=1}^{3} a_k^L (\tau_k^L - x_{2k-3})^2 (3 - 2\theta_k) \\ &\quad - \frac{1}{16(n-2)^4} \sum_{k=4}^{[n/2]} \theta_k^2 \end{aligned}$$

(for the last inequality we have used $a_k^L < 2/m$ for $4 \le k \le [n/2]$). We estimate the last sum using Lemma 2 to obtain

$$\sum_{k=4}^{[n/2]} \theta_k^2 < \sum_{k=4}^{\infty} \left(\frac{4}{32^{2^{k-2}}} \right)^2 = 16 \sum_{k=4}^{\infty} \frac{1}{32^{2^{k-1}}} < 16 \sum_{k=1}^{\infty} \frac{1}{32^{8k}} = \frac{16}{32^8 - 1}.$$

Finally, the lower bound for $c_{3,\infty}(Q_n^L)$ is obtained by subtracting $1/((32^8-1)(n-2)^4)$ from the upper bound in Theorem 4. □

In [5] the slightly weaker estimate $c_{3,\infty}(Q_n^L) \le 1/(192(n-2)^4)$ was obtained as a consequence of the inequality $|K_3(Q_n^L;t)| \le |G_{3,n}(t)|$, $t \in [0,1]$ where

$$G_{3,n}(t) = -\frac{1}{(n-2)^3} B_3^*((n-2)t + 1/2)$$

and $B_3^*(t)$ is the one-periodic extension of the Bernoulli polynomial (18). Even for small n the graphs of $K_3(Q_n^L;t)$ and $G_{3,n}(t)$ are indistinguishable (see Fig. 2).

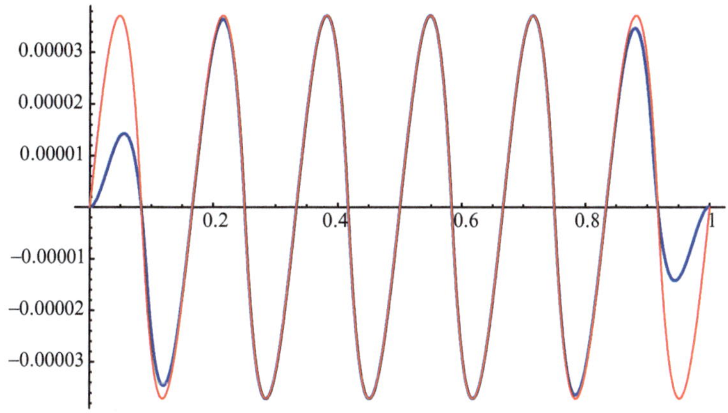

Fig. 2 The graphs of $K_3(Q_8^L;t)$ (*thick*) and $G_{3,8}(t)$ (*thin*)

5 Concluding Remarks

1. The algorithm for the construction of the n-point left Radau quadrature formula $Q_n^{R,l}$ is a hybrid between those for the Gauss and Lobatto quadratures. Namely, we apply the Lobatto scheme to obtain the nodes from the left-hand side and the Gauss scheme to find the nodes from the right-hand side. The number of the nodes obtained through the Lobatto and the Gauss scheme is the same (in the case of even n), or, in the case of odd n, we have one more node obtained through the Lobatto scheme. As in the Gauss and the Lobatto cases, there is a slight modification of the formulae for the nodes around the middle of the integration interval and in the corresponding weights.

Needles to say, the estimates for the weights and the nodes in the Gauss and the Lobatto case are valid here, too. Moreover, a compact formula for the error constant $c_{3,\infty}(Q_n^{R,l})$ can be obtained in the same manner, and then used for the derivation of sharp lower and upper bounds for $c_{3,\infty}(Q_n^{R,l})$. We omit the details.

Figure 3 depicts the third Peano kernel of $Q_5^{R,l}$ and the associated adjusted Bernoulli monospline $F_{3,5}(t)$, which satisfy the inequality $|K_3(Q_5^{R,l};t)| \leq |F_{3,5}(t)|$ for every $t \in [0,1]$. It is seen that, except for some neighborhoods of the endpoints of the integration interval, the graphs of the two functions are indistinguishable.

2. As was shown in Lemmas 1 and 2, the sequences of the nodes and the weights of the Gauss-type quadratures associated with the spaces of parabolic splines with equidistant knots converge extremely fast. This observation allows us to propose

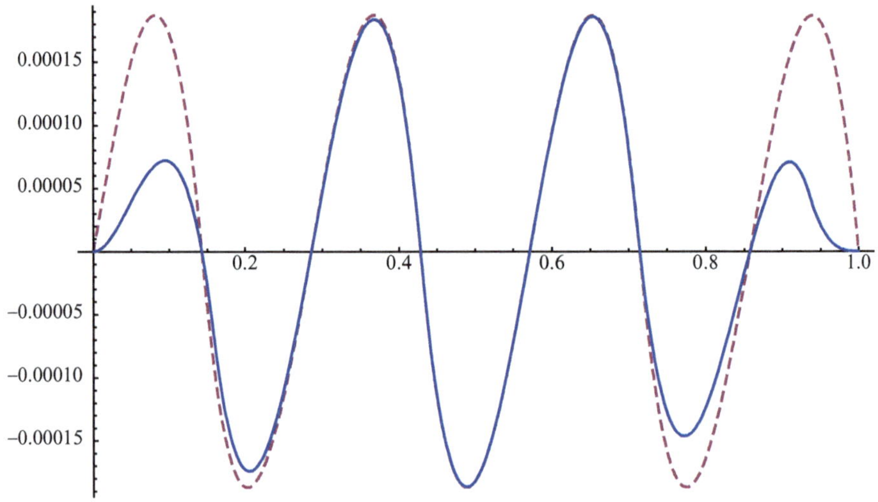

Fig. 3 The graphs of $K_3(Q_5^{R,l};t)$ (*solid*) and $F_{3,5}(t)$ (*dashed*)

some modifications of these quadratures, in which almost all nodes are equidistant and almost all weights are equal. For instance, for $n \geq 9$ the quadrature formula Q_n^G can be replaced by

$$\widetilde{Q}_n^G[f] := \sum_{k=1}^{4} a_{k,n}^G[f(\tau_{k,n}^G) + f(1-\tau_{k,n}^G)] + \frac{1}{n-1}\sum_{k=5}^{n-4} f\left(\frac{k-1}{n-1}\right),$$

where $\{a_{k,n}^G\}_{k=1}^4$ and $\{\tau_{k,n}^G\}_{k=1}^4$ are the first four weights and nodes of Q_n^G. Their numerical values are

$$a_{1,n}^G = \frac{0.5571909584179366}{n-1}, \quad \tau_{1,n}^G = \frac{0.2265409196609864}{n-1},$$

$$a_{2,n}^G = \frac{0.9432633913217402}{n-1}, \quad \tau_{2,n}^G = \frac{1.015518997179282}{n-1},$$

$$a_{3,n}^G = \frac{0.9995456760850675}{n-1}, \quad \tau_{3,n}^G = \frac{2.000113658441225}{n-1},$$

$$a_{4,n}^G = \frac{0.9999999741752557}{n-1}, \quad \tau_{4,n}^G = \frac{3.000000006456186}{n-1}.$$

On using Lemma 1, one can bound the error of $\widetilde{Q}_n^G$ for integrands in $C^3[0,1]$ as follows:

$$|I[f] - \widetilde{Q}_n^G[f]| \leq c_{3,\infty}(Q_n^G)||f'''||_{C[0,1]} + \frac{2}{128^8(n-1)}\left[||f'||_{C[0,1]} + 4(n-8)||f||_{C[0,1]}\right].$$

Similarly, one can obtain a modification of Q_n^L along with an appropriate error bound based on Lemma 2.

Acknowledgements The first-named author was supported by the Swiss National Science Foundation (SCOPES Joint Research Project no. IB7320–111079 “New Methods for Quadrature”), and by the Sofia University Research Fund through Contract no. 135/2008.

References

1. Bojanov, B.D.: Uniqueness of the monosplines of least deviation. In: Numerische Integration (G. Hämmerlin, Ed.), ISNM 45, Birkhäuser, Basel, 67–97 (1979)
2. Bojanov, B.D.: Existence and characterization of monosplines of least L_p deviation. In: Constructive Function Theory ’77 (Bl. Sendov and D. Vačov, Eds.), Sofia, BAN, 249–268 (1980)
3. Braß, H.: Quadraturverfahren. Vandenhoeck & Ruprecht, Göttingen (1977)
4. Karlin, S., Micchelli, C.A.: The fundamental theorem of algebra for monosplines satisfying boundary conditions. Israel J. Math. **11**, 405–451 (1972)
5. Köhler, P., Nikolov, G.P.: Error bounds for Gauss type quadrature formulae related to spaces of splines with equidistant knots. J. Approx. Theory **81**(3), 368–388 (1995)
6. Ligun, A.A.: Exact inequalities for splines and best quadrature formulas for certain classes of functions. Mat. Zametki **19**, 913–926 (1976) (in Russian); English Translation in: Math. Notes **19**, 533–541 (1976)
7. Motornii, V.P.: On the best quadrature formula of the form $\sum p_k f(x_k)$ for some classes of differentiable periodic functions. Izv. Akad. Nauk SSSR Ser. Mat. **38**, 583–614 (1974) (in Russian); English Translation in: Math. USSR Izv. **8**, 591–620 (1974)
8. Nikolov, G.: Gaussian quadrature formulae for splines. In: Numerische Integration, IV (G. Hämmerlin and H. Brass, Eds.), ISNM Vol. 112, Birkhäuser, Basel, 267–281 (1993)
9. Zhensykbaev, A.: Best quadrature formulae for some classes of periodic differentiable functions. Izv. Akad. Nauk SSSR Ser. Mat. **41** (1977) (in Russian); English Translation in: Math. USSR Izv. **11**, 1055–1071 (1977)
10. Zhensykbaev, A.: Monosplines and optimal quadrature formulae for certain classes of non-periodic functions. Anal. Math. **5**, 301–331 (1979) (in Russian)

Approximation of the Hilbert Transform on the Real Line Using Freud Weights

Incoronata Notarangelo

Dedicated to Professor Gradimir V. Milovanović on his 60th birthday

1 Introduction

Let us consider the Cauchy principal value integral

$$\mathcal{H}(G,y) := \int_{\mathbb{R}} \frac{G(x)}{x-y}\,\mathrm{d}x = \lim_{\varepsilon\to 0^+} \int_{|x-y|>\varepsilon} \frac{G(x)}{x-y}\,\mathrm{d}x,$$

where $y \in \mathbb{R}$. If the limit exists, we call $\mathcal{H}(G)$ the Hilbert transform of the function G. It is well known that $\mathcal{H}$ is a bounded operator in $L^p(\mathbb{R})$ for $1 < p < \infty$, while it is, in general, unbounded in the space of continuous functions. Nevertheless, if the function G satisfies the Dini-type condition

$$\int_0^1 \frac{\omega(G,t)}{t}\,\mathrm{d}t < \infty,$$

where ω is its usual modulus of smoothness with step t, then its Hilbert transform $\mathcal{H}(G)$ is continuous on $\mathbb{R}$ (see [12, p. 218]).

The numerical approximation of the Hilbert transform on $\mathbb{R}$ has interested several authors (see, for instance, [1, 3–5, 13, 14, 23–25]). To be more precise, to this end the zeros of Hermite and Markov–Sonin polynomials have been used in [4] and [5], respectively.

Incoronata Notarangelo
PhD student "International Doctoral Seminar entitled J. Bolyai",
Dipartimento di Matematica e Informatica, Università degli Studi della Basilicata,
Via dell'Ateneo Lucano 10, I-85100 Potenza, Italy,
incoronata.notarangelo@unibas.it

W. Gautschi et al. (eds.), *Approximation and Computation: In Honor of Gradimir V. Milovanović*, Springer Optimization and Its Applications 42,
DOI 10.1007/978-1-4419-6594-3_15,

In this paper, we want to compute integrals of the form

$$\mathcal{H}(fw,y) = \int_{\mathbb{R}} \frac{f(x)w(x)}{x-y}\,\mathrm{d}x\,,$$

where $w(x) = \mathrm{e}^{-|x|^\alpha}$, $\alpha > 1$, is a Freud weight. Concerning the study of the smoothness of the function $\mathcal{H}(fw)$, we refer the reader to [19]. We remark that integrals of this form appear in Cauchy singular integral equations, and the numerical treatment of these equations usually requires the approximation of Hilbert transforms (see [19]).

For this purpose, we suggest some simple quadrature rules obtained from a Gauss-type formula based on the zeros of Freud polynomials. These rules, in some aspects, are different from those used in [4,5]. We will consider two different cases: the first when y is sufficiently small, and the second otherwise. The main effort in this paper will be to prove the stability and the convergence of the proposed rules.

The paper is organized as follows. In Sect. 2, we recall some basic facts. In Sect. 3, we introduce the quadrature rules and state our main results. In Sect. 4, some numerical examples are described. In Sect. 5, we prove our main results. Finally, the Appendix deals with the computation of the Hilbert transform of a Freud weight.

2 Preliminary Results and Notations

In the sequel, C will stand for a positive constant that can assume different values in each formula, and we shall write $\mathrm{C} \neq \mathrm{C}(a,b,\ldots)$ when C is independent of $a,b,\ldots$. Furthermore, $A \sim B$ will mean that if A and B are positive quantities depending on some parameters, then there exists a positive constant C independent of these parameters such that $(A/B)^{\pm 1} \leq \mathrm{C}$.

2.1 Function Spaces

Let us consider the weight function u defined by

$$u(x) = (1+|x|)^\beta\,\mathrm{e}^{-|x|^\alpha}, \quad \beta \geq 0,\ \alpha > 1\,,$$

for $x \in \mathbb{R}$.

We denote by C_u the following set of continuous functions,

$$C_u = \left\{ f \in C^0(\mathbb{R}) : \lim_{x\to\pm\infty} f(x)u(x) = 0 \right\},$$

equipped with the norm

$$\|fu\| := \|fu\|_\infty = \sup_{x\in\mathbb{R}} |f(x)u(x)|\,.$$

In the sequel, we will write

$$\|fu\|_E = \sup_{x\in E} |f(x)u(x)|$$

for any $E \subset \mathbb{R}$. We note that the Weierstrass theorem implies the limit conditions in the definition of C_u.

Subspaces of C_u are the Sobolev spaces, defined by

$$W_r(u) = \left\{ f \in C_u : \ f^{(r-1)} \in AC(\mathbb{R}), \ \|f^{(r)}u\| < \infty \right\}, \quad r \in \mathbb{Z}^+,$$

where $AC(\mathbb{R})$ denotes the set of all functions which are absolutely continuous on every closed subset of $\mathbb{R}$. We equip these spaces with the norm

$$\|f\|_{W_r(u)} = \|fu\| + \|f^{(r)}u\|.$$

For any $f \in C_u$, we consider the following main part of the r-th modulus of smoothness (see [6])

$$\Omega^r(f,t)_u = \sup_{0<h\leq t} \|\Delta_h^r(f)\,u\|_{\mathcal{I}_{r,h}}, \quad r \in \mathbb{Z}^+,$$

where $\mathcal{I}_{r,h} = \left[-\mathrm{A}rh^{-1/(\alpha-1)}, \mathrm{A}rh^{-1/(\alpha-1)}\right]$, $\mathrm{A} > 0$ is a constant, and

$$\Delta_h f(x) = f\left(x+\frac{h}{2}\right) - f\left(x-\frac{h}{2}\right), \quad \Delta^r = \Delta\left(\Delta^{r-1}\right).$$

The r-th modulus of smoothness is given by

$$\omega^r(f,t)_u = \Omega^r(f,t)_u + \inf_{P\in\mathbb{P}_{r-1}} \|(f-P)\,u\|_{(-\infty,-\mathrm{A}rt^*)} + \inf_{P\in\mathbb{P}_{r-1}} \|(f-P)\,u\|_{(\mathrm{A}rt^*,+\infty)},$$

with step $t < t_0$ (t_0 sufficiently small) and $t^* := t^{-1/(\alpha-1)}$.

This modulus of smoothness is equivalent to the following K-functional

$$K(f,t^r)_u = \inf_{g\in W_r(u)} \left\{ \|(f-g)\,u\| + t^r \|g^{(r)}u\| \right\},$$

namely $\omega^r(f,t)_u \sim K(f,t^r)_u$. It follows that

$$\omega^r(f,t)_u \leq \mathrm{C}t^r \|f^{(r)}u\| \tag{1}$$

for any $f \in W_r(u)$, with $\mathrm{C} \neq \mathrm{C}(f,t)$.

Let us denote by $\mathbb{P}_m$ the set of all algebraic polynomials of degree at most m and by $E_m(f)_u = \inf_{P\in\mathbb{P}_m} \|(f-P)\,u\|$ the error of best polynomial approximation of $f \in C_u$. The following Jackson and Stechkin-type inequalities hold true (see [6]):

$$E_m(f)_u \leq \mathrm{C}\,\omega^r\left(f, \frac{a_m}{m}\right)_u, \quad r < m, \tag{2}$$

$$\omega^r\left(f,\frac{a_m}{m}\right)_u \le \mathrm{C}\left(\frac{a_m}{m}\right)^r \sum_{k=1}^{m}\left(\frac{k}{a_k}\right)^r \frac{E_k(f)_u}{k}, \tag{3}$$

where $f \in C_u$, $a_m := a_m(u) \sim m^{1/\alpha}$ is the Mhaskar–Rahmanov–Saff (M–R–S) number related to the weight u; in both cases, C is a positive constant independent of f and m, and $\lfloor a \rfloor$ stands for the largest integer smaller than or equal to $a \in \mathbb{R}^+$.

An estimate, weaker than (2), for the error of best polynomial approximation is given by

$$E_m(f)_u \le \mathrm{C}\int_0^{a_m/m} \frac{\Omega^r(f,t)_u}{t}\,\mathrm{d}t \tag{4}$$

with $\mathrm{C} \ne \mathrm{C}(m,f)$.

By means of the modulus of smoothness, we can define the Zygmund spaces

$$Z_s(u) = \left\{ f \in C_u : \sup_{t>0} \frac{\omega^r(f,t)_u}{t^s} < \infty,\ r > s \right\}, \quad s \in \mathbb{R}^+,$$

with the norm

$$\|f\|_{Z_s(u)} = \|fu\| + \sup_{t>0} \frac{\omega^r(f,t)_u}{t^s}.$$

We remark that, by inequalities (4) and (3), $\sup_{t>0} \Omega^r(f,t)_u t^{-s} < \infty$ implies $\sup_{t>0} \omega^r(f,t)_u t^{-s} < \infty$. Therefore, in the definition of the Zygmund space, $\omega^r(f,t)_u$ can be replaced by $\Omega^r(f,t)_u$.

Finally, in the sequel we will write $\omega(f,t)_u$ and $\Omega(f,t)_u$ in place of $\omega^1(f,t)_u$ and $\Omega^1(f,t)_u$.

2.2 *Orthonormal Polynomials and Gaussian Rule*

Let us consider the Freud weight $w(x) = \mathrm{e}^{-|x|^\alpha}$, $\alpha > 1, x \in \mathbb{R}$, and its related M-R-S number a_m, given by (see for instance [15])

$$a_m = a_m(w) = \left[\frac{2\pi}{\alpha B((\alpha+1)/2, 1/2)}\right]^{1/\alpha} m^{1/\alpha},$$

where B is the beta function.

Let $\{p_m(w)\}_{m\in\mathbb{N}}$ be the corresponding sequence of orthonormal polynomials with positive leading coefficient and degree m. We denote by $x_k := x_{m,k}$, $1 \le k \le \lfloor m/2 \rfloor$, the positive zeros of $p_m(w)$ and by $x_{-k} := x_{m,-k} = -x_{m,k}$ the negative ones, both ordered increasingly. If m is odd then $x_{m,0} = 0$ is a zero of $p_m(w)$. These zeros satisfy (see for instance [16])

$$-a_m(w) < x_{m,-\lfloor m/2 \rfloor} < \cdots < x_{m,1} < x_{m,2} < \cdots < x_{m,\lfloor m/2 \rfloor} < a_m(w).$$

For a fixed $\theta \in (0,1)$, we define an index $j = j(m)$ such that

$$x_j = \min_{1 \le k \le \lfloor m/2 \rfloor} \{x_k : x_k \ge \theta a_m(w)\}. \tag{5}$$

Hence, letting $\Delta x_k := \Delta x_{m,k} = x_{m,k+1} - x_{m,k}$, $k \in \{-\lfloor m/2 \rfloor, \dots, \lfloor m/2 \rfloor - 1\}$, we have (see [18])

$$\Delta x_{m,k} \sim \frac{a_m}{m}, \quad |k| \le j. \tag{6}$$

The following proposition will be useful in the sequel.

Proposition 1. *Let $x_{m+1,k}$, $|k| \le \lfloor (m+1)/2 \rfloor$, be the zeros of $p_{m+1}(w)$. If $x_{m+1,k+1}$, $x_{m,k} \in (-\theta a_m, \theta a_m)$, with a fixed $\theta \in (0,1)$, we have*

$$x_{m+1,k+1} - x_{m,k} \sim \frac{a_m}{m},$$

where the constants in "$\sim$" are independent of m and k.

Finally, we will need the following "truncated" Gaussian quadrature rule,

$$\int_{\mathbb{R}} f(x)w(x)\,\mathrm{d}x = \sum_{|k| \le j} \lambda_{m,k}(w) f(x_{m,k}) + \rho_m(f), \tag{7}$$

where $\rho_m(f)$ is the remainder term, $\lambda_k := \lambda_{m,k}(w)$ are the coefficients of the usual Gaussian rule and $x_{m,k}$ are the zeros of $p_m(w)$. An estimate of the remainder term is given by the following proposition, proved in [18].

Proposition 2. *Let $f \in C_u$, where $u(x) = (1+|x|)^\beta \mathrm{e}^{-|x|^\alpha}$, $\alpha > 1$. If $\beta > 1$, then there holds*

$$|\rho_m(f)| \le \mathcal{C}\left\{E_M(f)_u + \mathrm{e}^{-\mathcal{A}m}\|fu\|\right\},$$

where $M = \lfloor (\theta/(\theta+1))^\alpha m/2 \rfloor \sim m$, $\mathcal{C}$ and $\mathcal{A}$ are positive constants independent of m and f.

3 Main Results

To compute the Hilbert transform

$$\mathcal{H}(fw,y) = \int_{\mathbb{R}} \frac{f(x)w(x)}{x-y}\,\mathrm{d}x,$$

where the integral is understood in the Cauchy principal value sense, we use the well-known decomposition

$$\mathcal{H}(fw,y) = \int_{\mathbb{R}} \frac{f(x)-f(y)}{x-y} w(x)\,\mathrm{d}x + f(y)\int_{\mathbb{R}} \frac{w(x)}{x-y}\,\mathrm{d}x. \tag{8}$$

We assume that the second integral can be calculated with the required precision (in the Appendix, we will give some examples of how this can be done). The other is an ordinary improper integral, and so we apply a quadrature rule to compute it. The presence of the weight w in the first integral in (8) leads us to use the Gauss-type rule (7). Hence, we get

$$\begin{aligned}
\mathcal{H}(fw,y) &= \sum_{|k|\le j} \lambda_{m,k}(w)\frac{f(x_{m,k})-f(y)}{x_{m,k}-y} + f(y)\int_{\mathbb{R}} \frac{w(x)}{x-y}\,\mathrm{d}x + e_m(f,y) \\
&= \sum_{|k|\le j} \lambda_{m,k}(w)\frac{f(x_{m,k})}{x_{m,k}-y} + \\
&\quad + f(y)\left[\int_{\mathbb{R}} \frac{w(x)}{x-y}\,\mathrm{d}x - \sum_{|k|\le j}\frac{\lambda_{m,k}(w)}{x_{m,k}-y}\right] + e_m(f,y) \qquad (9) \\
&=: \sum_{|k|\le j} \lambda_{m,k}(w)\frac{f(x_{m,k})}{x_{m,k}-y} + f(y)A_m(y) + e_m(f,y) \qquad (10) \\
&=: H_m(f,y) + e_m(f,y),
\end{aligned}$$

$e_m(f,y)$ being the remainder term, and assuming $x_{m,k} \neq y$, $|k| \le j$.

Since the quantities $x_{m,k}-y$ could be "too small", the rule $H_m(f,y)$ is essentially unstable. Nevertheless, it can be productively used, making some "careful choices".

First of all, we observe that if, for fixed y and m, there holds $|y| \ge x_{m,j}+1$, then (9) can be replaced by the simpler formula

$$\int_{\mathbb{R}} \frac{f(x)w(x)}{x-y}\,\mathrm{d}x = \sum_{|k|\le j} \lambda_{m,k}(w)\frac{f(x_{m,k})}{x_{m,k}-y} + \rho_m(f,y), \qquad (11)$$

where the remainder term $\rho_m(f,y)$ can be estimated using the following theorem.

Theorem 1. *Assume $|y| \ge x_{m,j}+1$. For any $f \in W_r(u)$, with $\beta > 1$, and for a sufficiently large m (say, $m \ge m_0$), we have*

$$|\rho_m(f,y)| \le \mathrm{C}\left(\frac{a_M}{M}\right)^r \|f\|_{W_r(u)}, \qquad (12)$$

where $M = \lfloor(\theta/(\theta+1))^\alpha m/2\rfloor \sim m$ and C is a positive constant independent of m and f.

Now we observe that, for every fixed y and for $m \ge m_0$ ($m_0 = m_0(y,\theta)$), we have $|y| \le x_{m,j}$. Therefore, let us consider the rule $H_m(f,y)$ under the assumption $|y| \le x_{m,j}$. We remark that the term in (9) producing instability is the one related to the knot $x_{m,d}$ closest to y, i.e., $\lambda_{m,d}(w)/(x_{m,d}-y)$. But for $y \in [-x_{m,j}, x_{m,j}]$ there holds

$$\frac{\lambda_{m,d}(w)}{x_{m,d}-y} \sim \frac{a_m}{m}\frac{w(x_{m,d})}{x_{m,d}-y}.$$

Thus, for a fixed $y \in [-x_{m,j}, x_{m,j}]$, we are going to construct a sequence of integers $\{m^*\} \subset \mathbb{N}$, and then a sequence $\{H_{m^*}(f,y)\}_{m^* \in \mathbb{N}}$, such that

$$|x_{m^*,d} - y| \sim \frac{a_m}{m}.$$

This is possible by virtue of Proposition 1, and our choice is determined as follows.

Assume that $x_{m,d} \leq y < x_{m,d+1}$ for some $d \in \{-j,\dots,j-1\}$. Because of the interlacing properties of the zeros of $p_{m+1}(w)$ with those of $p_m(w)$, two cases are possible:

$$\text{(a)} \quad x_{m,d} \leq y < x_{m+1,d+1}; \quad \text{(b)} \quad x_{m+1,d+1} \leq y < x_{m,d+1}.$$

In the case (a), if $y < (x_{m,d} + x_{m+1,d+1})/2$, then we choose $m^* = m+1$ and we use the quadrature rule $H_{m+1}(f,y)$, otherwise we choose $m^* = m$ and we use $H_m(f,y)$. We make a similar choice in the case (b).

Thus, for every fixed $y \in [-x_{m,j}, x_{m,j}]$ we can define a sequence $\{H_{m^*}(f,y)\}$, $m^* \in \{m, m+1\}$. The next theorem proves the stability and the convergence of this sequence.

Theorem 2. *Let $y \in [-x_{m^*,j}, x_{m^*,j}]$ be fixed, with j defined by (5). For any $f \in C_u$, with $\beta > 1$, we have*

$$|H_{m^*}(f,y)| \leq \mathrm{C}\,\|fu\| \log m. \tag{13}$$

Moreover, if f is such that

$$\int_0^1 \frac{\omega^r(f,t)_u}{t}\,\mathrm{d}t < \infty, \quad r \in \mathbb{Z}^+,$$

there holds

$$|e_{m^*}(f,y)| \leq \mathrm{C}\left\{\log m \int_0^{\frac{a_M}{M}} \frac{\omega^r(f,t)_u}{t}\,\mathrm{d}t + \mathrm{e}^{-\mathrm{A}m}\|fu\|\right\}. \tag{14}$$

In both inequalities, C *and* A *are positive constants independent of m, y and f, while $M = \lfloor (\theta/(\theta+1))^\alpha m/2 \rfloor \sim m$.*

In particular, by Theorem 2, for $m \geq m_0$ and for $\beta > 1$, if $f \in Z_s(u)$, $s > 0$, we have

$$|e_{m^*}(f,y)| \leq \mathrm{C} \log m \left(\frac{a_m}{m}\right)^s \|f\|_{Z_s(u)},$$

while if $f \in W_r(u)$ we get

$$|e_{m^*}(f,y)| \leq \mathrm{C} \log m \left(\frac{a_m}{m}\right)^r \|f\|_{W_r(u)}, \tag{15}$$

using inequality (1).

Thus, Theorem 2 shows that the rule $H_{m^*}(f,y)$ is stable and its remainder term converges with the same order of the best polynomial approximation, apart from an extra factor $\log m$.

From a numerical point of view, we remark that we have to compute only $2j$ (or $2j+1$ if m^* is odd) Christoffel numbers, zeros of $p_{m^*}(w)$, and values of the function f. Moreover, if $\alpha = 2$, then $\{p_{m^*}(w)\}_{m\in\mathbb{N}}$ is the sequence of Hermite polynomials, that is the simplest case, and one can use the routine "gaussq" (see [10, 11]) or the routines "recur" and "gauss" (see [9]). In the case $\alpha \neq 2$, we can use the Mathematica Package "`OrthogonalPolynomials`" (see [2]).

Remark 1. We can use the same method to evaluate

$$\mathcal{H}(fw_\delta, y) = \int_{\mathbb{R}} \frac{f(x)w_\delta(x)}{x-y}\,\mathrm{d}x,$$

where $w_\delta(x) = \mathrm{e}^{-\delta|x|^\alpha}$, $\delta > 0$.

In fact, since $\{p_m(w_\delta, x)\}_{m\in\mathbb{N}} = \{\delta^{1/(2\alpha)} p_m(w, \delta^{1/\alpha}x)\}_{m\in\mathbb{N}}$, we have

$$p_m(w_\delta, z_k) = 0 \Leftrightarrow z_k = \frac{x_{m,k}(w)}{\delta^{1/\alpha}}, \quad |k| \leq \left\lfloor \frac{m}{2} \right\rfloor,$$

and the same relation holds for the Christoffel numbers.

4 Numerical Examples

In this section, we show some approximate values for the integral $\mathcal{H}(fw, y)$, $y \in \mathbb{R}$, obtained by using the algorithms described in Sect. 3.

Since the exact value of the integral is not known, in all the tables we report the digits which are correct according to the results obtained for $m = 400$.

Moreover, $J = J(m)$ will denote the number of the points that we use in the quadrature rules (9) and (11). To be more precise, J will be equal to $2j$, with j given by (5), if m^* is even, and it will be equal to $2j+1$ otherwise.

All the computations have been done in double-precision arithmetic (ca. 16 decimal digits).

Example 1. We want to evaluate the following integral

$$\int_{\mathbb{R}} \frac{|x-1|^4 \mathrm{e}^{-|x|^3}}{x-y}\,\mathrm{d}x.$$

Since the function $f(x) = |x-1|^4 \in W_4(u)$, with $u(x) = (1+|x|)^\beta \mathrm{e}^{-|x|^3}$ and $\beta > 1$, the theoretical error behaves like $m^{-8/3}\log m$ by using (15) and $a_m \sim m^{1/3}$. In Table 1, we approximate the above integral by using quadrature rules (9) and (11), choosing $\theta = 3/5$ in (5). We can see that the numerical results agree with the theoretical ones.

Table 1 Approximate values obtained for $\theta = 3/5$

m	J	$y = 1$	J	$y = 10$
16	10	−3.7	10	−0.5
32	21	−3.7859	20	−0.592
64	42	−3.785959023	42	−0.592611694
128	83	−3.78595902313849	82	−0.592611694730986

Example 2. Let us consider the integral

$$\int_{\mathbb{R}} \frac{\cosh x \, \mathrm{e}^{-x^4}}{x-y} \, \mathrm{d}x .$$

Since the function $f(x) = \cosh x = (\mathrm{e}^x + \mathrm{e}^{-x})/2$ is an analytic function, we obtain very accurate results. In Table 2, we report the results obtained for $\theta = 3/5$.

We remark that one obtains the same approximations for $m = 384$ and $\theta = 3/5$ and for $m = 32$ and $\theta = 0.95$. Therefore, the parameter θ influences the numerical results, notably for very smooth functions and small values of m, but the appropriate choice of this parameter is not yet totally clear.

Table 2 Approximate values obtained for $\theta = 3/5$

m	J	$y = 0.5$	J	$y = -6$
16	11	−1.16	10	0.3
32	18	−1.167	18	0.360
64	39	−1.167487	38	0.36052
128	75	−1.16748708017	76	0.36052114077
256	150	−1.16748708017153	150	0.360521140776152
384	225	−1.167487080171531	224	0.360521140776152

Example 3. Now we want to evaluate the integral

$$\int_{\mathbb{R}} \frac{\mathrm{e}^{-|x|^3}}{(1+x^2)^7 (x-y)} \, \mathrm{d}x .$$

As in Example 2, the function $f(x) = (1+x^2)^{-7}$ is very smooth and one could observe the influence of the parameter θ in the numerical results. In Table 3, we have chosen $\theta = 1/2$.

Example 4. Finally, we consider the integral

$$\int_{\mathbb{R}} \frac{\mathrm{d}x}{(1+x^2)^7 (x-y)} = \int_{\mathbb{R}} \frac{f(x) \, \mathrm{e}^{-|x|^3}}{x-y} \, \mathrm{d}x .$$

Table 3 Approximate values obtained for $\theta = 1/2$

m	J	$y = 0.2$	J	$y = 7$
16	9	-1.553	8	$-9.72\,E-2$
32	17	-1.553	16	$-9.7383\,E-2$
64	35	-1.5536242	34	$-9.738383637\,E-2$
128	68	-1.5536242711590	68	$-9.7383836377130\,E-2$
256	135	-1.55362427115905	134	$-9.7383836377130\,E-2$
384	201	-1.553624271159052	202	$-9.738383637713081\,E-2$

Since $f(x) = (1+x^2)^{-7}\mathrm{e}^{|x|^3} \in W_6(u)$, with $u(x) = (1+|x|)^2\mathrm{e}^{-|x|^3}$, the theoretical error behaves like $m^{-4}\log m$. Note that $\|f^{(6)}u\| = \mathcal{O}(10^5)$, and this influences in a negative way the numerical results in Table 4, obtained for $\theta = 1/2$.

Table 4 Approximate values obtained for $\theta = 1/2$

m	J	$y = 0.2$	J	$y = 7$
16	9	-1.5	8	-0.101
32	17	-1.519	16	-0.1014
64	35	-1.5192	34	-0.10143
128	68	-1.5192	68	-0.101431
256	135	-1.51924	134	-0.1014318
384	201	-1.5192406	202	-0.1014318

5 Proofs

First of all, we recall a well-known inequality (see [16]). Let $1 \le p \le \infty$ and let $u(x) = (1+|x|)^\beta \mathrm{e}^{-|x|^\alpha}$, with $\alpha > 1$, $\beta \ge 0$. For every $P \in \mathbb{P}_m$ and $\theta > 0$, there holds

$$\|Pu\|_{L^p(J_m)} \le \mathrm{C}\mathrm{e}^{-\mathrm{A}m}\|Pu\|_p, \tag{16}$$

where $J_m = \{x \in \mathbb{R} : |x| > (1+\theta)a_m(u)\}$, $a_m(u) \sim m^{1/\alpha} \sim a_{m/2}(w)$, C and A are positive constants independent of m and P.

By (16), it easily follows that (see for instance [21]) the inequality

$$\|fu\|_{J_m} \le \mathrm{C}\left\{E_m(f)_u + \mathrm{e}^{-\mathrm{A}m}\|fu\|\right\} \tag{17}$$

holds for any $f \in C_u$.

Proof of Proposition 1. An analog of this proposition was proved in [4] (Lemma 2.1) for Hermite zeros. Similar arguments apply in the case of Freud zeros, taking into account (6). Therefore, we omit the proof. □

We recall the Posse–Markov–Stieltjes inequalities (see [8], p. 33 and [17]). If a function g is such that $g^{(k)}(x) \geq 0$, $k = 0,1,\ldots,2m-1$, $m > 1$, for $x \in (-\infty, x_d]$, $d = -\lfloor m/2 \rfloor + 1, \ldots, \lfloor m/2 \rfloor$, then we have

$$\sum_{k=-\lfloor m/2 \rfloor}^{d-1} \lambda_k g(x_k) \leq \int_{-\infty}^{x_d} g(x) w(x)\,\mathrm{d}x \leq \sum_{k=-\lfloor m/2 \rfloor}^{d} \lambda_k g(x_k). \tag{18}$$

On the other hand, if $(-1)^k g^{(k)}(x) \geq 0$, $k = 0,1,\ldots,2m-1$, $m > 1$, for $x \in [x_d, +\infty)$, $d = -\lfloor m/2 \rfloor, \ldots, \lfloor m/2 \rfloor - 1$, then we have

$$\sum_{k=d+1}^{\lfloor m/2 \rfloor} \lambda_k g(x_k) \leq \int_{x_d}^{+\infty} g(x) w(x)\,\mathrm{d}x \leq \sum_{k=d}^{\lfloor m/2 \rfloor} \lambda_k g(x_k). \tag{19}$$

Proposition 3. *Let $A_m(y)$ and j be given by (10) and (5), respectively. Let us assume that $|y| \leq x_j$ and $\min_k |x_k - y| \geq \mathrm{c}\, a_m/m$, $\mathrm{c} \neq \mathrm{c}(m)$. Then we have*

$$|A_m(y)| \leq \mathrm{C}\, w(y) \log m, \tag{20}$$

where C *is a positive constant independent of m and y.*

Proof. Let $x_{d-1} < x_d \leq y < x_{d+1}$, $|d| \leq j$. We can write

$$|A_m(y)| \leq \left| \mathcal{H}(w,y) - \sum_{|k| \leq \lfloor m/2 \rfloor} \frac{\lambda_k}{x_k - y} \right| + \left| \sum_{|k| > j} \frac{\lambda_k}{x_k - y} \right| =: |B_1| + |B_2|. \tag{21}$$

We can decompose the integral in B_1 as

$$\int_{\mathbb{R}} \frac{w(x)}{x-y}\,\mathrm{d}x = \left\{ \int_{-\infty}^{x_{d-1}} + \int_{x_{d-1}}^{x_{d+1}} + \int_{x_{d+1}}^{+\infty} \right\} \frac{w(x)}{x-y}\,\mathrm{d}x.$$

Using the Posse–Markov–Stieltjes inequalities (18) for the first integral, with $g(x) = 1/(y-x)$, and (19) for the last integral, with $g(x) = 1/(x-y)$, we get

$$\int_{x_{d-1}}^{x_{d+1}} \frac{w(x)}{x-y}\,\mathrm{d}x - \frac{\lambda_{d+1}}{x_{d+1}-y} - \frac{\lambda_d}{x_d - y} \leq B_1 \leq \int_{x_{d-1}}^{x_{d+1}} \frac{w(x)}{x-y}\,\mathrm{d}x - \frac{\lambda_{d-1}}{x_{d-1}-y} - \frac{\lambda_d}{x_d - y}.$$

It follows that

$$|B_1| \leq \left| \int_{x_{d-1}}^{x_{d+1}} \frac{w(x)}{x-y}\,\mathrm{d}x \right| + \left| \frac{\lambda_d}{x_d - y} \right| + \max\left\{ \left| \frac{\lambda_{d-1}}{x_{d-1}-y} \right|, \left| \frac{\lambda_{d+1}}{x_{d+1}-y} \right| \right\}. \tag{22}$$

Since $y - x_d \sim a_m/m \sim x_{d+1} - x_d$, $|d| \leq j$, we have (see for instance [18])

$$\frac{\lambda_d}{|x_d - y|} \sim \frac{\Delta x_d\, w(x_d)}{|x_d - y|} \leq \mathrm{C}\, w(x_d) \sim w(y), \tag{23}$$

using

$$|x-y| \le \mathcal{C}\frac{a_m}{m} \quad \Rightarrow \quad w(x) \sim w(y). \tag{24}$$

Analogously, we obtain

$$\frac{\lambda_{d\pm 1}}{|x_{d\pm 1}-y|} \le \mathcal{C}w(y). \tag{25}$$

On the other hand, by the mean value theorem, we have

$$\begin{aligned}\left|\int_{x_{d-1}}^{x_{d+1}} \frac{w(x)}{x-y}\,\mathrm{d}x\right| &\le \left|\int_{x_{d-1}}^{x_{d+1}} \frac{w(x)-w(y)}{x-y}\,\mathrm{d}x\right| + w(y)\left|\int_{x_{d-1}}^{x_{d+1}} \frac{\mathrm{d}x}{x-y}\right| \\ &\le \alpha\int_{x_{d-1}}^{x_{d+1}} |\xi_x|^{\alpha-1} w(\xi_x)\,\mathrm{d}x + w(y)\left|\log\frac{x_{d+1}-y}{y-x_{d-1}}\right|, \end{aligned} \tag{26}$$

for some $\xi_x \in (x,y)$. Since $x_{d+1}-y \sim y-x_{d-1}$, by (24) and (6), the right-hand side of (26) is equivalent to

$$\int_{x_{d-1}}^{x_{d+1}} \mathrm{d}w(x) + w(y) \sim w(y) \tag{27}$$

Combining (23), (25)–(27) in (22), we get

$$|B_1| \le \mathcal{C}w(y). \tag{28}$$

Now let us consider the term B_2. Since $w(x_k) \le w(y)$ for $|k| > j$, we have

$$|B_2| \le \mathcal{C}\sum_{|k|>j} \frac{\Delta x_k w(x_k)}{|x_k-y|} \le \mathcal{C}w(y)\sum_{|k|>j}\frac{\Delta x_k}{|x_k-y|} \le \mathcal{C}w(y)\log m. \tag{29}$$

Combining (28) and (29) in (21), we obtain (20). □

Proof of Theorem 1. We will consider only the case $f \in W_1(u)$, since the case $r \ge 2$ can be proved by iteration.

Given $\Psi \in C^\infty(\mathbb{R})$ an arbitrary nondecreasing function such that

$$\Psi(x) = \begin{cases} 0, & x \le 0, \\ 1, & x \ge 1, \end{cases}$$

and x_j defined by (5), we set

$$\Psi_j(x) = \Psi\left(\frac{|x|-x_j}{\Delta x_j}\right) = \begin{cases} 1, & |x| \ge x_{j+1}, \\ 0, & |x| \le x_j. \end{cases}$$

Then we can write

$$\frac{f(x)}{x-y} = \frac{[1-\Psi_j(x)]\,f(x)}{x-y} + \frac{\Psi_j(x)f(x)}{x-y} =: F_1(x) + F_2(x),$$

and so

$$\int_{\mathbb{R}} \frac{f(x)}{x-y}\,\mathrm{d}x = \int_{\mathbb{R}} F_1(x)\,\mathrm{d}x + \int_{\mathbb{R}} F_2(x)\,\mathrm{d}x.$$

Using the truncated Gaussian rule (7), we obtain

$$\int_{\mathbb{R}} F_1(x)w(x)\,\mathrm{d}x = \sum_{|k|\le j} \frac{\lambda_k f(x_k)}{x_k - y} + \rho_m(F_1)\,.$$

Setting

$$\rho_m(f,y) = \rho_m(F_1) + \int_{\mathbb{R}} F_2(x)\,\mathrm{d}x, \tag{30}$$

we have

$$\int_{\mathbb{R}} \frac{f(x)}{x-y}\,\mathrm{d}x = \sum_{|k|\le j} \frac{\lambda_k f(x_k)}{x_k - y} + \rho_m(f,y)\,.$$

Let us consider the first term on the right of (30). By Proposition 2, since $\beta > 1$, we get

$$\begin{aligned} |\rho_m(F_1)| &\le \mathrm{C}\left\{E_M(F_1)_u + \mathrm{e}^{-\mathrm{A}m}\|F_1 u\|\right\} \\ &\le \mathrm{C}\left\{\frac{a_M}{M}\|F_1' u\| + \mathrm{e}^{-\mathrm{A}m}\|fu\|\right\} \le \mathrm{C}\frac{a_M}{M}\|f\|_{W_1(u)}, \end{aligned} \tag{31}$$

by Jackson's inequality (2) and (1). In fact for $m \ge m_0$ and $|x| \le x_{j+1}$, we have $|x-y| \ge 1 - \Delta x_j \ge 1/2$ and then $\|F_1 u\| \le 2\|fu\|$. Moreover, we can write

$$\begin{aligned} F_1'(x) &= -\frac{\Psi_j'(x) f(x)}{x-y} + \frac{[1-\Psi_j(x)]\,f'(x)}{x-y} - \frac{[1-\Psi_j(x)]\,f(x)}{(x-y)^2} \\ &=: G_1(x) + G_2(x) + G_3(x)\,. \end{aligned} \tag{32}$$

There holds

$$\|G_3 u\| \le \sup_{|x|\le x_{j+1}} \frac{|fu|(x)}{(x-y)^2} \le 4\,\|fu\| \tag{33}$$

and

$$\|G_2 u\| \le \sup_{|x|\le x_{j+1}} \frac{|f'u|(x)}{|x-y|} \le 2\,\|f'u\|\,, \tag{34}$$

since $|y| \ge x_j + 1$ and $|x-y| \ge 1 - \Delta x_j \ge 1/2$. Furthermore, using inequalities (17), (2) and (1), we get

$$\begin{aligned} \|G_1 u\| &= \sup_{x_j \le |x| \le x_{j+1}} \frac{|fu|(x)}{\Delta x_j |x-y|} \le \mathrm{C}\frac{m}{a_m} \sup_{x_j \le |x| \le x_{j+1}} |fu|(x) \\ &\le \mathrm{C}\left\{\frac{m}{a_m} E_M(f)_u + \mathrm{e}^{-\mathrm{A}m}\|fu\|\right\} \le \mathrm{C}\,\|f\|_{W_1(u)}, \end{aligned} \tag{35}$$

since $M \sim m$. Therefore, by (33)–(35) and (32), we obtain (31).

To handle the term in (30) containing F_2, we assume for simplicity $y>0$, since the other case is similar. We have

$$\int_{\mathbb{R}} F_2(x)w(x)\,\mathrm{d}x = \left\{\int_{x_j}^{+\infty} + \int_{-\infty}^{-x_j}\right\} \frac{\Psi_j(x)f(x)w(x)}{x-y}\,\mathrm{d}x =: I_1 + I_2 . \tag{36}$$

Using (17) we get

$$\begin{aligned} |I_2| &\le \int_{-\infty}^{-x_j} \left|\frac{f(x)w(x)}{x-y}\right| \mathrm{d}x \le \|fu\|_{(-\infty,-x_j]} \int_{-\infty}^{-x_j} \frac{\mathrm{d}x}{(y-x)(1+|x|)^{\beta}} \\ &\le \mathrm{C}\left\{E_M(f)_u + \mathrm{e}^{-\mathrm{A}m}\|fu\|\right\} \le \mathrm{C}\frac{a_M}{M}\|f\|_{W_1(u)}, \end{aligned} \tag{37}$$

since $\beta>1$ and $M=\lfloor(\theta/(\theta+1))^{\alpha}m/2\rfloor$.

However, setting $\varepsilon = a_m/m$, we have

$$I_1 = \left\{\int_{x_j}^{y-\varepsilon} + \int_{y-\varepsilon}^{y-\varepsilon} + \int_{y+\varepsilon}^{+\infty}\right\} \frac{\Psi_j(x)f(x)w(x)}{x-y}\,\mathrm{d}x =: J_1 + J_2 + J_3 . \tag{38}$$

The integrals J_1 and J_3 are nonsingular and using the same arguments as in the estimate of I_2, we obtain

$$|J_1| + |J_3| \le \mathrm{C}\frac{a_M}{M}\|f\|_{W_1(u)} . \tag{39}$$

To estimate J_2, we can write

$$\begin{aligned} |J_2| &= \left|\int_{y-\varepsilon}^{y+\varepsilon} \frac{f(x)w(x)}{x-y}\,\mathrm{d}x\right| \\ &\le w(y)\left|\int_{y-\varepsilon}^{y+\varepsilon} \frac{f(x)-f(y)}{x-y}\,\mathrm{d}x\right| + \left|\int_{y-\varepsilon}^{y+\varepsilon} f(x)\frac{w(x)-w(y)}{x-y}\,\mathrm{d}x\right| \\ &=: A_1 + A_2 . \end{aligned} \tag{40}$$

Using the mean value theorem, with $\xi, \xi_x \in (y-\varepsilon, y+\varepsilon)$, and by (24), since $\beta>1$, we have

$$A_1 \le 2\varepsilon\, w(y)|f'(\xi)| \le \mathrm{C}\frac{a_m}{m}\|f'u\| \tag{41}$$

and

$$\begin{aligned} A_2 &\le |f(\xi)| \int_{y-\varepsilon}^{y+\varepsilon} |w'(\xi_x)|\,\mathrm{d}x \\ &\sim |f(\xi)| \int_{y-\varepsilon}^{y+\varepsilon} -\mathrm{d}w(x) = |f(\xi)|\,[w(y-\varepsilon)-w(y+\varepsilon)] \\ &\le \mathrm{C}\|fu\|_{[x_j,+\infty)} \le \mathrm{C}\left\{E_M(f)_u + \mathrm{e}^{-\mathrm{A}m}\|fu\|\right\} \\ &\le \mathrm{C}\frac{a_M}{M}\|f\|_{W_1(u)}, \end{aligned} \tag{42}$$

by (17). Combining (41) and (42) in (40), we get

$$|J_2| \leq \mathrm{C}\frac{a_M}{M}\|f\|_{W_1(u)}.$$

It follows that recalling (39) and (38),

$$|I_1| \leq \mathrm{C}\frac{a_M}{M}\|f\|_{W_1(u)}. \tag{43}$$

By (43), (37), and (36) we obtain

$$\left|\int_{\mathbb{R}} F_2(x)w(x)\,\mathrm{d}x\right| \leq \mathrm{C}\frac{a_M}{M}\|f\|_{W_1(u)}. \tag{44}$$

Finally, combining (31) and (44) in (30), we get (12). □

To prove Theorem 2 we need the following lemmas.

Lemma 1. *Let* $w(x) = \mathrm{e}^{-|x|^\alpha}$, $\alpha > 1$, *and* $u(x) = (1+|x|)^\beta w(x)$, $\beta > 0$. *For* $\theta \in (0,1)$ *and* $|y| \leq \theta a_m$ *we have*

$$\left|\int_{\mathbb{R}} \frac{f(x)w(x)}{x-y}\,\mathrm{d}x\right| \leq \mathrm{C}\left\{\|fu\|\log m + \int_0^1 \frac{\Omega(f,t)_u}{t}\,\mathrm{d}t\right\}, \tag{45}$$

with C *a constant independent of* f, m *and* y.

Proof. We assume $y > 0$, since the other case can be treated in a similar way.

Letting $\varepsilon = a_m/m$, we can write

$$\begin{aligned}\int_{\mathbb{R}} \frac{f(x)w(x)}{x-y}\,\mathrm{d}x &= \left\{\int_{-\infty}^{-2a_m} + \int_{-2a_m}^{y-\varepsilon} + \int_{y-\varepsilon}^{y+\varepsilon} + \int_{y+\varepsilon}^{2a_m} + \int_{2a_m}^{+\infty}\right\} \frac{f(x)w(x)}{x-y}\,\mathrm{d}x \\ &=: I_1 + I_2 + I_3 + I_4 + I_5.\end{aligned} \tag{46}$$

Since $\beta > 0$, we get

$$|I_5| \leq \|fu\| \int_{2a_m}^{+\infty} \frac{\mathrm{d}x}{(x-y)(1+|x|)^\beta} \leq \mathrm{C}\|fu\| \int_{2a_m}^{+\infty} \frac{\mathrm{d}x}{x(1+|x|)^\beta} \leq \mathrm{C}\|fu\| \tag{47}$$

and also

$$|I_1| \leq \mathrm{C}\|fu\|. \tag{48}$$

Moreover, we have

$$|I_4| \leq \|fu\| \int_{y+\varepsilon}^{2a_m} \frac{\mathrm{d}x}{x-y} \leq \mathrm{C}\|fu\| \log\frac{2a_m - y}{\varepsilon} \leq \mathrm{C}\|fu\|\log m. \tag{49}$$

Analogously, we obtain

$$|I_2| \leq \mathrm{C}\|fu\|\log m. \tag{50}$$

Now let us consider the term I_3. We can write

$$|I_3| = \left| \int_{y-\varepsilon}^{y+\varepsilon} \frac{f(x)w(x) - f(y)w(y)}{x-y} \, \mathrm{d}x \right| = \left| \int_0^{2\varepsilon} \frac{[fw](y+t/2) - [fw](y-t/2)}{t} \, \mathrm{d}t \right|.$$

Since $w(y+t/2) < w(y) < u(y)$, we obtain

$$\begin{aligned} |I_3| &\le \int_0^{2\varepsilon} \frac{|\Delta_t f(y)|\, u(y)}{t} \, \mathrm{d}t + \|fu\| \int_0^{2\varepsilon} \frac{|\Delta_t w(y)|}{t\, w(y-t/2)} \, \mathrm{d}t \\ &\le \int_0^1 \frac{\Omega(f,t)_u}{t} \, \mathrm{d}t + \|fu\| \int_0^{2\varepsilon} \frac{|\Delta_t w(y)|}{t\, w(y-t/2)} \, \mathrm{d}t. \end{aligned} \tag{51}$$

As in the proof of Proposition 3, using the mean value theorem, (24) and (6), we have

$$\int_0^{2\varepsilon} \frac{|\Delta_t w(y)|}{t\, w(y-t/2)} \, \mathrm{d}t \le \int_0^{2\varepsilon} \frac{|\xi_t|^{\alpha-1} w(\xi_t)}{w(y-t/2)} \, \mathrm{d}t \le \mathrm{C} \int_0^{2\varepsilon} |\xi_t|^{\alpha-1} \, \mathrm{d}t \le \mathrm{C} \frac{a_m}{m} a_m^{\alpha-1} \le \mathrm{C}. \tag{52}$$

Combining (47)–(52) in (46) we get (45). □

Lemma 2. *For any function f such that*

$$\int_0^1 \frac{\omega^r(f,t)_u}{t} \, \mathrm{d}t < \infty, \quad r \in \mathbb{Z}^+,$$

and with $P \in \mathbb{P}_m$ the polynomial of best approximation of f, we have

$$\int_0^1 \frac{\Omega(f-P,t)_u}{t} \, \mathrm{d}t \le \mathrm{C} \log m \int_0^{a_m/m} \frac{\omega^r(f,t)_u}{t} \, \mathrm{d}t, \tag{53}$$

with C a positive constant independent of m, f and P.

Proof. By inequality (4), we have

$$\begin{aligned} \int_0^1 \frac{\Omega(f-P,t)_u}{t} \, \mathrm{d}t &= \int_0^{a_m/m} \frac{\Omega(f-P,t)_u}{t} \, \mathrm{d}t + \int_{a_m/m}^1 \frac{\Omega(f-P,t)_u}{t} \, \mathrm{d}t \\ &\le \int_0^{a_m/m} \frac{\Omega(f-P,t)_u}{t} \, \mathrm{d}t + E_m(f)_u \log m \\ &\le \int_0^{a_m/m} \frac{\Omega(f-P,t)_u}{t} \, \mathrm{d}t + \mathrm{C} \log m \int_0^{a_m/m} \frac{\Omega^r(f,t)_u}{t} \, \mathrm{d}t. \end{aligned}$$

By using the Jackson and Stechkin-type inequalities (2) and (3), and proceeding as in the proof of Proposition 4.2 in [20], we obtain

$$\int_0^{a_m/m} \frac{\Omega(f-P,t)_u}{t} \, \mathrm{d}t \le \mathrm{C} \int_0^{a_m/m} \frac{\omega^r(f,t)_u}{t} \, \mathrm{d}t,$$

from which we get (53). □

Proof of Theorem 2. Let us first prove (13). By Proposition 3, with $\beta > 1$, we have

$$
\begin{aligned}
|H_{m^*}(f,y)| &\le |f(y)|\,|A_{m^*}(y)| + \left| \sum_{|k|\le j} \lambda_{m^*,k}(w) \frac{f(x_{m^*,k})}{x_{m^*,k}-y} \right| \\
&\le \mathrm{C}\,\|fu\| \log m + \lambda_{m^*,d}(w) \left| \frac{f(x_{m^*,d})}{x_{m^*,d}-y} \right| \\
&\quad + \left| \sum_{|k|\le j, k\ne d} \lambda_{m^*,k}(w) \frac{f(x_{m^*,k})}{x_{m^*,k}-y} \right| \\
&=: \mathrm{C}\,\|fu\| \log m + |B_1| + |B_2|, \qquad (54)
\end{aligned}
$$

$x_{m^*,d}$ being a zero closest to y.

Since $\beta > 1$, by (23), we get

$$|B_1| \le \mathrm{C}\|fu\|. \qquad (55)$$

Moreover, we have

$$|B_2| \le \mathrm{C}\,\|fu\| \sum_{|k|\le j, k\ne d} \frac{\Delta x_{m^*,k}}{|x_{m^*,k}-y|} \le \mathrm{C}\,\|fu\| \log m. \qquad (56)$$

Combining (55) and (56) in (54), we obtain (13).

Now, we prove (14). Letting $P \in \mathbb{P}_M$ be the polynomial of best approximation of f of degree $M = \lfloor (\theta/(\theta+1))^\alpha m/2 \rfloor \sim m$, we have

$$|e_{m^*}(f,y)| \le |e_{m^*}(f-P,y)| + |e_{m^*}(P,y)|. \qquad (57)$$

Since the ordinary Gaussian rule is exact for polynomials of degree at most $2m-1$, we get

$$
\begin{aligned}
|e_{m^*}(P,y)| &= \left| \sum_{|k|>j} \lambda_{m^*,k}(w) \frac{P(x_{m^*,k}) - P(y)}{x_{m^*,k}-y} \right| \\
&\le \left| \sum_{|k|>j} \lambda_{m^*,k}(w) \frac{P(x_{m^*,k})}{x_{m^*,k}-y} \right| + |P(y)| \sum_{|k|>j} \frac{\lambda_{m^*,k}(w)}{|x_{m^*,k}-y|} \\
&=: |S_1| + |S_2|. \qquad (58)
\end{aligned}
$$

By (29), since $\beta > 1$, and by (16), we have

$$|S_1| \le \mathrm{C} \sum_{|k|>j} \Delta x_{m^*,k} \frac{|P(x_{m^*,k})|\, w(x_{m^*,k})}{|x_{m^*,k}-y|} \le \mathrm{C}\,\mathrm{e}^{-\mathrm{A}m} \|Pu\|. \qquad (59)$$

Moreover, we have

$$|S_2| \le \mathrm{C}\,\|Pu\|_{\{x\in\mathbb{R}:|x|\ge|y|\}} \sum_{|k|>j} \frac{\Delta x_{m^*,k}\, \mathrm{e}^{|y|^\alpha - |x_{m^*,k}|^\alpha}}{|x_{m^*,k}-y|}.$$

Two cases are possible. If $|y| \le x_{m^*,j+1}/2^{1/\alpha}$, then

$$|y|^\alpha - |x_{m^*,k}|^\alpha \le -\frac{x_{m^*,j+1}^\alpha}{2} \le -\frac{(\theta a_m)^\alpha}{2} \le -\frac{m}{2}.$$

Otherwise, by (16), we get

$$\|Pu\|_{\{x\in\mathbb{R}:|x|\ge|y|\}} \le \|Pu\|_{\{x\in\mathbb{R}:|x|\ge x_{m^*,j+1}/2\}} \le \mathrm{C}\mathrm{e}^{-\mathrm{A}m}\|Pu\|.$$

In both cases, we obtain

$$|S_2| \le \mathrm{C}\mathrm{e}^{-\mathrm{A}m}\|Pu\|. \tag{60}$$

Combining (59) and (60) in (58), we have

$$|e_{m^*}(P,y)| \le \mathrm{C}\mathrm{e}^{-\mathrm{A}m}\|Pu\| \le \mathrm{C}\left\{E_M(f)_u + \mathrm{e}^{-\mathrm{A}m}\|fu\|\right\}. \tag{61}$$

On the other hand, using Proposition 3 and proceeding as in the proof of inequality (13), we have

$$\begin{aligned}|e_{m^*}(f-P,y)| &\le |f(y)-P(y)|\,|A_{m^*}(y)| \\ &\quad + \left|\sum_{|k|\le j}\lambda_{m^*,k}(w)\frac{[f-P]\,(x_{m^*,k})}{x_{m^*,k}-y}\right| + \left|\int_{\mathbb{R}}\frac{[f-P]\,(x)w(x)}{x-y}\,\mathrm{d}x\right| \\ &\le \mathrm{C}E_M(f)_u\log m + \left|\int_{\mathbb{R}}\frac{[f-P]\,(x)w(x)}{x-y}\,\mathrm{d}x\right|.\end{aligned}$$

By using Lemmas 1 and 2, we obtain

$$\begin{aligned}\left|\int_{\mathbb{R}}\frac{[f-P]\,(x)w(x)}{x-y}\,\mathrm{d}x\right| &\le \mathrm{C}\left\{E_M(f)_u\log m + \int_0^1\frac{\Omega(f-P,t)_u}{t}\,\mathrm{d}t\right\} \\ &\le \mathrm{C}\left\{E_M(f)_u\log m + \log m\int_0^{a_M/M}\frac{\omega^r(f,t)_u}{t}\,\mathrm{d}t\right\}.\end{aligned}$$

Therefore, we have

$$|e_{m^*}(f-P,y)| \le \mathrm{C}\log m\left\{E_M(f)_u + \int_0^{a_M/M}\frac{\omega^r(f,t)_u}{t}\,\mathrm{d}t\right\}. \tag{62}$$

By inequalities (61), (62) and (57), using (4), we get (14). □

Acknowledgements The author is very grateful to Professor Giuseppe Mastroianni for his helpful remarks and suggestions.

Appendix

We want to show a simple case for computing

$$\mathcal{H}(w,y) = \int_{\mathbb{R}} \frac{\mathrm{e}^{-|x|^{\alpha}}}{x-y}\,\mathrm{d}x, \quad \alpha > 1,$$

that is if $\alpha \in \mathbb{N}$. We recall that for the case $\alpha = 2$ one can use the error function (see for instance [22]).

Let us assume $y > 0$, since $\mathcal{H}(w)$ is an odd function. We can write

$$\begin{aligned}\int_{-\infty}^{+\infty} \frac{\mathrm{e}^{-|x|^{\alpha}}}{x-y}\,\mathrm{d}x &= \int_{0}^{+\infty} \frac{\mathrm{e}^{-x^{\alpha}}}{x-y}\,\mathrm{d}x - \int_{0}^{+\infty} \frac{\mathrm{e}^{-x^{\alpha}}}{x+y}\,\mathrm{d}x \\ &= \frac{1}{\alpha}\left[\int_{0}^{+\infty} \frac{x^{1/\alpha-1}\mathrm{e}^{-x}}{x^{1/\alpha}-y}\,\mathrm{d}x - \int_{0}^{+\infty} \frac{x^{1/\alpha-1}\mathrm{e}^{-x}}{x^{1/\alpha}+y}\,\mathrm{d}x\right]. \qquad (63)\end{aligned}$$

By (63), if α is even, we get

$$\int_{\mathbb{R}} \frac{\mathrm{e}^{-x^{\alpha}}}{x-y}\,\mathrm{d}x = \frac{2}{\alpha}\sum_{k=1}^{\alpha/2} y^{2k-1} \int_{0}^{+\infty} \frac{x^{(1-2k)/\alpha}\mathrm{e}^{-x}}{x-y^{\alpha}}\,\mathrm{d}x, \qquad (64)$$

while if α is odd we have

$$\begin{aligned}\int_{\mathbb{R}} \frac{\mathrm{e}^{-|x|^{\alpha}}}{x-y}\,\mathrm{d}x = \frac{1}{\alpha}\sum_{k=1}^{\alpha} y^{k-1} \Bigg[&\int_{0}^{+\infty} \frac{x^{(1-k)/\alpha}\mathrm{e}^{-x}}{x-y^{\alpha}}\,\mathrm{d}x \\ &+(-1)^{k-1}\int_{0}^{+\infty} \frac{x^{(1-k)/\alpha}\mathrm{e}^{-x}}{x+y^{\alpha}}\,\mathrm{d}x\Bigg]. \qquad (65)\end{aligned}$$

The Cauchy principal value integrals in (64) and (65) are confluent hypergeometric functions, while, for the Stieltjes transforms in (65), for any $k = 1, \ldots, \alpha$, there holds (see for instance [7])

$$y^{k-1}\int_{0}^{+\infty} \frac{x^{(1-k)/\alpha}\mathrm{e}^{-x}}{x+y^{\alpha}}\,\mathrm{d}x = \mathrm{e}^{y^{\alpha}}\Gamma\left(1+\frac{1-k}{\alpha}\right)\Gamma\left(\frac{k-1}{\alpha}, y^{\alpha}\right),$$

where Γ stands for the Gamma and the incomplete Gamma functions.

Finally, we remark that a similar method can be applied also if α is rational.

References

1. Bialecki, B.: Sinc Quadrature for Cauchy Principal Value Integrals. In: Numerical Integration, Recent Developtments, Software and Applications. (T.O. Espelid and A. Genz, eds.), NATO ASI Series C: Mathematical and Physical Sciences, vol. 357, Kluwer, Dordrecht (1992)

2. Cvetković, A.S., Milovanović, G.V.: The Mathematica package "Orthogonal Polynomials", Facta Univ. Ser. Math. Inform. **19**, 17–36 (2004)
3. Davis, P.J., Rabinowitz, P.: Methods of Numerical Integration (2nd ed.), Computer Science and Applied Mathematics, Academic Press, Orlando (1984)
4. De Bonis, M.C., Della Vecchia, B., Mastroianni, G.: Approximation of the Hilbert transform on the real line using Hermite zeros. Math. Comp. **71**(239), 1169–1188 (2002)
5. De Bonis, M.C., Mastroianni, G.: Some simple quadrature rules for evaluating the Hilbert transform on the real line. Arch. Inequalities Appl. **1**, 475–494 (2003)
6. Ditzian, Z., Totik, V.: Moduli of Smoothness. Springer Series in Computational Mathematics, vol. 9, Springer, New York (1987)
7. Erdélyi, A., Magnus, W., Oberhettinger, F., Tricomi, F.G.: Tables of Integral Transforms. Bateman Manuscript Project California Institute of Technology, vol. 2, MacGraw-Hill (1954)
8. Freud, G.: Orthogonal Polynomials. Akadémiai Kiadó/Pergamon, Budapest (1971)
9. Gautschi, W.: Algorithm 726: ORTHPOL–a package of routines for generating orthogonal polynomials and Gauss-type quadrature rules. ACM Trans. Math. Software **20**, 21–62 (1994)
10. Golub, G.H.: Some modified matrix eigenvalue problems. SIAM Review **15**, 318–334 (1973)
11. Golub, G.H., Welsch, J.H.: Calculation of Gaussian quadrature rules. Math. Comp. **23**, 221–230 (1969)
12. Khavin, V.P., Nikol'skij, N.K.: Commutative Harmonic Analysis I. Encyclopedia of Mathematical Sciences, vol. 15, Springer, Berlin (1991)
13. Kress, V.R., Martensen, E.: Anwendung der Rechteckregel auf die reelle Hilberttransformation mit unendlichem Intervall. ZAMM **50**, T 61–T 64 (1970)
14. Kumar, S.: A note on quadrature formulae for Cauchy principal value integrals. J. Inst. Math. Appl. **26**, 447–451 (1980)
15. Levin, A.L., Lubinsky, D.S.: Christoffel functions, orthogonal polynomials, and Nevai's conjecture for Freud weights. Constr. Approx. **8**(4), 463–535 (1992)
16. Levin, A.L., Lubinsky, D.S.: Orthogonal polynomials for exponential weights. CMS Books in Mathematics/Ouvrages de Mathématiques de la SMC, vol. 4, Springer, New York (2001)
17. Lubinsky, D.S., Rabinowitz, P.: Rates of convergence of Gaussian quadrature for singular integrands. Math. Comp. **43**, 219–242 (1984)
18. Mastroianni, G., Notarangelo, I.: A Lagrange-type projector on the real line. Math. Comp. **79**, 327–352 (2010)
19. Mastroianni, G., Notarangelo, I.: A Nyström method for Fredholm integral equations on the real line to appear in J. Integral Equations Appl.
20. Mastroianni, G., Russo, M.G.: Lagrange interpolation in weighted Besov spaces. Constr. Approx. **15**, 257–289 (1999)
21. Mastroianni, G., Szabados, J.: Direct and converse polynomial approximation theorems on the real line with weights having zeros. In: Frontiers in Interpolation and Approximation. (N.K. Govil, H.N. Mhaskar, R.N. Mohpatra, Z. Nashed and J. Szabados, eds.), Chapman & Hall, Boca Raton, pp. 287–306 (2007)
22. Poppe, G.P.M., Wijers, M.J.: Algorithm 680: Evaluation of the complex error function. ACM Trans. Math. Software **16**, 47 (1990)
23. Stenger, F.: Approximations via Whittaker's cardinal function. J. Approx. Theory **17**, 222–240 (1976)
24. Stenger, F.: Numerical methods based on Whittaker cardinal or sinc functions. SIAM Review **23**, 165–224 (1981)
25. Stenger, F.: Numerical Methods Based on Sinc and Analytic Functions. Springer, Berlin (1993)

The Remainder Term of Gauss–Turán Quadratures for Analytic Functions

Miodrag M. Spalević and Miroslav S. Pranić

Dedicated to the 60th birthday of Professor Gradimir V. Milovanović

1 Introduction

We consider quadrature rules with multiple nodes over a finite interval, taken to be $[-1,1]$,

$$\int_{-1}^{1} f(t)\,w(t)\,\mathrm{d}t = \sum_{\nu=1}^{n}\sum_{i=0}^{2s} A_{i,\nu} f^{(i)}(\tau_\nu) + R_{n,s}(f), \tag{1}$$

involving a positive weight function w, assumed integrable over $[-1,1]$. In order to achieve the highest possible degree of algebraic precision $2(s+1)n-1$, the nodes τ_ν in (1) must be zeros of the corresponding s-orthogonal polynomials $\pi_n = \pi_{n,s}$ satisfying the following orthogonality conditions

$$\int_{-1}^{1} \pi_n(t)^{2s+1} t^k w(t)\,\mathrm{d}t = 0, \quad k = 0,1,\ldots,n-1. \tag{2}$$

The conditions (2) mean that the monic $\pi_{n,s}$ is a monic minimal polynomial with respect to the $L^{2s+2}(w(t)\mathrm{d}t)$ norm. The weights $A_{i,\nu,n}$ can be represented in the form

$$A_{i,\nu,n} = \int_{-1}^{1} h_{i,\nu,n}\, w(t)\,\mathrm{d}t,$$

Miodrag M. Spalević
Department of Mathematics, Faculty of Mechanical Engineering, University of Beograd, Kraljice Marije 16, 11000 Beograd, Serbia, e-mail: mspalevic@mas.bg.ac.rs

Miroslav S. Pranić
Department of Mathematics and Informatics, Faculty of Science, University of Banja Luka, M. Stojanovića 2, 51000 Banja Luka, Bosnia and Herzegovina, e-mail: pranic77m@yahoo.com

W. Gautschi et al. (eds.), *Approximation and Computation: In Honor of Gradimir V. Milovanović*, Springer Optimization and Its Applications 42, DOI 10.1007/978-1-4419-6594-3_16, © Springer Science+Business Media, LLC 2011

where the $h_{i,\nu,n}$ are the fundamental polynomials of Hermite interpolation. Under the assumption $\omega(\cos\varphi) \in C_{2\pi}^{m+\alpha}$ with $m+\alpha > (2s+2)^2$, $\omega(t) = \pi w(t)\sqrt{1-t^2}$, they have the following asymptotic representation (see [29])

$$A_{2s-2i,\nu,n,s} = \frac{a_{i,s}(1-\tau_{\nu,n}^2)^{s-j}\omega(\tau_{\nu,n})}{(s!)^2 n^{2s+1-2i}}\left[1+O\left(\frac{1}{n}\right)\right]$$

and

$$A_{2s-2i-1,\nu,n,s} = \left(\frac{-(2s-2i-1)}{2}\frac{\tau_{\nu,n}}{1-\tau_{\nu,n}^2} + b_{j,s}\frac{\omega'(\tau_{\nu,n})}{\omega(\tau_{\nu,n})}\right)(2s-2i)A_{2s-2i,\nu,n,s},$$

where $m \in \mathbb{N}$, $\alpha \in [0,1)$, $C_{2\pi}^{m+\alpha}$ denotes the space of 2π-periodic functions which are m-times differentiable on $[0,2\pi]$ with the m-th derivative in Lip α on $[0,2\pi]$, $2^{-2s}\prod_{i=1}^{s}(t+(2i)^2) = \sum_{i=0}^{s} a_{i,s}t^{s-i}$, and $b_{i,s}$ is an unknown constant which depends on i and s only, in particular not on the weight function ω ($b_{j,s} = 1$ for $j = 0,1$ and all $s \in \mathbb{N}_0$, and the same is conjectured for all j).

Numerically stable methods for constructing the nodes τ_ν and weights $A_{i,\nu}$ can be found in [22] and [32]. Some interesting results concerning this theory and its applications can be found in [31, and references therein] and [11, 14, 20, 29].

An elementary and old technique of deriving derivative-free error estimates for quadrature rules is based on Cauchy's integral formula. This technique has already been used in 1878 by Hermite [8], and in 1881 by Heine [7], to derive error estimates for polynomial interpolation; these, when integrated, yield error estimates for interpolatory quadrature rules.

Let Γ be a simple closed curve in the complex plane surrounding the interval $[-1,1]$ and let D be its interior. If the integrand f is analytic on D then the remainder term $R_{n,s}$ in (1) admits the contour integral representation

$$R_{n,s}(f) = \frac{1}{2\pi i}\oint_\Gamma K_{n,s}(z;w)f(z)\,\mathrm{d}z. \tag{3}$$

The *kernel* is given by

$$K_{n,s}(z;w) = \frac{q_{n,s}(z;w)}{[\pi_{n,s}(z)]^{2s+1}}, \quad z \notin [-1,1],$$

where

$$q_{n,s}(z;w) = \int_{-1}^{1}\frac{[\pi_{n,s}(t)]^{2s+1}}{z-t}w(t)\,\mathrm{d}t.$$

The modulus of the kernel is symmetric with respect to the real axis, that is, $|K_{n,s}(\overline{z})| = |K_{n,s}(z)|$. If the weight function w is even, the modulus of the kernel is symmetric with respect to both axes, that is, $|K_{n,s}(-\overline{z})| = |K_{n,s}(z)|$ (see [15, Lemma 2.1]).

The integral representation (3) leads to a general error estimate, by using Hölder's inequality,

$$|R_{n,s}(f)| = \frac{1}{2\pi}\left|\oint_\Gamma K_{n,s}(z;w)f(z)\,\mathrm{d}z\right|$$
$$\leq \frac{1}{2\pi}\left(\oint_\Gamma |K_{n,s}(z;w)|^r|\mathrm{d}z|\right)^{1/r}\left(\oint_\Gamma |f(z)|^{r'}|\mathrm{d}z|\right)^{1/r'},$$

that is,

$$|R_{n,s}(f)| \leq \frac{1}{2\pi}\|K_{n,s}\|_r\|f\|_{r'}, \tag{4}$$

where $1 \leq r \leq +\infty$, $1/r + 1/r' = 1$, and

$$\|f\|_r := \begin{cases} \left(\oint_\Gamma |f(z)|^r|\mathrm{d}z|\right)^{1/r}, & 1 \leq r < +\infty, \\ \max_{z\in\Gamma}|f(z)|, & r = +\infty. \end{cases}$$

Important features of the estimate (4), besides being derivative-free, are its sharpness, its natural conduciveness to a comparison of different quadrature processes, and the neat separation it expresses between the influence of the quadrature rule (given by $\|K_{n,s}\|_r$) and the function to which it is applied (given by $\|f\|_{r'}$). The following choices of r are natural. The case $r = +\infty$ ($r' = 1$) gives

$$|R_{n,s}(f)| \leq \frac{1}{2\pi}\left(\max_{z\in\Gamma}|K_{n,s}(z;w)|\right)\|f\|_1, \tag{5}$$

whereas for $r = 1$ ($r' = +\infty$) we have

$$|R_{n,s}(f)| \leq \frac{1}{2\pi}\left(\oint_\Gamma |K_{n,s}(z;w)||\mathrm{d}z|\right)\|f\|_\infty. \tag{6}$$

We focus on weight functions which admit explicit Gauss–Turán quadrature formulae, that is, on cases when explicit formulae for the corresponding s-orthogonal polynomials are known. There are only a couple of them.

In 1930, S. Bernstein [1] showed that the monic Chebyshev polynomial $\hat{T}_n(t) = T_n(t)/2^{n-1}$ minimizes all integrals of the form

$$\int_{-1}^{1}\frac{|\pi_n(t)|^{k+1}}{\sqrt{1-t^2}}\,\mathrm{d}t \quad (k \geq 0).$$

This means that the Chebyshev polynomials T_n are s-orthogonal on $(-1,1)$ for each $s \geq 0$. Ossicini and Rosati [28] found three other weight functions $w_k(t)$ ($k = 2,3,4$),

$$w_2(t) = (1-t^2)^{1/2+s}, \quad w_3(t) = \frac{(1+t)^{1/2+s}}{(1-t)^{1/2}}, \quad w_4(t) = \frac{(1-t)^{1/2+s}}{(1+t)^{1/2}},$$

for which the s-orthogonal polynomials can be identified with Chebyshev polynomials of the second, third, and fourth kind: U_n, V_n, and W_n, which are defined by

$$U_n(\cos\theta) = \frac{\sin(n+1)\theta}{\sin\theta}, \quad V_n(\cos\theta) = \frac{\cos(n+\frac{1}{2})\theta}{\cos\frac{1}{2}\theta}, \quad W_n(\cos\theta) = \frac{\sin(n+\frac{1}{2})\theta}{\sin\frac{1}{2}\theta},$$

respectively. These weight functions depend on s. It is easy to see that $W_n(-t) = (-1)^n V_n(t)$, so that in the investigation it is sufficient to study only the first three Jacobi measures $w_k(t)$, $k = 1,2,3$.

Recently, Gori and Micchelli [5] have introduced for each n a class of weight functions defined on $[-1,1]$ for which explicit Gauss–Turán quadrature formulae of all orders can be found. In other words, these weight functions do not depend on s, but depend on n. This class includes certain generalized Jacobi weight functions $w_{n,\mu}(t) = |U_{n-1}(t)/n|^{2\mu+1}(1-t^2)^{\mu}$, where $\mu > -1$. In this case, the Chebyshev polynomials T_n appear as s-orthogonal polynomials. In this chapter, we consider the case $\mu = \ell - 1/2$, $\ell \in \mathbb{N}$, that is,

$$w(t) = w_{n,\ell}(t) = \frac{U_{n-1}^{2\ell}(t)}{n^{2\ell}}(1-t^2)^{\ell-1/2} \quad (\ell \in \mathbb{N}). \tag{7}$$

The bounds in (5) and (6) depend on the contour Γ, while the contour Γ is arbitrary and $R_{n,s}(f;w)$ does not depend on Γ. Therefore, there is a possibility for optimizing these bounds over suitable families of contours. Two choices of the contours Γ have been most frequently used: concentric circles $C_r = \{z \in \mathbb{C} : |z| = r\}$, $r > 1$, and confocal ellipses

$$\mathcal{E}_\rho = \left\{z \in \mathbb{C} : z = \frac{1}{2}\left(u + u^{-1}\right),\ 0 \le \theta \le 2\pi\right\}, \quad u = \rho e^{i\theta},\ \rho > 1, \tag{8}$$

having foci at ± 1 and the sum of semi-axes equal to ρ. We limit ourselves to elliptic contours, which we consider to be more flexible than circular ones. Indeed, they can be chosen to snuggle tightly around the interval $[-1,1]$ by selecting ρ sufficiently close to 1, thereby avoiding possible singularities or excessive growth of f. The circular contours are used in [18] and [21].

2 Error Bounds of Type (5)

In this section, we consider the behavior of the kernel on elliptic contours and make an attempt to determine where exactly on the contour the modulus of the kernel attains its maximum.

We discuss in some detail the particularly interesting case of the first Chebyshev weight function $w = w_1$. Here, one finds [15], for $z \in \mathcal{E}_\rho$, the explicit formula

$$|K_{n,s}(z;w_1)| = \frac{2^{1-s}\pi}{\rho^n} \frac{\left|Z_{n,s}^{(1)}\left(\rho e^{i\theta}\right)\right|}{(a_2 - \cos 2\theta)^{1/2}(a_{2n} + \cos 2n\theta)^{1/2+s}}, \tag{9}$$

where

$$a_j = a_j(\rho) = \frac{1}{2}\left(\rho^j + \rho^{-j}\right), \quad j \in \mathbb{N}, \tag{10}$$

and

$$Z_{n,s}^{(1)}(u) = \sum_{k=0}^{s} \binom{2s+1}{s-k} u^{-2nk}. \tag{11}$$

Theorem 1. *There exists a $\rho_0 = \rho_0(n,s)$ such that*

$$\max_{z\in\mathcal{E}_\rho} |K_{n,s}(z;w_1)| = K_{n,s}\left(\frac{1}{2}(\rho+\rho^{-1});w_1\right) \tag{12}$$

for each $\rho \geq \rho_0$.

Proof. The inequality $|Z_{n,s}^{(1)}(\rho e^{i\theta})| \leq Z_{n,s}^{(1)}(\rho)$ immediately follows from (11). Because of that and (9), it is sufficient to prove

$$\frac{1}{(a_2-\cos 2\theta)^{1/2}(a_{2n}+\cos 2n\theta)^{1/2+s}} \leq \frac{1}{(a_2-1)^{1/2}(a_{2n}+1)^{1/2+s}} \tag{13}$$

for a sufficiently large ρ $(\rho \geq \rho_0(n,s))$ and $\theta \in (0,\pi/2]$, where the a_j are given by (10).

By squaring (13), it is reduced to

$$(a_2-1)(a_{2n}+1)^{2s+1} \leq (a_2-\cos 2\theta)(a_{2n}+\cos 2n\theta)^{2s+1}. \tag{14}$$

The following identity will be used

$$a_2 - \cos 2\theta = (a_2-1) + 2\sin^2\theta. \tag{15}$$

Further, we will use

$$\begin{aligned}(a_{2n}+\cos 2n\theta)^{2s+1} &= \left[(a_{2n}+1) - 2\sin^2 n\theta\right]^{2s+1}\\ &= (a_{2n}+1)^{2s+1} + \sum_{k=1}^{2s+1} (-2)^k \binom{2s+1}{k} (a_{2n}+1)^{2s+1-k}\sin^{2k} n\theta,\end{aligned}$$

that is,

$$(a_{2n}+\cos 2n\theta)^{2s+1} = (a_{2n}+1)^{2s+1} - 2(\sin^2 n\theta)E_{n,s}(\rho,\theta), \tag{16}$$

where

$$E_{n,s}(n,\theta) = \sum_{k=1}^{2s+1} (-2)^{k-1} \binom{2s+1}{k} (a_{2n}+1)^{2s+1-k}\sin^{2k-2} n\theta \quad (\geq 0).$$

It is easy to see that $E_{n,s}(\rho,\theta)$ can be represented in the form

$$E_{n,s}(\rho,\theta) = (2s+1)(a_{2n}+1)^{2s} + \sum_{k=2}^{2s+1} (-2)^{k-1} \binom{2s+1}{k} (a_{2n}+1)^{2s+1-k}\sin^{2k-2} n\theta,$$

that is,

$$E_{n,s}(\rho,\theta) = (2s+1)(a_{2n}+1)^{2s} - \sum_{k=1}^{s} 2^{2k-1}\binom{2s+1}{2k}(a_{2n}+1)^{2s-2k+1}\sin^{4k-2} n\theta + \sum_{k=1}^{s} 2^{2k}\binom{2s+1}{2k+1}(a_{2n}+1)^{2s-2k}\sin^{4k} n\theta. \tag{17}$$

By virtue of (15) and (16), the inequality (14) reduces to

$$(a_2-1)(a_{2n}+1)^{2s+1} \le \left[(a_2-1)+2\sin^2\theta\right]\left[(a_{2n}+1)^{2s+1} - 2(\sin^2 n\theta)E_{\rho,s}(n,\theta)\right],$$

that is, $2\sin^2\theta\,(a_{2n}+1)^{2s+1} - 2\sin^2 n\theta\,[(a_2-1)+2\sin^2\theta]\,E_{\rho,s}(n,\theta) \ge 0$. Dividing this inequality by $2\sin^2\theta$, it becomes

$$(a_{2n}+1)^{2s+1} - \frac{\sin^2 n\theta}{\sin^2\theta}\left[(a_2-1)+2\sin^2\theta\right] E_{n,s}(\rho,\theta) \ge 0. \tag{18}$$

By using the well-known fact $|\sin n\theta/\sin\theta| \le n$, it is easy to see that

$$\frac{\sin^2 n\theta}{\sin^2\theta}\left[(a_2-1)+2\sin^2\theta\right] = (a_2-1)\frac{\sin^2 n\theta}{\sin^2\theta} + 2\sin^2 n\theta \le (a_2-1)n^2+2. \tag{19}$$

In view of (17), we conclude that

$$E := E_{n,s}(\rho,\theta) - (2s+1)(a_{2n}+1)^{2s} = \sum_{k=1}^{s} \frac{4^k(2s+1)!}{(2k)!(2s-2k)!}(a_{2n}+1)^{2s-2k} \times \left[\frac{\sin^2 n\theta}{2k+1} - \frac{a_{2n}+1}{2(2s-2k+1)}\right]\sin^{4k-2} n\theta.$$

Since $\sin^{4k-2} n\theta \le 1$ and

$$\frac{\sin^2 n\theta}{2k+1} - \frac{a_{2n}+1}{2(2s-2k+1)} \le \frac{1}{2k+1} - \frac{a_{2n}+1}{2(2s-2k+1)},$$

from the previous equality we obtain

$$E \le \sum_{k=1}^{s} 4^k (a_{2n}+1)^{2s-2k}\left[\binom{2s+1}{2k+1} - \frac{a_{2n}+1}{2}\binom{2s+1}{2k}\right].$$

Therefore,

$$E_{n,s}(\rho,\theta) \le \sum_{k=0}^{s} 4^k\binom{2s+1}{2k+1}(a_{2n}+1)^{2s-2k} - \frac{1}{2}\sum_{k=1}^{s} 4^k\binom{2s+1}{2k}(a_{2n}+1)^{2s-2k+1}.$$

Using the last inequality and (19), we conclude that the left-hand side of (18) is greater than or equal to $F(\rho) \equiv F_{n,s}(\rho)$, where

$$F_{n,s}(\rho) := (a_{2n}+1)^{2s+1} - \left[(a_2-1)n^2+2\right]$$
$$\times\left[\sum_{k=0}^{s} 4^k \binom{2s+1}{2k+1}(a_{2n}+1)^{2s-2k} - \frac{1}{2}\sum_{k=1}^{s} 4^k \binom{2s+1}{2k}(a_{2n}+1)^{2s-2k+1}\right].$$

Since $F_{n,s}(\rho)$ (n,s – are fixed) is continuous on $\mathbb{R}$ and $\lim_{\rho\to+\infty} F_{n,s}(\rho) = +\infty$, it follows that $F_{n,s}(\rho) > 0$, for each $\rho > r$, where r is the largest zero of $F_{n,s}(\rho)$. For ρ_0, we can take r. □

The numerical results presented in Table 1 show that the ellipse has to be quite slim for (12) to fail.

Table 1 Optimal ρ_0 and its approximations r

	$s=2$		$s=6$	
n	r	Optimal ρ_0	r	Optimal ρ_0
2	4.2925	3.9534	7.1045	6.8474
3	2.3836	2.2505	3.1389	3.0605
4	1.8777	1.7964	2.2804	2.2361
5	1.6433	1.5854	1.9119	1.8816
7	1.4197	1.3834	1.5780	1.5598
10	1.2760	1.2529	1.3728	1.3616
15	1.1758	1.1615	1.2342	1.2274
30	1.0842	1.0776	1.1107	1.1076

For the generalized Chebyshev weight function of the second kind, $w = w_2$, one finds for $z \in \mathcal{E}_\rho$ that

$$|K_{n,s}^{(2)}(z)| = \frac{\pi}{4^s\rho^{n+1}}\left[\frac{a_2-\cos 2\theta}{a_{2n+2}-\cos(2n+2)\theta}\right]^{s+1/2} |Z_{n,s}^{(2)}(\rho e^{i\theta})|,$$

where

$$Z_{n,s}^{(2)}(u) = \sum_{k=0}^{s}(-1)^k\binom{2s+1}{s-k}u^{-2(n+1)k}.$$

The maximum of $|K_{n,s}^{(2)}(z)|$ on $\mathcal{E}_\rho$ is attained on the imaginary axis, if n is odd. If n is even, this is true only for ρ sufficiently large. The precise result is as follows [15,23].

Theorem 2. *There holds*

$$\max_{z\in\mathcal{E}_\rho} |K_{n,s}(z;w_2)| = \left|K_{n,s}\left(\frac{i}{2}(\rho-\rho^{-1});w_2\right)\right|$$

for all $\rho > 1$, if n is odd, and for all $\rho > \rho_0(n)$, if n is even, where $\rho_0(n)$ is the unique root of

$$\frac{\rho+\rho^{-1}}{\rho^{n+1}+\rho^{-(n+1)}} = \frac{1}{n+1}, \quad \rho > 1.$$

As in the previous case, $\rho_0(n)$ tends to 1 as $n \to \infty$.

For the generalized Chebyshev weight function $w = w_3$, we have the following explicit formula on $\mathcal{E}_\rho$

$$|K_{n,s}^{(3)}(z)| = \frac{2^{1-s}\pi}{\rho^{n+1/2}} \frac{(a_1+\cos\theta)^{s+1}|Z_{n,s}^{(3)}(\rho e^{i\theta})|}{(a_2-\cos 2\theta)^{1/2}[a_{2n+1}+\cos(2n+1)\theta]^{1/2+s}},$$

where

$$Z_{n,s}^{(3)}(u) = \sum_{k=0}^{s} \binom{2s+1}{s-k} u^{-(2n+1)k},$$

which is very similar to (9), and an argument similar to the one in the proof of Theorem 1 will establish:

Theorem 3. *There exists $\rho_0 = \rho_0(n,s)$ such that*

$$\max_{z\in\mathcal{E}_\rho} |K_{n,s}(z;w_3)| = K_{n,s}\left(\frac{1}{2}(\rho+\rho^{-1});w_3\right)$$

for each $\rho \geq \rho_0$.

The behavior of $K_{n,s}(z;w_{n,\ell})$ on $\mathcal{E}_\rho$, where $w_{n,\ell}$ is the Gori-Micchelli weight, is considerably more difficult to analyze, taking into account the formula

$$|K_{n,s}(z;w_{n,\ell})| = \frac{\pi}{2^{s+2k+1/2}n^{2k+2}\rho^{2n}} \frac{(a_{2n}-\cos 2n\theta)^{1/2}}{(a_2-\cos 2\theta)^{1/2}(a_{2n}+\cos 2n\theta)^{s+1/2}} |V_{n,s,k}(u)|,$$

where $k = \ell - 1$ and

$$V_{n,s,k}(u) = \sum_{\lambda=0}^{s+k} F_{s,k}(\lambda)\frac{1}{u^{2\lambda n}},$$

with

$$F_{s,k}(\lambda) = \sum_{j+p=\lambda} (-1)^p \binom{2k+1}{k-p}\binom{2s+1}{s-j} + \sum_{|p-j|=\lambda+1} (-1)^p \operatorname{sign}(p-j)\binom{2k+1}{k-p}\binom{2s+1}{s-j},$$

$j = 0,1\ldots,s,\ p = 0,1,\ldots,k.$

Computation shows that $|K_{n,s}(z;w_{n,\ell})|$, $z \in \mathcal{E}_\rho$, attains its maximum on the real axis ($z = \pm(\rho+\rho^{-1})/2$) if $\rho > \rho_0(n,s,\ell)$. This empirical observation can be verified asymptotically as $\rho \to \infty$. A lengthy calculation reveals that

$$|K_{n,s}(z;w_{n,\ell})| \sim \frac{\pi F_{s,k}(0)}{n^{2k+2}4^k\rho^{2n(s+1)+1}}\left(1+\frac{2\cos 2\theta}{\rho^2}\right)^{1/2}, \quad \rho\to\infty.$$

For more details see [15, 23, 25]. The case $s=0$ is considered in [3, 4].

3 Error Bounds of the Type (6)

In this section, we study the quantities $\frac{1}{2\pi}\oint_{\mathcal{E}_\rho}|K_{n,s}(z;w)|\,|\mathrm{d}z|$ and obtain bounds for them, which are very sharp if ρ is not very close to 1.

As in the previous section, we discuss in detail the case $w=w_1$. We need the following two equalities: $|\mathrm{d}z| = 2^{-1/2}\sqrt{a_2-\cos 2\theta}\,\mathrm{d}\theta$ and

$$|Z_{n,s}^{(1)}(\rho \mathrm{e}^{\mathrm{i}\theta})| = \left[\rho^{\,2ns}\sum_{j=0}^{s}A_j\cos 2jn\theta\right]^{1/2}$$

with

$$A_0 = \frac{1}{\rho^{2ns}}\sum_{\nu=0}^{s}\binom{2s+1}{\nu}^2\rho^{4n\nu},$$

$$A_j = \frac{2}{\rho^{2n(s-j)}}\sum_{\nu=0}^{s-j}\binom{2s+1}{\nu}\binom{2s+1}{\nu+j}\rho^{4n\nu}, \quad j=1,\dots,s.$$

The first equality obviously follows from $z=\frac{1}{2}(u+u^{-1})$, $u=\rho\mathrm{e}^{\mathrm{i}\theta}$, whereas the second one follows from [17, Lemma 4.1]. According to (9) we have that

$$\oint_{\mathcal{E}_\rho}|K_{n,s}(z;w_1)|\,|\mathrm{d}z| = \frac{\pi}{2^{s-1/2}\rho^{n(s+1)}}\int_0^{2\pi}\sqrt{\frac{\sum_{k=0}^{s}A_k\cos 2kn\theta}{(a_{2n}+\cos 2n\theta)^{2s+1}}}\,\mathrm{d}\theta\,.$$

Using the periodicity of the integrand and applying Cauchy's inequality, we get

$$\frac{1}{2\pi}\oint_{\mathcal{E}_\rho}|K_{n,s}(z;w_1)|\,|\mathrm{d}z| \le \frac{\sqrt{\pi}}{2^{s-1/2}\rho^{(s+1)n}}\left(\sum_{k=0}^{s}A_kJ_k(a_{2n})\right)^{1/2},$$

where (see [6, (3.616.7)])

$$J_k(a) = \int_0^{\pi}\frac{\cos k\theta}{(a+\cos\theta)^{2s+1}}\,\mathrm{d}\theta$$

$$= \frac{(-1)^k\pi 2^{2s+1}x^{s-(k-1)/2}}{(x-1)^{4s+1}}\sum_{\nu=0}^{2s}\binom{2s+\nu}{\nu}\binom{2s+k}{\nu+k}(x-1)^{2s-\nu},$$

with $a=(x+1)/(2\sqrt{x})$, $x>1$.

In the same way, we get the following bounds:

$$\frac{1}{2\pi}\oint_{\mathcal{E}_\rho} |K_{n,s}(z;w_2)||dz| \leq \frac{\sqrt{\pi}}{2^{2s+1/2}\rho^{(s+1)(n+1)}}\left(M_{2s+2}(\rho^2)\sum_{k=0}^{s} A_k J_k(a_{2n+2})\right)^{1/2},$$

$$\frac{1}{2\pi}\oint_{\mathcal{E}_\rho} |K_{n,s}(z;w_3)||dz| \leq \frac{\sqrt{\pi}}{2^{s-1/2}\rho^{(s+1)(n+1/2)}}\left(M_{2s+2}(\rho)\sum_{k=0}^{s} A_k J_k(a_{2n+1})\right)^{1/2},$$

and

$$\frac{1}{2\pi}\oint_{\mathcal{E}_\rho} |K_{n,s}(z;w_{n,\ell})|\,|dz| \leq \frac{\pi^{1/2}}{2^{s+2\ell-1}n^{2\ell}\rho^{n(s+\ell+1)}} \times\left\{\sum_{j=0}^{s+\ell-1}\bar{A}_j\left[a_{2n}J_j(a_{2n})-\frac{1}{2}(J_{j+1}(a_{2n})+J_{|j-1|}(a_{2n}))\right]\right\}^{1/2},$$

where

$$M_n(\rho) = \frac{1}{\pi}\int_0^{\pi}(a_1 \pm \cos\theta)^n\,d\theta = \left(\frac{\rho-\rho^{-1}}{2}\right)^n P_n\left(\frac{\rho+\rho^{-1}}{\rho-\rho^{-1}}\right) = (2\rho)^{-n}\sum_{\nu=0}^{n}\binom{n}{\nu}^2\rho^{2\nu},$$

P_n is the Legendre polynomial of degree n, $k=\ell-1$, and

$$\bar{A}_0 = \frac{1}{\rho^{2n(s+k)}}\sum_{\nu=0}^{s+k}\left[F_{s,k}(s+k-\nu)\right]^2\rho^{4n\nu},$$

$$\bar{A}_j = \frac{2}{\rho^{2n(s+k-j)}}\sum_{\nu=0}^{s+k-j}F_{s,k}(s+k-\nu)F_{s,k}(s+k-\nu-j)\rho^{4n\nu}, \quad j=1,\ldots,s.$$

For more details, see [17, 25]. The case $s=0$ is considered in [9]. The cases of Gauss–Turán quadratures of Radau or Lobatto type are considered in [16].

4 Practical Error Estimates

A very popular method for obtaining a practical error estimate in numerical integration is to use two quadrature formulae A and B, where the nodes used by formula B form a proper subset of those used by formula A, and where rule A is also of higher degree of precision. Kronrod originated this method (see [10]), which has been used many times to date. For more details, see, for example, [12, 26, 27]. The difference

$|A(f)-B(f)|$, that is, $|R^{(A)}(f)-R^{(B)}(f)|$, where f is the integrand, is usually quite a good estimate of the error for the rule B. Following this idea, taking Gauss–Turán quadratures (GT) as rule B, we derive the error estimates for Gauss–Turán quadratures which have the same form (const $\cdot \|f\|_\infty$) as the bounds (6).

As rule A we take Kronrod extensions of Gauss–Turán quadratures (K), first introduced in [13],

$$\int_{-1}^{1} w(t)f(t)\,\mathrm{d}t = \sum_{\nu=1}^{n}\sum_{i=0}^{2s}\sigma_{i,\nu}f^{(i)}(\tau_\nu)+\sum_{\mu=1}^{n+1}K_\mu f(\hat{\tau}_\mu)+R_{n,s}(f),$$

which is exact for all algebraic polynomials of degree less than or equal to $2(s+1)n+n+1$. The nodes τ_ν are the same as in (1), whereas the new nodes $\hat{\tau}_\mu$ and new weights $\sigma_{i,\nu}$, K_μ are chosen to maximize the degree of exactness. The new nodes must be zeros of the generalized Stieltjes polynomial $\hat{\pi}_{n+1}$ satisfying the conditions

$$\int_{-1}^{1} w(t)t^m[\pi_n(t)]^{2s+1}\hat{\pi}_{n+1}(t)\,\mathrm{d}t = 0, \quad m=0,1,\ldots,n.$$

The existence and uniqueness of monic $\hat{\pi}_{n+1}$ is proved, but its zeros are not necessarily real and distinct. For instance, in [30] it is proved that classical ($s=0$), real, and positive Gauss–Kronrod quadrature formulae with respect to the Gegenbauer weight function $(1-t^2)^{\lambda-1/2}$ do not exist for $\lambda > 3$ (i.e., w_2 for $s>2$) and sufficiently large n. Numerical tests (using routines from [2]) suggest the same for Gauss–Kronrod quadratures with respect to w_3 when $s=2$, $n\geq 6$ and when $s=3,4,\ldots,10$, $n\geq 2$. In all these cases, Gauss–Turán–Kronrod quadratures can be regarded as alternatives of the standard Gauss–Kronrod rules because of (see [19]): $\hat{\pi}_{n+1}(t;w_1)=(1-t^2)U_{n-1}(t)$, $\hat{\pi}_{n+1}(t;w_2)=T_{n+1}(t)$, and $\hat{\pi}_{n+1}(t;w_3)=(1-t)W_n(t)$.

The corresponding kernels and error bounds of the types (5) and (6) for Gauss–Turán–Kronrod quadratures with respect to the weight functions w_1, w_2, and w_3 are derived in [19] and [24]. The kernels exhibit the same behavior on elliptic contours as the kernels of Gauss–Turán quadratures with respect to the same weight function (see Theorems 1, 2, and 3). Although the kernel's representations are rather complicated, it is possible to derive a quite simple representation of their difference, which is used for obtaining the bounds on $|R^{(K)}(f;w_i)-R^{(GT)}(f;w_i)|$, $i=1,2,3$.

We demonstrate this for the case $w=w_1$. Here one finds [24], for $z\in\mathcal{E}_\rho$, explicit formulae

$$|K_{n,s}^{(K)}(z;w_1)| = \frac{\pi}{2^{s-1/2}\rho^{2n}}\,\frac{|B_{n,s}^{(1)}(\rho e^{i\theta})|}{(a_2-\cos 2\theta)^{1/2}(a_{2n}-\cos 2n\theta)^{1/2}(a_{2n}+\cos 2n\theta)^{s+1/2}}$$

and

$$K_{n,s}^{(K)}(z;w_1) = K_{n,s}^{(GT)}(z;w_1)+\frac{\dfrac{\pi}{2^{2s+1}}\dbinom{2s+1}{s}}{(1-z^2)\,[T_n(z)]^{2s+1}\,U_{n-1}(z)},$$

where

$$B_{n,s}^{(1)}(u) = \sum_{j=0}^{s} k(j)\,\frac{1}{u^{2jn}},$$

$$k(j) = \begin{cases} \binom{2s+1}{s-j} - \binom{2s+1}{s-j-1}, & j = 0,1,\ldots,s-1, \\ 1, & j = s. \end{cases}$$

We need the following result:

$$\begin{aligned} \tilde{J}_0(a) &= \int_0^{\pi} \frac{1}{(a+\cos\theta)^{2s+1}(a-\cos\theta)}\,\mathrm{d}\theta \\ &= \frac{2\pi x^{s+1}}{(x-1)^2}\left[\frac{x-1}{(x+1)^{2s+1}} + \left(\frac{2}{x-1}\right)^{2s}\psi(s,x)\right], \end{aligned}$$

where

$$\psi(s,x) = \sum_{\nu=1}^{s} \frac{(-1)^{\nu}}{2^{\nu}}\binom{2s+1}{\nu+1}\sum_{\ell=0}^{\nu}(-1)^{\ell}\binom{2s+\ell}{\ell}\left(\frac{2}{x-1}\right)^{\ell}\left[1-\left(\frac{2}{x+1}\right)^{\nu-\ell+1}\right],$$

and $a = (x+1)/(2\sqrt{x})$, $x > 1$.

Theorem 4. *There holds*

$$|R_{n,s}^{(K)}(f;w_i) - R_{n,s}^{(GT)}(f;w_i)| \le V_{n,s,\rho}^{(i)}\,\|f\|_\infty$$

and

$$|R_{n,s}^{(K)}(f;w_i) - R_{n,s}^{(GT)}(f;w_i)| \le W_{n,s,\rho}^{(i)}\,\|f\|_\infty,$$

where $i = 1,2,3$, *and*

$$\begin{aligned} V_{n,s,\rho}^{(1)} &= \frac{\sqrt{\pi}}{2^s}\binom{2s+1}{s}\sqrt{\tilde{J}_0(a_{2n})}, \\ W_{n,s,\rho}^{(1)} &= \binom{2s+1}{s}\frac{2\pi}{(\rho^n+\rho^{-n})(\rho^n-\rho^{-n})^{2s+1}}, \\ V_{n,s,\rho}^{(2)} &= \frac{\sqrt{\pi}}{2^{2s+1}}\binom{2s+1}{s}\sqrt{M_{2s+2}(\rho^2)}\sqrt{\tilde{J}_0(a_{2n+2})}, \end{aligned}$$

$$\begin{aligned} W_{n,s,\rho}^{(2)} &= \frac{\pi}{2^s}\binom{2s+1}{s}\frac{M_{s+1}(\rho^2)}{(\rho^{n+1}+\rho^{-(n+1)})(\rho^{n+1}-\rho^{-(n+1)})^{2s+1}}, \\ V_{n,s,\rho}^{(3)} &= \frac{\sqrt{\pi}}{2^s}\binom{2s+1}{s}\sqrt{M_{2s+2}(\rho)}\sqrt{\tilde{J}_0(a_{2n+1})}, \\ W_{n,s,\rho}^{(3)} &= 2\pi\binom{2s+1}{s}\frac{M_{s+1}(\rho)}{(\rho^{n+1/2}+\rho^{-(n+1/2)})(\rho^{n+1/2}-\rho^{-(n+1/2)})^{2s+1}}. \end{aligned}$$

Proof. We prove the case $i = 1$. According to (3), we obtain

$$
\begin{aligned}
|R_{n,s}^{(K)}(f;w_1) - R_{n,s}^{(GT)}(f;w_1)| &= \frac{1}{2\pi}\left|\oint_{\Gamma}\left[K_{n,s}^{(K)}(z;w_1) - K_{n,s}^{(GT)}(z;w_1)\right] f(z)\,\mathrm{d}z\right| \\
&\le \frac{1}{2\pi}\|K_{n,s}^{(K)}(z;w_1) - K_{n,s}^{(GT)}(z;w_1)\|_1\,\|f\|_\infty .
\end{aligned}
$$

Further we get

$$
\begin{aligned}
&\oint_{\mathcal{E}_\rho} |K_{n,s}^{(K)}(z;w_1) - K_{n,s}^{(GT)}(z;w_1)|\,|\mathrm{d}z| \\
&\quad = \frac{\pi}{2^{2s+1}}\binom{2s+1}{s}\oint_{\mathcal{E}_\rho}\frac{1}{|1-z^2|\,|T_n(z)|^{2s+1}\,|U_{n-1}(z)|}|\mathrm{d}z| \\
&\quad = \frac{\pi}{2^s}\binom{2s+1}{s}\int_0^{2\pi}\frac{1}{(a_{2n}-\cos 2n\theta)^{1/2}(a_{2n}+\cos 2n\theta)^{s+1/2}}\mathrm{d}\theta \\
&\quad = \frac{\pi}{2^{s-1}}\binom{2s+1}{s}\int_0^{\pi}\sqrt{\frac{1}{(a_{2n}-\cos\theta)(a_{2n}+\cos\theta)^{2s+1}}}\,\mathrm{d}\theta .
\end{aligned}
$$

We can continue applying Hölder's inequality with different choices of parameters r and r', such that $1/r + 1/r' = 1$. Then the case $r = \infty$, $r' = 1$ gives the bound with the factor $W_{n,s,\rho}^{(1)}$, whereas the case $r = r' = 2$ gives the bound with the factor $V_{n,s,\rho}^{(1)}$. □

References

1. Bernstein, S: Sur les polynomes orthogonaux relatifs à un segment fini. J. Math. Pure Appl. **9**, 127–177 (1930)
2. Gautschi, W.: Orthogonal Polynomials: Computation and Approximation. Numerical Mathematics and Scientific Computation. Oxford University Press, Oxford (2004)
3. Gautschi, W., Varga, R.S.: Error bounds for Gaussian quadrature of analytic functions. SIAM J. Numer. Anal. **20**, 1170–1186 (1983)
4. Gautschi, W., Tychopoulos, E., Varga, R.S.: A note on the contour integral representation of the remainder term for a Gauss–Chebyshev quadrature rule. SIAM J. Numer. Anal. **27**, 219–224 (1990)
5. Gori, L., Micchelli, C.A.: On weight functions which admit explicit Gauss–Turán quadrature formulas. Math. Comp. **65**, 1567–1581 (1996)
6. Gradshteyn, I.S., Ryzhik, I.M.: Table of Integrals, Series, and Products. Academic Press, New York (1980)
7. Heine, E.: Anwendungen der Kugelfunctionen und der verwandten Functionen. 2nd ed. Reimer, Berlin (1881) [I. Theil: Mechanische Quadratur, 1–31]
8. Hermite, C.: Sur la formule d'interpolation de Lagrange. J. Reine Angew. Math. **84**, 70–79 (1878) [Oeuvres III, 432–443]
9. Hunter, D.B.: Some error expansions for Gaussian quadrature. BIT **35**, 64–82 (1995)
10. Kronrod, A.S.: Nodes and Weights for Quadrature Formulae. Sixteen Place Tables. Nauka, Moscow (1964) (Translation by Consultants Bureau, New York, 1965)

11. Kroó, A., Peherstorfer, F.: Asymptotic representation of L_p-minimal polynimials, $1 < p < \infty$. Constr. Approx. **25**, 29–39 (2007)
12. Laurie, D.P.: Anti-Gaussian quadrature formulas. Math. Comp. **65**, 739–747 (1996)
13. Li, S.: Kronrod extension of Turán formula. Studia Sci. Math. Hungar. **29**, 71–83 (1994)
14. Milovanović, G.V.: Quadratures with multiple nodes, power orthogonality, and moment-preserving spline approximation. J. Comput. Appl. Math. **127**, 267–286 (2001)
15. Milovanović, G.V., Spalević, M.M.: Error bounds for Gauss–Turán quadrature formulae of analytic functions. Math. Comp. **72**, 1855–1872 (2003)
16. Milovanović, G.V., Spalević, M.M: Error analysis in some Gauss–Turán–Radau and Gauss–Turán-Lobatto quadratures for analytic functions. J. Comput. Appl. Math. **164–165**, 569–586 (2004)
17. Milovanović, G.V., Spalević, M.M.: An error expansion for Gauss–Turán quadratures and L^1-estimates of the remainder term. BIT **45**, 117–136 (2005)
18. Milovanović, G.V., Spalević, M.M.: Bounds of the error of Gauss–Turán-type quadratures. J. Comput. Appl. Math. **178**, 333–346 (2005)
19. Milovanović, G.V., Spalević, M.M.: Gauss–Turán quadratures of Kronrod type for generalized Chebyshev weight functions. Calcolo **43**, 171–195 (2006)
20. Milovanović, G.V., Spalević, M.M.: Quadrature rules with multiple nodes for evaluating integrals with strong singularities. J. Comput. Appl. Math. **189**, 689–702 (2006)
21. Milovanović, G.V., Spalević, M.M.: On monotony of the error in Gauss–Turán quadratures for analytic functions. ANZIAM J. **48**, 567–581 (2007)
22. Milovanović, G.V., Spalević, M.M., Cvetković, A.S.: Calculation of Gaussian type quadratures with multiple nodes. Math. Comput. Modelling **39**, 325–347 (2004)
23. Milovanović, G.V, Spalević, M.M., Pranić, M.S.: Maximum of the modulus of kernels in Gauss–Turán quadratures. Math. Comp. **77**, 985–994 (2008)
24. Milovanović, G.V., Spalević, M.M., Pranić, M.S.: Error estimates for Gauss–Turán quadratures and their Kronrod extensions. IMA J. Numer. Anal. **29**, 486–507 (2009)
25. Milovanović, G.V., Spalević, M.M., Pranić, M.S.: Bounds of the error of Gauss–Turán-type quadratures II. Appl. Numer. Math. **60**, 1–9 (2010)
26. Monegato, G.: Stieltjes polynomials and related quadrature rules. SIAM Rev. **24**, 137–158 (1982)
27. Monegato, G.: An overview of the computational aspects of Kronrod quadrature rules. Numer. Algorithms **26**, 173–196 (2001)
28. Ossicini, A., Rosati, F.: Funzioni caratteristiche nelle formule di quadratura gaussiane con nodi multipli. Bull. Un. Mat. Ital. **11**, 224–237 (1975)
29. Peherstorfer, F.: Gauss–Turán quadrature formulas: asymptotics of weights. SIAM J. Numer. Anal. **47**, 2638–2659 (2009)
30. Peherstorfer, F., Petras, K.: Ultraspherical Gauss–Kronrod quadrature is not possible for $\lambda > 3$. SIAM J. Numer. Anal **37**, 927–948 (2000)
31. Shi, Y.G.: Christoffel type functions for m-orthogonal polynomials. J. Approx. Theory **137**, 57–88 (2005)
32. Shi, Y.G., Xu, G.: Construction of σ-orthogonal polynomials and Gaussian quadrature formulas. Adv. Comput. Math. **27**, 79–94 (2007)

Towards a General Error Theory of the Trapezoidal Rule

Jörg Waldvogel

Dedicated to the 60th birthday of Gradimir V. Milovanović

1 Introduction

The (composite) trapezoidal rule, being the simplest and probably oldest algorithm for numerical quadrature, is often dismissed as insufficient for the purpose of high-precision quadrature, mainly due to its low order of precision for integrating generic integrands over a finite interval. However, for integrals over the real line $\mathbb{R}$ of functions analytic in an open strip containing $\mathbb{R}$, the discretization error of the trapezoidal rule has been proven to be exponentially small in the reciprocal step size. Hence, in some cases, the trapezoidal rule is among the most powerful algorithms for numerical quadrature of analytic functions.

By means of appropriate analytic transformations of the integration variable, general intervals may be mapped to the real line, and the decay rate of the integrand may be enhanced, which is desirable to reduce the number of terms in infinite trapezoidal sums. Since such transformations may generate new singularities in the complex plane, no strip of analyticity may exist for the transformed integrand. To the knowledge of this author, no general theory of the discretization error of the trapezoidal rule in the absence of a strip of analyticity exists.

The trapezoidal rule in connection with quadrature of analytic functions has already been studied by Goodwin in 1949 [7], by Schwartz in 1969 [12] and later by Stenger [14]. The idea of using transformations of the integration variable has been mentioned in [12] and by Takahasi and Mori in 1973 [16]; see also [13, Chap. 8], for a summary. The entire topic is comprehensively treated in the textbooks [4] and [15].

Jörg Waldvogel
Seminar für Angewandte Mathematik, ETH Zürich, Zürich, Switzerland,
e-mail: waldvogel@math.ethz.ch

W. Gautschi et al. (eds.), *Approximation and Computation: In Honor of Gradimir V. Milovanović*, Springer Optimization and Its Applications 42,
DOI 10.1007/978-1-4419-6594-3_17,

Here, we will first summarize the well-known error theory of the classical trapezoidal rule. We then consider the particular cases with an exponentially small discretization error, mainly using integrals over the real line $\mathbb{R}$. Finally, we will present examples of integrals with singularities arbitrarily close to $\mathbb{R}$, where the trapezoidal rule still yields powerful high-precision integrators of somewhat slower convergence, still with an exponentially small discretization error.

2 The Classical Trapezoidal Rule

In its classical form the (composite) trapezoidal rule is stated as an algorithm for numerically approximating the integral I of the integrable function f over the interval $a \le x \le b$,

$$I := \int_a^b f(x)\,\mathrm{d}x, \tag{1}$$

using the $n+1 \ge 2$ equally spaced points $x_j := a + jh$, $j = 0, 1, \ldots, n$, $h := (b-a)/n$, where h is the step size. The trapezoidal sum $T(h)$ is then defined as

$$T(h) := h \sum_{j=0}^{n} w_j f(x_j), \tag{2}$$

where the weights w_j are $w_0 = w_n = 1/2$, $w_1 = w_2 = \cdots = w_{n-1} = 1$.

If f has at least $2N-1 > 0$ continuous derivatives, the *discretization error* of a trapezoidal sum is given by the Euler–Maclaurin summation formula (for a good account see, e.g., [8]) as

$$\begin{aligned} T(h) - I = {} & \frac{h^2}{12}\left[f'(b) - f'(a)\right] - \frac{h^4}{720}\left[f'''(b) - f'''(a)\right] + \cdots \\ & + \frac{h^{2N} B_{2N}}{(2N)!}\left[f^{(2N-1)}(b) - f^{(2N-1)}(a)\right] + R_{2N}, \end{aligned} \tag{3}$$

where B_{2N} is the Bernoulli number of the even order $2N$,

$$B_0 = 1, \quad B_2 = \frac{1}{6}, \quad B_4 = -\frac{1}{30}, \quad B_6 = \frac{1}{42}, \quad B_8 = -\frac{1}{30}, \quad B_{10} = \frac{5}{66}, \; \ldots,$$

and R_{2N} is the remainder term. R_{2N} can be specified [8], but this is rather involved. Since the above series is usually divergent for any fixed step $h > 0$, it must be truncated at a finite value of N.

In general, the trapezoidal rule converges very slowly with respect to step refinement. Convergence is of *second* order: Halving of the step (i.e., doubling the computational effort) reduces the discretization error by a factor of 4, that is, merely yields 0.6 additional digits of accuracy. This does not allow to obtain more than a few digits of accuracy.

In the following, we restrict ourselves to integrands f *analytic* in the *open* interval of integration, that is, integrable boundary singularities are permitted, and x will be understood as a *complex* variable. Of particular interest are the cases with the property that the terms of the series (3) vanish up to arbitrary values of N. Then, the discretization error is given by the remainder R_{2N}. We distinguish three cases:

1. f is periodic with no singularities on the real line $\mathbb{R}$; the interval $[a,b]$ is a full period (or an integer number of periods).
2. f is integrable over the real line $\mathbb{R}$, and I is its integral over $\mathbb{R}$, that is, $a=-\infty$, $b=\infty$.
3. f is a flat function at both boundaries, that is, all derivatives of f at $x=a$ and $x=b$ vanish.

In all three cases, convergence with respect to step refinement is faster than any finite order, referred to as *exponential convergence*. Case 3 corresponds to using the *IMT rule* proposed in 1969 by Iri, Moriguti and Takasawa (see, e.g., [10]). Case 2 may be considered as a particular instance of Case 3; however, integration over the entire real line is more fundamental and often leads to more efficient algorithms. We will mainly concentrate on integrals over $\mathbb{R}$, but we begin with a short discussion of Case 1.

3 The Periodic Case

With no loss of generality, we assume $a=0$, $b=2\pi$. Let f be 2π-periodic and analytic on the entire real axis, and consider the integral

$$I := \int_0^{2\pi} f(x)\,\mathrm{d}x\,.$$

With $n \geq 1$ subintervals of equal length $h=2\pi/n$ the trapezoidal sum (2) becomes

$$T(h) = h\sum_{\ell=0}^{n-1} f(\ell h)\,. \tag{4}$$

The Fourier series of f (in a somewhat unconventional notation),

$$f(x) = \frac{1}{2\pi}\sum_{k=-\infty}^{\infty} c_k \mathrm{e}^{\mathrm{i}kx} \quad \text{with } c_k = \int_0^{2\pi} f(x)\mathrm{e}^{-\mathrm{i}kx}\,\mathrm{d}x\,, \tag{5}$$

implies $I=c_0$. Hence, the trapezoidal sum (4) becomes

$$T(h) = \frac{1}{n}\sum_{k=-\infty}^{\infty} c_k \sum_{\ell=0}^{n-1}\exp\left(\mathrm{i}\frac{2\pi}{n}k\ell\right) = \sum_{j=-\infty}^{\infty} c_{jn}\,,$$

and we obtain the discretization error $E_n := T(h) - I$ with $n = 2\pi/h$ as

$$\boxed{E_n = c_n + c_{-n} + c_{2n} + c_{-2n} + \cdots}. \tag{6}$$

Therefore, the discretization error of the trapezoidal rule for small steps is governed by the decay rate of the Fourier coefficients c_n of the integrand for large absolute values of the index n.

A more informative statement may be formulated if f is analytic in the symmetric strip $|\mathrm{Im}\, x| < \gamma$, $\gamma > 0$. Then, the theory of Fourier series [4, 11] states

$$c_n = O\left(\mathrm{e}^{-(\gamma-\varepsilon)|n|}\right) \quad \text{for any } \varepsilon \in (0,\gamma) \quad \text{as } n \to \pm\infty.$$

Combining this with the error formula (6) yields the following result:

Theorem 1. *Let f be a 2π-periodic function, analytic in the strip $|\mathrm{Im}\, x| < \gamma$, $\gamma > 0$, and let I be the integral of f over a full period. Then the discretization error $E_n := T(h) - I$ of the trapezoidal sum $T(h)$, with $n \geq 1$ subintervals of equal length $h = 2\pi/n$, decays as $O\left(\mathrm{e}^{-(\gamma-\varepsilon)|n|}\right)$ as $n \to \pm\infty$ for any $\varepsilon \in (0,\gamma)$.*

4 Integrals Over the Real Line

We now consider Case 2 of Sect. 2, that is, (1) is specialized to

$$I := \int_{-\infty}^{\infty} f(x)\,\mathrm{d}x\,, \tag{7}$$

where f is analytic on $\mathbb{R}$ and integrable over $\mathbb{R}$. By slightly generalizing the definition (2), we will approximate I by the *shifted* trapezoidal sum with step h and offset s,

$$T(h,s) = h \sum_{j=-\infty}^{\infty} f(s + jh)\,, \tag{8}$$

where the doubly infinite sum converges as a consequence of the integrability of f. As a function of s, $T(h,s)$ is periodic with period h, $T(h,s) = T(h,s+h)$, and the relation

$$T\left(\frac{h}{2}, s\right) = \frac{1}{2}\left[T(h,s) + T\left(h, s + \frac{h}{2}\right)\right]$$

may be used for an efficient transition from step h to step $h/2$.

Aside from the approximation of I, the evaluation of the infinite sum (8) may be a problem in itself. Here, we will not discuss this aspect; instead, we assume a sufficiently fast decay of $|f(x)|$ as $x \to \pm\infty$, such that simple truncation rules for the infinite sum (8) may be used. In the next section, we will motivate this approach by discussing transformations of the integration variable to obtain quickly decaying integrands.

As in the periodic case of Sect. 3, the discretization error may be obtained by Fourier theory. The basic tool is the Poisson summation formula (see, e.g., [9]),

$$T(h,s) = PV \sum_{k=-\infty}^{\infty} \widehat{f}(kr)\, \mathrm{e}^{\mathrm{i}skr} \quad \text{with } r := \frac{2\pi}{h}, \tag{9}$$

which expresses the shifted trapezoidal sum (8) of f with step h again as a (weighted) trapezoidal sum, but of the Fourier transform $\widehat{f}$, and with step $r := 2\pi/h$. Here, PV stands for the Cauchy principal value of the sum, that is, the sum over all integers k must be taken as the limit of the symmetric sum from $-K$ to K as $K \to \infty$. The Fourier transform $\widehat{f}$ of f is defined as

$$\widehat{f}(\omega) := \int_{-\infty}^{\infty} \mathrm{e}^{-\mathrm{i}\omega x} f(x)\,\mathrm{d}x, \quad \omega \in \mathbb{R}, \tag{10}$$

which immediately yields $I = \widehat{f}(0)$.

In the following, we restrict ourselves to the particular case of vanishing offset, $s = 0$, using the notation $T(h) := T(h,0)$. In view of the definition $r := 2\pi/h$ in (9), we introduce the discretization error $E(r)$ as a function of r,

$$E(r) := T(h) - I, \quad h := \frac{2\pi}{r}.$$

Equation (9) now yields the *error formula*

$$\boxed{E(r) = \widehat{f}(r) + \widehat{f}(-r) + \widehat{f}(2r) + \widehat{f}(-2r) + \cdots}. \tag{11}$$

Therefore, in complete analogy with (6), with r taking the role of the index n, the discretization error of the trapezoidal rule for small steps h is governed by the decay rate of the Fourier transform $\widehat{f}$ of the integrand for large values of $\omega = \pm r := \pm 2\pi/h$.

A more informative statement may be formulated if f is analytic in the symmetric strip $|\mathrm{Im}\, x| < \gamma$, $\gamma > 0$ and integrable along any path $\mathrm{Im}\, x = \gamma_0 = \text{const}$ with $|\gamma_0| < \gamma$. Then, we have

$$\widehat{f}(\omega) = O\left(\mathrm{e}^{-(\gamma-\varepsilon)|\omega|}\right) \quad \text{for any } \varepsilon \in (0,\gamma) \quad \text{as } \omega \to \pm\infty \tag{12}$$

([3, Chap. 3], [11, Theorem IX.14], or [4]). Combining (11) and (12) yields the following theorem:

Theorem 2. *Let f be analytic in the strip $|\mathrm{Im}\, x| < \gamma$, $\gamma > 0$ and integrable along any path $\mathrm{Im}\, x = \gamma_0 = \text{const}$ with $|\gamma_0| < \gamma$, and let I be the integral of f over $\mathbb{R}$. Then the discretization error $E_n := T(h) - I$ of the trapezoidal sum $T(h)$ with step $h = 2\pi/r$ decays as $O\left(\mathrm{e}^{-(\gamma-\varepsilon)|r|}\right)$ as $r \to \pm\infty$ for any $\varepsilon \in (0,\gamma)$.*

5 Transforming the Integration Variable

The trapezoidal rule for integrals over $\mathbb{R}$ is particularly attractive for quickly decaying integrands (at least exponential decay). Then, the truncation of the infinite trapezoidal sums can be easily handled. A simple truncation rule for an infinite sum is *truncation if the contribution of the current term is below a given tolerance* $\varepsilon > 0$. If necessary, this simple rule can be made more robust by truncating only if two (or three) consecutive terms do not contribute. Accumulation of doubly infinite sums has to be done upwards and downwards from an interior point.

To validate such a truncation procedure, the tail of the truncated sum needs to be estimated. Ideally, the contribution of the truncated tail should be $< \varepsilon$. We will use the function

$$f(x) := \exp(-\mathrm{e}^{\alpha x}), \quad \alpha > 0, \tag{13}$$

as a simple model of a doubly exponentially decaying integrand as $x \to +\infty$. The tail of the sum, truncated at $x = X > 0$, will be modeled by the integral

$$R_X := \int_X^\infty f(x)\,\mathrm{d}x,$$

which can be expressed in terms of the exponential integral $E_1(x)$ (see [1]) as $R_X = \alpha^{-1}\, E_1(\exp(\alpha X))$. Asymptotic expansion then yields

$$R_X \sim \frac{1}{\alpha} \exp\left(-\mathrm{e}^{\alpha X}\right) \left(\mathrm{e}^{-\alpha X} - \mathrm{e}^{-2\alpha X} + 2!\mathrm{e}^{-3\alpha X} + \cdots\right).$$

The suggested truncation limit X, given by $f(X) = \varepsilon$, is found to be

$$X = \frac{1}{\alpha} \log\log\frac{1}{\varepsilon};$$

therefore

$$R_X = -\frac{\varepsilon}{\alpha}\left\{\frac{1}{\log\varepsilon} + O\left[(\log\varepsilon)^{-2}\right]\right\}.$$

Truncation is safe even for $\alpha \ll 1$, since $|R_X| < \varepsilon$ if ε is sufficiently small.

Our strategy is to use the simple tool of transformations of the integration variable to transform a given integral into one over $\mathbb{R}$ with a quickly decaying integrand. Beginning with the integral I of (1), we use an appropriate transformation $x = \phi(t)$, $t \in \mathbb{R}$, where ϕ is a monotonically increasing, differentiable function $\phi : t \in \mathbb{R} \mapsto x = \phi(t) \in (a,b)$. The result is

$$I = \int_{-\infty}^{\infty} F(t)\,\mathrm{d}t \quad \text{with } F(t) := f(\phi(t))\,\phi'(t). \tag{14}$$

There is a wide range for the choice of ϕ. Since we are working with analytic integrands we only consider analytic transformation functions. In view of numerical applications, we will choose ϕ as a combination of elementary functions, for which reliable implementations in arbitrary precision are readily available.

In the following table, we list a few standard intervals $[a,b]$ for the integral (1), together with suggested elementary transformations $x = \phi(t)$. Possibly, a composition of several of the suggested mappings may have to be used to achieve doubly exponential decay of the transformed integrand $F(t)$ as $t \to \pm\infty$. In all cases (including the cases with finite boundaries), integrable boundary singularities are allowed. However, in the case of finite intervals care must be taken to accurately transmit the distances to both interval boundaries.

Interval: $x \in (a,b)$	Transformation: $x = \phi(t) =$
1. Finite interval $(-1,1)$:	$\tanh(t/2)$
2. Finite interval $(0,1)$:	$1/(1+\exp(-t))$
3. Semi-infinite interval $(0,\infty)$:	$\exp(t)$
4. Real line $\mathbb{R}$, accelerate decay as $x \to \pm\infty$	$\sinh(t)$
5. Real line $\mathbb{R}$, accelerate decay as $x \to +\infty$	$t+\exp(t)$
6. Real line $\mathbb{R}$, accelerate decay as $x \to -\infty$	$t-\exp(-t)$.

6 Error Theory of Integrals Over $\mathbb{R}$

In the remaining sections, we will go back to the problem of Sect. 4, the integral I over $\mathbb{R}$ of (7), to be approximated by the infinite trapezoidal sum of (8), $T(h) = T(h,0)$ with step h. In the case of f being analytic in a symmetric strip of half-width $\gamma > 0$ containing $\mathbb{R}$, the discretization error of $T(h)$ is exponentially small in h according to Theorem 2, more precisely,

$$E(r) = O\left(\mathrm{e}^{-(\gamma-\varepsilon)|r|}\right) \quad \text{as } r := \pm 2\pi/h \to \pm\infty, \quad \forall \varepsilon \in (0,\gamma). \tag{15}$$

This also applies to integrals one would hardly attempt to directly approximate by the trapezoidal rule, for example, $I = \int_{-\infty}^{\infty} \mathrm{d}x/(1+x^2) = \pi$. In such cases, it is advisable to enhance the decay of the integrand according to Sect. 5 before invoking the trapezoidal rule. However, transformations enhancing the decay of the integrand may bear the danger of generating new singularities in the complex plane (possibly closing in on the real line), such that Theorem 2 is no longer applicable. We will consider this situation in more detail in Sect. 7.

The essential ingredient for an error theory of the infinite trapezoidal sum is the Fourier transform $\widehat{f}(\omega)$ of the integrand, defined in (10); then, the discretization error is given by the error formula (11). We will now develop an error formula for transformed integrals. Consider the integral I over $\mathbb{R}$ of (7), and assume $\phi : t \in \mathbb{R} \mapsto x = \phi(t) \in \mathbb{R}$ is used to modify the decay rate of the integrand; the transformed integral is given by (14). The Fourier transform $\widehat{F}(\omega)$ of $F(t)$ becomes

$$\widehat{F}(\omega) = \int_{-\infty}^{\infty} \mathrm{e}^{-\mathrm{i}\omega t} f(\phi(t))\, \phi'(t)\, \mathrm{d}t\,.$$

Going back to the original variable x by using the *inverse* transformation $\phi^{[-1]} : x \in \mathbb{R} \mapsto t = \phi^{[-1]}(x) \in \mathbb{R}$ and the relation $\phi'(t)\,\mathrm{d}t = \mathrm{d}x$ yields

$$\widehat{F}(\omega) = \int_{-\infty}^{\infty} \mathrm{e}^{-\mathrm{i}\omega\phi^{[-1]}(x)} f(x)\,\mathrm{d}x\,, \tag{16}$$

a *modified "Fourier" transform* of f, where the factor x in the exponent of (10) is replaced by $\phi^{[-1]}(x)$. Combining this with the error formula (11) yields the following result:

Theorem 3. *Let $\widetilde{T}(h)$ be the infinite trapezoidal sum with step h, obtained from the integral $I = \int_{-\infty}^{\infty} f(x)\,\mathrm{d}x$ after having transformed the integration variable by means of the isomorphism $\phi:\ t \in \mathbb{R} \mapsto x = \phi(t) \in \mathbb{R}$,*

$$\widetilde{T}(h) = h \sum_{k=-\infty}^{\infty} f(\phi(kh))\,\phi'(kh)\,.$$

Then the discretization error $\widetilde{E}(r) := \widetilde{T}(h) - I$ of $\widetilde{T}(h)$ is given by

$$\widetilde{E}(r) = \sum_{k \neq 0} \widehat{F}(kr), \qquad r = \frac{2\pi}{h}\,,$$

where $\widehat{F}(\omega)$ is given by (16).

7 Asymptotics

According to the error formula (11), the discretization error of the infinite trapezoidal rule for small steps h is governed by the Fourier transform $\widehat{f}(r)$ of the integrand at large arguments $r = \pm 2\pi/h$. Therefore, an error theory of the trapezoidal rule amounts to an asymptotic theory of the Fourier transform of the integrand. Such a theory is complicated and has not been fully developed so far.

In the simple case of a strip of analyticity (of half-width γ) with the symmetry axis $\mathbb{R}$, Theorem 2 establishes the discretization error to be of order $O(\exp((\varepsilon-\gamma)|r|))$. However, if a transformation $x = \phi(t)$ is necessary to achieve a sufficient decay rate of the integrand, the strip of analyticity may be lost (i.e., the singularities may fill the area outside a "funnel" closing in on the real axis, see Fig. 1). Then, Theorem 2 is no longer applicable, and an asymptotic theory of the modified Fourier transform (16) is needed. In several examples, conjectured decay rates of order $O(\exp(-\gamma\omega/\log(c\omega))$ or $O\left(\exp(-\gamma\sqrt{\omega})\right)$ with $\gamma > 0$, $c > 0$ were observed.

7.1 An Introductory Example

We begin with the discussion of the explicit example mentioned in connection with (15), subjected to repeated sinh transformations of the integration variable. Adapting the notation by using $t_0 := x$, $I_0 := I$, consider

$$I_k = \int_{-\infty}^{\infty} f_k(t_k)\,\mathrm{d}t_k = \pi,\ k = 0,1,2,3 \quad \text{with } f_0(t) := \frac{1}{1+t^2},$$

and the transformations

$$t_k = \sinh(t_{k+1}), \quad f_{k+1}(t) = f_k(\sinh(t))\cosh(t), \quad k = 0,1,2.$$

Hence, the integrands

$$f_1(t) = \frac{1}{\cosh(t)}, \quad f_2(t) = \frac{\cosh(t)}{\cosh(\sinh(t))}, \quad f_3(t) = \frac{\cosh\big(\sinh(t)\big)\cosh(t)}{\cosh\big(\sinh(\sinh(t))\big)}$$

show single, double, or triple exponential decay. The singularities (poles) t_0, t_1, t_2 of f_0, f_1, f_2, respectively, are given by

$$\begin{aligned}
t_0 &= \pm\mathrm{i},\\
t_1 &= \mathrm{i}\,u\frac{\pi}{2}, \quad u \text{ odd},\\
t_2 &= \operatorname{sign}(u)\,\operatorname{Acosh}\left(|u|\frac{\pi}{2}\right) + \mathrm{i}v\frac{\pi}{2}, \quad u,v \text{ odd}.
\end{aligned}$$

The poles t_2 of f_2 still leave a strip of analyticity $|\mathrm{Im}\,t| < \pi/2$, whereas the poles of f_3 close in on the real axis as $t \to \pm\infty$, see Fig. 1.

In the cases $k = 0$ and $k = 1$, the Fourier transforms $\widehat{f_k}(\omega)$ and the discretization errors $E_k(r)$, see (10, 11), may be expressed explicitly as

$$\begin{aligned}
\widehat{f_0}(\omega) &= \pi\,\mathrm{e}^{-|\omega|}, & E_0(r) &= \frac{2\pi}{\mathrm{e}^r - 1}, \quad r = 2\pi/h,\\
\widehat{f_1}(\omega) &= \frac{\pi}{\cosh(\omega\,\pi/2)}, & E_1(r) &= \sum_{j=1}^{\infty} \frac{2\pi}{\cosh(j\,r\,\pi/2)}.
\end{aligned}$$

In the case $k = 2$, Theorem 3 yields the Fourier transform $\widehat{f_2}(\omega)$ as the integral

$$\widehat{f_2}(\omega) = \int_{-\infty}^{\infty} \frac{\mathrm{e}^{-\mathrm{i}\omega \operatorname{Asinh}(t)}}{\cosh(t)}\,\mathrm{d}t \tag{17}$$

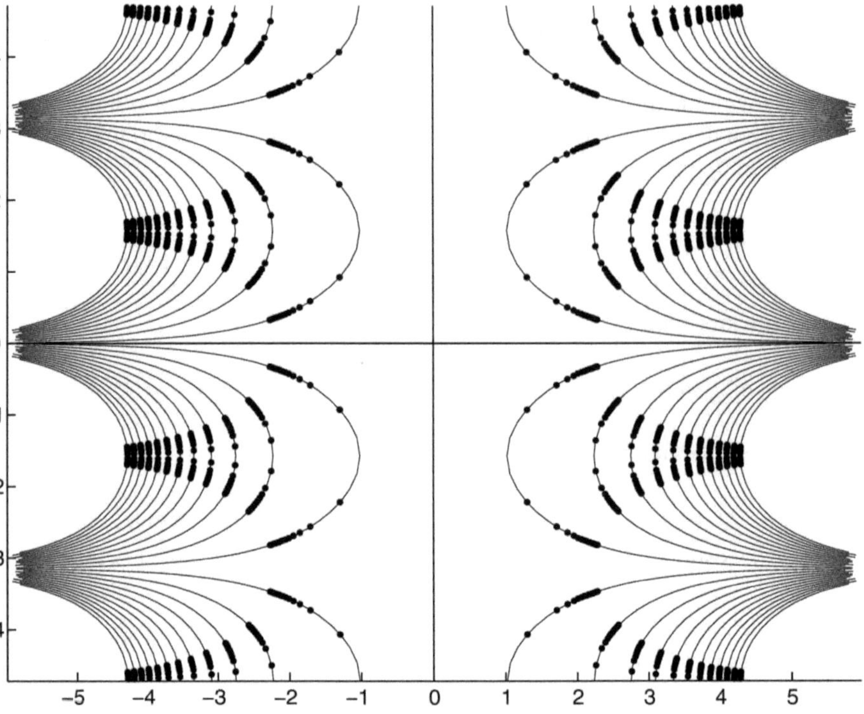

Fig. 1 Singularities of the integrand f_3 of $\int_{-\infty}^{\infty} \mathrm{d}x/(1+x^2)$ after three sinh transformations. The *dots on the curves*, to be continued in increasing density, mark singularities. The *arc-shaped areas* to the left and to the right are densely filled with singularities not marked in the figure

which does not have an obvious closed form. However, according to Theorem 2 we have

$$\widehat{f_2}(\omega) = O\left(\mathrm{e}^{(\varepsilon-\pi/2)\,|\omega|}\right) \quad \text{as } \omega \to \pm\infty \quad \text{for any } \varepsilon \in (0,\pi/2)\,. \tag{18}$$

Evaluation of the integral (17) by a branch cut integral along the imaginary axis or by the saddle point method ([5, 6] for a worked example) corroborates (18). In Fig. 2, the exponentially amplified discretization error, $\exp(r\pi/2)E_2(r)$, is plotted versus $r \in (0.5, 70)$, where $E_2(r) \approx 2\mathrm{Re}\,\widehat{f_2}(r)$ according to the error formula (11). The complicated irregular behavior seen in Fig. 2 reflects the fact that infinitely many singularities or saddles contribute to the integral (17). The slow (sub-exponential) growth of the amplitude in Fig. 2 exemplifies the necessity of the parameter ε in Theorem 2.

The case $k = 3$ is even more complicated. In spite of a "mine field" of singularities, the real axis does not hit a single pole, and the integrand on $\mathbb{R}$ becomes exceedingly small. Nevertheless, in the range of Fig. 3, the discretization error seems to be exponentially small of the conjectured order

$$E_3(r) \approx O\left(\mathrm{e}^{-\gamma r/\log(cr)}\right) \quad \text{with } \gamma \doteq 2.285\,,\ c \doteq 6.05\,. \tag{19}$$

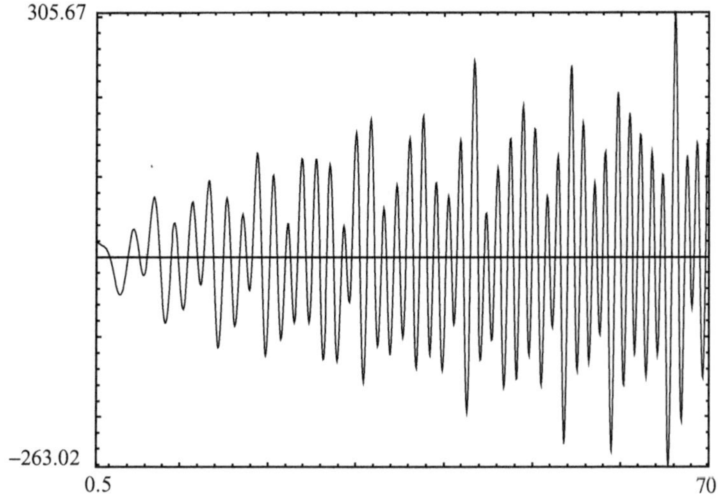

Fig. 2 Exponentially amplified discretization error, $e^{r\pi/2}E_2(r)$, versus $r = 2\pi/h$, $0.5 \leq r \leq 70$. High-precision computations done with the software package PARI [2]

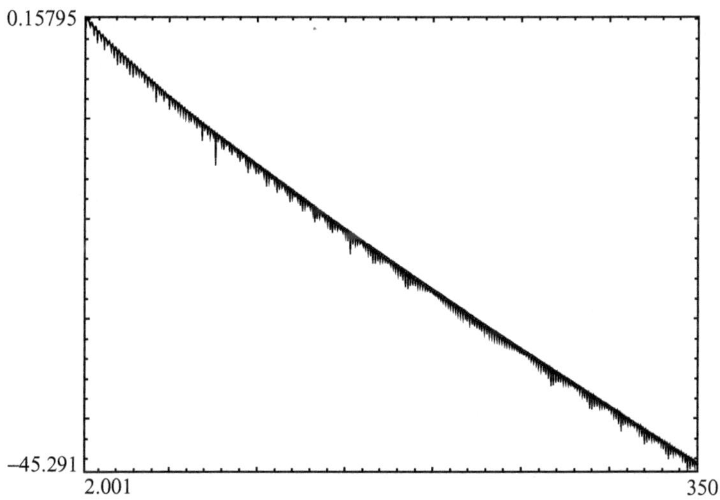

Fig. 3 The integral $\int_{-\infty}^{\infty} \mathrm{d}x/(1+x^2)$ with three sinh transformations. Plot of $\log_{10}(|E_3(r)|) \in (-45.3, 0.2)$ versus $r \in (2, 350)$, including an empirical upper envelope

Figure 3 shows a plot of $\log_{10}(|E_3(r)|)$, that is, a plot of minus the number of correct digits, versus $r = 2\pi/h$, together with the empirical envelope

$$\log_{10}(9\exp(-\gamma r/\log(cr)))$$

with γ, c from (19). In spite of the singularities, halving the step size almost doubles the number of correct digits, at least up to an accuracy of 45 digits.

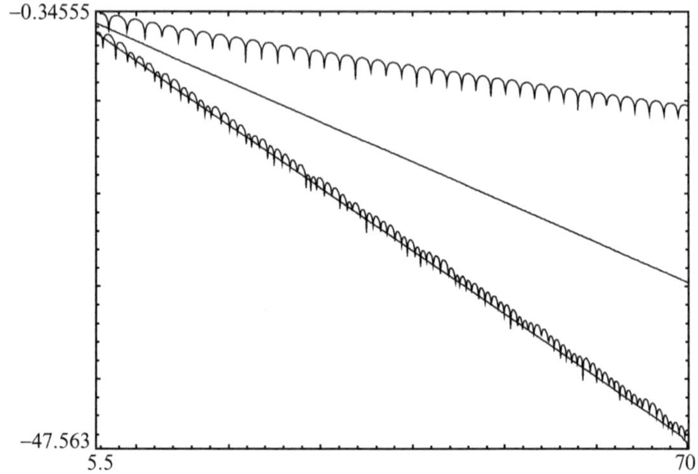

Fig. 4 Plots of $\log_{10}(|E_k(r)|) \in (-47,0)$ for $k = 1,2,0,3$ (bottom to top) versus $r = 2\pi/h \in (5.5,70)$

As a summary, Fig. 4 shows $\log_{10}(|E_k(r)|)$ for $k = 1,2,0,3$ (bottom to top) versus $r = 2\pi/h$ in the interval $5.5 \le r \le 70$. The algorithm with $k = 0$ is impractical due to the slow decay of the integrand (very long trapezoidal sums). $k = 3$ works with short trapezoidal sums but, as a consequence of the small slope in Fig. 4, requires a much smaller step size for reaching the same accuracy. The best performance results from $k = 2$, featuring an integrand of doubly exponential decay in a strip of analyticity.

7.2 *A Difficult Integral*

Slowly decaying analytic functions with many singularities may be difficult to integrate over $\mathbb{R}$ by the algorithms discussed here. Transformations to accelerate the decay of the integrand (Sect. 5) may breed more singularities (Fig. 1) and slow down the decay rate of the discretization error with respect to step refinement. Consider, for example, the integral

$$I(a) := \int_{-\infty}^{\infty} \frac{\mathrm{d}x}{\cosh(x)\,(\cosh(a) + \cos(x))}\,. \tag{20}$$

Explicit values may be obtained from calculus of residues using the poles $x_1 = \mathrm{i}u\pi/2$, $x_2 = v\pi \pm \mathrm{i}\, a$, u, v odd:

$$I(a) = 2\pi\left(\sum_{k=0}^{\infty} \frac{(-1)^k}{\cosh(a) + \cosh(k+1/2)\pi} + \frac{2\cos(a)}{\sinh(a)} \sum_{u=1,3,\ldots} \frac{\cosh(u\pi)}{\cosh^2(u\pi) - \sin^2(a)}\right).$$

For $a = 1$, the quickly convergent series easily yields the 50D approximation

$$I(1) = 1.94734\,99863\,38691\,95445\,99206\,53366\,23422\,62265\,42793\,65329$$

to be used as a reference value.

For a numerical high-precision evaluation by the trapezoidal rule with step h, an integrand of doubly exponential decay may be desirable; therefore, we transform the integration variable by means of $x = \sinh(t)$.

The behaviour of the discretization error $E(r)$ of the transformed integral as function of $r := 2\pi/h$ turns out to be complicated. Here, we formulate a conjecture concerning the decay rate of $E(r)$:

Conjecture 1.

$$E(r) = O\left((ar)^{1/4}\mathrm{e}^{-2\sqrt{ar}}\right), \quad r := \frac{2\pi}{h}. \tag{21}$$

In the following, we give a heuristic derivation of the above conjecture. In view of the error formula (11), $E(r) \sim \widehat{F}(r) + \widehat{F}(-r)$ and by Theorem 3, it suffices to investigate the modified Fourier transform

$$\widehat{F}(\omega) = \int_{-\infty}^{\infty} \frac{\mathrm{e}^{-\mathrm{i}\omega \mathrm{A}\sinh(z)}}{\cosh(z)\,(\cosh(a) + \cos(z))}\,\mathrm{d}z$$

for large $|\omega|$. For this purpose, we deform ("pull down") the path of integration $\mathbb{R}$ and consider the following contributions F_k, $k = 1,2,3,4$, to the integral $\widehat{F}(\omega)$:

1. Lower semi-circle of radius R: $F_1 \to 0$ as $R \to \infty$
2. Branch cut along the imaginary axis, $\operatorname{Im} z < -1$: $F_2 = O\left(\mathrm{e}^{-|\omega|\pi/2}\right)$, negligible as $\omega \to \pm\infty$
3. Poles on branch cut, $z_\nu := \mathrm{i}\pi/2\ \nu$, $\nu < 0$, odd: $F_3 = O\left(\mathrm{e}^{-|\omega|\pi/2}\right)$, negligible as $\omega \to \pm\infty$
4. Poles $z_u := u\pi - \mathrm{i}a$, u odd: the *dominant* contributions F_4

Summation over the poles z_u yields

$$F_4 = \frac{4\pi}{\sinh(a)} \sum_{u=1,3,\ldots} \mathrm{Re}\,\frac{\mathrm{e}^{-\mathrm{i}\omega \mathrm{A}\sinh(u\pi - \mathrm{i}a)}}{\cosh(u\pi - \mathrm{i}a)}.$$

This is simplified by approximating the sum by the corresponding integral and replacing the terms of the sum by the expressions

$$2\mathrm{Re}\left\{\exp\left[-\mathrm{i}\omega\left(\log(2u\pi) - \frac{\mathrm{i}a}{u\pi}\right) + \mathrm{i}a - u\pi\right]\right\},$$

which represent the terms of the sum asymptotically as $u \to \infty$. The result is

$$F_4 \approx \frac{4}{\sinh(a)} \int_0^\infty \exp\left(-\frac{a\omega}{v} - v\right)\cos\left(\omega \log(2v) - a\right)\mathrm{d}v,$$

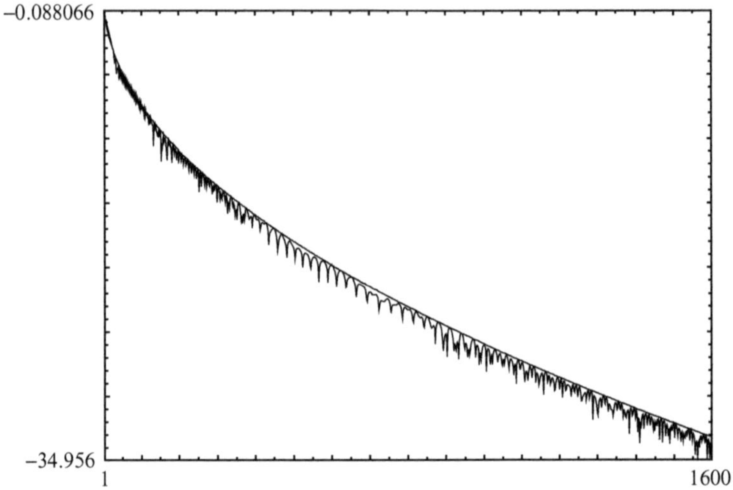

Fig. 5 Plot of $\log_{10}(|E(r)|/2) \in (-35,0)$ versus $r = 2\pi/h \in (1,1600)$ for the integral (20) with one sinh transformation, together with the conjectured upper bound from (23)

where $v = u\pi$. A crude upper bound for $|F_4|$ is therefore

$$|F_4| \dot{<} \frac{4}{\sinh(a)} \int_0^\infty \exp\left(-\frac{a\omega}{v} - v\right) \mathrm{d}v. \tag{22}$$

The substitution $v = \sqrt{a\omega}\,\mathrm{e}^t$ and (9.6.24) of [1] results in the following closed form of the integral in (22) in terms of a *modified Bessel* function:

$$\int_0^\infty \exp\left(-\frac{a\omega}{v} - v\right) \mathrm{d}v = 2\sqrt{a\omega}\mathrm{K}_1\left(2\sqrt{a\omega}\right).$$

Asymptotics of K_1 [1, (9.7.2)] yields the approximate inequality

$$|\widehat{F}(\omega)| \dot{<} c(a\omega)^{1/4}\mathrm{e}^{-2\sqrt{a\omega}}, \quad c = \frac{4\sqrt{\pi}}{\sinh(a)}. \tag{23}$$

Figures 5 and 6 show this bound in comparison with half the error of the trapezoidal sum with step $h = 2\pi/\omega$. The agreement is striking!

The error law (21) implies that doubling the accuracy requires quartering the step; this is still exponential convergence, but at a slower rate. For this reason, the trapezoidal rule applied to the original integral (20) turns out to be more efficient, in spite of the longer trapezoidal sums. With the software package PARI [2] and a working precision of 57 digits (resulting in about 54 correct digits), the computation time for the direct approach is 45% of the time taken for the transformed integral.

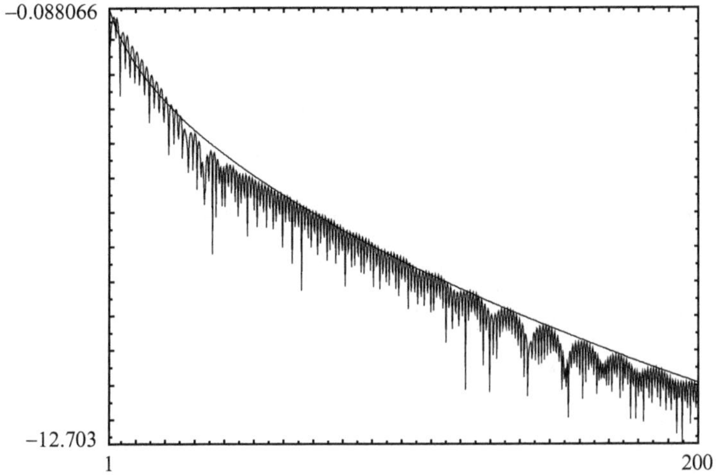

Fig. 6 Detail of Fig. 5 in the range $r \in (1,200)$ in higher resolution. As expected, (23) does not give a strict bound

8 Conclusions

We have proposed to use the trapezoidal rule on the entire real line $\mathbb{R}$ as the standard algorithm for numerical quadrature of analytic functions. Finite and semi-infinite intervals as well as boundary singularities may be handled by using a new integration variable t, thus transforming the original variable x as $x = \phi(t)$, where in practice ϕ may be chosen as an appropriate combination of elementary functions. Slow decay at infinity may be accelerated by the transformation $x = \sinh(t)$.

If the transformed integrand is analytic in a strip containing $\mathbb{R}$, the discretization error is known to be exponentially small in $r := 2\pi/h$, h being the step size. If the integrand has singularities arbitrarily close to $\mathbb{R}$, as it often happens in integrands transformed with sinh, the behaviour of the discretization error may be very complicated; no general theory is known. Decay rates $O(\exp(-\gamma r/\log(cr)))$ and $O(\exp(-\gamma\sqrt{r}))$ with $\gamma > 0$, $c > 0$, have been observed; in both cases, the trapezoidal rule is still a powerful algorithm.

References

1. Abramowitz, M., Stegun, I.A.: Handbook of Mathematical Functions. US Government Printing Office, Washington D.C. (1964)
2. Batut, C., Belabas, K., Bernardi, D., Cohen, H., Olivier, M.: The software package PARI (freeware). http://pari.math.u-bordeaux.fr/
3. Bornemann, F., Laurie, D., Wagon, S., Waldvogel, J.: The SIAM 100-Digit Challenge. SIAM, Philadelphia, 306 pp. (2004)

4. Davis, P.J., Rabinowitz, P.: Methods of Numerical Integration, 2nd edition. Academic Press, San Diego, 512 pp. (1984)
5. Erdélyi, A.: Asymptotic Expansions. Dover, New York (1956)
6. Gautschi, W., Waldvogel, J.: Computing the Hilbert transform of the generalized Laguerre and Hermite weight functions. BIT **41**, 490–503 (2001)
7. Goodwin, E.T.: The evaluation of integrals of the form $\int_{-\infty}^{\infty} f(x) \exp(-x^2)\, dx$. Proc. Cambr. Philos. Soc. **45**, 241–245 (1949)
8. Graham, R.L., Knuth, D.E., Patashnik, O.: Concrete Mathematics. Addison-Wesley, Reading, MA, 625 pp. (1989)
9. Henrici, P.: Applied and Computational Complex Analysis, Vol. 2, in particular p. 270 ff. Wiley, New York, 662 pp. (1977)
10. Iri, M., Moriguti, S., Takasawa, Y.: On a certain quadrature formula. J. Comp. Appl. Math. **17**, 3–20 (1987)
11. Reed, M., Simon, B.: Methods of Modern Mathematical Physics. II. Fourier Analysis, Self-Adjointness. Academic Press, New York (1975)
12. Schwartz, C.: Numerical integration of analytic functions. J. Comput. Phys. **4**, 19–29 (1969)
13. Schwarz, H.R.: Numerical Analysis. Wiley, New York, 517 pp. (1989). In particular: Numerical Quadrature, 330–350, by J. Waldvogel. Original edition: Numerische Mathematik, Teubner, Leipzig, 496 pp. (1986)
14. Stenger, F.: Integration formulae based on the trapezoidal formula. J. Inst. Math. Appl. **12**, 103–114 (1973)
15. Stenger, F.: Numerical Methods Based on sinc and Analytic Functions. Springer Series in Computational Mathematics, Vol. 20. Springer, New York (1993)
16. Takahasi, H., Mori, M.: Quadrature formulas obtained by variable transformation. Numer. Math. **21**, 206–219 (1973)

Part IV
Differential Equations

Finite Difference Method for a Parabolic Problem with Concentrated Capacity and Time-Dependent Operator

Dejan R. Bojović and Boško S. Jovanović

Dedicated to Professor Gradimir V. Milovanović on the occasion of his 60th birthday

1 Introduction

The finite difference method is one of the basic tools for the numerical solution of partial differential equations. In the case of problems with discontinuous coefficients and concentrated factors (Dirac delta functions, free boundaries, etc.) the solution has weak global regularity and it is impossible to establish convergence of finite difference schemes using the classical Taylor series expansion. Often, the Bramble-Hilbert lemma takes the role of the Taylor formula for functions from the Sobolev spaces [5, 6, 11].

Following Lazarov et al. [11], a convergence rate estimate of the form

$$\|u-v\|_{W_{2,h}^k} \leq Ch^{s-k}\|u\|_{W_2^s}, \quad s>k,$$

is called **compatible** with the smoothness (regularity) of the solution u of the boundary-value problem. Here v is the solution of the discrete problem, h the spatial mesh step, W_2^s and $W_{2,h}^k$ are Sobolev spaces of functions with continuous and discrete argument, respectively, and C is a constant which does not depend on u and h. For the parabolic case, typical estimates are of the form

$$\|u-v\|_{W_{2,h\tau}^{k,k/2}} \leq C(h+\sqrt{\tau})^{s-k}\|u\|_{W_2^{s,s/2}}, \quad s>k,$$

Dejan R. Bojović
Faculty of Science, University of Kragujevac, R. Domanovića 12, 34000 Kragujevac, Serbia,
e-mail: bojovicd@ptt.rs

Boško S. Jovanović
Faculty of Mathematics, University of Belgrade, Studentski trg 16, 11000 Belgrade, Serbia,
e-mail: bosko@matf.bg.ac.rs

W. Gautschi et al. (eds.), *Approximation and Computation: In Honor of Gradimir V. Milovanović*, Springer Optimization and Its Applications 42,
DOI 10.1007/978-1-4419-6594-3_18,

where τ is the time step. In the case of equations with variable coefficients the constant C in the error bounds depends on the norm of the coefficients (see, for example, [1,2,6,17]).

One interesting class of parabolic problems models processes in heat-conducting media with concentrated capacity, in which the heat capacity coefficient contains a Dirac delta function, or equivalently, the jump of the heat flow in the singular point is proportional to the time-derivative of the temperature [13]. Such problems are nonstandard, and the classical tools of the theory of finite difference schemes are difficult to apply to their convergence analysis.

In this paper a finite difference scheme, approximating the one-dimensional initial-boundary value problem for the heat equation with concentrated capacity and time-dependent operator, is derived. A special energetic Sobolev norm (corresponding to the norm $W_2^{2,1}$ for the classical heat conduction problem) is constructed. In this norm, a convergence rate estimate, compatible with the smoothness of the solution of the boundary value problem, is obtained.

An analogous result for a parabolic problem with concentrated capacity and a constant coefficient multiplying the spatial derivative is obtained in [8]. Note that the convergence to classical solutions is studied in [4] and [19]. A parabolic problem with a variable operator (without a Dirac delta function) is studied in [1] and [2]. The convergence of the finite difference method in the $\widetilde{W}_2^{1,1/2}$ norm for a parabolic problem with concentrated capacity and a spatially variable operator is proved in [3]. The convergence of difference schemes for a hyperbolic problem with concentrated mass has been studied in [9].

2 Preliminary Results

Let H be a real separable Hilbert space endowed with inner product $(\cdot,\cdot)$ and norm $\|\cdot\|$, and S an unbounded self-adjoint positive definite linear operator with domain $D(S)$ dense in H. It is easy to see that the product $(u,v)_S = (Su,v)$ $(u,v \in D(S))$ satisfies the axioms of inner product. The closure of $D(S)$ in the norm $\|u\|_S = (u,u)_S^{1/2}$ is a Hilbert space $H_S \subset H$. The inner product (u,v) continuously extends to $H_S^* \times H_S$, where $H_S^* = H_{S^{-1}}$ is the dual space for H_S. The spaces H_S, H, and $H_{S^{-1}}$ form a Gelfand triple $H_S \subset H \subset H_{S^{-1}}$ with continuous imbeddings. The operator S extends to the map $S: H_S \to H_S^*$. There exists an unbounded self-adjoint positive definite linear operator $S^{1/2}$ such that $D(S^{1/2}) = H_S$ and $(u,v)_S = (Su,v) = (S^{1/2}u, S^{1/2}v)$. We also define the Sobolev spaces $W_2^s(a,b;H)$, $W_2^0(a,b;H) = L_2(a,b;H)$ of the functions $u = u(t)$ mapping the interval $(a,b) \subset \mathbb{R}$ into H (see [12,20]).

Let A and B be unbounded self-adjoint positive definite linear operators, $A = A(t)$, $B \neq B(t)$, in the Hilbert space H, in general noncommutative, with $D(A)$ dense in H and $H_A \subset H_B$. We consider the following abstract Cauchy problem (cf. [15,20]):

$$B\frac{\mathrm{d}u}{\mathrm{d}t} + Au = f(t), \quad 0 < t < T; \quad u(0) = u_0, \tag{1}$$

where $f(t)$ and u_0 are given and $u(t)$ is an unknown function with values in H. Let us also assume that $A_0 \leq A(t) \leq cA_0$, where $c = \text{const} > 1$ and A_0 is a constant self-adjoint positive definite linear operator in H. We also assume that $A(t)$ is a decreasing operator in the variable t:

$$\left(\frac{\mathrm{d}A(t)}{\mathrm{d}t}u,u\right) < 0, \quad \forall u \in H. \tag{2}$$

The following proposition holds.

Lemma 1. *The solution of the problem* (1) *satisfies the a priori estimate:*

$$\int_0^T \left(\|Au(t)\|_{B^{-1}}^2 + \left\|\frac{\mathrm{d}u(t)}{\mathrm{d}t}\right\|_B^2\right)\mathrm{d}t \leq C\left(\|u_0\|_{A_0}^2 + \int_0^T \|f(t)\|_{B^{-1}}^2\,\mathrm{d}t\right), \tag{3}$$

provided that $u_0 \in H_A$ *and* $f \in L_2(0,T;H_{B^{-1}})$.

Proof. It follows from the abstract theory of parabolic initial-value problems (see [15, 20]) that for $u_0 \in H_B$ and $f \in L_2(0,T;H_{A_0^{-1}})$ the problem (1) has a unique solution $u \in L_2(0,T;H_{A_0})$ with $\mathrm{d}u/\mathrm{d}t \in L_2(0,T;H_{BA_0^{-1}B})$. We take the inner product of (1) with $2\mathrm{d}u/\mathrm{d}t$, and estimate the right-hand side by the Cauchy–Schwarz inequality:

$$2\left\|\frac{\mathrm{d}u}{\mathrm{d}t}\right\|_B^2 + 2\left(Au,\frac{\mathrm{d}u}{\mathrm{d}t}\right) = 2\left(f,\frac{\mathrm{d}u}{\mathrm{d}t}\right) \leq \left\|\frac{\mathrm{d}u}{\mathrm{d}t}\right\|_B^2 + \|f\|_{B^{-1}}^2.$$

By (2) this implies that

$$\frac{\mathrm{d}}{\mathrm{d}t}(\|u\|_A^2) \leq 2\left(Au,\frac{\mathrm{d}u}{\mathrm{d}t}\right) \quad \forall u \in H.$$

Furthermore, we have

$$2\left\|\frac{\mathrm{d}u}{\mathrm{d}t}\right\|_B^2 + \frac{\mathrm{d}}{\mathrm{d}t}(\|u\|_A^2) \leq \left\|\frac{\mathrm{d}u}{\mathrm{d}t}\right\|_B^2 + \|f\|_{B^{-1}}^2.$$

Integration with respect to t gives

$$\int_0^T \left\|\frac{\mathrm{d}u}{\mathrm{d}t}\right\|_B^2 \mathrm{d}t + \|u(T)\|_{A(T)}^2 \leq \|u_0\|_{A(0)}^2 + \int_0^T \|f(t)\|_{B^{-1}}^2\,\mathrm{d}t. \tag{4}$$

We also take the inner product of (1) with $2B^{-1}Au$ and obtain

$$2\|Au\|_{B^{-1}}^2 + \frac{\mathrm{d}}{\mathrm{d}t}(\|u\|_A^2) \leq \|Au\|_{B^{-1}}^2 + \|f\|_{B^{-1}}^2.$$

Again, integration with respect to t gives

$$\int_0^T \|Au(t)\|_{B^{-1}}^2\,\mathrm{d}t + \|u(T)\|_{A(T)}^2 \leq \|u_0\|_{A(0)}^2 + \int_0^T \|f(t)\|_{B^{-1}}^2\,\mathrm{d}t. \tag{5}$$

Finally, from (4) and (5) we deduce the a priori estimate (3). □

Analogous results hold for operator-difference schemes. Let H_h be a finite-dimensional real Hilbert space with inner product $(\cdot,\cdot)_h$ and norm $\|\cdot\|_h$. Let $A_h = A_h(t)$ and $B_h \neq B_h(t)$ be self-adjoint positive linear operators defined on H_h, in general noncommutative. By H_{S_h}, where $S_h = S_h^* > 0$, we denote the space with inner product $(y,v)_{S_h} = (S_h y, v)_h$ and norm $\|y\|_{S_h} = (S_h y, y)_h^{1/2}$.

Let ω_τ be a uniform mesh on $(0,T)$ with stepsize $\tau = T/m$, $\omega_\tau^- = \omega_\tau \cup \{0\}$, $\omega_\tau^+ = \omega_\tau \cup \{T\}$, and $\overline{\omega}_\tau = \omega_\tau \cup \{0,T\}$. In the sequel we shall use standard notation from the theory of difference schemes [16, 17]. In particular we set

$$v_{\bar{t}} = v_{\bar{t}}(t) = \frac{v(t) - v(t-\tau)}{\tau}, \quad v_t = v_t(t) = \frac{v(t+\tau) - v(t)}{\tau} = v_{\bar{t}}(t+\tau).$$

We will consider the simplest implicit operator-difference scheme

$$B_h v_{\bar{t}} + A_h v = \varphi(t), \ \ t \in \omega_\tau^+; \quad v(0) = v_0, \tag{6}$$

where v_0 is a given element of H_h, $\varphi(t)$ is known, and $v(t)$ is an unknown mesh function with values in H_h. Analogously as in the previous case, we assume that $A_{0h} \leq A_h(t) \leq cA_{0h}$, where $c = \text{const} > 1$ and A_{0h} is a constant self-adjoint positive linear operator in H_h. Also, we assume that $((A_h(t+\tau) - A_h(t))u, u) < 0$. The following analogue of Lemma 1 is true (cf. [6, 7]).

Lemma 2. *For the solution of the problem* (6) *the following estimate holds:*

$$\tau \sum_{t\in\overline{\omega}_\tau}{}' \|A_h v(t)\|_{B_h^{-1}}^2 + \tau \sum_{t\in\omega_\tau^+} \|v_{\bar{t}}(t)\|_{B_h}^2 \leq C\Big(\|v_0\|_{A_{0h}}^2 + \tau\|A_h v_0\|_{B_h^{-1}}^2 + \tau \sum_{t\in\omega_\tau^+} \|\varphi(t)\|_{B_h^{-1}}^2\Big),$$

where we denote

$$\sum_{t\in\overline{\omega}_\tau}{}' w(t) = \frac{w(0)}{2} + \sum_{t\in\omega_\tau} w(t) + \frac{w(T)}{2}.$$

We also need the following result (see [14]):

Lemma 3. *For $f \in W_2^1(0,1)$ and $\varepsilon \in (0,1)$ the following estimate holds:*

$$\|f\|_{L_2(0,\varepsilon)} \leq C\varepsilon^{1/2} \|f\|_{W_2^1(0,1)}.$$

3 Heat Equation with Concentrated Capacity

Let us consider the initial-boundary-value problem for the heat equation in the presence of a concentrated capacity at the interior point $x = \xi$:

$$[1+K\delta(x-\xi)]\frac{\partial u}{\partial t} - \frac{\partial}{\partial x}\left(a(x,t)\frac{\partial u}{\partial x}\right) = f(x,t), \quad (x,t) \in Q,$$
$$u(0,t) = 0, \;\; u(1,t) = 0, \quad 0 < t < T, \tag{7}$$
$$u(x,0) = u_0(x), \quad x \in (0,1),$$

where $Q = (0,1) \times (0,T)$, K is a positive constant, $\delta(x)$ is the Dirac delta generalized function, and equality is considered in the sense of distributions [18]. Our aim is to investigate the singularity of the solution of the problem (7) caused by the presence of the singular coefficient $K\delta(x-\xi)$; therefore, we restrict ourselves to the simplest Dirichlet boundary conditions. An analogous problem with constant coefficient a is considered in [8].

In a standard manner one obtains the following weak form of the initial-boundary-value problem (7):

$$\int_0^T\int_0^1 \frac{\partial u}{\partial t} v\,\mathrm{d}x\mathrm{d}t + \int_0^T \frac{\partial u}{\partial t}(\xi,t)v(\xi,t)\,\mathrm{d}t + \int_0^T\int_0^1 a(x,t)\frac{\partial u}{\partial x}\frac{\partial v}{\partial x}\,\mathrm{d}x\mathrm{d}t = \int_0^T\int_0^1 fv\,\mathrm{d}x\mathrm{d}t,$$
$$\forall v \in \overset{\circ}{W}{}_2^{1,0}(Q) = \{v \in W_2^{1,0}(Q) \,:\, v = 0 \text{ on } \{0,1\} \times (0,T)\}. \tag{8}$$

The same weak form (8) corresponds to the following initial-boundary-value problem:

$$\frac{\partial u}{\partial t} - \frac{\partial}{\partial x}\left(a(x,t)\frac{\partial u}{\partial x}\right) = f(x,t), \quad (x,t) \in Q_1 \cup Q_2,$$
$$[u]_{x=\xi} \equiv u(\xi+0,t) - u(\xi-0,t) = 0, \quad \left[a\frac{\partial u}{\partial x}\right]_{x=\xi} = K\frac{\partial u}{\partial t}(\xi,t), \tag{9}$$
$$u(0,t) = 0, \;\; u(1,t) = 0, \;\; 0 < t < T, \quad u(x,0) = u_0(x), \;\; x \in (0,1),$$

where $Q_1 = (0,\xi) \times (0,T)$ and $Q_2 = (\xi,1) \times (0,T)$. In this sense, the initial-boundary-value problems (7) and (9) are equivalent.

If $H = L_2(0,1)$ it is easy to see that the initial-boundary-value problem (7) can be written in the form (1), where

$$Au = -\frac{\partial}{\partial x}\left(a(x,t)\frac{\partial u}{\partial x}\right), \quad Bu = [1+K\delta(x-\xi)]u(x,t),$$

or

$$(Av,w) = \int_0^1 a(x,t)v'(x)w'(x)\,\mathrm{d}x,$$

for $v,w \in \overset{\circ}{W}{}_2^1(0,1)$ and

$$(Bv,w) = \int_0^1 v(x)w(x)\,\mathrm{d}x + Kv(\xi)w(\xi).$$

Assuming that $a \in L_\infty(Q)$, $0 < c_1 \le a(x,t) \le c_2$, a.e. in Q, we immediately obtain

$$\|w\|_A^2 = \int_0^1 a(x,t)|w'(x)|^2\,\mathrm{d}x \asymp \|w\|^2_{W_2^1(0,1)}, \quad w \in \overset{\circ}{W}{}_2^1(0,1),$$

so we can put $H_A = \overset{\circ}{W}{}_2^1(0,1)$ and $H_{A^{-1}} = W_2^{-1}(0,1)$.

The operator B is defined on the subset H_B of functions in $L_2(0,1)$ with finite norm

$$\|w\|_B^2 = \|w\|^2_{L_2(0,1)} + Kw^2(\xi) \asymp \|w\|^2_{L_2(0,1)} + w^2(\xi) = \|w\|^2_{\widetilde{L}_2(0,1)},$$

so we can put $H_B = \widetilde{L}_2(0,1) =$ closure of the set $C[0,1]$ in the norm $\|\cdot\|_{\widetilde{L}_2(0,1)}$. Obviously, $H_A = \overset{\circ}{W}{}_2^1(0,1) \subset C[0,1] \subset \widetilde{L}_2(0,1) = H_B$. The "negative" norm $\|w\|_{B^{-1}}$ satisfies the relation

$$\|w\|_{B^{-1}} = (B^{-1}w,w)^{1/2} = \sup_{0 \neq v \in H_B} \frac{|(w,v)|}{\|v\|_B}.$$

4 The Difference Problem

Let ω_h be a uniform mesh on $(0,1)$ with stepsize $h = 1/n$, $\omega_h^- = \omega_h \cup \{0\}$, and $\overline{\omega}_h = \omega_h \cup \{0,1\}$. Suppose for simplicity that ξ is a rational number. Then one can choose the step h so that $\xi \in \omega_h$. Also, we assume that the condition $c_1 h^2 \le \tau \le c_2 h^2$ is satisfied. Define finite differences in the usual way:

$$\begin{aligned}
v_{\bar{t}}(x,t) &= \frac{v(x,t) - v(x,t-\tau)}{\tau} = v_t(x,t-\tau),\\
v_{\bar{x}}(x,t) &= \frac{v(x,t) - v(x-h,t)}{h} = v_x(x-h,t),\\
v_{x\bar{x}}(x,t) &= (v_x(x,t))_{\bar{x}} = \frac{v(x+h,t) - 2v(x,t) + v(x-h,t)}{h^2}.
\end{aligned}$$

The problem (7) can be approximated on the mesh $\overline{Q}_{h\tau} = \overline{\omega}_h \times \overline{\omega}_\tau$ by the following implicit difference scheme with averaged right-hand side:

$$\begin{aligned}
&[1 + K\delta_h(x-\xi)]\, v_{\bar{t}} - (\tilde{a}v_x)_{\bar{x}} = T_x^2 T_t^- f, \quad (x,t) \in \omega_h \times \omega_\tau^+,\\
&v(0,t) = 0, \;\; v(1,t) = 0, \quad t \in \omega_\tau^+, \qquad\qquad (10)\\
&v(x,0) = u_0(x), \quad x \in \overline{\omega}_h,
\end{aligned}$$

where $\tilde{a}(x,t) = 0.5[a(x+0,t) + a(x+h-0,t)]$,

$$\delta_h(x-\xi) = \begin{cases} 0, & x \in \omega_h \setminus \{\xi\}, \\ 1/h, & x = \xi \end{cases}$$

is the mesh Dirac function, and T_x, T_t^- are Steklov averaging operators [18], defined as follows

$$T_x f(x,t) = T_x^- f(x+h/2,t) = T_x^+ f(x-h/2,t) = \frac{1}{h}\int_{x-h/2}^{x+h/2} f(x',t)\,\mathrm{d}x',$$

$$T_t^- f(x,t) = T_t^+ f(x,t-\tau) = \frac{1}{\tau}\int_{t-\tau}^{t} f(x,t')\,\mathrm{d}t'.$$

Notice that these operators commute and map derivatives into finite differences; for example,

$$T_x^2 \frac{\partial^2 u}{\partial x^2} = u_{x\bar{x}}, \quad T_t^- \frac{\partial u}{\partial t} = u_{\bar{t}}.$$

Let H_h denote the set of functions defined on the mesh $\overline{\omega}_h$ and equal to zero at $x=0$ and $x=1$. We define the following inner product and norms by:

$$(v,u)_h = h\sum_{x\in\omega_h} v(x)u(x), \quad \|v\|_h = \|v\|_{L_{2,h}} = (v,v)_h^{1/2}, \quad |[v\|_h = \left(h\sum_{x\in\omega_h^-} v^2(x)\right)^{1/2}.$$

The difference scheme (10) can be reduced to the form (6) by setting $A_h v = -(\tilde{a}v_x)_{\bar{x}}$ and $B_h v = [1 + K\delta_h(x-\xi)]\,v$. For each $v \in H_h$ we have

$$\|v\|_{A_h}^2 = (A_h v, v)_h \asymp |[v_x\|_h^2, \quad \|v\|_{B_h}^2 = \|v\|_h^2 + Kv^2(\xi),$$

$$\|v\|_{B_h^{-1}}^2 = (B_h^{-1}v,v)_h = h\sum_{x\in\omega\setminus\{\xi\}} v^2(x) + \frac{h^2}{K+h}v^2(\xi),$$

$$\|v\|_{\widetilde{W}_{2,h}^2}^2 = \|v_{x\bar{x}}\|_{B_h^{-1}}^2 + |[v_x\|_h^2 + \|v\|_{B_h}^2.$$

Note that the norms $\|A_h v\|_{B_h^{-1}}$ and $\|v\|_{\widetilde{W}_{2,h}^2}$ are equivalent [10].

We also define the discrete $\widetilde{W}_2^{2,1}$ norm by:

$$\|v\|_{\widetilde{W}_2^{2,1}(Q_{h\tau})}^2 = \tau\sum_{t\in\overline{\omega}_\tau}{}' \|v(\cdot,t)\|_{\widetilde{W}_{2,h}^2}^2 + \tau\sum_{t\in\omega_\tau^+} \|v_{\bar{t}}(\cdot,t)\|_{B_h}^2.$$

5 Convergence of the Difference Scheme

In this section we shall prove the convergence of the difference scheme (10) in the $\widetilde{W}_2^{2,1}$ norm. Let u be the solution of the boundary-value problem (7) and v the solution of the difference problem (10). The error $z = u - v$ satisfies the finite difference scheme

$$\begin{aligned} &[1+K\delta_h(x-\xi)]\,z_{\bar t}+A_h z=\varphi \text{ on } Q_{h\tau},\\ &z(0,t)=z(1,t)=0, \quad t\in\omega_\tau^+,\\ &z(x,0)=0, \quad x\in\overline{\omega}_h, \end{aligned} \tag{11}$$

where $\varphi=\varphi_1+\varphi_2$ with

$$\varphi_1=T_t^-\frac{\partial u}{\partial t}-T_x^2T_t^-\frac{\partial u}{\partial t}, \quad \varphi_2=T_x^2T_t^-\frac{\partial}{\partial x}\left(a\frac{\partial u}{\partial x}\right)-(\tilde a u_x)_{\bar x}.$$

In the sequel we shall assume that

$$a\in W_2^{3,3/2}(Q_1)\cap W_2^{3,3/2}(Q_2)\subset L_\infty(Q), \quad f\in W_2^{2,1}(Q)$$

and

$$u\in W_2^{4,2}(Q_1)\cap W_2^{4,2}(Q_2), \quad u(\xi,\cdot)\in W_2^2(0,T).$$

Also, we assume that $0<c_1\le a(x,t)\le c_2$ and that $a(x,t)$ is a decreasing function in the variable t.

Using Lemma 2, we directly obtain the following a priori estimate for the solution of the difference scheme (11):

$$\|z\|_{\widetilde W_2^{2,1}(Q_{h\tau})}\le\left(\tau\sum_{t\in\omega_\tau^+}\|\varphi(\cdot,t)\|^2_{B_h^{-1}}\right)^{1/2}. \tag{12}$$

Therefore, in order to estimate the rate of convergence of the difference scheme (10), it is sufficient to estimate the right-hand side of the inequality (12).

The term φ_1 is estimated in [8]:

$$\left(\tau\sum_{t\in\omega_\tau^+}\|\varphi_1(\cdot,t)\|^2_{B_h^{-1}}\right)^{1/2}\le Ch^2\Big(\|u\|_{W_2^{4,2}(Q_1)}+\|u\|_{W_2^{4,2}(Q_2)}+\|u(\xi,\cdot)\|_{W_2^2(0,T)}\Big). \tag{13}$$

The following estimate for φ_2 directly follows from [1]:

$$\begin{aligned}\left(h\tau\sum_{x\in\omega_h\setminus\{\xi\}}\sum_{t\in\omega_\tau^+}\varphi_2^2(x,t)\right)^{1/2}&\le Ch^2\\ &\times\left(\|a\|_{W_2^{3,3/2}(Q_1)}\|u\|_{W_2^{4,2}(Q_1)}+\|a\|_{W_2^{3,3/2}(Q_2)}\|u\|_{W_2^{4,2}(Q_2)}\right).\end{aligned} \tag{14}$$

At the point $x=\xi$ we have that $\varphi_2=\eta_{\bar x}$, where

$$\eta=T_x^+T_t^-\left(a\frac{\partial u}{\partial x}\right)-\tilde a u_x$$

and

$$\frac{h^2}{K+h}\varphi_2^2(\xi,t) \le C(|\eta(\xi,t)|^2 + |\eta(\xi-h,t)|^2).$$

We decompose $\eta = \eta_1 + \eta_2 + \eta_3$, where

$$\eta_1 = T_x^+ T_t^- \left(a\frac{\partial u}{\partial x}\right) - (T_x^+ T_t^- a)\left(T_x^+ T_t^- \frac{\partial u}{\partial x}\right),$$
$$\eta_2 = (T_x^+ T_t^- a - \tilde{a})\left(T_x^+ T_t^- \frac{\partial u}{\partial x}\right),$$
$$\eta_3 = \tilde{a}\left(T_x^+ T_t^- \frac{\partial u}{\partial x} - u_x\right).$$

Further, we decompose $\eta_1 = \eta_{1,1} + \eta_{1,2} + \eta_{1,3}$, where

$$\eta_{1,1}(\xi,t) = \frac{1}{2h^2\tau^2}\int_\xi^{\xi+h}\int_\xi^{\xi+h}\int_{t-\tau}^{t}\int_{t-\tau}^{t}\int_{t''}^{t'} \frac{\partial a(x',t''')}{\partial t} \times \left[\frac{\partial u(x',t'')}{\partial x} - \frac{\partial u(x'',t')}{\partial x}\right] \mathrm{d}t'''\mathrm{d}t''\mathrm{d}t'\mathrm{d}x''\mathrm{d}x',$$

$$\eta_{1,2}(\xi,t) = \frac{1}{2h^2\tau^2}\int_\xi^{\xi+h}\int_\xi^{\xi+h}\int_{t-\tau}^{t}\int_{t-\tau}^{t}\int_{t'}^{t''} [a(x',t'') - a(x'',t'')] \times \frac{\partial^2 u(x',t''')}{\partial t\partial x} \mathrm{d}t'''\mathrm{d}t''\mathrm{d}t'\mathrm{d}x''\mathrm{d}x',$$

$$\eta_{1,3}(\xi,t) = \frac{1}{2h^2\tau^2}\int_\xi^{\xi+h}\int_\xi^{\xi+h}\int_{t-\tau}^{t}\int_{t-\tau}^{t}\int_{x'}^{x''}\int_{x'}^{x''} \frac{\partial a(x''',t'')}{\partial x} \times \frac{\partial^2 u(x'''',t')}{\partial x^2} \mathrm{d}x''''\mathrm{d}x'''\mathrm{d}t''\mathrm{d}t'\mathrm{d}x''\mathrm{d}x'.$$

From the first integral representation we have

$$|\eta_{1,1}(\xi,t)| \le Ch^{1/2}\left\|\frac{\partial a}{\partial t}\right\|_{L_2(g)}\left\|\frac{\partial u}{\partial x}\right\|_{\mathbb{C}(\overline{g})},$$

where $g = (\xi,\xi+h)\times(t-\tau,t)$. Summing over the mesh ω_τ^+ we get

$$\tau\sum_{t\in\omega_\tau^+}|\eta_{1,1}(\xi,t)|^2 \le Ch^3\left\|\frac{\partial a}{\partial t}\right\|^2_{L_2(Q_2^h)}\left\|\frac{\partial u}{\partial x}\right\|^2_{\mathbb{C}(\overline{Q}_2)},$$

where $Q_2^h = (\xi,\xi+h)\times(0,T)$. Applying Lemma 3, we have

$$\left\|\frac{\partial a}{\partial t}\right\|_{L_2(Q_2^h)} \le Ch^{1/2}\left(\left\|\frac{\partial a}{\partial t}\right\|_{L_2(Q_2)} + \left\|\frac{\partial^2 a}{\partial t\partial x}\right\|_{L_2(Q_2)}\right) \le Ch^{1/2}\|a\|_{W_2^{3,3/2}(Q_2)}.$$

From the preceding inequality and the imbedding $W_2^{3,3/2}(Q_2) \subset \mathbb{C}(\overline{Q}_2)$ we obtain

$$\tau \sum_{t\in\omega_\tau^+} |\eta_{1,1}(\xi,t)|^2 \le Ch^4 \|a\|^2_{W_2^{3,3/2}(Q_2)} \|u\|^2_{W_2^{4,2}(Q_2)}.$$

By the same technique we can estimate the terms $\eta_{1,2}$ and $\eta_{1,3}$. We obtain

$$\tau \sum_{t\in\omega_\tau^+} |\eta_1(\xi,t)|^2 \le Ch^4 \|a\|^2_{W_2^{3,3/2}(Q_2)} \|u\|^2_{W_2^{4,2}(Q_2)}. \tag{15}$$

Furthermore,

$$\eta_2 = \left(T_x^+ T_t^- a - \tilde{a}\right) \left(T_x^+ T_t^- \frac{\partial u}{\partial x}\right) = \eta_{2,1} \left(T_x^+ T_t^- \frac{\partial u}{\partial x}\right).$$

The term $\eta_{2,1}$ is a bounded linear functional of the argument $a \in W_2^{2,1}(g)$. Further, $\eta_{2,1} = 0$ whenever a is a polynomial of degree one in x and a constant function in t. Applying the Bramble–Hilbert Lemma [5], we get

$$|\eta_{2,1}(\xi,t)| \le Ch^{1/2} |a|_{W_2^{2,1}(g)} \quad \text{and} \quad |\eta_2(\xi,t)| \le Ch^{1/2} |a|_{W_2^{2,1}(g)} \left\| \frac{\partial u}{\partial x} \right\|_{\mathbb{C}(\overline{Q}_2)}.$$

Summing over the mesh ω_τ^+, we get

$$\tau \sum_{t\in\omega_\tau^+} |\eta_2(\xi,t)|^2 \le Ch^3 |a|^2_{W_2^{2,1}(Q_2^h)} \left\| \frac{\partial u}{\partial x} \right\|^2_{\mathbb{C}(\overline{Q}_2)}.$$

Applying Lemma 3, we have

$$|a|_{W_2^{2,1}(Q_2^h)} \le Ch^{1/2} \|a\|_{W_2^{3,3/2}(Q_2)},$$

whereby

$$\tau \sum_{t\in\omega_\tau^+} |\eta_2(\xi,t)|^2 \le Ch^4 \|a\|^2_{W_2^{3,3/2}(Q_2)} \|u\|^2_{W_2^{4,2}(Q_2)}. \tag{16}$$

Representing the term η_3 as:

$$\eta_3(\xi,t) = \frac{1}{2h\tau} [a(\xi+0,t) + a(\xi+h,t)] \int_\xi^{\xi+h} \int_{t-\tau}^{t} \int_t^{t'} \frac{\partial^2 u(x',t'')}{\partial x \partial t} \,\mathrm{d}t''\mathrm{d}t'\mathrm{d}x',$$

we have

$$|\eta_3(\xi,t)| \le h^{1/2} \|a\|_{\mathbb{C}(\overline{Q}_2)} \left\| \frac{\partial^2 u}{\partial x \partial t} \right\|_{L_2(g)}.$$

Summing over the mesh ω_τ^+, we get

$$\tau \sum_{t\in\omega_\tau^+} |\eta_3(\xi,t)|^2 \le Ch^3 \|a\|^2_{\mathbb{C}(\overline{Q}_2)} \left\| \frac{\partial^2 u}{\partial x \partial t} \right\|^2_{L_2(Q_2^h)}.$$

Applying Lemma 3, we get

$$\left\| \frac{\partial^2 u}{\partial x \partial t} \right\|_{L_2(Q_2^h)} \le Ch^{1/2} \|u\|_{W_2^{4,2}(Q_2)},$$

whereby

$$\tau \sum_{t\in\omega_\tau^+} |\eta_3(\xi,t)|^2 \le Ch^4 \|a\|^2_{W_2^{3,3/2}(Q_2)} \|u\|^2_{W_2^{4,2}(Q_2)}. \tag{17}$$

From (15), (16), and (17) we obtain

$$\tau \sum_{t\in\omega_\tau^+} |\eta(\xi,t)|^2 \le Ch^4 \|a\|^2_{W_2^{3,3/2}(Q_2)} \|u\|^2_{W_2^{4,2}(Q_2)}. \tag{18}$$

By the same technique we obtain

$$\tau \sum_{t\in\omega_\tau^+} |\eta(\xi-h,t)|^2 \le Ch^4 \|a\|^2_{W_2^{3,3/2}(Q_1)} \|u\|^2_{W_2^{4,2}(Q_1)}. \tag{19}$$

From (14), (18), and (19) we get

$$\left(\tau \sum_{t\in\omega_\tau^+} \|\varphi_2(\cdot,t)\|^2_{B_h^{-1}} \right)^{1/2} \le Ch^2 \times \left(\|a\|_{W_2^{3,3/2}(Q_1)} \|u\|_{W_2^{4,2}(Q_1)} + \|a\|_{W_2^{3,3/2}(Q_2)} \|u\|_{W_2^{4,2}(Q_2)} \right). \tag{20}$$

Finally, from the a priori estimate (12) and estimates (13) and (20) we obtain the following main result of this paper:

Theorem 1. *The solution of the difference scheme* (10) *converges in* $\widetilde{W}_2^{2,1}(Q_{h\tau})$ *to the solution of the differential problem* (7)*, and the following estimate is valid:*

$$\|u-v\|_{\widetilde{W}_2^{2,1}(Q_{h\tau})} \le Ch^2 \left(\|a\|_{W_2^{3,3/2}(Q_1)} + \|a\|_{W_2^{3,3/2}(Q_2)} + 1 \right) \times \left(\|u\|_{W_2^{4,2}(Q_1)} + \|u\|_{W_2^{4,2}(Q_2)} + \|u(\xi,\cdot)\|_{W_2^2(0,T)} \right).$$

This estimate is compatible with the smoothness of the coefficient and the solution of the differential problem (7).

Acknowledgements This research was supported by the Ministry of Science of the Republic of Serbia under project # 144005A.

References

1. Bojović, D.: Convergence of finite difference method for parabolic problem with variable operator. Lect. Notes Comp. Sci. **1988**, 110–116 (2001)
2. Bojović, D.: Convergence in $W_2^{1,1/2}$ norm of finite difference method for parabolic problem. Comput. Methods Appl. Math. **3**, No 1, 45–58 (2003)
3. Bojović, D., Jovanović, B.S.: Convergence of finite difference method for the parabolic problem with concentrated capacity and variable operator. J. Comp. Appl. Math. **189**, 286–303 (2006)
4. Braianov, I.: Convergence of a Crank-Nicolson difference scheme for heat equations with interface in the heat flow and concentrated heat capacity. Lect. Notes Comp. Sci. **1196**, 58–65 (1997)
5. Bramble, J.H., Hilbert, S.R.: Bounds for a class of linear functionals with applications to Hermite interpolation. Numer. Math. **16**, 362–369 (1971)
6. Jovanović, B.S.: Finite difference method for boundary value problems with weak solutions. Posebna izdanja Mat. Instituta 16, Belgrade (1993)
7. Jovanović, B.S., Matus, P.P.: On the strong stability of operator-difference schemes in time-integral norms. Comput. Methods Appl. Math. **1**, 72–85 (2001)
8. Jovanović, B.S., Vulkov, L.G.: On the convergence of finite difference schemes for the heat equation with concentrated capacity. Numer. Math. **89**, No 4, 715–734 (2001)
9. Jovanović, B.S., Vulkov, L.G.: On the convergence of difference schemes for hyperbolic problems with concentrated data. SIAM J. Numer. Anal. **41**, No 2, 516–538 (2003)
10. Jovanović, B.S., Vulkov, L.G.: On the convergence of difference schemes for parabolic problems with concentrated data. Intern. J. Numer. Anal. Model. **5**, No 3, 386–406 (2008)
11. Lazarov, R.D., Makarov, V.L., Samarskii, A.A.: Application of exact difference schemes for constructing and investigating difference schemes on generalized solutions. Math. Sbornik **117**, 469–480 (1982) (Russian)
12. Lions, J.L., Magenes, E.: Non Homogeneous Boundary Value Problems and Applications. Springer-Verlag, Berlin and New York (1972)
13. Lykov, A.V.: Heat and Mass Transfer. Nauka, Moscow (1989) (Russian)
14. Oganesyan, L.A., Rukhovets, L.A.: Variational-Difference Method for Solving Eliptic Equations. AN Arm. SSR (1979) (Russian)
15. Renardy, M., Rogers, R.C.: An Introduction to Partial Differential Equations. Springer-Verlag, Berlin and New York (1993)
16. Samarskii, A.A.: Theory of Difference Schemes. Nauka, Moscow (1989) (Russian; English edition: Pure and Appl. Math., Vol. **240**, Marcel Dekker, Inc. (2001))
17. Samarskii, A.A., Lazarov, R.D., Makarov, V.L.: Difference Schemes for Differential Equations with Generalized Solutions. Vysshaya Shkola, Moscow (1987) (Russian)
18. Vladimirov, V.S.: Equations of Mathematical Physics. Nauka, Moscow (1988) (Russian)
19. Vulkov, L.G.: Applications of Steklov-type eigenvalue problems to convergence of difference schemes for parabolic and hyperbolic equations with dynamical boundary conditions. Lect. Notes Comp. Sci. **1196**, 1309–1364 (1997)
20. Wloka, J.: Partial Differential Equations. Cambridge University Press, Cambridge (1987)

Adaptive Finite Element Approximation of the Francfort–Marigo Model of Brittle Fracture

Siobhan Burke, Christoph Ortner and Endre Süli

Dedicated to Gradimir V. Milovanović on his 60th birthday

1 Introduction

Beginning with the work of Francfort and Marigo [13], the variational theory of quasi-static brittle fracture mechanics has experienced a rapid and successful development. Upon recasting Griffith's idea of balancing energy release rate with a fictitious surface energy [15] as an energy minimization problem, Francfort and Marigo were able to formulate a model that was free of the usual constraints of fracture mechanics such as a predefined and piecewise smooth crack path. With the help of the theory of free-discontinuity problems [2], this model was soon shown to be well-posed in a surprisingly general setting [9, 10, 12]. We briefly review the model in Sect. 1.1.

The model of Francfort and Marigo is posed in terms of the minimization of a highly irregular energy functional, which is also used in image segmentation, where it is known as the Mumford-Shah functional [17]. Several methods have been proposed in the literature, which regularize this energy in order to render the problem

Siobhan Burke
Mathematical Institute, 24–29 St. Giles', Oxford, OX1 3LB, UK
e-mail: burke@maths.ox.ac.uk

Christoph Ortner
Mathematical Institute, 24–29 St. Giles', Oxford, OX1 3LB, UK
e-mail: ortner@maths.ox.ac.uk

Endre Süli
Mathematical Institute, 24–29 St. Giles', Oxford, OX1 3LB, UK
e-mail: suli@maths.ox.ac.uk

W. Gautschi et al. (eds.), *Approximation and Computation: In Honor of Gradimir V. Milovanović*, Springer Optimization and Its Applications 42,
DOI 10.1007/978-1-4419-6594-3_19,

accessible to numerical simulation [5]. Such methods typically use the theory of Γ-convergence to construct approximating functionals whose minimizers converge to those of the original functional.

In our experience, the Ambrosio-Tortorelli approximation [1,5] is one of the most promising approaches. A particularly nice feature of the Ambrosio-Tortorelli functional is that its minimization can be reduced to the solution of elliptic boundary value problems that are straightforward to discretize, for example, by a finite element method. This approach has been used successfully by Bourdin, Francfort and Marigo [4] and Bourdin [3] for the simulation of problems that are usually inaccessible to classical methods. A brief review of the Ambrosio-Tortorelli approximation is given in Sect. 1.2.

The Ambrosio-Tortorelli approximation can be understood as a phase field model for the crack set. To resolve the phase field variable, the mesh near the crack set has to be significantly finer than the mesh that would be required to resolve the elastic deformation away from the crack set. As we do not know the crack set in advance, it is a natural idea to use an adaptive finite element method.

We shall formulate an optimization algorithm that is well-defined in the function space in which the minimization of the Ambrosio-Tortorelli functional is posed. Each step of the algorithm requires the solution of a linear self-adjoint second-order elliptic boundary value problem by an adaptive finite element algorithm. The adaptive algorithm can be controlled by adjusting the refinement tolerance, to yield a convergent adaptive optimization scheme with guaranteed convergence to a critical point of the Ambrosio-Tortorelli functional.

In order to lay out the main ideas, our analysis in this work is restricted to linearized elasticity in anti-plane displacement, and to linear finite elements. We will extend our results to more general approximations and a wider range of models in future work.

1.1 The Francfort–Marigo Model of Brittle Fracture

In order to introduce the Francfort–Marigo model of brittle fracture, we briefly define the space of special functions of bounded variation [2]. Detailed knowledge of the properties of this function space is not necessary in order to follow the main ideas contained in the paper.

Let Ω be a domain in $\mathbb{R}^N$. For $p \in [1,\infty]$ we use $\mathrm{L}^p(\Omega)$ to denote the standard L^p-spaces and $\mathrm{H}^1(\Omega)$ to denote the standard Hilbertian Sobolev space of square-integrable functions on Ω whose distributional gradient is square-integrable on Ω. The N-dimensional Lebesgue and Hausdorff measures are denoted by $\mathcal{L}^N$ and $\mathcal{H}^N$, respectively.

We say that a function $f \in \mathrm{L}^1(\Omega)$ is a special function of bounded variation (or $f \in \mathrm{SBV}(\Omega)$) if its distributional gradient Du is a measure of bounded variation that has the form:

$$Df = \nabla f \mathcal{L}^N + (f^+(x) - f^-(x)) \otimes \nu_f(x) \mathcal{H}^{N-1} \lfloor J(f).$$

Here, $\nabla f \in \mathrm{L}^1(\Omega)^N$ is called the approximate gradient of f, $J(f)$ the jump set, ν_f is the unit normal to $J(f)$, and $f^{\pm}$ are the inner and outer traces of f on $J(u)$ with respect to ν_f.

The crack-free reference configuration of a linearly elastic body is denoted by a bounded Lipschitz domain $\Omega \subset \mathbb{R}^N$. For each $u \in \mathrm{SBV}(\Omega)$, and for each Hausdorff measurable set Γ, the energy functional of the Francfort–Marigo model of brittle fracture [13] is defined by:

$$E(u,\Gamma) := \begin{cases} \|\nabla u\|^2_{\mathrm{L}^2(\Omega)} + \mathcal{H}^{N-1}(\Gamma), & \text{if } \mathcal{H}^{N-1}(J(u)\setminus\Gamma) = 0, \\ +\infty, & \text{otherwise.} \end{cases}$$

The energy functional $E(u,\Gamma)$ reflects Griffith's principle that, to create a crack, one has to *spend* an amount of elastic energy that is proportional to the area of the crack created [15] (here, the constant of proportionality is set to one). The crack set Γ and the jump set $J(u)$ are decoupled in the definition of the total energy in order to be able to impose irreversibility of the crack evolution.

We wish to study how the body evolves in time under the action of a varying load $g(t)$, $t \in [0,T]$, with $T > 0$, which is applied on an open subset $\Omega_{\mathrm{D}} \subset \Omega$ of positive N-dimensional Lebesgue measure. We assume that $g \in \mathrm{L}^\infty(0,T;\mathrm{W}^{1,\infty}(\Omega)) \cap \mathrm{W}^{1,1}(0,T;\mathrm{H}^1(\Omega))$, and we define

$$\mathcal{A}(t) := \left\{u \in \mathrm{SBV}(\Omega) : u|_{\Omega_{\mathrm{D}}} = g(t)|_{\Omega_{\mathrm{D}}}\right\}, \quad t \in [0,T].$$

The fact that the Dirichlet boundary condition is imposed on a set of positive N-dimensional Lebesgue measure is mostly technical and ensures that the jump set on the Dirichlet boundary $\partial\Omega_{\mathrm{D}} \cap \Omega$ is well-defined. We call $\partial\Omega_{\mathrm{N}} := \partial\Omega \setminus \partial\Omega_{\mathrm{D}}$ the Neumann boundary.

The Francfort–Marigo model of irreversible brittle fracture is formulated as follows: Find a trajectory $(u(t),\Gamma(t))_{t\in[0,T]}$ such that the following conditions are satisfied:

1. **Irreversibility:**

$$\Gamma(s) \subset \Gamma(t) \quad \text{for all } \ s,t \in [0,T] \ \text{ such that } \ s \leq t;$$

2. **Global stability:**

$$E(u(t),\Gamma(t)) \leq E(v,J(v)\cup\Gamma(t)) \quad \text{for all } \ v \in \mathcal{A}(t);$$

3. **Conservation of energy:**

$$\frac{\mathrm{d}}{\mathrm{d}t}E(u(t),\Gamma(t)) = \int_\Omega \nabla\dot{g}(t)\cdot\nabla u(t)\,\mathrm{d}x.$$

Existence of solutions to this model, in its full generality, was first established in the paper of Francfort & Larsen [12].

1.2 The Ambrosio-Tortorelli Approximation

The ability to predict complicated crack paths is the greatest strength of the Francfort–Marigo model. However, this generality makes the numerical approximation of the model particularly challenging. A promising approach is to work with the Ambrosio-Tortorelli regularization of $E(u,\Gamma)$, which represents the crack set by a phase field variable v, and which is easily discretized using standard numerical methods. The regularized functional is chosen to approximate $E(u,\Gamma)$ in the sense of Γ-convergence [1,5]. Consequently, minimizers of the approximating functional converge to minimizers of $E(u,\Gamma)$ together with convergence of the minimized energy.

The Ambrosio-Tortorelli functional $I_\varepsilon : \mathrm{H}^1(\Omega;\mathbb{R}) \times \mathrm{H}^1(\Omega;[0,1]) \to \mathbb{R}$ is defined for $0 < \eta \ll \varepsilon \ll 1$ as follows:

$$I_\varepsilon(u,v) := \int_\Omega (v^2+\eta)|\nabla u|^2\,\mathrm{d}x + \int_\Omega \left[\frac{1}{4\varepsilon}(1-v)^2 + \varepsilon|\nabla v|^2\right]\mathrm{d}x.$$

The Ambrosio-Tortorelli functional regularizes the Francfort–Marigo model in space. In addition, we discretize the evolution in time. Let $0 = t_0 < t_1 < \cdots < t_K = T$ be a discretization of the time interval $[0,T]$, with $\Delta t := \max_{k=1,\dots,K}(t_k - t_{k-1})$. At time $t = t_0$, a crack field $v(0) := v(\cdot,0)$, with $0 \le v(x,0) \le 1$ for all $x \in \Omega$, is prescribed. The corresponding elastic field $u(0) := u(\cdot,0)$ is the (unique) minimizer of $I_\varepsilon(v,u)$ with $v = v(0)$ fixed, i.e., $u(0) = \operatorname{argmin}\{I_\varepsilon(\hat{u},v(0)) : \hat{u} \in \mathrm{H}^1(\Omega), \hat{u}|_{\Omega_\mathrm{D}} = g(0)\}$.

At subsequent times t_k, $k = 1,\dots,K$, we compute $(u_\varepsilon(t_k), v_\varepsilon(t_k))$ satisfying

$$\begin{aligned}(u_\varepsilon(t_k), v_\varepsilon(t_k)) \in \operatorname{argmin}\{I_\varepsilon(\hat{u},\hat{v}) \,:\; &\hat{u} \in \mathrm{H}^1(\Omega), \hat{u} = g(t_k) \text{ on } \Omega_\mathrm{D};\\ &\hat{v} \in \mathrm{H}^1(\Omega), \hat{v} \le v_\varepsilon(t_{k-1})\}.\end{aligned} \tag{1}$$

It was shown by Giacomini [14] that an evolution satisfying (1) converges in an appropriate sense, as $\Delta t, \varepsilon \to 0$, to a solution of the Francfort–Marigo model. We will, therefore, restrict our consideration to the problem of minimizing the Ambrosio-Tortorelli functional at a fixed moment in time t_k, and for fixed values of ε and η.

Note that the condition $\hat{v} \le v_\varepsilon(t_{k-1})$ enforces the irreversibility of the crack. In practice, however, we choose to implement the irreversibility criterion through the following equality constraint introduced by Bourdin [3]. At each time t_k, $k = 1,\dots,K$, we define the set

$$\mathtt{CR}(t_k) := \{x \in \overline{\Omega} : v_\varepsilon(x,t_{k-1}) < \mathtt{CRTOL}\}$$

for some small specified tolerance $\mathtt{CRTOL} > 0$, and we fix

$$v_\varepsilon(x,t_k) = 0 \text{ for all } x \in \mathtt{CR}(t_k).$$

Thus, if at a particular time, $t = t_k$, $v_\varepsilon(x,t_k)$ is close enough to zero to indicate that the point x lies on the crack path, then v_ε is set to zero at that point for all subsequent time steps. This considerably simplifies the minimization over $\hat{v}$ by allowing the use of an unconstrained minimization algorithm. However, it has yet to be shown that this modification of the irreversibility condition $\hat{v} \leq v_\varepsilon(t_{k-1})$ is equivalent to irreversibility of the crack as $\Delta t, \varepsilon \to 0$. We will address the question of imposing irreversibility via this pointwise monotonicity condition in future work.

1.3 Critical Points

In the Ambrosio-Tortorelli model, Lipschitz regularity of the domain is not required. As, in practice, it is more convenient to model a pre-existing crack by a slit domain than by the initial crack field $v(0)$, we will not assume that Ω is a Lipschitz domain. Instead, motivated by the fact that we will need to partition Ω for the purpose of defining a finite element approximation, we shall assume that Ω is a polyhedral domain. By this, we simply mean that Ω possesses a finite partition into non-degenerate open N-simplices: there exist open, pairwise disjoint, non-degenerate simplices $T_1, \ldots, T_K \subset \Omega$ such that $\mathcal{L}^N(\Omega \setminus \cup_k T_k) = 0$ (see also Sect. 2). This assumption guarantees that many of the usual trace and embedding theorems for Sobolev spaces hold on the domain Ω, while admitting domains with slits.

As we consider the minimization of the Ambrosio-Tortorelli functional at the fixed time $t = t_k$, $k \in \{1, \ldots, K\}$, it is useful to define the function spaces

$$\mathrm{H}^1_{\mathrm{D}}(\Omega) := \left\{\varphi \in \mathrm{H}^1(\Omega) : \varphi = 0 \text{ on } \Omega_{\mathrm{D}}\right\}, \quad \text{and}$$
$$\mathrm{H}^1_{\mathsf{CR}(t_k)}(\Omega) := \left\{\psi \in \mathrm{H}^1(\Omega) : \psi = 0 \text{ on } \mathsf{CR}(t_k)\right\}.$$

Fixing ε and η throughout, we relabel the Ambrosio-Tortorelli functional $I : \mathrm{H}^1(\Omega)^2 \to \mathbb{R} \cup \{+\infty\}$, where

$$I(u,v) := \int_\Omega \left[(v^2+\eta)|\nabla u|^2 + \alpha(1-v)^2 + \varepsilon|\nabla v|^2\right] \mathrm{d}x \quad \text{and } \alpha = 1/(4\varepsilon).$$

It can be seen, using a truncation argument, that any local minimizer (u,v) of I (in the $\mathrm{H}^1(\Omega)^2$ topology) satisfies $0 \leq v(x) \leq 1$ a.e. in Ω. Thus, all relevant test functions for v lie in the space $\mathrm{L}^\infty(\Omega)$. As such, in the following discussion of differentiability of I, we work with test functions for v from the space $\mathrm{H}^1(\Omega) \cap \mathrm{L}^\infty(\Omega)$.

It is easy to see that I is Fréchet-differentiable in $\mathrm{H}^1(\Omega) \times (\mathrm{H}^1(\Omega) \cap \mathrm{L}^\infty(\Omega))$; however, we note that $I(u,v)$ is not finite for all $(u,v) \in \mathrm{H}^1(\Omega)^2$, and thus I is *not* Gateaux-differentiable in $\mathrm{H}^1(\Omega)^2$. This motivates the following definition of a critical point.

Definition 1. Let $k \in \{1,\dots,K\}$. For all $(u,v) \in \mathrm{H}^1(\Omega) \times (\mathrm{H}^1(\Omega) \cap \mathrm{L}^\infty(\Omega))$, let $I'(u,v;\varphi,\psi)$ denote the Fréchet derivative of I at (u,v) in the direction $(\varphi,\psi) \in \mathrm{H}^1(\Omega) \times (\mathrm{H}^1(\Omega) \cap \mathrm{L}^\infty(\Omega))$. We say that

$$(u,v) \in (g(t_k) + \mathrm{H}^1_D(\Omega)) \times (\mathrm{H}^1_{\mathsf{CR}(t_k)}(\Omega) \cap \mathrm{L}^\infty(\Omega))$$

is a critical point of I if $I'(u,v;\varphi,\psi) = 0$ for all $\varphi \in \mathrm{H}^1_\mathrm{D}(\Omega)$ and for all $\psi \in \mathrm{H}^1_{\mathsf{CR}(t_k)}(\Omega)$.

Proposition 1. *Let $k \in \{1,\dots,K\}$. If $(u,v) \in \mathrm{H}^1(\Omega) \times (\mathrm{H}^1_{\mathsf{CR}(t_k)}(\Omega) \cap \mathrm{L}^\infty(\Omega))$ is a critical point of I, then $0 \le v(x) \le 1$ for a.e. $x \in \Omega$.*

2 Adaptive Finite Element Discretization

2.1 The Alternating Minimization Algorithm

The minimization of the functional I is a particularly challenging task as the term $(v^2+\eta)|\nabla u|^2$ renders the functional nonconvex. A number of minimization schemes can be employed; we note however that none would in general be able to find global minimizers, at least not easily. Instead, we must be satisfied with being able to locate *local* minimizers.

The minimization is achieved using an alternate minimization algorithm proposed by Bourdin et al. [4]. We state the algorithm for the minimization of I over $(g(t_k) + \mathrm{H}^1_\mathrm{D}(\Omega)) \times (\mathrm{H}^1_{\mathsf{CR}(t_k)}(\Omega) \cap \mathrm{L}^\infty(\Omega))$ at time $t = t_k$, $k \in \{1,\dots,K\}$. The main observation is that, although the functional I is nonconvex with respect to the pair (u,v), it is convex in each variable, taking the other variable fixed. Thus, it is straightforward to minimize with respect to one variable at a time.

Algorithm 1. Alternating Minimization

1. Let $v^0 \in \mathrm{H}^1_{\mathsf{CR}(t_k)}(\Omega) \cap \mathrm{L}^\infty(\Omega)$ be given
   (normally the crack field $v(t_{k-1})$ from the previous, $(k-1)$st, time step)
2. For $n = 1,2,3,\dots$ do:

   2.1. $u^n := \operatorname{argmin}\{I(z,v^{n-1}) : z \in g(t_k) + \mathrm{H}^1_\mathrm{D}(\Omega)\}$
   2.2. $v^n := \operatorname{argmin}\{I(u^n,w) : w \in \mathrm{H}^1_{\mathsf{CR}(t_k)}(\Omega)\}$

3. Set $u(t_k) = \lim_{n\to\infty} u^n$ and $v(t_k) = \lim_{n\to\infty} v^n$.

In practice, we terminate the iteration once $\|v^n - v^{n-1}\|_{\mathrm{L}^\infty(\Omega)}$ falls below a prescribed tolerance. Local convergence to isolated minimizers was established in [3, Theorem 2], while the first global convergence proof of alternating minimization was given in [6]. As this proof is well-hidden inside more general results, we shall outline the main ideas.

Proposition 2. *Let $k \in \{1,\dots,K\}$. There exists a subsequence $((u^{n_\ell},v^{n_\ell}))_{\ell=1}^\infty$ in $(g(t_k)+\mathrm{H}^1_{\mathrm{D}}(\Omega)) \times (\mathrm{H}^1_{\mathrm{CR}(t_k)}(\Omega) \cap \mathrm{L}^\infty(\Omega))$, and a critical point (u,v) of I in the same set, such that $(u^{n_\ell},v^{n_\ell}) \to (u,v)$ strongly in $\mathrm{H}^1(\Omega)^2$ as $\ell \to \infty$.*

Proof. Using the fact that the energy I is nonincreasing along the sequence $((u^n,v^n))_{n=1}^\infty$, and that I is coercive in $\mathrm{H}^1(\Omega)^2$, we may extract a subsequence $n_\ell \nearrow \infty$ such that, as $\ell \to \infty$,

$$(u^{n_\ell},v^{n_\ell}) \rightharpoonup (u,v) \quad \text{weakly in } \mathrm{H}^1(\Omega)^2 \quad \text{and } v^{n_\ell -1} \rightharpoonup \bar{v} \quad \text{weakly in } \mathrm{H}^1(\Omega),$$

where $u \in g(t_k)+\mathrm{H}^1_D(\Omega)$ and $v,\bar{v} \in \mathrm{H}^1_{\mathrm{CR}(t_k)}(\Omega) \cap \mathrm{L}^\infty(\Omega)$. Because $0 \le v^n \le 1$ for all n (slightly strengthening Proposition 1) it follows that also $0 \le v \le 1$ a.e. in Ω. The task now is to verify that (u,v) is a critical point of I.

Owing to the compactness of the embedding $\mathrm{H}^1(\Omega) \subset \mathrm{L}^2(\Omega)$ (we note that, although possibly non-Lipschitz, a domain Ω that is polyhedral in our sense satisfies the internal cone property, and therefore compactness of the embedding follows), $v^{n_\ell-1} \to \bar{v}$ strongly in $\mathrm{L}^2(\Omega)$ as $\ell \to \infty$. From this, and the fact that $\partial_u I(u^{n_\ell},v^{n_\ell-1}) = 0$ for all ℓ, it follows fairly easily that $\partial_u I(u,\bar{v}) = 0$. To prove that also $\partial_v I(u,v) = 0$, we first verify that $\nabla u^{n_\ell} \to \nabla u$ strongly in $\mathrm{L}^2(\Omega)$. This can be deduced by noting that

$$\begin{aligned}
\eta \|\nabla(u-u^{n_\ell})\|^2_{\mathrm{L}^2(\Omega)} &\le \int_\Omega ([v^{n_\ell-1}]^2+\eta)|\nabla(u-u^{n_\ell})|^2 \,\mathrm{d}x \\
&= \int_\Omega (\bar{v}^2+\eta)\nabla u \cdot \nabla(u-u^{n_\ell}) \,\mathrm{d}x \\
&\quad - \int_\Omega ([v^{n_\ell-1}]^2+\eta)\nabla u^{n_\ell} \cdot \nabla(u-u^{n_\ell}) \,\mathrm{d}x \\
&\quad + \int_\Omega ([v^{n_\ell-1}]^2-\bar{v}^2)\nabla u \cdot \nabla(u-u^{n_\ell}) \,\mathrm{d}x.
\end{aligned}$$

As $\partial_u I(u,\bar{v}) = 0$ and $\partial_u I(u^{n_\ell},v^{n_\ell-1}) = 0$, the first and second term on the right vanish. The third term can be easily shown to converge to zero in the limit of $\ell \to \infty$ using Lebesgue's Dominated Convergence Theorem.

Having shown that $\nabla u^{n_\ell} \to \nabla u$ strongly in $\mathrm{L}^2(\Omega)$, it is now an easy exercise to prove that $\partial_v I(u,v) = 0$ and that $v^{n_\ell} \to v$ strongly in $\mathrm{H}^1(\Omega)$ as $\ell \to \infty$. Owing to the boundedness of v^{n_ℓ} in $\mathrm{L}^\infty(\Omega)$, an application of Fatou's Lemma shows that $I(u^{n_\ell},v^{n_\ell}) \to I(u,v)$.

To prove that (u,v) is a critical point of I, it suffices to show that $\bar{v} = v$. To see this, note that monotonicity of the energy along the computed sequence and, in particular, $I(u^{n_\ell},v^{n_\ell-1}) \le I(u^{n_{\ell-1}},v^{n_{\ell-1}})$ implies

$$I(u,\bar{v}) \le \liminf_{\ell\to\infty} I(u^{n_\ell},v^{n_\ell-1}) \le \liminf_{\ell\to\infty} I(u^{n_{\ell-1}},v^{n_{\ell-1}}) = I(u,v).$$

Because $\partial_v I(u,v) = 0$ and I is strictly convex for fixed u, it follows that $\bar{v} = v$, and therefore that (u,v) is a critical point of I. □

In what follows, we shall solve each step of the alternating minimization algorithm by an adaptive finite element method.

2.2 Finite Element Discretization

For simplicity, we shall now restrict our presentation to the physically relevant case of $N \in \{2,3\}$. In particular, Ω is assumed to be a polygonal ($N=2$) or polyhedral ($N=3$) domain in $\mathbb{R}^N$, in the sense that it has a finite partition into pairwise disjoint nondegenerate open N-simplices, the complement (relative to Ω) of whose union is of N-dimensional Lebesgue measure zero. We consider a sequence $\{\mathcal{T}_j\}_{j\in J}$ of regular partitions $\mathcal{T}_j$ of Ω, where $J := \mathbb{N}_{\geq 0} \cup (1/2 + \mathbb{N}_{\geq 0})$. (The index set describes the half-steps in the alternating minimization algorithm.) We also assume that Ω_{D} is partitioned exactly by $\mathcal{T}_j$ for all $j \in J$. The associated P1 finite element space is defined as follows:

$$X_j := \left\{ w \in \mathrm{C}(\overline{\Omega}) : w \text{ is piecewise affine with respect to } \mathcal{T}_j \right\}.$$

For simplicity, we shall assume that $g(t_k) \in X_j$ for all $j \in J$ and $k \in \{1,\dots,K\}$. We also define the discrete test spaces for the variables u and v, respectively, as

$$X_{j,\mathrm{D}} := X_j \cap \mathrm{H}^1_D(\Omega) \quad \text{and } X_{j,\mathsf{CR}(t_k)} := X_j \cap \mathrm{H}^1_{\mathsf{CR}(t_k)}(\Omega).$$

Here, and henceforth, $\mathsf{CR}(t_k)$ denotes the union of the closures $\overline{e}$ of all (open) faces e of elements T in the finite element partition $\mathcal{T}_j$ of $\overline{\Omega}$ such that $v(x,t_{k-1}) < \mathtt{CRTOL}$ for all $x \in \overline{e}$. Note that the discrete trial space for the u variable at $t = t_k$ is given by $g(t_k) + X_{j,D}$, while the discrete trial space for the v variable at $t = t_k$ is simply $X_{j,\mathsf{CR}(t_k)}$.

In our subsequent analysis, we require the discrete formulation to satisfy the analogue of Proposition 1. In order to accomplish this we use a 'mass-lumping' approximation [19, Chap. 11] for I together with the following additional hypothesis on the mesh regularity.

Hypothesis A. We assume that all off-diagonal elements of the finite element stiffness matrix, associated with the finite element space X_j on the partition $\mathcal{T}_j$, $j \in J$, are non-positive.

This condition has been studied in detail in two dimensions [8], [18, p.78] as well as in three dimensions [16]. A sufficient condition for Hypothesis A to hold is that $\mathcal{T}_j$, $j \in J$, is a nonobtuse partition of Ω in the sense that, for $N = 2$, each internal angle of each $T \in \mathcal{T}_j$ is $\leq \pi/2$; and, for $N = 3$, each internal dihedral angle of each $T \in \mathcal{T}_j$ is $\leq \pi/2$. We note that Hypothesis A is only adopted for technical purposes, and in practice we do not expect it to be necessary for the conclusions of Proposition 3 later (in whose proof, in [6], Hypothesis A is crucially used) to hold.

The 'mass-lumping' approximation of I is the functional $I_j : X_j \times X_j \to \mathbb{R}$, $j \in J$, defined by

$$I_j(u,v) := \int_\Omega \left[\left(P_j(v^2) + \eta\right) |\nabla u|^2 + \alpha P_j\left((v-1)^2\right) + \varepsilon |\nabla v|^2 \right] \mathrm{d}x,$$

where $P_j : \mathrm{C}(\overline{\Omega}) \to X_j$ is the standard nodal interpolation operator on $\mathcal{T}_j$.

Definition 2. Let $k \in \{1,\dots,K\}$. We say that $(u,v) \in (g(t_k)+X_{j,\mathrm{D}}) \times X_{j,\mathsf{CR}(t_k)}$ is a critical point of I_j if $I_j'(u,v;\varphi,\psi)=0$ for all $\varphi \in X_{j,\mathrm{D}}$ and $\psi \in X_{j,\mathsf{CR}(t_k)}$, where

$$\begin{aligned} I_j'(u,v;\varphi,\psi) &= 2a_j(v;u,\varphi)+2b_j(u;v,\psi),\\ a_j(v;u,\varphi) &:= \int_\Omega \big(P_j(v^2)+\eta\big)\nabla u\cdot\nabla\varphi\,\mathrm{d}x, \quad \text{and}\\ b_j(u;v,\psi) &:= \int_\Omega \big[P_j(v\psi)|\nabla u|^2+\alpha P_j(((v-1)\psi))+\varepsilon\nabla v\cdot\nabla\psi\big]\,\mathrm{d}x. \end{aligned}$$

Proposition 3. *Let $k \in \{1,\dots,K\}$. Suppose that $(u,v) \in (g(t_k)+X_{j,D}) \times X_{j,\mathsf{CR}(t_k)}$ satisfies $b_j(u;v,\psi)=0$ for all $\psi \in X_{j,\mathsf{CR}(t_k)}$; then, $0 \le v(x) \le 1$ for all $x \in \Omega$.*

2.3 Adaptive Alternating Minimization

Steps 2.1 and 2.2 in the alternating minimization algorithm require the solution of elliptic boundary-value problems that arise from the criticality conditions for the respective minimization problems.

We shall modify the algorithm, requiring that the criticality conditions in each step are only satisfied approximately. The theory of residual-based a-posteriori error estimation for finite element methods provides powerful tools for achieving approximate criticality, up to a prescribed tolerance. In particular, we have the following residual-based a-posteriori estimates (cf. [6]).

Proposition 4 (Residual Estimates). *Suppose that $k \in \{1,\dots,K\}$.*
(i) Let $u \in g(t_k)+X_{j,D}, v \in X_{j,\mathsf{CR}(t_k)}$ be such that $a_j(v;u,\varphi)=0$ for all $\varphi \in X_{j,D}$; then,

$$|\partial_u I(u,v;\varphi)| \le C\mu_j(u,v)\|\nabla\varphi\|_{\mathrm{L}^2(\Omega)} \quad \forall\varphi \in \mathrm{H}^1_D(\Omega),$$

where

$$\begin{aligned}[\mu_j(u,v)]^2 := \sum_{\substack{T\in\mathcal{T}_j\\ T\cap\Omega_{\mathrm{D}}=\emptyset}} \Big[h_T^4\|\nabla v\|^4_{\mathrm{L}^\infty(T)}\|\nabla u\|^2_{\mathrm{L}^2(T)} + h_T^2\|2v(\nabla v\cdot\nabla u)\|^2_{\mathrm{L}^2(T)}\\ + h_T\|(v^2+\eta)[\nabla u]\|^2_{\mathrm{L}^2(\partial T\setminus\overline{\Omega}_{\mathrm{D}})}\Big] =: \sum_{\substack{T\in\mathcal{T}_j\\ T\cap\Omega_{\mathrm{D}}=\emptyset}} [\mu_j(u,v;T)]^2.\end{aligned}$$

(ii) Let $u \in g(t_k)+X_{j,D}, v \in X_{j,\mathsf{CR}(t_k)}$ be such that $b_j(u;v,\psi)=0$ for all $\psi \in X_{j,\mathsf{CR}(t_k)}$; then,

$$|\partial_v I(u,v;\psi)| \le C\nu_j(u,v)\|\nabla\psi\|_{\mathrm{L}^2(\Omega)} \quad \forall\psi \in \mathrm{H}^1_{\mathsf{CR}(t_k)}(\Omega),$$

where

$$[\nu_j(u,v)]^2 := \sum_{\substack{T\in\mathcal{T}_j\\ T\cap \mathsf{CR}(t_k)=\emptyset}} \Big[h_T^4\|\nabla v\|_{\mathrm{L}^\infty(T)}^2\|\alpha+|\nabla u|^2\|_{\mathrm{L}^2(T)}^2 + \varepsilon^2 h_T\|[\nabla v]\|_{\mathrm{L}^2(\partial T\setminus \mathsf{CR}(t_k))}^2$$

$$+ h_T^2\|(\alpha+|\nabla u|^2)v-\alpha\|_{\mathrm{L}^2(T)}^2\Big] =: \sum_{\substack{T\in\mathcal{T}_j\\ T\cap \mathsf{CR}(t_k)=\emptyset}} [\nu_j(u,v;T)]^2.$$

We note that the constant C in the preceding proposition depends on the mesh quality, which must be controlled during mesh refinement.

The residual estimates motivate the following formulation of our adaptive algorithm, where TOL_n is a sequence of residual tolerances, which satisfy $\mathrm{TOL}_n \searrow 0$ as $n \nearrow \infty$.

Algorithm 2. Adaptive Alternating Minimization

1. Let $\mathcal{T}_0$, v^0, $g(t_k)$ and $\mathsf{CR}(t_k)$ be given.
2. For $n = 1,2,3,\ldots$ do

   2.1. Construct a sub-mesh $\mathcal{T}_{n-1/2}$ of $\mathcal{T}_{n-1}$ such that the solution $u^n \in g(t_k) + X_{n-1/2,D}$ of

   $$a_{n-1/2}(v^{n-1};u^n,\varphi) = 0 \quad \forall\varphi\subset X_{n-1/2,D},$$

   satisfies $\mu_{n-1/2}(u^n,v^{n-1}) \le \mathrm{TOL}_n$.

   2.2. Construct a sub-mesh $\mathcal{T}_n$ of $\mathcal{T}_{n-1/2}$ such that the solution $v^n \in X_{n,\mathsf{CR}(t_k)}$ of

   $$b_n(u^n;v^n,\psi) = 0 \quad \forall\psi\in X_{n,\mathsf{CR}(t_k)}$$

   satisfies $\nu_n(u^n,v^n) \le \mathrm{TOL}_n$.

In practice, we terminate the algorithm once $\|v^n - v^{n-1}\|_{\mathrm{L}^\infty(\Omega)}$ is sufficiently small.

For details of the implementation of the adaptive mesh-refinement in steps 2.1 and 2.2 in Algorithm 2 we refer to the adaptive finite element literature, for example, [7, 11, 20], and to our extended paper [6]. We note that for Algorithm 2 to be well-posed one needs to prove convergence of the adaptive finite element algorithm that we use for solving the linear self-adjoint second-order elliptic boundary-value problems with 'mass-lumping' in steps 2.1 and 2.2. Convergence results of this type have been established in the literature, though, to the best of our knowledge, only in the absence of 'mass-lumping'. Having said this, the extension of these convergence results to the case of a 'mass-lumped' approximation is not foreseen to be technically demanding.

Provided that each step of Algorithm 2 can be executed, we obtain the following convergence theorem. Its proof follows essentially along the lines of the proof of Proposition 2, requiring only minor modifications (see, [6]).

Theorem 1. *Let $((u^n, v^n))_{n=1}^{\infty}$ be a sequence generated by Algorithm 2. Then, assuming convergence of the adaptive finite element approximation of the linear self-adjoint second-order elliptic problems with 'mass-lumping' in steps 2.1 and 2.2, there exists a subsequence $((u^{n_\ell}, v^{n_\ell}))_{\ell=1}^{\infty}$ and a critical point (u,v) of the functional I such that $(u^{n_\ell}, v^{n_\ell}) \to (u,v)$ strongly in $\mathrm{H}^1(\Omega)^2$ as $\ell \to \infty$.*

3 A Computational Example

We now briefly illustrate the performance of the adaptive alternating minimization algorithm in practice with a numerical example. Consider the rectangular domain $(-0.5, 6.5) \times (0,6)$ with two initial slits along $\{2\} \times [0,2]$ and $\{4\} \times [4,6]$. The domain is shown in Fig. 1, where the shaded region is Ω_{D}, together with the initial mesh. The time-dependent displacement imposed on Ω_{D} is given as follows:

$$g(x,t) := \begin{cases} -t, & \text{on } (-0.5,0) \times (0,6), \\ t, & \text{on } (6,6.5) \times (0,6), \end{cases}$$

where $t = 0.01s$, for $s = 0, 1, \ldots, 160$. We set the following values for the parameters in the computation: $\varepsilon = 0.05$, $\eta = 1 \times 10^{-5}$, $\mathtt{CRTOL} = 1 \times 10^{-4}$ and $\mathtt{TOL}_n = 0.3$. For each $s = 0, 1, \ldots, 160$, we terminate the adaptive alternating minimization algorithm when $\|v^n - v^{n-1}\|_{\mathrm{L}^\infty(\Omega)} < 1 \times 10^{-3}$.

Figure 2 shows the final mesh and crack path, together with the change in bulk, surface and total energies over time. As the incremental displacement is applied to Ω_{D} the bulk energy increases. There is an initial period in which this increase is not sufficient to cause the initial slits in the domain to extend. At time $t \sim 0.7$ the body starts to experience a period of steady crack growth in which both slits grow simultaneously. This continues until time $t = 1.43$ when the two cracks grow rapidly, meeting at the centre of the domain, thus causing the body to fracture into two pieces.

It can be seen that the adaptive algorithm refines the mesh around the growing cracks whilst retaining coarser elements elsewhere. The number of elements in the final mesh is 76,937. The smallest and largest elements have diameters of size 1×10^{-3} and 0.25, respectively. In particular, a (quasi-)uniform mesh with elements of diameter 1×10^{-3} would require roughly 8×10^7 elements.

Conclusion

We have presented an adaptive algorithm for computing finite element approximations of local minimizers of the Ambrosio-Tortorelli functional. We have primarily focused on convergence results for the algorithm. We have been able to show that, provided the associated residuals converge to zero, the algorithm generates a

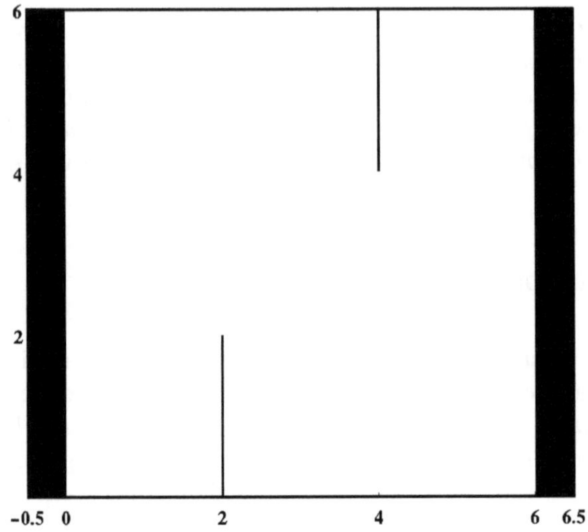

(a) Domain

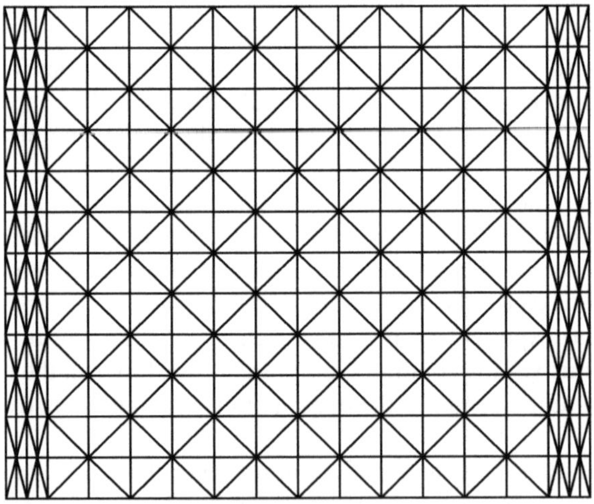

(b) Initial mesh

Fig. 1 Computational domain and initial mesh on the domain

sequence of numerical solutions that converges to a critical point of the energy functional I (as the sequence of termination tolerances tends to zero). Our preliminary computational results demonstrate the potential of using this method. In particular, we believe that the algorithm enables us to accurately and reliably compute the evolution of the crack path using considerably fewer elements compared to simulations with uniform meshes. We believe that the results presented are easily extendable to the cases of planar and three-dimensional elasticity. We are currently working on extending the theory and implementation to these models.

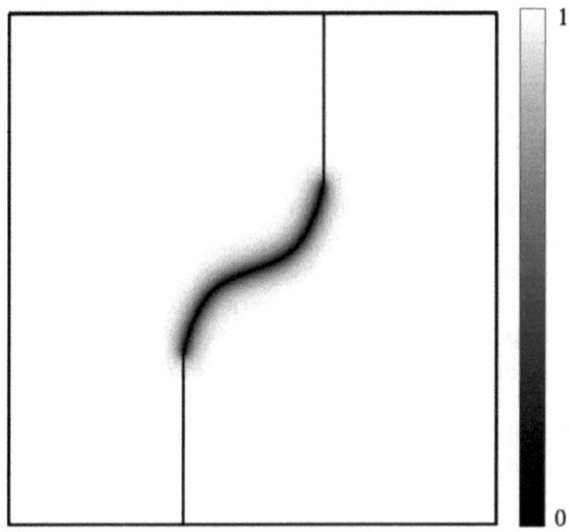

(a) Crack path: plot of v

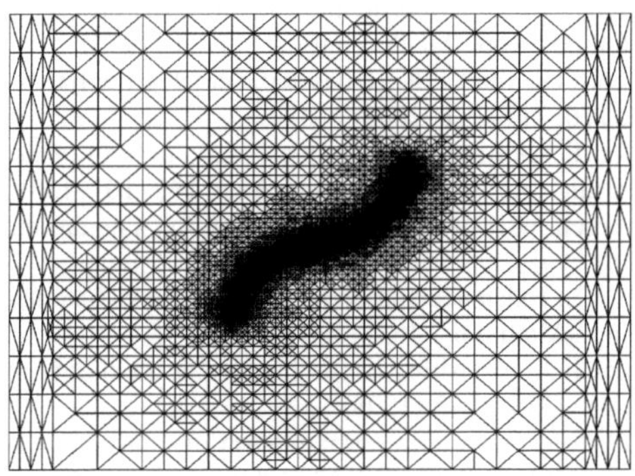

(b) Final mesh

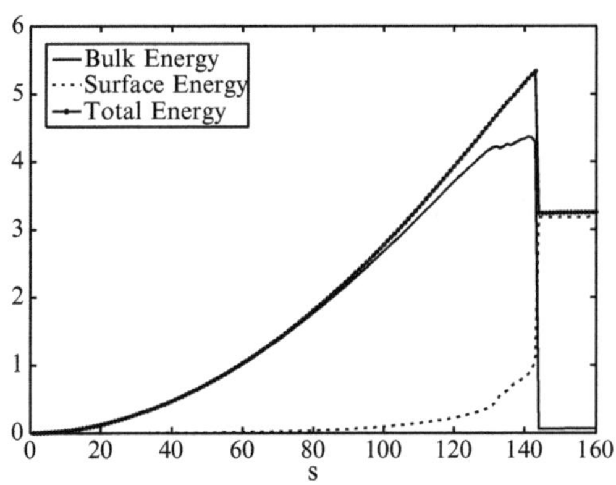

(c) Bulk and surface energies

Fig. 2 Evolution of the body: The figure at the *top* shows v, indicating the crack-path at the final time, $T = 1.6$. The *middle figure* depicts the final mesh, with 76,937 elements. The figure at the *bottom* shows the evolution of bulk, surface and total energies, for $t \in [0, 1.6]$

References

1. Ambrosio, L., Tortorelli, V.M.: On the approximation of free discontinuity problems. Boll. Un. Mat. Ital. B (7), **6**, 105–123 (1992)
2. Ambrosio, L., Fusco, N., Pallara, D.: Functions of Bounded Variation and Free Discontinuity Problems. Oxford University Press, Oxford (2000)
3. Bourdin, B.: Numerical implementation of the variational formulation for quasi-static brittle fracture. Interf. Free Bound. **9**, 411–430 (2007)
4. Bourdin, B., Francfort, G., Marigo, J.-J.: Numerical experiments in revisited brittle fracture. J. Mech. Phys. Solids. **48**, 797–826 (2000)
5. Braides, A.: Γ-Convergence for Beginners. Vol. 22 of Oxford Lecture Series in Mathematics and its Applications, Oxford University Press, Oxford (2002)
6. Burke, S., Ortner, C., Süli, E.: An adaptive finite element approximation of a variational model of brittle fracture. SIAM J. Numer. Anal., Vol. 48, No. 3. Published online 1 July 2010. DOI: 10.1137/080741033
7. Cascon, J., Kreuzer, C., Nochetto, R., Siebert, K.: Quasi-optimal convergence rate for an adaptive finite element method. SIAM J. Numer. Anal. **46**, 2524–2550 (2008)
8. Ciarlet, P., Raviart, P.-A.: Maximum principle and uniform convergence for the finite element method. Comput. Methods Appl. Mech. Engrg. **2**, 17–31 (1973)
9. Dal Maso, G., Francfort, G., Toader, R.: Quasistatic crack growth in nonlinear elasticity. Arch. Ration. Mech. Anal. **176**, 111–111 (2005)
10. Dal Maso, G., Toader, R.: A model for the quasi-static growth of brittle fractures: existence and approximation results. Arch. Ration. Mech. Anal. **162**, 101–135 (2002)
11. Dörfler, W.: A convergent adaptive algorithm for Poisson's equation. SIAM J. Numer. Anal. **33**, 1106–1124 (1996)
12. Francfort, G., Larsen, C.: Existence and convergence for quasi-static evolution in brittle fracture. Comm. Pure Appl. Math. **56**, 1465–1500 (2003)
13. Francfort, G., Marigo, J.-J.: Revisiting brittle fracture as an energy minimization problem. J. Mech. Phys. Solids. **46**, 1319–1342 (1998)
14. Giacomini, A.: Ambrosio-Tortorelli approximation of quasi-static evolution of brittle fractures. Calc. Var. Partial Diff. Eqs. **22**, 129–172 (2005)
15. Griffith, A.: The phenomena of rupture and flow in solids. Philosophical Trans. R. Soc. Lond. **221**, 163–198 (1921)
16. Korotov, S., Křížek, M., Neittaanmäki, P.: Weakened acute type condition for tetrahedral triangulations and the discrete maximum principle. Math. Comp. **70**, 107–119 (2001) (electronic)
17. Mumford, D., Shah, J.: Optimal approximation by piecewise smooth functions and associated variational problems. Comm. Pure Appl. Math. **42**, 577–685 (1989)
18. Strang, G., Fix, G.: An Analysis of the Finite Element Method. Prentice-Hall Inc., Englewood Cliffs, NJ, USA (1973)
19. Thomée, V.: Galerkin Finite Element Methods for Parabolic Problems. Vol. 1054 of Lecture Notes in Mathematics, Springer-Verlag, Berlin, New York (1984)
20. Verfürth, R.: A Review of A Posteriori Error Estimation and Adaptive Mesh-Refinement Techniques, Wiley & Teubner, New York (1996)

A Nyström Method for Solving a Boundary Value Problem on $[0,\infty)$*

Carmelina Frammartino

Dedicated to Professor Gradimir V. Milovanović on his 60-th birthday

1 Introduction

This paper deals with the numerical treatment of a special second-order boundary value problem on the real semiaxis,

$$\begin{cases} f''(x) - \mu a(x) f(x) = g(x), \\ f(0) = f(\infty) = 0, \end{cases} \tag{1}$$

where g and a are known functions and μ is a real parameter.

Let $(Af)(x) = \int_0^\infty G(x,t) f(t)\,\mathrm{d}t$ where

$$G(x,t) = -\begin{cases} \mathrm{e}^{-t} \sinh x, & 0 \le x \le t, \\ \mathrm{e}^{-x} \sinh t, & t < x < \infty. \end{cases}$$

Then the following identity holds

$$(Af'')(x) - (Af)(x) = f(x)\,. \tag{2}$$

We rewrite the differential equation in (1) as follows

$$f''(x) - f(x) = f(x)[\mu a(x) - 1] + g(x)\,;$$

Carmelina Frammartino
Department of Mathematics and Computer Science, University of Basilicata, Via dell'Ateneo Lucano 10, 85100 Potenza, Italy, e-mail: carmelina.frammartino@unibas.it

* The work is supported by the research project PRIN 2006 "Numerical methods for structured linear algebra and applications".

W. Gautschi et al. (eds.), *Approximation and Computation: In Honor of Gradimir V. Milovanović*, Springer Optimization and Its Applications 42,
DOI 10.1007/978-1-4419-6594-3_20,

then, multiplying both sides by the function $G(x,t)$, integrating from 0 to ∞ and applying (2), we obtain the following Fredholm integral equation

$$f(t) - \int_0^\infty G(x,t)a^*(x)f(x)\,\mathrm{d}x = \widetilde{g}(t)\,, \tag{3}$$

where $a^*(x) = [\mu a(x) - 1]$ and $\widetilde{g}(t) = \int_0^\infty G(x,t)g(x)\,\mathrm{d}x$.

If $a^*(t) \neq 0$ in $[0,\infty)$ we can multiply both sides of (3) by the function $a^*(t)\mathrm{e}^t$ obtaining the following Fredholm integral equation

$$\overline{f}(t) - \overline{a}(t)\int_0^\infty G(x,t)\overline{f}(x)\mathrm{e}^{-x}\,\mathrm{d}x = \overline{g}(t)\,, \tag{4}$$

where $\overline{f}(t) = f(t)a^*(t)\mathrm{e}^t$, $\overline{a}(t) = \mathrm{e}^t a^*(t)$ and $\overline{g}(t) = \widetilde{g}(t)a^*(t)\mathrm{e}^t$.

In this way we have reduced the problem (1) to solving an equivalent Fredholm integral equation on the real semiaxis.

Some other authors were interested in this kind of equation (see for example [4]), but our procedure is based on a product rule instead of a Gaussian quadrature rule. This kind of choice is strictly connected to the properties of the kernel, i.e., to the properties of the Green function $G(x,t)$.

The paper is organized as follows. In Sect. 2 we introduce some definitions and preliminary results. In Sect. 3 we describe the proposed numerical procedure and we analyze its convergence and stability. In Sect. 4 we provide some numerical examples in order to illustrate the accuracy of the method. Finally, in Sect. 5, we give the proofs of the main results.

2 Function Spaces and Preliminary Results

We introduce some notations and the spaces of functions that we need for studying the numerical method we will propose.

With the weight $u(x) = (1+x)^\lambda x^\gamma \mathrm{e}^{-x/2}$, $0 < \gamma < 1/4$, $1/2 < \lambda \leq 1$, $x > 0$, we define the space

$$C_u = \left\{ f \in C^0((0,\infty)) \,:\, \lim_{\substack{x\to 0\\ x\to\infty}} (fu)(x) = 0 \right\}$$

equipped with the norm

$$\|f\|_{C_u} = \|fu\|_\infty = \sup_{x\geq 0} |(fu)(x)|\,.$$

We also introduce the Sobolev-type space of order $r \in \mathbb{N}$, $r \geq 1$, defined by:

$$W_r(u) = \{f \in C_u \,:\, f^{(r-1)} \in AC(\mathbb{R}^+) \text{ and } \|f^{(r)}\varphi^r u\|_\infty < \infty\}\,,$$

where $\varphi(x) = \sqrt{x}$ and $AC(\mathbb{R}^+)$ is the set of all absolutely continuous functions on $\mathbb{R}^+$. In $W_r(u)$ we introduce the norm

$$\|f\|_{W_r(u)} = \|fu\|_\infty + \|f^{(r)}\varphi^r u\|_\infty .$$

Let

$$E_m(f)_{u,\infty} = \inf_{P\in\mathbb{P}_m} \|(f-P)u\|_\infty$$

be the error of best polynomial approximation in C_u. In [2] the following estimate was proved,

$$E_m(f)_{u,\infty} \le \mathcal{C}\left(\frac{\sqrt{a_m}}{m}\right)^r \|f\|_{W_r(u)} \tag{5}$$

for all $f \in W_r(u)$, with $a_m = a_m(u)$ the Mhaskar-Rakhmanov-Saff number with respect to the weight u. $\mathcal{C}$ is a positive constant.

We introduce the following interpolation process.

Let $w(x) = e^{-x}$ and let $\{p_m(w)\}_m$ be the related system of orthonormal polynomials having positive leading coefficient. Let $\mathcal{C}/m \le x_1 < \cdots < x_m \le 4m\left(1-\mathcal{C}/m^{2/3}\right)$ be the zeros of $p_m(w)$. For every $f \in C_u$ we define the Lagrange polynomial

$$L^*_{m+2}(w,f;x) = \sum_{k=0}^{m+1} l^*_k(x)f(x_k)$$

interpolating f at the nodes $x_0 = 0, x_1, \ldots, x_m, x_{m+1} = a_m$, where

$$l^*_k(x) = \frac{x(a_m-x)}{x_k(a_m-x_k)} l_k(x), \quad k = 1,\ldots,m,$$

with

$$l_k(x) = \frac{p_m(w,x)}{p'_m(w,x_k)(x-x_k)}, \quad l^*_0(x) = \frac{p_m(w,x)(a_m-x)}{p_m(w,0)a_m}, \quad l^*_{m+1}(x) = \frac{xp_m(w,x)}{a_m p_m(w,a_m)}.$$

We observe that in our case $f(0) = 0$; therefore, the Lagrange polynomial will be $L^*_{m+2}(w,f;x) = \sum_{k=1}^{m+1} l^*_k(x)f(x_k)$. Now let m be sufficiently large (say $m > m_0$) and the integer $j = j(m)$ be defined by:

$$x_j = x_{j(m)} = \min_{1\le k\le m}\{x_k : x_k \ge \theta a_m\}, \tag{6}$$

where $\theta \in (0,1)$ is fixed. Then we introduce the following truncated interpolation process

$$L^{**}_{m+2}(w,f;x) = \sum_{k=1}^{j} l^*_k(x)f(x_k) = L^*_{m+2}(w,f_j;x), \quad f \in C_u,$$

where $f_j = \Phi_j f$ and Φ_j is the characteristic function of the interval $[0,x_j]$.

We state the following theorem that will be useful in the sequel.

Theorem 1. *If the weights $w(x) = x^{\alpha}\mathrm{e}^{-x^{\beta}}$ and $u(x) = (1+x)^{\lambda}x^{\gamma}\mathrm{e}^{-x^{\beta}/2}$ with $\alpha > -1$ and $\beta > 1/2$ are such that the parameter α, γ, and λ satisfy the conditions*

$$\frac{1}{2} < \lambda \leq 1 \quad \textit{and} \quad \max\left(0, \frac{\alpha}{2} - \frac{3}{4}\right) < \gamma < \frac{\alpha}{2} + \frac{1}{4}, \tag{7}$$

then, for each $f \in L^{\infty}_{u}$, we have

$$\left\| L^{**}_{m+2}(w,f)\frac{w}{u} \right\|_1 \leq \mathcal{C}\|fu\|_{L^{\infty}([0,x_j])}, \tag{8}$$

with $\mathcal{C} \neq \mathcal{C}(m,f)$ and j defined as in (6)*. Consequently, under the conditions* (7)*, for each $f \in W_r(u)$ we have*

$$\left\| (f - L^{**}_{m+2}(w,f))\frac{w}{u} \right\|_1 \leq \mathcal{C}\frac{\|f\|_{W_r(u)}}{M^{(1-1/(2\beta))r}} \tag{9}$$

with $\mathcal{C} \neq \mathcal{C}(m,f)$ and $M = \left[m(\theta/(1+\theta))^{\beta} \right]$.

3 Numerical Method

Let

$$(Kf)(t) = \overline{a}(t)\int_0^{\infty} G(x,t)f(x)\mathrm{e}^{-x}\,\mathrm{d}x.$$

We first prove that, under a suitable hypothesis on the function $a^*(x)$, the operator $K : C_u \to C_u$ is compact. In fact we state the following

Theorem 2. *Let $a^*(t) \sim 1/t^{\alpha}$ for $t \to \infty$ with $\alpha \geq 1/2$. Then $K : C_u \to C_u$ is a compact operator.*

Then for (4) the Fredholm alternative is true.

At this point we are able to introduce the following sequence of operators,

$$(K_m f)(t) = \overline{a}(t)\int_0^{\infty} G(x,t)L^{**}_{m+2}(w,f;x)\mathrm{e}^{-x}\,\mathrm{d}x \quad (m > m_0), \tag{10}$$

and, in order to apply a Nyström method to approximate the integral equation (4) we consider the following equations

$$\overline{f}_m(t) - (K_m f_m)(t) = \overline{g}(t) \quad (m > m_0). \tag{11}$$

Multiplying both sides of (11) by the weight $u(t)$ and collocating the last equation at the zeros x_i, $i = 1,\ldots,j$, we obtain the linear system

$$\sum_{k=1}^{j}\left[\delta_{ik} - \frac{u(x_i)}{u(x_k)}\overline{a}(x_i)\overline{M}_k(x_i)\right]c_k = b_i, \quad i = 1,\ldots,j, \tag{12}$$

where $c_k = \overline{f}_m(x_k)u(x_k)$, $k = 1,\dots,j$, $b_i = u(x_i)\overline{g}(x_i)$, $i = 1,\dots,j$, and $\overline{M}_k(x_i) = \int_0^\infty G(x_i,x)l_k^*(x)\mathrm{e}^{-x}\,\mathrm{d}x$. To compute the quantities $\overline{M}_k(t)$ we propose a recurrence relation (see Appendix).

If $(c_1,\dots,c_j)$ is the unique solution of the linear system (12), we define the Nyström interpolant

$$(\overline{f}_m u)(t) = u(t)\overline{a}(t)\sum_{k=1}^{j}\frac{\overline{M}_k(t)}{u(x_k)}c_k + u(t)\overline{g}(t)\,. \tag{13}$$

The following theorem gives us the conditions for stability and convergence of the method.

Theorem 3. *If $a^*(t) \sim 1/t^\alpha$ for $t \to \infty$ with $\alpha \geq 2(\lambda+\gamma)+1/2$ and $g \in W_r(u)$, $r \geq 1$, then, for sufficiently large m, say $m > m_0$, (11) has a unique solution, and the condition number of the matrix of the system (12) is independent of m.*
Then, if we denote by $\overline{f}^*$ and $\overline{f}_m^*$ the solution of (4) and (11), respectively, we have

$$\|(\overline{f}^* - \overline{f}_m^*)u\|_\infty \leq \mathcal{C}\frac{\|\overline{f}^*\|_{W_r(u)}}{M^{r/2}},$$

where $\mathcal{C} \neq \mathcal{C}(m,f)$ and $M = [m\theta/(1+\theta)]$.

Theorem 3 follows in the usual way from the following

Lemma 1. *Let K_m be defined in (10), and assume that the function $a^*(t)$ satisfies the same conditions as in Theorem 3. Then $K_m : C_u \to C_u$ is a continuous operator. Moreover,*

$$(\forall f \in C_u) \quad \lim_m \|(K-K_m)f\|_{u,\infty} = 0 \tag{14}$$

and

$$\lim_m \|(K-K_m)K_m\|_{C_u\to C_u} = 0. \tag{15}$$

Remark 1. If it is not possible to have an analytic representation of $\widetilde{g}(t)$ we can approximate it by using a Gaussian quadrature rule in the following way:

$$\widetilde{g}(t) = \int_0^\infty G(x,t)g(x)\,\mathrm{d}x = -\mathrm{e}^{-t}\int_0^t \sinh x\, g(x)\,\mathrm{d}x - \sinh t\int_t^\infty \mathrm{e}^{-x}g(x)\,\mathrm{d}x$$

$$= -\frac{t\mathrm{e}^{-t}}{2}\int_{-1}^1 \sinh\left(\frac{(y+1)t}{2}\right)g\left(\frac{(y+1)t}{2}\right)\mathrm{d}y - \mathrm{e}^{-t}\sinh t\int_0^\infty g(y+t)\mathrm{e}^{-y}\,\mathrm{d}y$$

by setting $y = 2x/t - 1$ and $y = x - t$, respectively, in the first and second integral. Now approximating these two last integrals by means of Gaussian quadrature rules with respect to the Legendre resp. Laguerre weight, we replace $\widetilde{g}(t)$ by the following

$$\widetilde{g}^*(t) = -\frac{t\mathrm{e}^{-t}}{2}\sum_{k=0}^{m}\overline{\lambda}_k \sinh\left(\frac{(\overline{x}_k+1)t}{2}\right)g\left(\frac{(\overline{x}_k+1)t}{2}\right) - \mathrm{e}^{-t}\sinh t\sum_{k=1}^{m}\lambda_k g(x_k+t),$$

where $\overline{x}_k$, $k=1,\dots,m$, are the zeros of the m-th Legendre polynomial and x_k those of the m-th Laguerre polynomial. In this case the sequence (13) is replaced by:

$$(\overline{f}_m u)(t) = u(t)\overline{a}(t)\sum_{k=1}^{j}\frac{\overline{M}_k(t)}{u(x_k)}c_k + u(t)\widetilde{g}^*(t)\,. \tag{16}$$

4 Numerical Examples

In this section we show, by some examples, that our theoretical results are confirmed by numerical tests. If the solution is unknown, we consider as exact the approximate solutions obtained for $m=300$ and, in all the tables, we show only the digit that are correct accordingly. All the computations were performed in 16-digits arithmetic.

In any example, *cond* denotes the condition number in the infinity norm of the matrix of the system (12).

Example 1. We consider the following BVP:

$$\begin{cases} f''(x) - \left[\dfrac{2}{(1+x)^2}+1\right]f(x) = 2\mathrm{e}^{-x}(x+1)(1-3x), \\ f(0)=f(\infty)=0. \end{cases}$$

Then $\widetilde{g}(x) = x\mathrm{e}^{-x}(x+1)(2x+3)/2$ and the exact solution is $f(x)=(x+1)^2x\mathrm{e}^{-x}$.

As can be seen in Table 1 it is sufficient to solve a linear system of order 27 to obtain a maximum absolute error of the order 10^{-15}.

Table 1 $\theta = 1/2$, $\lambda = 5/8$, $\gamma = 1/8$

m	j	maxerr
8	8	2.59e-7
16	14	1.38e-10
32	27	2.77e-15

Example 2. Now we consider the following BVP:

$$\begin{cases} f''(x) - \left[\dfrac{3\log(x+2)}{(1+x)^{11/4}}+1\right]f(x) = \mathrm{e}^{-x}(1+x^2), \\ f(0)=f(\infty)=0. \end{cases}$$

Then $\widetilde{g}(x) = -x\mathrm{e}^{-x}(2x^2+3x+9)/12$.

In Table 2 we represent the values of the weighted approximate solution $(f_m^* u)(x)$ at the points $x = 1,5,20$ when m increases. In this case, to obtain 14 exact decimal digits, we have to solve a well-conditioned truncated linear system of order 157 (see Table 3).

Table 2 $(\overline{f}_m^* u)(x)$, $\lambda = 3/4$, $\gamma = 1/8$, $\theta = 1/4$

m	j	$x=1$	$x=5$	$x=20$
32	20	−4.7205e-1	−4.6758e-1	−1.983e-3
64	40	−4.720599e-1	−4.67582e-1	−1.9838e-3
128	79	−4.720599878e-1	−4.67582716e-1	−1.9838488e-3
256	157	−4.7205998789815e-1	−4.67582716368695e-1	−1.98384885929e-3

Table 3 Condition number

m	j	cond
32	20	1.22004
64	40	1.22044
128	79	1.22011
164	135	1.21998
256	157	1.21976

In Fig. 1 we represent the graph of the weighted approximate solution obtained with $m = 256$.

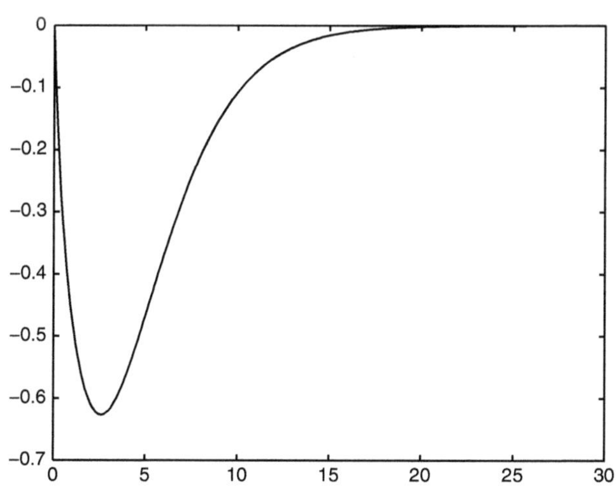

Fig. 1 The graph of the weighted solution $\overline{f}_{256}^*(x)u(x)$

Remark 2. If we apply (16) instead of (13), the numerical results are similar, as one can see in the next example.

Example 3. We consider the following BVP,

$$\begin{cases} f''(x) - \left[\dfrac{\arctan(x+3)}{(7+x)^3} + 1\right] f(x) = \mathrm{e}^{-x/2}\dfrac{\log(x+1)}{3}, \\ f(0) = f(\infty) = 0. \end{cases}$$

In Table 4 we represent the values of the weighted approximate solution $(f_m^* u)(x)$ at the points $x = 1, 5, 10, 20$ when m increases. Here also, to obtain 14 exact decimal digits, we have to solve a well-conditioned truncated linear system of order 157.

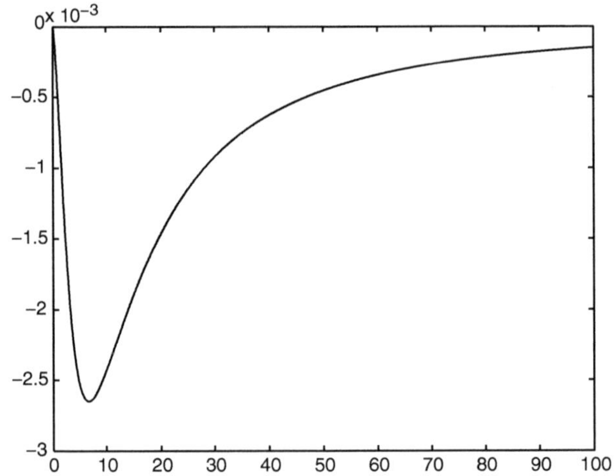

Fig. 2 The graph of the weighted solution $\overline{f}^*_{256}(x)u(x)$

Table 4 $(\overline{f}^*_m u)(x)$, $\lambda = 3/4$, $\gamma = 1/8$, $\theta = 1/4$

m	j	$x = 1$	$x = 5$	$x = 10$	$x = 20$
32	20	−5.6658426e-4	−2.532938e-3	−2.43043e-3	−1.46202e-3
64	40	−5.6658426e-4	−2.5329380e-3	−2.43043122e-3	−1.4620270e-3
128	79	−5.6658426e-4	−2.532938022e-3	−2.430431223e-3	−1.46202701e-3
256	157	−5.66584266699e-4	−2.532938022434e-3	−2.43043122322e-3	−1.4620270190e-3

Table 5 Condition number

m	j	cond
32	20	1.00161
64	40	1.00158
128	79	1.00156
256	157	1.00154

Also in this case, we solved well-conditioned linear systems of equations (see Table 5). In Fig. 2 we represent the graph of the weighted approximate solution obtained with $m = 256$.

5 Proofs

Proof of Theorem 1. Set $\sigma(x) = w(x)/u(x) = x^{\alpha-\gamma}e^{-x^\beta/2}/(1+x)^\lambda$. By the Remez-type inequality we have

$$\|L^{**}_{m+2}(w,f)\sigma\|_1 = \|L^*_{m+2}(w,f_j)\sigma\|_1 \le \mathcal{C}\|L^*_{m+2}(w,f_j)\sigma\|_{L^1(x_1,a_m)}$$
$$= \mathcal{C}\int_{x_1}^{a_m}\left(\sum_{k=1}^{j} f(x_k)l^*_k(x)\right)\widetilde{f}(x)\sigma(x)\,dx =: \mathcal{C}A(f),$$

with $\widetilde{f} = \operatorname{sgn}\left(L^{**}_{m+2}(w,f)\right)$. In order to estimate $A(f)$, we recall the following inequality,

$$\frac{1}{|p'_m(w,x_k)\sqrt{w(x_k)}|} \sim \sqrt[4]{a_m x_k}\Delta x_k\sqrt{1-\frac{x_k}{a_m}+\frac{1}{m^{2/3}}}, \quad k=1,\dots,m, \tag{17}$$

with constants in $\sim$ independent of m and k. Applying (17) to $A(f)$, we have

$$|A(f)| \le \mathcal{C}\|fu\|_{L^\infty[x_1,x_j]}a_m^{-1}\sum_{k=1}^{j}\frac{\sqrt[4]{a_m}\Delta x_k x_k^{\alpha/2-3/4-\gamma}}{(1+x_k)^\lambda}|\pi(x_k)|,$$

where

$$\pi(t) = \int_{x_1}^{a_m}\frac{x(a_m-x)p_m(w,x)q(x)-t(a_m-t)p_m(w,t)q(t)}{x-t}\frac{\widetilde{f}(x)\sigma(x)}{q(x)}\,dx$$

and q is an arbitrary polynomial of degree $lm-1$ (l fixed). Then $\pi \in \mathbb{P}_{lm+m}$ and, by using a Marcinkiewicz-type inequality, we have

$$|A(f)| \le \mathcal{C}\|f_j u\|_\infty a_m^{-1}\int_{x_1}^{a_m}\frac{\sqrt[4]{a_m}t^{\alpha/2-3/4-\gamma}}{(1+t)^\lambda}|\pi(t)|\,dt.$$

Now, we set $F(x) = x(a_m-x)p_m(w,x)\sqrt[4]{a_m}\widetilde{f}(x)\sigma(x)$; then

$$I := a_m^{-1}\int_{x_1}^{a_m}\frac{\sqrt[4]{a_m}t^{\alpha/2-3/4-\gamma}}{(1+t)^\lambda}|\pi(t)|\,dt = a_m^{-1}\int_{x_1}^{a_m}\frac{t^{\alpha/2-3/4-\gamma}}{(1+t)^\lambda}\left|\int_{x_1}^{a_m}\frac{F(x)}{x-t}\,dx\right.$$
$$\left. - \sqrt[4]{a_m}t(a_m-t)p_m(w,t)q(t)\int_{x_1}^{a_m}\frac{\sigma(x)\widetilde{f}(x)}{q(x)(x-t)}\,dx\right|\,dt$$

$$
\begin{aligned}
&\leq a_m^{-1}\int_{x_1}^{a_m}\left(\frac{t}{1+t}\right)^{\alpha/2-3/4-\gamma}(1+t)^{\alpha/2-3/4-\gamma-\lambda}\left|\int_{x_1}^{a_m}\frac{F(x)}{x-t}\,\mathrm{d}x\right|\,\mathrm{d}t\\
&\qquad+a_m^{-1}\int_{x_1}^{a_m}\frac{t^{\alpha/2-3/4-\gamma}}{(1+t)^{\gamma}}\sqrt[4]{a_m t}(a_m-t)|p_m(w,t)q(t)|\left|\int_{x_1}^{a_m}\frac{\sigma(x)\widetilde{f}(x)}{q(x)(x-t)}\,\mathrm{d}x\right|\,\mathrm{d}t\\
&=: I_1+I_2\,.
\end{aligned}
$$

The integrals are in the sense of principal value. Now we recall the following inequality (cf. [7, p. 440]):

$$
\begin{aligned}
&\int_0^{\infty}\left(\frac{t}{1+t}\right)^{r}(1+t)^{s}\left|\int_0^{\infty}\frac{F^*(x)}{x-t}\,\mathrm{d}x\right|\,\mathrm{d}t \qquad (18)\\
&\qquad\leq \mathcal{C}\left[1+\int_0^{\infty}\left(\frac{t}{1+t}\right)^{R}(1+t)^{S}|F^*(t)|\left(1+\log|F^*(t)|+\log^+ t\right)\,\mathrm{d}t\right],
\end{aligned}
$$

where $r>-1$, $R\leq 0$, $r\geq R$, $s<0$, $S\geq -1$, $s\leq S$, $\log^+ x=\log\max(1,x)$, and $\mathcal{C}\neq\mathcal{C}(f)$. In order to apply (18) to I_1, we set

$$
R=r=\frac{\alpha}{2}-\frac{3}{4}-\gamma \quad\text{and}\quad S=s=-\lambda+\frac{\alpha}{2}-\frac{3}{4}-\gamma
$$

and obtain

$$
\begin{aligned}
I_1\leq a_m^{-1}\Bigg[1+\int_0^{\infty}\left(\frac{t}{1+t}\right)^{\alpha/2-3/4-\gamma}&(1+t)^{-\lambda+\alpha/2-3/4-\gamma}\\
&\times|F(t)|(1+\log^+|F(t)|+\log^+ t)\,\mathrm{d}t\Bigg].
\end{aligned}
$$

As $|F(t)|\leq \mathcal{C}a_m t^{\alpha/2+3/4-\gamma}/(1+t)^{\lambda}$, by the assumptions on α, γ, and λ, we have

$$
I_1\leq \mathcal{C}\left[a_m^{-1}+\int_0^{\infty}\frac{t^{\alpha-2\gamma}}{(1+t)^{2\lambda}}\left(1+\log^+\frac{t^{\alpha/2+7/4-\gamma}}{(1+t)^{\lambda}}\right)\,\mathrm{d}t\right]\leq\mathcal{C}.
$$

In order to estimate I_2, we choose $q(x)\sim x^{\gamma}\mathrm{e}^{-x^{\beta}/2}$ (cf. [3, pp. 485–493], [6, pp. 200–208]) and apply (18) with $r=R=0$ and $s=S=-\lambda$. Then, by the assumptions on α, γ, and λ, we have

$$
\begin{aligned}
I_2&\leq \mathcal{C}\int_{x_1}^{a_m}\frac{1}{(1+t)^{\lambda}}\left|\int_{x_1}^{a_m}\frac{\sigma(x)\widetilde{f}(x)}{q(x)(x-t)}\,\mathrm{d}x\right|\,\mathrm{d}t\\
&\leq \mathcal{C}\left[1+\int_0^{\infty}\frac{t^{\alpha-2\gamma}}{(1+t)^{2\lambda}}\left(1+\log^+\frac{t^{\alpha-2\gamma+1}}{(1+t)^{\lambda}}\right)\,\mathrm{d}t\right]\leq\mathcal{C}
\end{aligned}
$$

and we conclude that (8) holds. In order to prove (9), we introduce the set

$$\mathbb{P}^*_{m+1} = \left\{ q \in \mathbb{P}_{m+1} \text{ s.t. } q(x_i) = 0,\, i > j \text{ and } q(a_m) = 0 \right\}$$

and observe that $\{l^*_0, l^*_1, \dots, l^*_j\}$ is a basis for $\mathbb{P}^*_{m+1}$ and $\forall q \in \mathbb{P}^*_{m+1}$, $L^*_{m+2}(w,q) = q$. Let $\widetilde{E}_m(f)_{u,p} = \inf_{P \in \mathbb{P}^*_{m+1}} \|(f-P)u\|_p$, and let $q \in \mathbb{P}^*_{m+1}$; then we have

$$\left\|(f - L^{**}_{m+2}(w,f))\frac{w}{u}\right\|_1 \le \left\|(f-q)\frac{w}{u}\right\|_1 + \left\|L^{**}_{m+2}(w,q-f)\frac{w}{u}\right\|_1 \le \mathcal{C}\|(f-q)u\|_\infty$$

by the assumptions on α, γ, and λ and by (8). Taking the infimum over $q \in \mathbb{P}^*_{m+1}$, we get

$$\left\|(f - L^{**}_{m+2}(w,f))\frac{w}{u}\right\|_1 \le \mathcal{C}\widetilde{E}_m(f)_{u,\infty}. \tag{19}$$

Now it is sufficient to observe that (see [5])

$$\widetilde{E}_m(f)_{u,p} \le \mathcal{C}(E_M(f)_{u,p} + \mathrm{e}^{-Am}\|fu\|_\infty),$$

where $M = [m(\theta/(1+\theta))^\beta]$. Applying (5) to the last inequality and recalling that $a_M \sim M^{1/\beta}$, (9) easily follows. □

Proof of Theorem 2. We first estimate $\|(Kf)u\|_\infty$. To this end we observe that

$$\begin{aligned}
|(Kf)(t)u(t)| &= \left|\mathrm{e}^t u(t) a^*(t) \int_0^\infty G(x,t) f(x) u(x) \frac{\mathrm{e}^{-x}}{u(x)}\,\mathrm{d}x\right| \\
&\le \mathcal{C}\|fu\|_\infty \mathrm{e}^t u(t)|a^*(t)| \left|\int_0^\infty \frac{G(x,t)}{x^\gamma(1+x)^\lambda}\mathrm{e}^{-x/2}\,\mathrm{d}x\right| \\
&\le \mathcal{C}\|fu\|_\infty \left[\mathrm{e}^t u(t)|a^*(t)|\mathrm{e}^{-t}\left|\int_0^t \frac{\sinh x}{x^\gamma(1+x)^\lambda}\mathrm{e}^{-x/2}\,\mathrm{d}x\right|\right. \\
&\quad \left. +\ \mathrm{e}^t u(t)|a^*(t)|\sinh t \int_t^\infty \frac{\mathrm{e}^{-3x/2}}{x^\gamma(1+x)^\lambda}\,\mathrm{d}x\right] \le \mathcal{C}\|fu\|_\infty,
\end{aligned}$$

because $\sinh x/(x^\gamma(1+x)^\lambda)$ is an increasing function. Taking the supremum over $t \ge 0$, we have

$$\|(Kf)u\|_\infty \le \mathcal{C}\|fu\|_\infty, \tag{20}$$

i.e., the continuity of the operator K in C_u. Now we estimate $\|(Kf)'\varphi u\|_\infty$. Thus,

$$\begin{aligned}
|(Kf)'(t)\varphi(t)u(t)| \le \mathcal{C}\|fu\|_\infty \varphi(t)u(t)\mathrm{e}^t &\left[|a^*(t)|\left|\int_0^\infty \frac{G(x,t)\mathrm{e}^{-x/2}}{x^\gamma(1+x)^\lambda}\,\mathrm{d}x\right|\right. \\
&\left. +|a'(t)|\left|\int_0^\infty \frac{G(x,t)\mathrm{e}^{-x/2}}{x^\gamma(1+x)^\lambda}\,\mathrm{d}x\right| + |a^*(t)|\left|\frac{\partial}{\partial t}\int_0^\infty \frac{G(x,t)\mathrm{e}^{-x/2}}{x^\gamma(1+x)^\lambda}\,\mathrm{d}x\right|\right].
\end{aligned}$$

In order to estimate the first two terms, we can proceed as in the previous case. For the last term we have to observe that

$$\left|\frac{\partial}{\partial t}\int_0^\infty \frac{G(x,t)\mathrm{e}^{-x/2}}{x^\gamma(1+x)^\lambda}\,\mathrm{d}x\right| = \left|\mathrm{e}^{-t}\int_0^t \frac{\sinh x\,\mathrm{e}^{-x/2}}{x^\gamma(1+x)^\lambda}\,\mathrm{d}x - \cosh t\int_t^\infty \frac{\mathrm{e}^{-3x/2}}{x^\gamma(1+x)^\lambda}\,\mathrm{d}x\right|$$

and proceed as in the previous case. Taking the supremum over $t \geq 0$, we get

$$\|(Kf)'\varphi u\|_\infty \leq \mathcal{C}\|fu\|_\infty. \tag{21}$$

By (20) and (21), it follows that

$$\|Kf\|_{W_1(u)} \leq \mathcal{C}\|fu\|_\infty. \tag{22}$$

Now we observe that, applying (5) with $r = 1$, we have

$$E_m(Kf)_{u,\infty} \leq \frac{\mathcal{C}}{m}\|Kf\|_{W_1(u)} \leq \frac{\mathcal{C}}{m}\|fu\|_\infty,$$

using (22) in the last inequality. Then

$$\lim_m \sup_{f\in C_u} \frac{E_m(Kf)_{u,\infty}}{\|fu\|_\infty} = 0,$$

from which it follows that the operator $K : C_u \to C_u$ is compact (see for instance [9, p. 44]). □

Proof of Lemma 1. In order to estimate $\|(K_m f)u\|_\infty$, we observe that, using the hypothesis on the function $a^*(t)$ and inequality (8), we have

$$\begin{aligned}
|(K_m f)(t)u(t)| &= \left|u(t)\mathrm{e}^t a^*(t)\int_0^\infty G(x,t)L^{**}_{m+2}(w,f;x)\mathrm{e}^{-x}\,\mathrm{d}x\right| \\
&\leq \mathcal{C}\mathrm{e}^t|a^*(t)|u(t)\int_0^\infty |G(x,t)|u(x)|L^{**}_{m+2}(w,f;x)|\frac{\mathrm{e}^{-x}}{u(x)}\,\mathrm{d}x \\
&= \mathcal{C}\left[|a^*(t)|u(t)\int_0^t \sinh x\mathrm{e}^{-x/2}x^\gamma(1+x)^\lambda|L^{**}_{m+2}(w,f;x)|\frac{\mathrm{e}^{-x}}{u(x)}\,\mathrm{d}x\right. \\
&\quad \left.+\,\mathrm{e}^t|a^*(t)|u(t)\sinh t\int_t^\infty \mathrm{e}^{-3x/2}x^\gamma(1+x)^\lambda|L^{**}_{m+2}(w,f;x)|\frac{\mathrm{e}^{-x}}{u(x)}\,\mathrm{d}x\right] \\
&\leq \mathcal{C}|a^*(t)|t^{2\gamma}(1+t)^{2\lambda}\|fu\|_{L^\infty([0,x_j])} \leq \mathcal{C}\|fu\|_\infty.
\end{aligned}$$

Taking the supremum over $t \geq 0$, we have

$$\|(K_m f)u\|_\infty \leq \mathcal{C}\|fu\|_\infty, \tag{23}$$

i.e., continuity of the operator $K_m : C_u \to C_u$.

In order to prove (14), we observe that

$$|(K-K_m)f(t)u(t)| = \left|\overline{a}(t)u(t)\int_0^\infty G(x,t)[f(x)-L^{**}_{m+2}(w,f;x)]\mathrm{e}^{-x}\,\mathrm{d}x\right|$$
$$\le \mathcal{C}\left\|[f-L^{**}_{m+2}(w,f)]\frac{w}{u}\right\|_1 \le \mathcal{C}\widetilde{E}_m(f)_{u,\infty}$$

by using the hypothesis on $a^*(t)$ and inequality (19). Taking the supremum over $t\ge 0$, we obtain

$$\|(K-K_m)f\|_{u,\infty} \le \mathcal{C}\widetilde{E}_m(f)_{u,\infty} \tag{24}$$

and then (14). In order to prove (15), we first prove that

$$\|K_m f\|_{W_1(u)} \le \mathcal{C}\|fu\|_\infty\,. \tag{25}$$

In fact,

$$|(K_m f)'(t)\varphi(t)u(t)| \le \left|u(t)\mathrm{e}^t[\mu a(t)-1]\varphi(t)\int_0^\infty G(x,t)L^{**}_{m+2}(w,f;x)\mathrm{e}^{-x}\,\mathrm{d}x\right|$$
$$+\left|u(t)\mathrm{e}^t\mu a'(t)\varphi(t)\int_0^\infty G(x,t)L^{**}_{m+2}(w,f;x)\mathrm{e}^{-x}\,\mathrm{d}x\right|$$
$$+\left|u(t)\mathrm{e}^t a^*(t)\frac{\partial}{\partial t}\int_0^\infty G(x,t)L^{**}_{m+2}(w,f;x)\mathrm{e}^{-x}\,\mathrm{d}x\right|.$$

As

$$\left|\frac{\partial}{\partial t}\int_0^\infty G(x,t)L^{**}_{m+2}(w,f;x)\mathrm{e}^{-x}\,\mathrm{d}x\right| = \left|\mathrm{e}^{-t}\int_0^t \sinh x L^{**}_{m+2}(w,f;x)\mathrm{e}^{-x}\,\mathrm{d}x\right.$$
$$\left. -\cosh t\int_t^\infty \mathrm{e}^{-x}L^{**}_{m+2}(w,f;x)\mathrm{e}^{-x}\,\mathrm{d}x\right|,$$

it is sufficient to apply inequality (8) and the hypothesis on $a^*(t)$ to obtain

$$\|(K_m f)'\varphi u\|_\infty \le \mathcal{C}\|fu\|_\infty\,. \tag{26}$$

Combining (23) and (26), (25) follows.

Now we can replace f by $K_m f$ in (24) and apply (5) with $r=1$. Then, we obtain

$$\|(K-K_m)K_m f\|_\infty \le \mathcal{C}\frac{\|K_m f\|_{W_1(u)}}{\sqrt{M}} \le \mathcal{C}\frac{\|f\|_{u,\infty}}{\sqrt{M}},$$

applying (25) in the last inequality. Now, taking the supremum on $\|f\|_{u,\infty}=1$, we conclude that

$$\sup_{\|f\|_{u,\infty}=1}\|(K-K_m)K_m f\|_\infty \le \frac{\mathcal{C}}{M},$$

from which (15) follows. □

Proof of Theorem 3. It is sufficient to apply Lemma 1, Theorem 4.1.1 from [1], and to observe that

$$\|(\overline{f}^* - \overline{f}_m^*)u\|_\infty \leq \mathcal{C}\|(I-K_m)^{-1}\|\|(K-K_m)\overline{f}^*\|_{u,\infty} \leq \mathcal{C}\frac{\|\overline{f}^*\|_{W_r(u)}}{M^{r/2}}$$

by using (24) and (5). □

Appendix

We propose a recurrence relation for calculating $\overline{M}_k(x) = \int_0^\infty G(x,t) l_k^*(x) \mathrm{e}^{-t}\,\mathrm{d}t$ by using the following three-term recurrence relation (see [8, pp. 101–102])

$$p_k(w,x) = \frac{x-2k+1}{k} p_{k-1}(w,x) - \frac{k-1}{k} p_{k-2}(w,x), \quad k = 2,\dots,m,$$

with $p_0(w,x) = 1$ and $p_1(w,x) = 1-x$, and the relation

$$\frac{\mathrm{d}}{\mathrm{d}x} p_k(w,x) = \frac{k}{x}\{p_k(w,x) - p_{k-1}(w,x)\}, \quad k = 1,\dots,m.$$

We proceed as follows:

$$\begin{aligned}
\overline{M}_k(x) &= \int_0^\infty G(x,t)\frac{l_k(t)t(a_m-t)}{x_k(a_m-x_k)}\mathrm{e}^{-t}\,\mathrm{d}t \\
&= \frac{1}{x_k}\int_0^\infty G(x,t)tl_k(t)\mathrm{e}^{-t}\,\mathrm{d}t - \frac{1}{x_k(a_m-x_k)}\int_0^\infty G(x,t)\frac{tp_m(w,t)}{p_m'(w,x_k)}\mathrm{e}^{-t}\,\mathrm{d}t \\
&= \frac{\lambda_k}{x_k}\sum_{l=0}^{m-1} p_l(w,x_k)M_l(x) - \frac{1}{(a_m-x_k)p_{m-1}(w,x_k)m}M_m(x),
\end{aligned}$$

where $M_k(x) = \int_0^\infty G(x,t)p_k(t)\mathrm{e}^{-t}\,\mathrm{d}t$. We observe that

$$M_{k+1}(x) = -\frac{\mathrm{e}^{-x}}{2}M_{k+1}^1(x) + \frac{\mathrm{e}^{-x}}{2}M_{k+1}^2(x) - \sinh x\, M_{k+1}^3(x), \quad k = 1,\dots,m,$$

where

$$M_{k+1}^1(x) = \int_0^x tp_{k+1}(w,t)\,\mathrm{d}t = \frac{x^2 p_{k+1}(w,x)}{k+3} - \frac{k+1}{k+3}M_k^1(x),$$

$$\begin{aligned}
M_{k+1}^2(x) &= \int_0^x \mathrm{e}^{-2t}tp_{k+1}(w,t)\,\mathrm{d}t \\
&= -\frac{\mathrm{e}^{-2x}x^2 p_k(w,x)}{2(k+1)} - \frac{3k}{2(k+1)}M_k^2(x) - \frac{k}{2(k+1)}M_{k-1}^2(x),
\end{aligned}$$

$$M_{k+1}^3(x) = \int_x^\infty \mathrm{e}^{-2t} t p_{k+1}(w,t)\,\mathrm{d}t$$
$$= \frac{\mathrm{e}^{-2x}x^2 p_k(w,x)}{2(k+1)} - \frac{3k}{2(k+1)} M_k^3(x) - \frac{k}{2(k+1)} M_{k-1}^3(x),$$

and

$$M_0^1(x) = \frac{x^2}{2}, \quad M_0^2(x) = \frac{1}{4} - \mathrm{e}^{-2x}\left(\frac{x}{2}+\frac{1}{4}\right),$$
$$M_1^2(x) = \frac{x^2\mathrm{e}^{-2x}}{2}, \quad M_0^3(x) = \mathrm{e}^{-2x}\left(\frac{x}{2}+\frac{1}{4}\right), \quad M_1^3(x) = \frac{x^2\mathrm{e}^{-2x}}{2}.$$

Acknowledgements The author is very grateful to Professor Giuseppe Mastroianni for his useful remarks and interesting discussions on the topic.

References

1. Atkinson, K.E.: The Numerical Solution of Integral Equations of the Second Kind. Cambridge Monographs on Applied and Computational Mathematics, Cambridge University Press, Cambridge (1997)
2. De Bonis, M.C., Mastroianni, G., Viggiani, M.: $K-$functionals, Moduli of Smoothness and Weighted Best Approximation on the semi-axis. In: Functions, Series, Operators, Alexits Memorial Conference, Edited by L. Leindler, F. Schipp, J. Szabados, János Bolyai Mathematical Society, 181–211 (2002)
3. Levin, A.L., Lubinsky, D.S.: Christoffel functions, orthogonal polynomials and Nevai's conjecture for Freud weights. Constr. Approx. **8**, no. 4, 463–535 (1992)
4. Mastroianni, G., Milovanović, G.V.: Some numerical methods for second kind Fredholm integral equation on the real semiaxis. IMA J. Numer. Anal. doi:10.1093/imanum/drn056
5. Mastroianni, G., Occorsio, D.: Some quadrature formulae with non standard weights. to appear in J. Comput. Appl. Math. DOI: 10.1016/j.cam.2010.06.011
6. Mhaskar, H.N.: Introduction to the Theory of Weighted Polynomial Approximation. World Scientific, Singapore, New Jersey, London, Hong Kong (1996)
7. Muckenhoupt, B.: Mean convergence of Hermite and Laguerre series II. Trans. Am. Math. Soc. **147**, 433–460 (1970)
8. Szegö, G.: Orthogonal Polynomials. American Mathematical Society Colloquium Publications, Volume XXIII (1939)
9. Timan, A.F.: Theory of Approximation of Functions of a Real Variable. Dover Publications, Inc. New York (1994)

Finite Difference Approximation of a Hyperbolic Transmission Problem

Boško S. Jovanović

Dedicated to Professor Gradimir V. Milovanović on the occasion of his 60th birthday

1 Introduction

It happens frequently in applications that the domain under consideration is occupied by several materials with different physical properties. Such problems can be modelled by partial differential equations whose input data and the solutions have discontinuities across one or several interfaces (lines, surfaces, etc.). Corresponding boundary value problems are commonly called transmission, diffraction or interface problems [1–4, 8].

In some cases, the domain in which we have to look for the solution is disjoint. For example, this situation occurs when the solution in the intermediate region is known, or can be determined from a simpler equation. The effect of the intermediate region can be taken into account by means of nonlocal jump conditions [4, 5].

A finite element approximation of an elliptic boundary value problem with nonlocal jump conditions is proposed in [9]. In [6] an initial boundary value problem for one-dimensional parabolic equation in two disjoint intervals is considered and a finite difference scheme for its solution is proposed and analyzed. A similar nonlinear problem is treated in [13]. Its well-posedness is proved and a monotone iterative method for its solution is proposed. An analogous hyperbolic problem is considered in [7]. In this paper we continue its investigation. The well-posedness of the initial boundary value problem and the convergence of the corresponding finite difference scheme are proved under weaker conditions.

Boško S. Jovanović
Faculty of Mathematics, University of Belgrade, Studentski trg 16, 11000 Belgrade, Serbia,
e-mail: bosko@matf.bg.ac.rs

W. Gautschi et al. (eds.), *Approximation and Computation: In Honor of Gradimir V. Milovanović*, Springer Optimization and Its Applications 42,
DOI 10.1007/978-1-4419-6594-3_21,

This paper is organized as follows. The investigated hyperbolic initial-boundary-value problem (IBVP) is formulated in Sect. 2. Section 3 is devoted to the analysis of the existence and the uniqueness of its weak solution. In Sect. 4 we introduce a finite difference scheme (FDS) approximating IBVP and investigate its convergence.

2 Formulation of the Initial Boundary Value Problem

We consider the following IBVP: Find functions $u_1(x,t)$ and $u_2(x,t)$ that satisfy the system of hyperbolic equations

$$\frac{\partial^2 u_1}{\partial t^2} - \frac{\partial}{\partial x}\left(p_1(x)\frac{\partial u_1}{\partial x}\right) = f_1(x,t), \quad x \in \Omega_1 \equiv (a_1,b_1),\ t > 0, \tag{1}$$

$$\frac{\partial^2 u_2}{\partial t^2} - \frac{\partial}{\partial x}\left(p_2(x)\frac{\partial u_2}{\partial x}\right) = f_2(x,t), \quad x \in \Omega_2 \equiv (a_2,b_2),\ t > 0, \tag{2}$$

where $-\infty < a_1 < b_1 < a_2 < b_2 < +\infty$, the internal boundary conditions of Robin-Dirichlet type

$$p_1(b_1)\frac{\partial u_1(b_1,t)}{\partial x} + \alpha_1 u_1(b_1,t) = \beta_1 u_2(a_2,t), \tag{3}$$

$$-p_2(a_2)\frac{\partial u_2(a_2,t)}{\partial x} + \alpha_2 u_2(a_2,t) = \beta_2 u_1(b_1,t), \tag{4}$$

the simplest external Dirichlet boundary conditions

$$u_1(a_1,t) = 0, \quad u_2(b_2,t) = 0, \tag{5}$$

and the initial conditions

$$u_1(x,0) = u_{10}(x), \quad u_2(x,0) = u_{20}(x), \tag{6}$$

$$\frac{\partial u_1}{\partial t}(x,0) = u_{11}(x), \quad \frac{\partial u_2}{\partial t}(x,0) = u_{21}(x). \tag{7}$$

Throughout the paper we assume that

$$p_i(x) \in L_\infty(\Omega_i), \quad 0 < p_{i0} \le p_i(x) \quad \text{a.e. in } \Omega_i, \quad i = 1,2. \tag{8}$$

We require p_1 to be continuous in a suitable left neighbourhood of b_1 and p_2 to be continuous in a suitable right neighbourhood of a_2. We also assume that

$$\alpha_i > 0,\ \beta_i > 0, \quad i = 1,2. \tag{9}$$

By c_j $(j = 0,1,\ldots)$ we denote positive constants, while C stands for a generic positive constant, independent of the solution of the IBVP and the mesh-sizes, which can take different values in different formulas.

3 Existence and Uniqueness of Weak Solution

Let the conditions (9) hold. We introduce the product space

$$L = L_2(\Omega_1) \times L_2(\Omega_2) = \{v = (v_1, v_2) \,|\, v_i \in L_2(\Omega_i)\},$$

endowed with the inner product and associated norm

$$(u,v)_L = \beta_2 (u_1, v_1)_{L_2(\Omega_1)} + \beta_1 (u_2, v_2)_{L_2(\Omega_2)}, \quad \|v\|_L = (v,v)_L^{1/2},$$

where $(u_i, v_i)_{L_2(\Omega_i)} = \int_{\Omega_i} u_i v_i \,\mathrm{d}x$. We also define the spaces

$$H^k = \{v = (v_1, v_2) \,|\, v_i \in H^k(\Omega_i)\}, \quad k = 1, 2, \ldots,$$

endowed with the inner products and norms

$$(u,v)_{H^k} = \beta_2 (u_1, v_1)_{H^k(\Omega_1)} + \beta_1 (u_2, v_2)_{H^k(\Omega_2)}, \quad \|v\|_{H^k} = (v,v)_{H^k}^{1/2},$$

where

$$(u_i, v_i)_{H^k(\Omega_i)} = \sum_{j=0}^{k} \left(\frac{\mathrm{d}^j u_i}{\mathrm{d}x^j}, \frac{\mathrm{d}^j v_i}{\mathrm{d}x^j} \right)_{L_2(\Omega_i)}.$$

In particular, we set

$$H_0^1 = \{v = (v_1, v_2) \in H^1 \,|\, v_1(a_1) = 0,\ v_2(b_2) = 0\}.$$

Finally, we define the following bilinear form:

$$\begin{aligned} A(u,v) = {} & \beta_2 \int_{\Omega_1} \left(p_1 \frac{\mathrm{d}u_1}{\mathrm{d}x} \frac{\mathrm{d}v_1}{\mathrm{d}x} \right) \mathrm{d}x + \beta_1 \int_{\Omega_2} \left(p_2 \frac{\mathrm{d}u_2}{\mathrm{d}x} \frac{\mathrm{d}v_2}{\mathrm{d}x} \right) \mathrm{d}x \\ & + \beta_2 \alpha_1 u_1(b_1) v_1(b_1) + \beta_1 \alpha_2 u_2(a_2) v_2(a_2) \\ & - \beta_1 \beta_2 \left[u_1(b_1) v_2(a_2) + u_2(a_2) v_1(b_1) \right]. \end{aligned} \tag{10}$$

Lemma 1. *Under the conditions* (8) *and* (9) *the bilinear form A, defined by* (10)*, is symmetric and bounded on* $H^1 \times H^1$*. Moreover, this form also satisfies Gårding's inequality on* H_0^1 *i.e., there exist positive constants* c_1 *and* k *such that*

$$A(u,u) + k \|u\|_L^2 \ge c_1 \|u\|_{H^1}^2, \quad \forall\, v \in H_0^1. \tag{11}$$

Proof. The symmetry of A is obvious, while its boundedness follows from (8) and the embeddings $H^1(\Omega_i) \subset C(\overline{\Omega}_i)$, $i = 1, 2$.

From the Poincaré type inequalities

$$\int_{\Omega_i} u_i^2(x) \,\mathrm{d}x \le \frac{b_i^2 - a_i^2}{2} \int_{\Omega_i} \left(\frac{\mathrm{d}u_i}{\mathrm{d}x}(x) \right)^2 \mathrm{d}x, \quad i = 1, 2,$$

there immediately follows that for $u \in H_0^1$

$$\beta_2 \int_{\Omega_1} p_1 \left(\frac{\mathrm{d}u_1}{\mathrm{d}x}\right)^2 \mathrm{d}x + \beta_1 \int_{\Omega_2} p_2 \left(\frac{\mathrm{d}u_2}{\mathrm{d}x}\right)^2 \mathrm{d}x \geq c_0 \|u\|_{H^1}^2,$$

where

$$c_0 = \max\left\{\frac{2p_{10}}{b_1^2 - a_1^2 + 2}, \frac{2p_{20}}{b_2^2 - a_2^2 + 2}\right\}.$$

Under the condition (9), for $\theta > 0$ we have

$$\beta_2 \alpha_1 u_1^2(b_1) + \beta_1 \alpha_2 u_2^2(a_2) - 2\beta_1 \beta_2 u_1(b_1) u_2(a_2) \geq -\beta_1 \beta_2 \theta u_1^2(b_1) - \frac{\beta_1 \beta_2}{\theta} u_2^2(a_2).$$

Further, for $\varepsilon > 0$,

$$\begin{aligned} u_1^2(b_1) &= \int_{\Omega_1} \frac{\mathrm{d}(u_1^2)}{\mathrm{d}x} \mathrm{d}x = \int_{\Omega_1} 2u_1 \frac{\mathrm{d}u_1}{\mathrm{d}x} \mathrm{d}x \leq \varepsilon \int_{\Omega_1} \left(\frac{\mathrm{d}u_1}{\mathrm{d}x}\right)^2 \mathrm{d}x + \frac{1}{\varepsilon} \int_{\Omega_1} u_1^2 \,\mathrm{d}x \\ &\leq \varepsilon \|u_1\|_{H^1(\Omega_1)}^2 + \frac{1}{\varepsilon} \|u_1\|_{L_2(\Omega_1)}^2, \end{aligned}$$

and analogously,

$$u_2^2(a_2) \leq \varepsilon \|u_2\|_{H^1(\Omega_2)}^2 + \frac{1}{\varepsilon} \|u_2\|_{L_2(\Omega_2)}^2.$$

Taking $\theta = \sqrt{\beta_2/\beta_1}$, from the previous inequalities we get

$$A(u,u) \geq \left(c_0 - \varepsilon\sqrt{\beta_1 \beta_2}\right) \|u\|_{H^1}^2 - \frac{\sqrt{\beta_1 \beta_2}}{\varepsilon} \|u\|_L^2,$$

from which for sufficiently small $\varepsilon > 0$ there follows (11). □

Let Ω be a domain in $\mathbb{R}^n$ and $u(t)$ a function mapping Ω into a Hilbert space H. In a standard manner (see [10]) we define the Sobolev space of vector-valued functions $H^k(\Omega, H)$, endowed with the inner product

$$(u,v)_{H^k(\Omega,H)} = \int_\Omega \sum_{|\alpha| \leq k} (D^\alpha u(t), D^\alpha v(t))_H \,\mathrm{d}t, \quad k = 0, 1, 2, \ldots .$$

For $k = 0$ we set $L_2(\Omega, H) = H^0(\Omega, H)$. We also introduce the space of continuous vector-valued functions $C(\overline{\Omega}, H)$, endowed with the norm

$$\|u\|_{C(\overline{\Omega},H)} = \max_{t \in \overline{\Omega}} \|u(t)\|_H.$$

Multiplying (1) by $v_1(x)$, with $v_1(a_1) = 0$, and integrating by parts, using condition (3), we obtain the identity

$$\left(\frac{\partial^2 u_1}{\partial t^2}(\cdot,t),v_1\right)_{L_2(\Omega_1)}+\int_{\Omega_1}\left(p_1\frac{\partial u_1}{\partial x}\frac{\mathrm{d}v_1}{\mathrm{d}x}\right)\mathrm{d}x$$
$$+\alpha_1 u_1(b_1,t)v_1(b_1)-\beta_1 u_2(a_2,t)v_1(b_1)=(f_1(\cdot,t),v_1)_{L_2(\Omega_1)}.$$

Analogously, multiplying (2) by $v_2(x)$, with $v_2(b_2)=0$, and integrating by parts, we obtain

$$\left(\frac{\partial^2 u_2}{\partial t^2}(\cdot,t),v_2\right)_{L_2(\Omega_2)}+\int_{\Omega_2}\left(p_2\frac{\partial u_2}{\partial x}\frac{\mathrm{d}v_2}{\mathrm{d}x}\right)\mathrm{d}x$$
$$+\alpha_2 u_2(a_2,t)v_2(a_2)-\beta_2 u_1(b_1,t)v_2(a_2)=(f_2(\cdot,t),v_2)_{L_2(\Omega_2)}.$$

Now multiplying the first of these identities by β_2, the second by β_1, and summing up, we get the weak form of (1)–(5):

$$\left(\frac{\partial^2 u}{\partial t^2}(\cdot,t),v\right)_L+A(u(\cdot,t),v)=(f(\cdot,t),v)_L,\quad \forall v\in H_0^1. \tag{12}$$

Applying Theorem 29.1 from [14] to (12), we immediately obtain the following assertion.

Theorem 1. *Let the assumptions* (8)–(9) *hold and suppose that* $u_0=(u_{10},u_{20})\in H_0^1$, $u_1=(u_{11},u_{21})\in L$, $f=(f_1,f_2)\in L_2((0,T),L)$. *Then the IBVP* (1)–(7) *has a unique weak solution* $u\in L_2((0,T),H_0^1)\cap H^1((0,T),L)$, *and it depends continuously on* f, u_0 *and* u_1. *For* $0<t<T<+\infty$ *the following a priori estimate holds:*

$$\left\|\frac{\partial u}{\partial t}(\cdot,t)\right\|_L^2+\|u(\cdot,t)\|_{H^1}^2\le c_2\mathrm{e}^{(2kT+1)t}\left(\|u_0\|_{H^1}^2+\|u_1\|_L^2+\int_0^T\|f(\cdot,t)\|_L^2\,\mathrm{d}t\right). \tag{13}$$

4 Finite Difference Approximation

4.1 Meshes and Finite Difference Operators

Let $\overline{\omega}_{1,h_1}$ be a uniform mesh in $\overline{\Omega}_1$ with the step size $h_1=(b_1-a_1)/n_1$, $\omega_{1,h_1}=\overline{\omega}_{1,h_1}\cap\Omega_1$, $\omega_{1,h_1}^-=\omega_{1,h_1}\cup\{a_1\}$, $\omega_{1,h_1}^+=\omega_{1,h_1}\cup\{b_1\}$. Analogously, we define a uniform mesh $\overline{\omega}_{2,h_2}$ in $\overline{\Omega}_2$ with the step size $h_2=(b_2-a_2)/n_2$ and its submeshes $\omega_{2,h_2}=\overline{\omega}_{2,h_2}\cap\Omega_2$, $\omega_{2,h_2}^-=\omega_{2,h_2}\cup\{a_2\}$, $\omega_{2,h_2}^+=\omega_{2,h_2}\cup\{b_2\}$. Finally, we introduce a uniform mesh $\overline{\omega}_\tau$ in $[0,T]$ with the step size $\tau=T/n$ and set $\omega_\tau=\overline{\omega}_\tau\cap(0,T)$, $\omega_\tau^-=\omega_\tau\cup\{0\}$, $\omega_\tau^+=\omega_\tau\cup\{T\}$. We shall consider vector-functions of the form $v=(v_1,v_2)$, where v_i is a mesh function defined on $\overline{\omega}_{i,h_i}\times\overline{\omega}_\tau$, $i=1,2$. We define difference quotients in the usual way (see [11]):

$$v_{i,x}(x,t)=\frac{v_i(x+h_i,t)-v_i(x,t)}{h_i}=v_{i,\bar{x}}(x+h_i,t),$$
$$v_{i,t}(x,t)=\frac{v_i(x,t+\tau)-v_i(x,t)}{\tau}=v_{i,\bar{t}}(x,t+\tau).$$

We shall use the following notational conventions:

$$v_i = v_i(x,t), \quad \overline{v}_i = \frac{1}{2}\left[v_i(x,t) + v_i(x,t+\tau)\right], \quad \check{v}_i = \frac{1}{2}\left[v_i(x,t-\tau) + v_i(x,t)\right],$$

$$v_i^{(\sigma)} = \sigma v_i(x,t+\tau) + (1-2\sigma)v_i(x,t) + \sigma v_i(x,t-\tau).$$

In this section we shall assume that u_i belongs to $H^4(Q_i)$, where $Q_i = \Omega_i \times (0,T)$, while $p_i \in H^3(\Omega_i)$, $i = 1,2$. We also assume that

$$h_1 \asymp h_2 \asymp \tau. \tag{14}$$

4.2 Finite Difference Scheme

We approximate (1) and (2) in the following manner:

$$v_{1,\bar{t}t} - \left(\overline{p}_1 v_{1,\bar{x}}^{(1/4)}\right)_x = f_1, \quad x \in \omega_{1,h_1}, \ t \in \omega_\tau, \tag{15}$$

$$v_{2,\bar{t}t} - \left(\overline{p}_2 v_{2,\bar{x}}^{(1/4)}\right)_x = f_2, \quad x \in \omega_{2,h_2}, \ t \in \omega_\tau, \tag{16}$$

where $\overline{p}_i(x) = [p_i(x) + p_i(x-h_i)]/2$, $i = 1,2$. To ensure the same order of approximation for $x = b_1$ and $x = a_2$ we set

$$\begin{aligned} v_{1,\bar{t}t}(b_1,t) &+ \frac{h_1}{3}\frac{1}{p_1(b_1)}\left[\widetilde{\alpha}_1 v_{1,\bar{t}t}(b_1,t) - \widetilde{\beta}_1 v_{2,\bar{t}t}(a_2,t)\right] + \frac{h_1}{6}\frac{p_1'(b_1)}{p_1(b_1)} v_{1,\bar{t}t}(b_1,t) \\ &+ \frac{2}{h_1}\left[\overline{p}_1(b_1) v_{1,\bar{x}}^{(1/4)}(b_1,t) + \widetilde{\alpha}_1 v_1^{(1/4)}(b_1,t) - \widetilde{\beta}_1 v_2^{(1/4)}(a_2,t)\right] \\ &= f_1(b_1,t) - \frac{h_1}{3} f_{1,\bar{x}}(b_1,t) + \frac{h_1}{6}\frac{p_1'(b_1)}{p_1(b_1)} f_1(b_1,t), \quad t \in \omega_\tau, \end{aligned} \tag{17}$$

$$\begin{aligned} v_{2,\bar{t}t}(a_2,t) &+ \frac{h_2}{3}\frac{1}{p_2(a_2)}\left[\widetilde{\alpha}_2 v_{2,\bar{t}t}(a_2,t) - \widetilde{\beta}_2 v_{1,\bar{t}t}(b_1,t)\right] - \frac{h_2}{6}\frac{p_2'(a_2)}{p_2(a_2)} v_{2,\bar{t}t}(a_2,t) \\ &- \frac{2}{h_2}\left[\overline{p}_2(a_2+h_2) v_{2,x}^{(1/4)}(a_2,t) - \widetilde{\alpha}_2 v_2^{(1/4)}(a_2,t) + \widetilde{\beta}_2 v_1^{(1/4)}(b_1,t)\right] \\ &= f_2(a_2,t) + \frac{h_2}{3} f_{2,x}(a_2,t) - \frac{h_2}{6}\frac{p_2'(a_2)}{p_2(a_2)} f_2(a_2,t), \quad t \in \omega_\tau, \end{aligned} \tag{18}$$

where $\widetilde{\alpha}_i = \alpha_i D_i$, $\widetilde{\beta}_i = \beta_i D_i$, $i = 1,2$, and

$$D_1 = \left[1 + \frac{h_1^2}{12}\left(\frac{p_1''(b_1)}{p_1(b_1)} + \frac{(p_1'(b_1))^2}{p_1^2(b_1)}\right)\right], \quad D_2 = \left[1 + \frac{h_2^2}{12}\left(\frac{p_2''(a_2)}{p_2(a_2)} + \frac{(p_2'(a_2))^2}{p_2^2(a_2)}\right)\right].$$

The Dirichlet boundary conditions (5) and initial conditions (6) can be satisfied exactly:

$$v_1(a_1,t)=0,\ \ v_2(b_2,t)=0,\quad t\in\overline{\omega}_\tau, \tag{19}$$
$$v_i(x,0)=u_{i0}(x),\quad x\in\omega^{\pm}_{i,h_i},\quad i=1,2, \tag{20}$$

while the initial conditions (7) we approximate by:

$$v_{i,t}(x,0)=u_{i1}(x)+\frac{\tau}{2}\left[\frac{\mathrm{d}}{\mathrm{d}x}\left(p_i\frac{\mathrm{d}u_{i0}}{\mathrm{d}x}\right)+f_i(x,0)\right],\quad x\in\omega^{\pm}_{i,h_i},\quad i=1,2. \tag{21}$$

At each time level $t=j\tau$ the FDS (15)–(21) reduces to a tridiagonal linear system with n_1+n_2 unknowns, which can be solved by the Thomas algorithm. In this way, the FDS (15)–(21) is computationally efficient.

4.3 Convergence of the Finite Difference Scheme

Let $u=(u_1,u_2)$ be the solution of the IBVP (1)–(7) and $v=(v_1,v_2)$ the solution of the FDS (15)–(21). Then the error $z=u-v$ satisfies the following FDS:

$$z_{1,\bar{t}t}-\left(\overline{p}_1 z^{(1/4)}_{1,\bar{x}}\right)_x=\varphi_1,\quad x\in\omega_{1,h_1},\ t\in\omega_\tau, \tag{22}$$

$$\begin{aligned}
&z_{1,\bar{t}t}(b_1,t)+\frac{h_1}{3}\frac{1}{p_1(b_1)}\left[\widetilde{\alpha}_1 z_{1,\bar{t}t}(b_1,t)-\widetilde{\beta}_1 z_{2,\bar{t}t}(a_2,t)\right]+\frac{h_1}{6}\frac{p_1'(b_1)}{p_1(b_1)}z_{1,\bar{t}t}(b_1,t)\\
&\quad+\frac{2}{h_1}\left[\overline{p}_1(b_1)z^{(1/4)}_{1,\bar{x}}(b_1,t)+\widetilde{\alpha}_1 z_1^{(1/4)}(b_1,t)-\widetilde{\beta}_1 z_2^{(1/4)}(a_2,t)\right]\\
&\quad=\varphi_1(b_1,t),\quad t\in\omega_\tau,
\end{aligned} \tag{23}$$

$$z_{2,\bar{t}t}-\left(\overline{p}_2 z^{(1/4)}_{2,\bar{x}}\right)_x=\varphi_2,\quad x\in\omega_{2,h_2},\ t\in\omega_\tau, \tag{24}$$

$$\begin{aligned}
&z_{2,\bar{t}t}(a_2,t)+\frac{h_2}{3}\frac{1}{p_2(a_2)}\left[\widetilde{\alpha}_2 z_{2,\bar{t}t}(a_2,t)-\widetilde{\beta}_2 z_{1,\bar{t}t}(b_1,t)\right]-\frac{h_2}{6}\frac{p_2'(a_2)}{p_2(a_2)}z_{2,\bar{t}t}(a_2,t)\\
&\quad-\frac{2}{h_2}\left[\overline{p}_2(a_2+h_2)z^{(1/4)}_{2,x}(a_2,t)-\widetilde{\alpha}_2 z_2^{(1/4)}(a_2,t)+\widetilde{\beta}_2 z_1^{(1/4)}(b_1,t)\right]\\
&\quad=\varphi_2(a_2,t),\quad t\in\omega_\tau,
\end{aligned} \tag{25}$$

$$z_1(a_1,t)=0,\quad z_2(b_2,t)=0,\quad t\in\overline{\omega}_\tau, \tag{26}$$
$$z_i(x,0)=0,\quad x\in\omega^{\pm}_{i,h_i},\quad i=1,2, \tag{27}$$
$$z_{i,t}(x,0)=\zeta_i(x),\quad x\in\omega^{\pm}_{i,h_i},\quad i=1,2, \tag{28}$$

where

$$\varphi_i=\psi_i+\chi_i,\quad x\in\omega_{i,h_i},\quad t\in\omega_\tau,\quad i=1,2,$$

$$\varphi_1(b_1,t) = \psi_1(b_1,t) + \frac{h_1}{6}\frac{p_1'(b_1)}{p_1(b_1)}\psi_1(b_1,t) + \frac{h_1}{3}\eta_1(b_1,t) + \widetilde{\chi}_1(b_1,t),$$

$$\varphi_2(a_2,t) = \psi_2(a_2,t) - \frac{h_2}{6}\frac{p_2'(a_2)}{p_2(a_2)}\psi_2(a_2,t) + \frac{h_2}{3}\eta_2(a_2,t) + \widetilde{\chi}_2(a_2,t),$$

$$\psi_i = u_{i,\bar{t}t} - \frac{\partial^2 u_i}{\partial t^2}, \quad \chi_i = \frac{\partial}{\partial x}\left(p_i\frac{\partial u_i}{\partial x}\right) - (\overline{p}_i u_{i,\bar{x}}^{(1/4)})_x,$$

$$\widetilde{\chi}_1(b_1,t) = \left\{\frac{\partial}{\partial x}\left(p_1\frac{\partial u_1}{\partial x}\right) - \frac{h_1}{3}\left[\frac{\partial}{\partial x}\left(p_1\frac{\partial u_1}{\partial x}\right)\right]_{\bar{x}} + \frac{h_1}{6}\frac{p_1'}{p_1}\frac{\partial}{\partial x}\left(p_1\frac{\partial u_1}{\partial x}\right)\right\}\Bigg|_{(b_1,t)}$$
$$+\frac{2}{h_1}\left(\overline{p}_1(b_1)u_{1,\bar{x}}^{(1/4)}(b_1,t) + \widetilde{\alpha}_1 u_1^{(1/4)}(b_1,t) - \widetilde{\beta}_1 u_2^{(1/4)}(a_2,t)\right),$$

$$\widetilde{\chi}_2(a_2,t) = \left\{\frac{\partial}{\partial x}\left(p_2\frac{\partial u_2}{\partial x}\right) + \frac{h_2}{3}\left[\frac{\partial}{\partial x}\left(p_2\frac{\partial u_2}{\partial x}\right)\right]_{\bar{x}} - \frac{h_2}{6}\frac{p_2'}{p_2}\frac{\partial}{\partial x}\left(p_2\frac{\partial u_2}{\partial x}\right)\right\}\Bigg|_{(a_2,t)}$$
$$-\frac{2}{h_2}\left(\overline{p}_2(a_2+h_2)u_{2,x}^{(1/4)}(a_2,t) - \widetilde{\alpha}_2 u_2^{(1/4)}(a_2,t) + \widetilde{\beta}_2 u_1^{(1/4)}(b_1,t)\right),$$

$$\eta_1(b_1,t) = \left(\frac{\partial^2 u_1}{\partial t^2}\right)_{\bar{x}}\Bigg|_{(b_1,t)} + \frac{1}{p_1(b_1)}\left[\widetilde{\alpha}_1 u_{1,\bar{t}t}(b_1,t) - \widetilde{\beta}_1 u_{2,\bar{t}t}(a_2,t)\right],$$

$$\eta_2(a_2,t) = -\left(\frac{\partial^2 u_2}{\partial t^2}\right)_{x}\Bigg|_{(a_2,t)} + \frac{1}{p_2(a_2)}\left[\widetilde{\alpha}_2 u_{2,\bar{t}t}(a_2,t) - \widetilde{\beta}_2 u_{1,\bar{t}t}(b_1,t)\right],$$

$$\zeta_i(x) = u_{i,t}(x,0) - u_{i1}(x) - \frac{\tau}{2}\left[\frac{\mathrm{d}}{\mathrm{d}x}\left(p_i\frac{\mathrm{d}u_{i0}}{\mathrm{d}x}\right) + f_i(x,0)\right], \quad x \in \omega_{i,h_i}^{\pm}, \quad i = 1,2.$$

To obtain the discrete analogue of the a priori estimate (13), we introduce the discrete inner products and associated norms:

$$(v,w)_{L_h} = \widetilde{\beta}_2 \sum_{x\in\omega_{1,h_1}^+} v_1 w_1 \hbar_1 + \widetilde{\beta}_1 \sum_{x\in\omega_{2,h_2}^-} v_2 w_2 \hbar_2, \quad \|v\|_{L_h}^2 = (v,v)_{L_h},$$

$$(v,w)_{L_{h'}} = \widetilde{\beta}_2 h_1 \sum_{x\in\omega_{1,h_1}^+} v_1 w_1 + \widetilde{\beta}_1 h_2 \sum_{x\in\omega_{2,h_2}^+} v_2 w_2, \quad \|v\|_{L_{h'}}^2 = (v,v)_{L_{h'}},$$

where we denoted $\hbar_i = \hbar_i(x) = h_i$, $x \in \omega_{i,h_i}$, $i = 1,2$, $\hbar_1(b_1) = h_1/2$, and $\hbar_2(a_2) = h_2/2$. We also define the following discrete norm

$$\|v\|_{H_h^1}^2 = \|v_{\bar{x}}\|_{L_{h'}}^2 + \|v\|_{L_h}^2.$$

Theorem 2. *Let $p_i \in C(\overline{\Omega}_i)$, for $i = 1,2$, and let the assumptions* (8), (9) *and* (14) *hold. Then the solution z of FDS* (22)–(28) *satisfies the a priori estimate*

$$\|z_t(\cdot,t)\|_{L_h}^2 + \|\bar{z}(\cdot,t)\|_{H_h^1}^2 \le C\left(\|z_t(\cdot,0)\|_{L_h}^2 + \|\bar{z}(\cdot,0)\|_{H_h^1}^2 + \tau\sum_{t'\in\omega_\tau}\|\varphi(\cdot,t')\|_{L_h}^2\right), \quad (29)$$

where $t \in \omega_\tau^+$ and $C = c_3 \mathrm{e}^{c_4 T t}$.

Proof. Taking the inner product of (22)–(25) with $\tau(z_t+z_{\bar t})$ and performing partial summation, one obtains

$$\|z_t\|^2_{L_h}-\|z_{\bar t}\|^2_{L_h}+A_h(\bar z,\bar z)-A_h(\check z,\check z)=R(\varphi,z_t,z_{\bar t}),$$

where

$$A_h(z,w)=(\overline{p}\,z_{\bar x},w_{\bar x})_{L_{h'}}+\widetilde\beta_2\widetilde\alpha_1 z_1(b_1)w_1(b_1)+\widetilde\beta_1\widetilde\alpha_2 z_2(a_2)w_2(a_2)$$
$$-\widetilde\beta_1\widetilde\beta_2\left[z_1(b_1)w_2(a_2)+z_2(a_2)w_1(b_1)\right]$$

and

$$R(\varphi,z_t,z_{\bar t})=\tau(\varphi,z_t)_{L_h}+\tau(\varphi,z_{\bar t})_{L_h}$$
$$-\widetilde\beta_2\frac{h_1^2}{12}\frac{p_1'(b_1)}{p_1(b_1)}\left[z^2_{1,t}(b_1,t)-z^2_{1,\bar t}(b_1,t)\right]+\widetilde\beta_1\frac{h_2^2}{12}\frac{p_2'(a_2)}{p_2(a_2)}\left[z^2_{2,t}(a_2,t)-z^2_{2,\bar t}(a_2,t)\right]$$
$$-\widetilde\alpha_1\widetilde\beta_2\frac{h_1^2}{6}\frac{1}{p_1(b_1)}\left[z^2_{1,t}(b_1,t)-z^2_{1,\bar t}(b_1,t)\right]-\widetilde\alpha_2\widetilde\beta_1\frac{h_2^2}{6}\frac{1}{p_2(a_2)}\left[z^2_{2,t}(a_2,t)-z^2_{2,\bar t}(a_2,t)\right]$$
$$+\widetilde\beta_1\widetilde\beta_2\left(\frac{h_1^2}{6}\frac{1}{p_1(b_1)}+\frac{h_2^2}{6}\frac{1}{p_2(a_2)}\right)\left[z_{1,t}(b_1,t)z_{2,t}(a_2,t)-z_{1,\bar t}(b_1,t)z_{2,\bar t}(a_2,t)\right]$$
$$-\widetilde\beta_1\widetilde\beta_2\left(\frac{h_1^2}{6}\frac{1}{p_1(b_1)}-\frac{h_2^2}{6}\frac{1}{p_2(a_2)}\right)\left[z_{1,t}(b_1,t)z_{2,\bar t}(a_2,t)-z_{1,\bar t}(b_1,t)z_{2,t}(a_2,t)\right].$$

Using (8), (14) and the Cauchy-Schwarz inequality, we immediately obtain

$$|R(\varphi,z_t,z_{\bar t})|\le\tau\|\varphi\|^2_{L_h}+c_5\tau\left(\|z_t\|^2_{L_h}+\|z_{\bar t}\|^2_{L_h}\right)$$

and after summation on t

$$\|z_t(\cdot,t)\|^2_{L_h}+A_h(\bar z(\cdot,t),\bar z(\cdot,t))\le\|z_t(\cdot,0)\|^2_{L_h}+A_h(\bar z(\cdot,0),\bar z(\cdot,0))$$
$$+\tau\sum_{t'=\tau}^{t}\|\varphi(\cdot,t')\|^2_{L_h}+2c_5\tau\sum_{t'=0}^{t}\|z_t(\cdot,t')\|^2_{L_h}.$$

Under the conditions (8) and (9) the bilinear form $A_h(v,w)$ satisfies inequalities analogous to its continuous counterpart $A(u,v)$:

$$A_h(v,v)+k\|v\|^2_{L_h}\ge c_1\|v\|^2_{H_h^1}\quad\text{and}\quad|A_h(v,w)|\le c_6\|v\|_{H_h^1}\|w\|_{H_h^1}.$$

In this way we obtain

$$\|z_t(\cdot,t)\|^2_{L_h}+c_1\|\bar z(\cdot,t)\|^2_{H_h^1}\le\|z_t(\cdot,0)\|^2_{L_h}+c_6\|\bar z(\cdot,0)\|^2_{H_h^1}+\tau\sum_{t'=\tau}^{T}\|\varphi(\cdot,t')\|^2_{L_h}$$
$$+k\|\bar z(\cdot,t)\|^2_{L_h}+2c_5\tau\sum_{t'=0}^{t}\|z_t(\cdot,t')\|^2_{L_h}.$$

Further,

$$
\begin{aligned}
\|\bar{z}(\cdot,t)\|^2_{L_h} &= \left\| \bar{z}(\cdot,0) + \frac{\tau}{2} z_t(\cdot,0) + \tau \sum_{t'=\tau}^{t-\tau} z_t(\cdot,t') + \frac{\tau}{2} z_t(\cdot,t) \right\|^2_{L_h} \\
&\le \left(\|\bar{z}(\cdot,0)\|_{L_h} + \frac{\tau}{2} \|z_t(\cdot,0)\|_{L_h} + \tau \sum_{t'=\tau}^{t-\tau} \|z_t(\cdot,t')\|_{L_h} + \frac{\tau}{2} \|z_t(\cdot,t)\|_{L_h} \right)^2 \\
&\le 2\|\bar{z}(\cdot,0)\|_{L_h} + 2T \left(\frac{\tau}{2} \|z_t(\cdot,0)\|^2_{L_h} + \tau \sum_{t'=\tau}^{t-\tau} \|z_t(\cdot,t')\|^2_{L_h} + \frac{\tau}{2} \|z_t(\cdot,t)\|^2_{L_h} \right) \\
&\le 2\|\bar{z}(\cdot,0)\|_{H^1_h} + 2T\tau \sum_{t'=0}^{t} \|z_t(\cdot,t')\|^2_{L_h}.
\end{aligned}
$$

Our assertion follows from the last two inequalities by using the discrete Grönwall lemma [12]. □

Therefore, in order to determine the convergence rate of the FDS (15)–(21), it is enough to estimate the terms on the right-hand side of the inequality (29). Such estimates are constructed in [7]:

$$
\begin{aligned}
&\left(\tau \sum_{t\in\omega_\tau} \sum_{x\in\omega^+_{i,h_i}} \hbar_i |\psi_i|^2 \right)^{1/2} \le C(\tau^2 + h_i^2) \|u_i\|_{H^4(Q_i)}, \\
&\left(\tau \sum_{t\in\omega_\tau} h_i \sum_{x\in\omega_{i,h_i}} |\chi_i|^2 \right)^{1/2} \le C(\tau^2 + h_i^2) \|p_i\|_{H^3(\Omega_i)} \|u_i\|_{H^4(Q_i)}, \\
&\left(\tau \sum_{t\in\omega_\tau} \frac{h_i}{2} |\tilde{\chi}_i(d_i,t)|^2 \right)^{1/2} \le C(\tau^2 + h_i^2) \|p_i\|_{H^3(\Omega_i)} \left(\|u_1\|_{H^4(Q_1)} + \|u_2\|_{H^4(Q_2)} \right), \\
&\left(\tau \sum_{t\in\omega_\tau} \frac{h_i}{2} |h_i \eta_i(d_i,t)|^2 \right)^{1/2} \le C(\tau^2 + h_i^2) \left(\|u_1\|_{H^4(Q_1)} + \|u_2\|_{H^4(Q_2)} \right), \\
&\left(\sum_{x\in\omega^+_{i,h_i}} |\bar{z}_i(x,0)|^2 \hbar_i \right)^{1/2} = \left(\sum_{x\in\omega^+_{i,h_i}} |\zeta_i|^2 \hbar_i \right)^{1/2} \le C(\tau^2 + h_i^2) \|u_i\|_{H^4(Q_i)}, \\
&\left(h_i \sum_{x\in\omega^+_{i,h_i}} |\bar{z}_{i,\bar{x}}(x,0)|^2 \right)^{1/2} = \left(h_i \sum_{x\in\omega^+_{i,h_i}} \left| \frac{\tau}{2} \zeta_{i,\bar{x}}(x) \right|^2 \right)^{1/2} \le C\tau^2 \|u_i\|_{H^4(Q_i)},
\end{aligned}
$$

where $i = 1,2$, $d_1 = b_1$ and $d_2 = a_2$. From these inequalities and (29) one obtains the next assertion.

Theorem 3. *Let* $p_i \in H^3(\Omega_i)$, $i = 1,2$, *and let the assumptions* (8)–(9) *hold. Let further the solution of IBVP* (1)–(7) *belong to the space* H^4. *Then the solution* v *of FDS* (15)–(21) *converges to the solution* u *of IBVP* (1)–(7), *and the following convergence rate estimate holds:*

$$\max_{t\in\omega_\tau}\left(\|z_t(\cdot,t)\|_{L_h}+\|\bar{z}(\cdot,t)\|_{H_h^1}\right)\leq C(h_1^2+h_2^2+\tau^2)\max_i\|p_i\|_{H^3(\Omega_i)}\|u\|_{H^4},$$

where $z = u - v$.

Acknowledgements This research was supported by the Ministry of Science of the Republic of Serbia under project # 144005A.

References

1. Angelova, I.T., Vulkov, L.G.: High-order finite difference schemes for elliptic problems with intersecting interfaces. Appl. Math. Comput. **187**, 824–843 (2007)
2. Caffarelli, L.: A monotonicity formula for heat functions in disjoint domains. In: Boundary-Value Problems for PDEs and Applications, RMA Res. Notes Appl. Math. **29**, Masson, Paris, 53–60 (1993)
3. Datta, A.K.: Biological and bioenvironmental heat and mass transfer. Marcel Dekker, New York (2002)
4. Givoli, D.: Exact representation on artificial interfaces and applications in mechanics. Appl. Mech. Rev. **52**, 333–349 (1999)
5. Givoli, D.: Finite element modeling of thin layers. Comput. Model. Eng. Sci. **5** (6), 497–514 (2004)
6. Jovanović, B.S., Vulkov, L.G.: Finite difference approximation of strong solutions of a parabolic interface problem on disconnected domains. Publ. Inst. Math. **83** (2008)
7. Jovanović, B.S., Vulkov, L.G.: Numerical solution of a hyperbolic transmission problem. Comput. Methods Appl. Math. **8**, No 4 (2008)
8. Kandilarov, J., Vulkov, L.G.: The immersed interface method for two-dimensional heat-diffusion equations with singular own sources. Appl. Numer. Math. **57**, 5–7, 486–497 (2007)
9. Koleva, M.: Finite element solution of boundary value problems with nonlocal jump conditions. Math. Model. Anal. **13**, No 3, 383–400 (2008)
10. Lions, J.L., Magenes, E.: Non homogeneous boundary value problems and applications. Springer, Berlin and New York (1972)
11. Samarskiĭ, A.A.: Theory of difference schemes. Marcel Dekker, New York and Basel (2001)
12. Thomée, V.: Galerkin finite element methods for parabolic problems. Springer Series in Computational Mathematics Vol. 25, Springer, Berlin etc. (1997)
13. Vulkov, L.G.: Well posedness and a monotone iterative method for a nonlinear interface problem on disjoint intervals. Am. Inst. Phys., Proc. Ser. 946 (2007)
14. Wloka, J.: Partial differential equations. Cambridge University Press, Cambridge (1987)

Homeomorphisms and Fredholm Theory for Perturbations of Nonlinear Fredholm Maps of Index Zero and of A-Proper Maps with Applications

P. S. Milojević

Dedicated to Prof. Gradimir V. Milovanović for his 60th birthday

1 Part I. Existence Theory

1.1 Perturbations of Homeomorphisms and Nonlinear Fredholm Alternatives

Throughout the paper, we assume that X and Y are infinite-dimensional Banach spaces. Define the set measure of noncompactness α by:

$$\alpha(T) = \sup\{\alpha(T(A))/\alpha(A) \mid A \subset X \text{ bounded}, \ \alpha(A) > 0\},$$
$$\beta(T) = \inf\{\alpha(T(A))/\alpha(A) \mid A \subset X \text{ bounded}, \ \alpha(A) > 0\}.$$

Here, $\alpha(T)$ and $\beta(T)$ are related to the properties of compactness and properness of the map T, respectively. Let U be an open subset of X and $T : U \to Y$. For $p \in U$, let $B_r(p)$ be the open ball in X centered at p with radius r. Suppose that $B_r(p) \subset U$ and set

$$\alpha\left(T|_{B_r(p)}\right) = \sup\{\alpha(T(A))/\alpha(A) \mid A \subset B_r(p) \text{ bounded}, \ \alpha(A) > 0\}.$$

This is nondecreasing as a function of r, and clearly $\alpha\left(T|_{B_r(p)}\right) \le \alpha(T)$. Hence, the following definition makes sense:

$$\alpha_p(T) = \lim_{r\to 0} \alpha\left(T|_{B_r(p)}\right).$$

P. S. Milojević
Department of Mathematical Sciences and CAMS, New Jersey Institute of Technology, Newark, NJ, USA, e-mail: milojevi@ADM.NJIT.EDU

W. Gautschi et al. (eds.), *Approximation and Computation: In Honor of Gradimir V. Milovanović*, Springer Optimization and Its Applications 42,
DOI 10.1007/978-1-4419-6594-3_22,

Similarly, we define $\beta_p(T)$. We have $\alpha_p(T) \leq \alpha(T)$ and $\beta_p(T) \geq \beta(T)$ for any p. If T is of class C^1, then $\alpha_p(T) = \alpha(T'(p))$ and $\beta_p(T) = \beta(T'(p))$ for any p ([5]). Note that for a Fredholm map $T : X \to Y$, $\beta_p(T) > 0$ for all $p \in X$.

To look at ball-condensing perturbations, we recall that the *ball-measure of noncompactness* of a bounded set $D \subset X$ is defined as $\chi(D) = \inf\{r > 0 \mid D \subset \cup_{i=1}^n B(x_i, r),\ x_i \in X,\ n \in \mathbb{N}\}$. Let ϕ denote either the set or the ball-measure of noncompactness. Then a map $T : D \subset X \to Y$ is said to be $k - \phi$*-contractive* (ϕ*-condensing*) if $\phi(T(Q)) \leq k\phi(Q)$ (respectively, $\phi(T(Q)) < \phi(Q)$) whenever $Q \subset D$ (with $\phi(Q) \neq 0$). We note that T is completely continuous if and only if it is $\alpha - 0$-contractive. Moreover, if T is Lipschitz continuous with constant k, then it is $\alpha - k$-contractive.

We say that a map $T : X \to Y$ satisfies condition $(+)$ if $\{x_n\}$ is bounded whenever $\{Tx_n\}$ converges. Recall that a map $T : X \to Y$ is a c-expansive map if $\|Tx - Ty\| \geq c\|x - y\|$ for all $x, y \in X$ and some $c > 0$. For a continuous map $F : X \to Y$, let Σ be the set of all points $x \in X$ where F is not locally invertible, and let $\operatorname{card} F^{-1}(f)$ be the cardinal number of the set $F^{-1}(f)$. Let ϕ be either the set or ball measure of noncompactness.

In [25], using Browder's theorem [4], we have shown that if $T : X \to Y$ is closed on bounded closed subsets of X and is a local homeomorphism, then it is a homeomorphism if and only if it satisfies condition $(+)$. The results in this subsection are contained in [25, 26].

Theorem 1 (Fredholm Alternative). *Let $T : X \to Y$ be a homeomorphism and $C : X \to Y$ be such that $\alpha(C) < \beta(T)$ (T being a c-expansive homeomorphism, and C being a $k - \phi$-contraction with $k < c$, respectively). Then either*

(i) *$T + C$ is injective (locally injective, respectively), in which case it is an open map, and $T + C$ is a homeomorphism if and only if either one of the following conditions holds:*
  (a) *$T + C$ is closed (in particular, proper, or satisfies condition $(+)$),*
  (b) *$T + C$ is injective and $R(T + C)$ is closed, or*

(ii) *$T + C$ is locally injective in the first case (not locally injective in either case, respectively), in which case, assuming additionally that $T + tC$ satisfies condition $(+)$, the equation $Tx + Cx = f$ is solvable for each $f \in Y$ with the cardinal number* $\operatorname{card}(T + C)^{-1}(f)$ *positive and finite for each $f \in Y$ (the cardinal number* $\operatorname{card}(T + C)^{-1}(f)$ *positive, constant, and finite on each connected component of the set $Y \setminus (T + C)(\Sigma)$, respectively).*

Remark 1. If X is an infinite-dimensional Banach space and $T : X \to X$ is a homeomorphism then T satisfies condition $(+)$ but it need not be coercive in the sense that $\|Tx\| \to \infty$ as $\|x\| \to \infty$ (cf. [7, Corollary 8]). We refer to [25] for a number of particular conditions on the maps that imply condition $(+)$.

Corollary 1. *Let $T : X \to Y$ be a homeomorphism and $C : X \to Y$ be continuous and uniformly bounded, i.e., $\|Cx\| \leq M$ for all x and some $M > 0$, and $\alpha(C) < \beta(T)$. Then either*

(i) *$T+C$ is injective, in which case $T+C$ is a homeomorphism, or*

(ii) *$T+C$ is locally injective and satisfies condition $(+)$, in which case the cardinal number* $\mathrm{card}(T+C)^{-1}$ *is positive and finite for each $f \in Y$, or*

(iii) *$T+C$ is not locally injective, in which case $T+C$ is surjective, $(T+C)^{-1}(f)$ is compact for each $f \in Y$, and the cardinal number* $\mathrm{card}(T+C)^{-1}(f)$ *is positive, constant, and finite on each connected component of the set $Y \setminus (T+C)(\Sigma)$.*

When T is Fredholm of index zero, then the injectivity of $T+C$ can be replaced by the local injectivity.

Theorem 2 (Fredholm Alternative). *Let $T : X \to Y$ be a Fredholm map of index zero and $C : X \to Y$ be such that $\alpha(C) < \beta(T)$. Then either*

(i) *$T+C$ is locally injective, in which case it is an open map and it is a homeomorphism if and only if one of the following conditions holds:*
  (a) *$T+C$ is closed (in particular, proper, or satisfies condition $(+)$),*
  (b) *$T+C$ is injective and $R(T+C)$ is closed, or*

(ii) *$T+C$ is not locally injective, in which case, assuming additionally that T is locally injective and $T+tC$ satisfies condition $(+)$, the equation $Tx+Cx=f$ is solvable for each $f \in Y$ with $(T+C)^{-1}(f)$ compact, and the cardinal number* $\mathrm{card}(T+C)^{-1}(f)$ *is positive, constant, and finite on each connected component of the set $Y \setminus (T+C)(\Sigma)$.*

Condition $\alpha(C) < \beta(T)$ in (i) can be replaced by $\alpha_p(C) < \beta_p(T)$ for each $p \in X$ since $T+C$ is also open in this case [5]. This condition always holds if T is a Fredholm map of index zero and C is compact. In view of this remark, we have the following extension of Tromba's homeomorphism result for proper locally injective Fredholm maps of index zero [35].

Corollary 2. *Let $T : X \to Y$ be a Fredholm map of index zero, $C : X \to Y$ be compact and $T+C$ be locally injective and closed. Then $T+C$ is a homeomorphism.*

Corollary 3. *Let $T : X \to Y$ be a locally injective closed Fredholm map of index zero, $C : X \to Y$ be continuous and uniformly bounded and $\alpha(C) < \beta(T)$. Then either*

(i) *$T+C$ is locally injective, in which case $T+C$ is a homeomorphism, or*

(ii) *$T+C$ is not locally injective, in which case $T+C$ is surjective, $(T+C)^{-1}(f)$ is compact for each $f \in Y$, and the cardinal number* $\mathrm{card}(T+C)^{-1}(f)$ *is positive, constant, and finite on each connected component of the set $Y \setminus (T+C)(\Sigma)$.*

Next, we shall discuss some nonlinear extensions of the Fredholm alternative to set contractive like perturbations of homeomorphisms as well as of Fredholm maps of index zero that are asymptotically close to positive k-homogeneous maps. Recall that a map T is positive k-homogeneous outside some ball if $T(\lambda x) = \lambda^k T(x)$ for some $k \geq 1$, all $\|x\| \geq R$ and all $\lambda \geq 1$. We say that T is asymptotically close to a positive k-homogeneous map A if

$$|T-A| = \limsup_{\|x\|\to\infty} \|Tx - Ax\|/\|x\|^k < \infty.$$

We note that T is asymptotically close to a positive k-homogeneous map A if there is a functional $c : X \to [0,a]$ such that $\|T(tx)/t^k - Ax\| \leq c(t)\|x\|^k$. In this case, $|T-A| \leq a$.

Theorem 3 (Fredholm alternative). *Let $T : X \to Y$ be a homeomorphism and $C, D : X \to Y$ be continuous maps such that $\alpha(D) < \beta(T) - \alpha(C)$ with $|D|$ sufficiently small (T being a c-expansive homeomorphism and C being a $k-\phi$-contraction with $k < c$, respectively), where*

$$|D| = \limsup_{\|x\|\to\infty} \|Dx\|/\|x\|^k < \infty.$$

Assume that $T+C$ is injective (locally injective, respectively) and either $\|Tx + Cx\| \geq c\|x\|^k - c_0$ for all $\|x\| \geq R$ for some R, c, and c_0, or $T+C$ asymptotically close to a continuous, closed (proper, in particular) on bounded and closed subsets of X positive k-homogeneous map A outside some ball in X, i.e., there are $k \geq 1$ and $R_0 > 0$ such that $A(\lambda x) = \lambda^k Ax$ for all $\|x\| \geq R_0$, all $\lambda \geq 1$ with $A^{-1}(0)$ bounded and $|T+C-A|$ sufficiently small. Then either

(i) *$T+C+D$ is injective, in which case $T+C+D$ is a homeomorphism, or*

(ii) *$T+C+D$ is not injective, in which case the solution set $(T+C+D)^{-1}(f)$ is nonempty and compact for each $f \in Y$, and the cardinal number* $\operatorname{card}(T+C+D)^{-1}(f)$ *is constant, finite, and positive on each connected component of the set $Y \setminus (T+C+D)(\Sigma)$.*

If $C = 0$ in Theorem 3, then the injectivity of $T+D$ can be weakened to local injectivity when T is c-expansive.

Corollary 4. *Let $T : X \to Y$ be a c-expansive homeomorphism and $D : X \to Y$ a continuous map such that $\alpha(D) < c$ and*

$$|D| = \limsup_{\|x\|\to\infty} \|Dx\|/\|x\| < c.$$

Then either

(i) *$T+D$ is locally injective, in which case $T+D$ is a homeomorphism, or*

(ii) *$T+D$ is not locally injective, in which case the solution set $(T+D)^{-1}(f)$ is nonempty and compact for each $f \in Y$, and the cardinal number* $\operatorname{card}(T+D)^{-1}(f)$ *is constant, finite, and positive on each connected component of the set $Y \setminus (T+D)(\Sigma)$.*

Next, we shall look at various conditions on a c-expansive map that make it a homeomorphism (see [6,12,14]). They came about when some authors tried to give a positive answer to the Nirenberg problem on surjectivity of a c-expansive map with $T(0) = 0$ and mapping a neighborhood of zero onto a neighborhood of zero.

Corollary 5. *Let $T : X \to Y$ be a c-expansive map and $D : X \to Y$ a continuous map such that $\alpha(D) < c$ and*

$$|D| = \limsup_{\|x\| \to \infty} \|Dx\|/\|x\| < c.$$

Suppose that either one of the following conditions holds:

(a) *Y is reflexive, T is Fréchet differentiable and*

$$\limsup_{x \to x_0} \|T'(x) - T'(x_0)\| < c \quad \text{for each } x_0 \in X;$$

(b) *$T : X \to X$ is Fréchet differentiable and such that the logarithmic norm $\mu(T'(x))$ of $T'(x)$ is strictly negative for all $x \in X$, where*

$$\mu(T'(x)) = \lim_{t \to 0^+} (\|I + tT'(x)\| - 1)/t;$$

(c) *$X = Y = H$ is a Hilbert space, T is Fréchet differentiable and such that either*

$$\inf_{\|h\|=1} \mathrm{Re}(T'(x)h, h) > 0 \quad \text{for all } x \in H,$$

or

$$\sup_{\|h\|=1} \mathrm{Re}(T'(x)h, h) < 0 \quad \text{for all } x \in H;$$

(d) *X is reflexive and $T : X \to X^*$ is a C^1 potential map.*

Then either

(i) *$T + D$ is locally injective, in which case $T + D$ is a homeomorphism, or*

(ii) *$T + D$ is not locally injective, in which case the solution set $(T + D)^{-1}(f)$ is nonempty and compact for each $f \in Y$, and the cardinal number $\mathrm{card}(T + D)^{-1}(f)$ is constant, finite, and positive on each connected component of the set $Y \setminus (T + D)(\Sigma)$.*

Recall that a map $T : X \to Y$ is expansive along rays if for each $y \in Y$, there is a $c(y) > 0$ such that $\|Tx - Ty\| \geq c(y)\|x - y\|$ for all $x, y \in T^{-1}([0, y])$, where $[0, y] = \{ty \mid 0 \leq t \leq 1\}$.

Theorem 4. *An expansive along rays local homeomorphism $T : X \to Y$ is a homeomorphism. A locally expansive local homeomorphism is a homeomorphism if $m = \inf_x c(x) > 0$. In general, a locally expansive local homeomorphism need not be a homeomorphism.*

Corollary 6. *Let $T: X \to Y$ be an expansive along rays local homeomorphism and $D: X \to Y$ be a continuous map such that $\alpha(D) < \beta(T)$ and $\|Dx\| \leq M$ for all $x \in X$ and some $M > 0$. Then either*

(i) *$T + D$ is injective, in which case $T + D$ is a homeomorphism, or*

(ii) *$T + D$ is not injective, in which case the solution set $(T+D)^{-1}(f)$ is nonempty and compact for each $f \in Y$, and the cardinal number* $\mathrm{card}(T+D)^{-1}(f)$ *is constant, finite, and positive on each connected component of the set $Y \setminus (T+D)(\Sigma)$.*

Next, we shall give another extension of the Fredholm Alternative to perturbations of nonlinear Fredholm maps of index zero.

Theorem 5 (Fredhom Alternative). *Let $T: X \to Y$ be a Fredholm map of index zero and $C, D: X \to Y$ be continuous maps such that $\alpha(D) < \beta(T) - \alpha(C)$ with $|D|$ sufficiently small, where*

$$|D| = \limsup_{\|x\| \to \infty} \|Dx\| / \|x\|^k < \infty.$$

Assume that either $\|Tx + Cx\| \geq c\|x\|^k - c_0$ for all $\|x\| \geq R$ for some R, c and c_0, or $T + C$ is asymptotically close to a continuous, closed (in particular, proper) on bounded and closed subsets of X positive k-homogeneous map A, outside some ball in X, i.e., there are $k \geq 1$ and $R_0 > 0$ such that $A(\lambda x) = \lambda^k Ax$ for all $\|x\| \geq R_0$, all $\lambda \geq 1$ and $(A)^{-1}(0)$ bounded with $|T + C - A|$ sufficiently small. Then either

(i) *$T + C + D$ is locally injective, in which case $T + C + D$ is a homeomorphism, or*

(ii) *$T + C + D$ is not locally injective, in which case, assuming additionally that $T + C$ is locally injective, the solution set $(T + C + D)^{-1}(f)$ is nonempty and compact for each $f \in Y$, and the cardinal number* $\mathrm{card}(T + C + D)^{-1}(f)$ *is constant, finite, and positive on each connected component of the set $Y \setminus (T + C + D)(\Sigma)$.*

1.2 Finite Solvability of Equations with Perturbations of Odd Fredholm Maps of Index Zero

In this subsection, we shall study perturbations of Fredholm maps of index zero, assuming that the maps are odd. The results on compact perturbations are based on the Fitzpatrick-Pejsachowisz-Rabier-Salter degree ([11, 29, 31]).

Theorem 6 (Generalized First Fredholm Theorem). *Let $T: X \to Y$ be a Fredholm map of index zero that is proper on bounded and closed subsets of X and $C, D: X \to Y$ be compact maps with $|D|$ sufficiently small, where*

$$|D| = \limsup_{\|x\| \to \infty} \|Dx\| / \|x\|^k < \infty.$$

Assume that $T+C$ is odd, asymptotically close to a continuous, closed (in particular, proper) on bounded and closed subsets of X positive k-homogeneous map A, outside some ball in X, i.e., there exists $R_0 > 0$ such that $A(\lambda x) = \lambda^k Ax$ for all $\|x\| \geq R_0$, for all $\lambda \geq 1$ and some $k \geq 1$, $\|x\| \leq M < \infty$ if $Ax = 0$ and $|T+C-A|$ is sufficiently small. Then the equation $Tx+Cx+Dx = f$ is solvable for each $f \in Y$ with $(T+C+D)^{-1}(f)$ compact, and the cardinal number $\mathrm{card}(T+C+D)^{-1}(f)$ *is constant, finite, and positive on each connected component of the set $Y \setminus (T+C+D)(\Sigma)$.*

Remark 2. Earlier generalizations of the first Fredholm theorem to condensing vector fields, maps of type (S_+), monotone-like maps and (pseudo) A-proper maps assumed the homogeneity of T with $Tx = 0$ only if $x = 0$ (see [13, 17–19, 28] and the references therein).

Next, we provide some generalizations of the Borsuk-Ulam principle for odd compact perturbations of the identity. The first result generalizes Theorem 6 when $D = 0$.

Theorem 7. *Let $T : X \to Y$ be a Fredholm map of index zero that is proper on closed bounded subsets of X and $C : X \to Y$ be compact such that $T+C$ is odd outside some ball $B(0,R)$. Suppose that $T+C$ satisfies condition $(+)$. Then $Tx+Cx = f$ is solvable, $(T+C)^{-1}(f)$ is compact for each $f \in Y$, and the cardinal number* $\mathrm{card}(T+C)^{-1}(f)$ *is positive and constant on each connected component of $Y \setminus (T+C)(\Sigma)$.*

The next result provides a more general version of Theorem 7.

Theorem 8. *Let $T : X \to Y$ be a Fredholm map of index zero that is proper on closed bounded subsets of X and $C_1, C_2 : X \to Y$ be compact such that $T+C_1$ is odd outside some ball $B(0,R)$. Suppose that $H(t,x) = Tx + C_1x + tC_2x - tf$ satisfies condition $(+)$. Then $Tx + C_1x + C_2x = f$ is solvable for each $f \in Y$ with $(T+C_1+C_2)^{-1}(f)$ compact, and the cardinal number* $\mathrm{card}(T+C_1+C_2)^{-1}(f)$ *is positive and constant on each connected component of $Y \setminus (T+C_1+C_2)(\Sigma)$.*

Next, we shall study k-set contractive perturbarions of Fredholm maps of index zero. Denote by $\deg_{\mathrm{BCF}}$ the degree of Benevieri-Calamai-Furi [1–3]. When $T+C$ is not locally injective, we have the following extension of Theorem 5.

Theorem 9 (Generalized First Fredholm Theorem). *Let $T : X \to Y$ be a Fredholm map of index zero and $C, D : X \to Y$ be continuous maps such that $\alpha(D) < \beta(T) - \alpha(C)$ with $|D|$ sufficiently small, where*

$$|D| = \limsup_{\|x\| \to \infty} \|Dx\| / \|x\|^k < \infty.$$

Assume that $T+C$ is asymptotically close to a continuous, closed (in particular, proper) on bounded and closed subsets of X positive k-homogeneous map A, outside some ball in X for all $\lambda \geq 1$ and some $k \geq 1$, $\|x\| \leq M < \infty$ if $Ax = 0$, $|T+C-A|$

is sufficiently small, and $\deg_{\mathrm{BCF}}(T+C, B(0,r), 0) \neq 0$ *for all large* r. *Then the equation* $Tx + Cx + Dx = f$ *is solvable for each* $f \in Y$ *with* $(T+C+D)^{-1}(f)$ *compact, and the cardinal number* $\mathrm{card}(T+C+D)^{-1}(f)$ *is constant, finite, and positive on each connected component of the set* $Y \setminus (T+C+D)(\Sigma)$.

Remark 3. Theorems 6 and 9 are valid if k-positive homogeneity of $T+C$ is replaced by $\|Tx+Cx\| \geq c\|x\|^k$ for all x outside some ball.

1.3 Applications to (Quasi) Linear Elliptic Nonlinear Boundary Value Problems

A. Potential problems with strongly nonlinear boundary value conditions. Consider the nonlinear BVP

$$\begin{aligned} \Delta\Phi &= 0 \quad \text{in } Q \subset \mathbb{R}^2, \\ -\partial_n\Phi &= b(x, \Phi(x)) - h \quad \text{on } \Gamma = \partial Q, \end{aligned} \tag{1}$$

where $b(x,u)$ is strongly nonlinear, Γ a simple smooth closed curve, and ∂_n is the outer normal derivative on Γ. The nonlinearity appears only in the boundary condition. By means of the Kirchhoff transformation, more general quasilinear equations can be transformed into this form. This kind of equations with various nonlinearities arise in many applications like steady-state heat transfer, electromagnetic problems with variable electrical conductivity of the boundary, heat radiation and heat transfer (cf. [33] and the references therein). Except for [32], the earlier studies assume that the nollinearities have at most a linear growth and were based on the boundary element method. We shall study BVP (1) using the theory developed in Sects. 1.1–1.2. We follow [26].

Assume $b(x,u) = b_0(x,u) + b_1(x,u)$ satisfies:

1. $b_0 : \Gamma \times \mathbb{R} \to \mathbb{R}$ is a Carathéodory function, i.e., $b_0(\cdot, u)$ is measurable for all $u \in \mathbb{R}$ and $b_0(x, \cdot)$ is continuous for a.e. $x \in \Gamma$;
2. $b_0(x, \cdot)$ is strictly increasing on $\mathbb{R}$;
3. For $p \geq 2$, there exist constants $a_1 > 0$, $a_2 \geq 0$, $c_1 > 0$, and $c_2 \geq 0$ such that

$$|b_0(x,u)| \leq a_1|u|^{p-1} + a_2, \quad b_0(x,u)u \geq c_1|u|^p + c_2;$$

4. b_1 satisfies the Carathéodory conditions and $|b_1(x,u)| \leq M$ for all $(x,u) \in \Gamma \times \mathbb{R}$ and some $M > 0$.

We shall reformulate the BVP (1) as a boundary integral equation. Recall that the single-layer operator V is defined by:

$$Vu(x) = -1/(2\pi) \int_\Gamma u(y) \log|x-y| \, ds_y, \quad x \in \Gamma,$$

and the double layer operator K is defined by:

$$Ku(x) = 1/(2\pi)\int_{\Gamma} u(y)\partial_n \log|x-y|\,\mathrm{d}s_y, \quad x \in \Gamma.$$

We shall make the ansatz: Find a boundary distribution u (in some space) such that

$$\Phi(x) = -1/(2\pi)\int_{\Gamma} u(y)\log|x-y|\,\mathrm{d}s_y, \quad x \in Q.$$

Then, by the properties of the normal derivative of the monopole potential ([8, 36]), we derive the nonlinear boundary integral equation ([32]) $(1/2I - K^*)u + B(Vu) = f$. This equation can be written in the form $Tu + Cu = f$, where we set $T = 1/2I - K^* + B_0V$, $Cu = B_1V$ with $B_iu = b_i(x,u)$, $i = 0,1$.

Theorem 10. *Let* (1)–(4) *hold and* $q = p/(p-1)$. *Then either*

(i) The BVP (1) is locally injective in $L_q(Q)$, *in which case it is uniquely solvable in* $L_q(Q)$ *for each* $h \in L_q(Q)$, *and the solution depends continuously on h, or*

(ii) *The BVP (1) is not locally injective, in which case it is solvable in* $L_q(Q)$ *for each* $h \in L_q(Q)$, *the solution set is compact, and the cardinality of the solution set is finite and constant on each connected component of* $L_q(Q) \setminus (T+C)(\Sigma)$, *where* $\Sigma = \{u \in L_q(Q) \mid BVP\ (1)\ \text{is not locally uniquely solvable}\}$.

Example 1. If $b_0(u) = |u|u^{p-2}$ and $b_1 = 0$, or if $b_0(u) = |u|u^{p-2}$ and $b_1(u) = \arctan u$ with p even, then part (i) of Theorem 10 holds. Note that $b(u)$ is strictly increasing in either case, and therefore the BVP (1) is injective as in the proof of Theorem 10. Part (ii) of Theorem 10 is valid if e.g., $b_1(u) = a\sin u + b\cos u$.

When the nonlinearities have a linear growth, we have the following result.

Theorem 11. *Let* $b(x,u) = b_0(x,u) + b_1(x,u)$ *be such that* $b_0(x,u)$ *is a Carathéodory function,* $b_0(x,\cdot)$ *is strictly increasing on* $\mathbb{R}$ *and*

(1)*There exist constants* $a_1 > 0$ *and* $a_2 \geq 0$ *such that* $|b_0(x,u)| \leq a_1|u| + a_2$;

(2)b_1 *satisfies the Carathéodory conditions and for some positive constants* c_1 *and* c_2 *with* c_1 *sufficiently small,* $|b_1(x,u)| \leq c_1|u| + c_2$ *for all* $(x,u) \in \Gamma \times \mathbb{R}$.

Assume that $I - K$ *is injective. Then either*

(i) *The BVP (1) is locally injective in* $L_2(Q)$, *in which case it is uniquely solvable in* $L_2(Q)$ *for each* $h \in L_2(Q)$, *and the solution depends continuously on h, or*

(ii) *The BVP (1) is not locally injective, in which case it is solvable in* $L_2(Q)$ *for each* $h \in L_2(Q)$, *the solution set is compact, and the cardinality of the solution set is finite and constant on each connected component of* $L_2(Q) \setminus (T+C)(\Sigma)$, *where* $\Sigma = \{u \in L_2(Q) \mid BVP\ (1)\ \text{is not locally uniquely solvable}\}$ *and* $T = I - K + B_0V$.

Remark 4. Theorem 11 is also valid when $b_0 = 0$. The injectivity of $I - K$ has been studied in [8, 36].

B. Semilinear elliptic equations with nonlinear boundary value conditions. Condider the nonlinear BVP

$$\Delta u = f(x,u,\nabla u) + g \quad \text{in } Q \subset \mathbb{R}^n, \tag{2}$$

$$-\partial_n u = b(x,u(x)) - h \quad \text{on } \Gamma = \partial Q, \tag{3}$$

where $Q \subset \mathbb{R}^n$, $n = 2$ or 3, is a bounded domain with smooth boundary Γ satisfying a scaling assumption $\operatorname{diam}(Q) < 1$ for $n = 2$, and ∂_n is the outer normal derivative on Γ.

Let $b = b_0 + b_1$. As in [33], assume that

1. $b_0(x,u)$ is a Carathéodory function such that $\frac{\partial}{\partial u} b_0(x,u)$ is Borel measurable and satisfies

$$0 < c \le \frac{\partial}{\partial u} b_0(x,u) \le C < \infty \quad \text{for almost all } x \in \Gamma \text{ and all } u \in \mathbb{R},$$

2. b_1 and f are Carathéodory functions such that

$$|b_1(x,u)| \le a(x) + c(1+|u|), \quad |f(x,u,v)| \le d(x) + c(1+|u|+|v|)$$

for almost all $x \in \Gamma$ and $u, v \in \mathbb{R}$, some functions $a(x) \in L_2(\Gamma)$ and $b(x) \in L_2(Q)$, and $c > 0$ sufficiently small.

Define the Nemitsky maps $B_i : L_2(\Gamma) \to L_2(\Gamma)$ by $B_i u(x) = b_i(x,u(x))$, $i = 0,1$, and $F : H^1(Q) \to L_2(Q)$ by $Fu(x) = f(x,u(x),\nabla u(x))$. Denote by $H^s(Q)$ and $H^s(\Gamma)$ the Sobolev spaces of order s in Q and on Γ, respectively. In particular, $H^{-s}(Q) = (\widetilde{H}^s(Q))^*$, where $\widetilde{H}^s$ is the completion of $C_0^\infty(Q)$ in $H^s(\mathbb{R}^n)$. We also have that ([33]) for each $0 \le s \le 1$, $B_0 : H^s(\Gamma) \to H^s(\Gamma)$ is bounded. Denote by $(\cdot,\cdot)$ the L_2 inner product. As in [10], inserting (2)–(3) into the Green formula

$$\int_Q \Delta u \cdot v \,\mathrm{d}x + \int_Q \Delta u \cdot \Delta v \,\mathrm{d}x - \int_\Gamma \frac{\partial u}{\partial n} v \,\mathrm{d}s_\Gamma = 0,$$

we obtain the weak formulation of the BVP (2)–(3): for a given $g \in \widetilde{H}^{-1}$, find $u \in H^1(Q)$ such that for all $v \in H^1(Q)$

$$\begin{aligned}(Au,v)_{H^1(Q)} &= (\nabla u, \nabla v)_Q + (B_0 u|_\Gamma, v|_\Gamma)_\Gamma - (B_1 u|_\Gamma, v|_\Gamma)_\Gamma \\ &\quad - (h, v|_\Gamma)_\Gamma - (Fu,v)_Q - (g,v)_Q = 0.\end{aligned}$$

Define $T, C : H^1(Q) \to H^1(Q)$ by:

$$(Tu,v)_{H^1(Q)} = (\nabla u, \nabla v)_Q + (B_0 u|_\Gamma, v|_\Gamma)_\Gamma$$

and

$$(Cu,v)_{H^1(Q)} = -(B_1 u|_\Gamma, v|_\Gamma)_\Gamma - (h, v|_\Gamma)_\Gamma - (Fu,v)_Q - (g,v)_Q.$$

Theorem 12. *Let* (1)–(2) *hold. Then either*

(i) *The problem* (2)–(3) *is locally injective in* $H^1(Q)$, *in which case it is uniquely solvable in* $H^1(Q)$ *for each* $g \in \widetilde{H}^{-1}(Q)$ *and* $h \in H^{-1/2}(\Gamma)$, *and the solution depends continuously on* (g,h), *or*

(ii) *The problem* (2)–(3) *is not locally injective in* $H^1(Q)$, *in which case it is solvable in* $H^1(Q)$ *for each* $(g,h) \in \widetilde{H}^{-1}(Q) \times H^{-1/2}(\Gamma)$, *the solution set is compact, and the cardinality of the solution set is finite and constant on each connected component of* $H^1(Q) \setminus (T+C)(\Sigma)$, *where* $\Sigma = \{u \in H^1(Q) \mid BVP\ (2)\text{–}(3)$ *is not locally uniquely solvable*$\}$.

Remark 5. The solvability of problem (2)–(3) with sublinear nonlinearities was proved by Efendiev, Schmitz, and Wendland [10] using a degree theory for compact perturbations of strongly monotone maps.

C. Semilinear elliptic equations with strong nonlinearities. Consider

$$\begin{aligned} &\Delta u + \lambda_1 u - f(u) + g(u) = h, \quad h \in L_2, \\ &u|_{\partial Q} = 0, \end{aligned} \tag{4}$$

where Q is a bounded domain in $\mathbb{R}^n$ and λ_1 is the smallest positive eigenvalue of Δ on Q.

Assume that f and $g = g_1 + g_2$ are Carathéodory functions such that

1. For $p \geq 2$ if $n = 2$ and $p \in [2, 2n/(n-2))$ if $n \geq 3$, there exist constants $a_1 > 0$, $a_2 \geq 0$, $c_1 > 0$ and $c_2 \geq 0$ such that
$$|f(u)| \leq a_1|u|^p + a_2, \quad f(u)u \geq c_1|u|^{p+1} + c_2;$$
2. f is differentiable;
3. $|g_1(u)| \leq b_1|u|^p + b_2$ with $b_1 \leq a_1$ and $g_1(u)u \geq c_1|u|^{p+1} + c_2$;
4. $|g_2(u)| \leq c_1|u|^p + c_2$ for all u with c_1 sufficiently small.

Define $X = \{u \in W_2^1(Q) \mid u = 0 \text{ on } \partial Q\}$. Note that X is compactly embedded into $L_p(Q)$ for each p as in (1) by the Sobolev embedding theorem. We shall look at weak solutions of (4), i.e., $u \in X$ such that $Tu + Cu = h$, where

$$(Tu,v)_{1,2} = (\nabla u, \nabla v) - \lambda_1(u,v) + (f(u),v), \quad (C_i u, v) = (g_i(u),v), \ i = 1,2,$$

$C = C_1 + C_2$, and $(\cdot,\cdot)$ is the L_2 inner product. In X, the derivative of T is $T'(u)v = \Delta v - \lambda_1 v - f'(u)v$. Since $T'(u)$ is a selfadjoint elliptic map in X, T is Fredholm of index zero in X.

Theorem 13. *Let* (1)–(4) *hold. Then either*

(i) *BVP* (4) *is locally injective in* X, *in which case it is uniquely solvable in* X *for each* $h \in L_2(Q)$, *and the solution depends continuously on* h, *or*

(ii) BVP (4) *is not locally injective, in which case it is solvable in X for each $h \in L_2(Q)$, and its solution set is compact. Moreover, the cardinality of the solution set is finite and constant on each connected component of $X \setminus (T+C)(\Sigma)$, where $\Sigma = \{u \in X \mid BVP$ (4) is not locally uniquely solvable$\}$.*

Theorem 14. *Let* (1)–(2) *hold with $f'(0) = 0$, $f'(u) > 0$, and $|g(u)| \leq M$ for some $M > 0$ and all u. Then either*

(i) *The BVP* (4) *is locally injective in X, in which case it is uniquely solvable in X for each $h \in L_2(Q)$, and the solution depends continuously on h, or*

(ii) BVP (4) *is not locally injective, in which case it is solvable in X for each $h \in L_2(Q)$, and the solution set is compact. Moreover, the cardinality of the solution set is finite and constant on each connected component of $X \setminus (T+C)(\Sigma)$, where $\Sigma = \{u \in X \mid BVP$ (4) is not locally uniquely solvable$\}$.*

Theorem 15. *Let* 1)– 4) *hold and f and g_1 be odd. Then the BVP* (4) *is solvable in X for each $h \in L_2(Q)$, and its solution set is compact. Moreover, the cardinality of the solution set is finite and constant on each connected component of $X \setminus (T+C)(\Sigma)$, where $\Sigma = \{u \in X \mid BVP$ (4) is not locally uniquely solvable$\}$.*

Remark 6. Applications to solvability of quasilinear elliptic BVP's with asymptotically positive homogeneous nonlinearities on $\mathbb{R}^n$, and on bounded domains, can be found in [25].

2 Part II. Constructive Theory

2.1 Constructive Homeomorphism Results and Error Estimates

Let $\{X_n\}$ and $\{Y_n\}$ be finite-dimensional subspaces of Banach spaces X and Y, respectively, such that $\dim X_n = \dim Y_n$ for each n and $\mathrm{dist}(x, X_n) \to 0$ as $n \to \infty$ for each $x \in X$. Let $P_n : X \to Y_n$ and $Q_n : Y \to Y_n$ be linear projections onto X_n and Y_n, respectively, such that $P_n x \to x$ for each $x \in X$ and $\delta = \max \|Q_n\| < \infty$. Then $\Gamma = \{X_n, P_n; Y_n, Q_n\}$ is a projection scheme for (X, Y).

A map $T : D \subset X \to Y$ is said to be *approximation-proper* (*A-proper* for short) with respect to Γ if (i) $Q_n T : D \cap X_n \to Y_n$ is continuous for each n and (ii) whenever $\{x_{n_k} \in D \cap X_{n_k}\}$ is bounded and $\|Q_{n_k} T x_{n_k} - Q_{n_k} f\| \to 0$ for some $f \in Y$, then a subsequence $x_{n_{k(i)}} \to x$ and $Tx = f$. A map T is said to be *pseudo A-proper* w.r.t. Γ if in (ii) above we do not require that a subsequence of $\{x_{n_k}\}$ converges to x for which $Tx = f$. If f is given in advance, we say that T is (pseudo) A-proper at f.

The first few results deal with approximation stable homeomorphisms. We say that T is locally p-Lipschitz for some $p > 0$ if for each $x \in X$ there are positive numbers r and M (depending on x) such that

$$\|Ty - Tz\| \leq M\|x - y\|^p \quad \text{for all } y, z \in \overline{B}(x, r).$$

Theorem 16 ([23, 27]). *Let $T : X \to Y$ be continuous, surjective and the function $c : \mathbb{R}^+ \to \mathbb{R}^+$ be such that $c(r)r^q \to \infty$ as $r \to \infty$, and assume that for some $q > 0$, all $r > 0$ and for each large n*

$$\|Q_nTx - Q_nTy\| \geq c(r)\|x-y\|^q \quad \text{for all } x,y \in \overline{B}(0,r) \cap X_n.$$

Then T is a homeomorphism. Moreover, if T is also locally p-Lipschitz for some $p > 0$, it is A-proper w.r.t. $\Gamma = \{X_n, P_n, Y_n, Q_n\}$, and for each $f \in Y$, $Tx = f$ is uniquely approximation solvable w.r.t. Γ, with the approximate solutions $x_n \in \overline{B}(0,r) \cap X_n$ for some r and satisfying for $n \geq n_0$

$$\|x_n - x\| \leq c_1\|Tx_n - f\|^{1/q}, \tag{5}$$
$$\|x_n - x\| \leq c_2\|x - P_nx\|^s \leq k_2 d(x, X_n)^s, \tag{6}$$

where d is the distance, $k_2 = c_2\delta$, $\delta = \sup\|Q_n\|$, with $s = 1$ if $p \geq q$ and $s = p/q$ if $0 < p/q < 1$.

Corollary 7. *Let $T : X \to Y$ be a continuous surjective c-strongly K-monotone map with a suitable K, i.e., for each $r > 0$*

$$(Tx - Ty, K(x-y)) \geq c(r)\|x-y\|^q\,\|K(x-y)\| \quad \text{for all } x,y \in \overline{B}(0,r),$$

and let $\Gamma = \{X_n, P_n, Y_n, Q_n\}$ be an approximation scheme for (X,Y) such that $Q_n^*Kx = Kx$ for all $x \in X_n$ and $n \geq 1$.

Then T is a homeomorphism. Moreover, if T is also locally p-Lipschitz for some $p > 0$, then T is A-proper w.r.t. $\Gamma = \{X_n, P_n, Y_n, Q_n\}$, and for each $f \in Y$, $Tx = f$ is uniquely approximation solvable w.r.t. Γ, with the unique approximate solutions $x_n \in \overline{B}(0,r) \cap X_n$ for some r and satisfying the error estimates (5)–(6).

Remark 7. If $Y = X$ and $K = J$, the normalized duality map, then $T : X \to X$ is called a c-strongly accretive map. It is known that a c-strongly accretive map with $c(r)$ =constant is surjective if it is continuous, or if it is demicontinuous and X^* is uniformly convex. This is proven using ordinary differential equations in Banach spaces. If T is only demicontinuous and c-strongly accretive, then Webb [37] showed that T is still A-proper and the equation $Tx = f$ is uniquely approximation solvable, but no error estimates are given. Here, for a scheme $\Gamma = \{X_n, P_n\}$ for X, we know that $P_n^*Jx = Jx$ for all $x \in X_n$. If $Y = X^*$ and $K = I$, then $T : X \to X^*$ is a c-strongly monotone map and thus surjective. In both cases, these choices of K satisfy the conditions in the above corollary.

Let us now discuss stongly K-monotone maps $T : X \to X$ with $K \neq J$ and $P_n^*Kx \neq Kx$ on X_n. They appear in some applications (see [34] and the references therein). Let X be a separable reflexive Banach space and (X, H, X^*) be in duality, i.e., there is a Hilbert space H such that $X \subset H \subset X^*$ with X dense in H and the embedding continuous. Let the bilinear continuous form $(\cdot,\cdot)$ be defined in $X^* \times X$ and in $X \times X^*$ such that it extends the inner product in H. Let $Z \subset X$ be another Banach space with continuous injection. Suppose that the linear map $K : X^* \to X$ is an isomorphism and $K : H \to Z$ is bounded. Let $(K^{-1})^* : X \to X^*$ be the adjoint of the map $K^{-1} : X \to X^*$ defined by the relation $\big(K^{-1}x^*, x\big) = \big(x^*, (K^{-1})^*x\big)$ for $x \in X$ and $x^* \in X^*$. Let $T : X \to X$ be c-strongly $(K^{-1})^*$-monotone, i.e.,

$$\big(Tx - Ty, (K^{-1})^*(x-y)\big) \geq c\|x-y\|^2 \quad \text{for all } x, y \in X.$$

Let $\Gamma = \{X_n, P_n\}$ be a projection scheme for X such that the P_n's have the following approximation property,

$$\|x - P_n x\| \leq \varepsilon_n \|x\|_Z, \quad n \geq 1, \tag{7}$$

with $\varepsilon_n \to 0$ as $n \to \infty$. Equivalently, this can be expressed as:

$$\|I - P_n\|_{L(Z,X)} \leq \varepsilon_n \to 0 \quad \text{as } n \to \infty,$$

where $\|\cdot\|_{L(Z,X)}$ denotes the norm of the map. The schemes satisfying (7) naturally appear in Galerkin and collocation methods that use smooth splines as trial functions (cf. [32–34]). For such maps we have the following result.

Corollary 8. *Let $T : X \to X$ be a c-strongly $(K^{-1})^*$-monotone map, $T : H \to H$ be Lipschitz continuous, and $\Gamma = \{X_n, P_n\}$ satisfy* (7)*. Then T is an A-proper w.r.t. Γ homeomorphism, and the unique approximate solutions $x_n \in X_n$ satisfy for $n \geq n_0$,*

$$\|x_n - x\| \leq c_1 \|Tx_n - f\|, \quad \|x_n - x\| \leq c_2 \|x - P_n x\| \leq k_2 d(x, X_n),$$

where d is the distance, $k_2 = c_2\delta$, and $\delta = \sup \|P_n\|$.

For k-contractive perturbations, we have the following extension of Theorem 16.

Theorem 17. *Let $T : X \to Y$ be continuous, surjective, locally p-Lipschitz for some $p > 0$ and, for some $c > 0$, and each large n, $\|Q_n Tx - Q_n Ty\| \geq c\|x - y\|$ for all $x, y \in X_n$. Let $C : X \to Y$ be q-Lipschitz with $c > \delta q$, and $T + C$ be locally injective. Then $T + C$ is A-proper w.r.t. $\Gamma = \{X_n, P_n, Y_n, Q_n\}$ and, for each $f \in Y$, $Tx + Cx = f$ is uniquely approximation solvable w.r.t. Γ, with the approximate solutions $x_n \in \overline{B}(0, r) \cap X_n$ for some r, and satisfying for $n \geq n_0$,*

$$\|x_n - x\| \leq c_1 \|Tx_n + Cx_n - f\|, \tag{8}$$

$$\|x_n - x\| \leq c_2 \|x - P_n x\| \leq k_2 d(x, X_n), \tag{9}$$

where d is the distance, $k_2 = c_2\delta$, for some $c_2 > 0$, and $\delta = \sup \|Q_n\|$.

Corollary 9. *Let $T : X \to X$ be a c-strongly $(K^{-1})^*$-monotone map, $T : H \to H$ be Lipschitz continuous and $\Gamma = \{X_n, P_n\}$ satisfy* (7)*, and $C : X \to Y$ be k-contractive with k sufficiently small. Then $T + C$ is A-proper w.r.t. $\Gamma = \{X_n, P_n\}$ and, for each $f \in X$, $Tx + Cx = f$ is uniquely approximation solvable w.r.t. Γ, with the unique approximate solutions $x_n \in X_n$ satisfying the error estimates* (8)–(9) *for $n \geq n_0$.*

Next, we shall use these results to discuss Gateaux differentiable and gradient homeomorphisms. We follow [27] (cf. also [20, 21]).

First, we shall look at $x - Nx = f$ with N Gateaux differentiable and such that $B_1 \leq N'(x) \leq B_2$ for all $x \in H$, where $B_1, B_2 : H \to H$ are selfadjoint maps with $B_1 \leq B_2$, i.e., $(B_1x, x) \leq (B_2x, x)$ for $x \in H$. A fixed point theory for such maps has been developed by Perov [30] and Krasnoselskii-Zabreiko (cf. [15]) assuming that $\{B_1, B_2\}$ is a regular pair. These maps have been studied extensively in the context of semilinear equations by the author [19–24].

The pair $\{B_1, B_2\}$ is said to be regular if 1 is not in the spectrum $\sigma(B_1) \cup \sigma(B_2)$, $\sigma(B_1) \cap (1, \infty) = \{\lambda_1, \ldots, \lambda_k\}$, $\sigma(B_2) \cap (1, \infty) = \{\mu_1, \ldots, \mu_m\}$, where the λ_i's and the μ_j's are eigenvalues of B_1 and B_2, respectively, of finite multiplicities and $\beta(B_1)$ =the sum of the multiplicities of the λ_i's is equal to $\beta(B_2)$ =the sum of multiplicities of the μ_j's. It has been shown in [15] that if $\{B_1, B_2\}$ is a regular pair, then there is a constant $c > 0$ such that for each selfadjoint map C with $B_1 \leq C \leq B_2$ we have that $\|x - Cx\| \geq c\|x\|$ for all $x \in H$.

Here, $c = \min\{1 - \lambda_-, \lambda_+ - 1\}$, where λ_- is the supremum of all spectral values of B_2 that are smaller than 1, and λ_+ is the minimum of all spectral values of B_1 that are larger than 1. If there are no spectral values larger than 1, we put $\lambda_+ = \infty$.

Theorem 18. *Let $N : H \to H$ be Gateaux differentiable, $N'(x)$ be selfadjoint and $B_1 \leq N'(x) \leq B_2$ for each $x \in H$ and some selfadjoint regular pair $\{B_1, B_2\}$. Then $I - N$ is a homeomorphism, the equation $x - Nx = f$ is uniquely approximation solvable for each $f \in H$, with the unique approximate solutions $x_n \in H_n$ of $x - P_nNx = P_nf$ satisfying the following error estimates for each large n,*

$$\|x_n - x\| \leq c_1\|x_n - Nx_n - f\|,$$
$$\|x_n - x\| \leq c_2\|x - P_nx\| \leq k_2 d(x, H_n),$$

where d is the distance, $k_2 = c_2\delta$, $\delta = \sup\|P_n\|$ and $c_1 > 0$.

The following results are proven by showing that the involved maps are strongly monotone and therefore Theorem 16 applies. For general gradient maps, we have the following result.

Theorem 19. *Let $A : H \to H$ be a continuous gradient map and suppose that there are closed subspaces H_1 and H_2 of H such that $H = H_1 + H_2$, selfadjoint maps $B_1, B_2 : H \to H$ and $q > 0$ such that*

(i) $A - B_1$ and $B_2 - A$ are monotone maps and
(ii) $(B_1x, x) \geq q\|x\|^2$ on H_1 and $(B_2x, x) \leq -q\|x\|^2$ on H_2.

Then A is a homeomorphism, the equation $Ax = f$ is uniquely approximation solvable for each $f \in H$, with the unique approximate solutions $x_n \in H_n$ of $P_nAx = P_nf$ satisfying the following error estimates,

$$\|x_n - x\| \leq c_1\|Ax_n - f\|, \quad \|x_n - x\| \leq c_2\|x - P_nx\| \leq k_2 d(x, H_n),$$

for each large n, where d is the distance, $k_2 = c_2\delta$, $\delta = \sup\|P_n\|$, and $c_1 > 0$.

Remark 8. If $B_1, B_2 : H \to H$ are selfadjoint maps such that $B_1 \le B_2$ and $A : H \to H$ is Gateaux differentiable, $A'(x)$ is selfadjoint and $B_1 \le A'(x) \le B_2$ for each $x \in H$, then $A - B_1$ and $B_2 - A$ are monotone maps.

The homeomorphism assertion in the next result was proved by Dancer [9] using different methods.

Corollary 10. *Let $B_1, B_2 : H \to H$ be selfadjoint, bounded invertible maps such that $B_1 \le B_2$ and $A : H \to H$ be Gateaux differentiable with $B_1 \le A'(x) \le B_2$ for each $x \in H$ and*
(i) $(B_1 x, x) \ge q\|x\|^2$ *on* H_1 *and* $(B_2 x, x) \le -q\|x\|^2$ *on* H_2.
Then A is a homeomorphism, the equation $Ax = f$ is uniquely approximation solvable for each $f \in H$, with the unique approximate solutions $x_n \in H_n$ of $P_n Ax = P_n f$ satisfying the following error estimates for each large n,

$$\|x_n - x\| \le \|K\|/q\|Ax_n - f\|, \quad \|x_n - x\| \le c_2\|x - P_n x\| \le k_2 d(x, X_n),$$

where d is the distance, $k_2 = c_2\delta$, and $\delta = \sup\|P_n\|$.

Remark 9. Condition (i) in Corollary 10 is equivalent to $R(P_1^+) + R(P_2^-) = H$ ([9]), where P_1^- is the spectral projection for B_1 corresponding to $(-\infty, 0)$, P_1^+ is the spectral projection for B_2 corresponding to $(0, \infty)$, and $R(L)$ is the range of L.

Recall that $N : H \to H$ is said to be regularly $\{B_1, B_2\}$-quasilinear if for each $x, y \in H$ there is a selfadjoint map $D(x-y)$ with $B_1 \le D(x-y) \le B_2$ and $Nx - Ny = D(x-y)(x-y)$. If N is a Gateaux differential gradient map, then it is regularly $\{B_1, B_2\}$-quasilinear if and only if $B_1 \le N'(x) \le B_2$ for each $x \in H$. Let H_0 be the null space of $B_2 - B_1$ and H_1 be its orthogonal complement in H. Then the square root $(B_2 - B_1)^{1/2}$ is also a positive definite selfadjoint map on H_1 with range R_1 and has the inverse $(B_2 - B_1)^{-1/2} : R_1 \to H_1$. Moreover, N is regularly $\{B_1, B_2\}$-quasilinear if and only if ([15]) $Nx - B_1 x$, $B_2 - Nx \in R_1$ for all $x \in H$ and

$$\big((B_2 - B_1)^{-1}(Nx - Ny - B_1(x-y)), B_2(x-y) - (Nx - Ny)\big) \ge 0 \quad \text{for all } x, y \in H.$$

It is clear that if N is regularly $\{B_1, B_2\}$-quasilinear then $N - B_1$ and $B_2 - N$ are monotone maps.

The differentiability condition on A can be further relaxed. Assume that $A : H \to H$ has the Gateaux differential $DA(x;h)$, i.e., for each $x, h \in H$ the limit

$$\lim_{t \to 0}(A(x + th) - Ax)/t = DA(x;h)$$

exists. For fixed x, the map $DA(x;h)$ is homogeneous in h.

We have the following extension of Theorem 2 in [9]. No regularity of the pair $\{B_1, B_2\}$ is required.

Theorem 20. *Let $A : H \to H$ be regularly $\{B_1, B_2\}$-quasilinear with $Ax - Ay = D(x-y)(x-y)$ for each $x, y \in H$, and $D(x-y)$ a selfadjoint map with $B_1 \le$*

$D(x-y) \leq B_2$. Assume that there is a $q>0$ such that $\|D^{-1}(x-y)\| \leq q$ for all $x,y \in H$. Then the range of A is closed. Let, in addition, the Gateaux differential $DA(x;h)$ have a dense range in H for each fixed $x \in H$, i.e., the set $\{DA(x;h) \mid h \in H\}$ is dense in H. Then A is a homeomorphism. If, in addition, A is A-proper w.r.t. Γ and if x_n is an approximate solution to x, then $\|x_n - x\| \leq q^{-1}\|Ax_n - f\|$ for each n.

Corollary 11. *Let $B_1, B_2 : H \to H$ be selfadjoint, bounded invertible maps such that $B_1 \leq B_2$ and $A : H \to H$ be regularly $\{B_1, B_2\}$-quasilinear. Suppose that there are a constant $q>0$ and closed subspaces H_1 and H_2 of H such that $H = H_1 + H_2$ and*
(i) $(B_1x,x) \geq q\|x\|^2$ on H_1 and $(B_2x,x) \leq -q\|x\|^2$ on H_2.
Then the range of A is closed. Let, in addition, the Gateaux differential $DA(x;h)$ have a dense range in H for each fixed $x \in H$. Then A is a homeomorphism. If, in addition, A is A-proper w.r.t. Γ and if x_n is an approximate solution to x with $Ax = f$, then $\|x_n - x\| \leq q^{-1}\|Ax_n - f\|$ for each n.

Note that in Theorem 20 we have that A is a closed map, mapping closed bounded subsets of H into closed sets and satisfying condition $(+)$. These properties imply that $R(A)$ is closed. In view of this, we have the following extension of Theorem 20.

Theorem 21. *Let X and Y be Banach spaces and $A : X \to Y$ be closed, mapping closed bounded subsets of X into closed subsets of Y and satisfying condition $(+)$. Let, in addition, the Gateaux differential $DA(x;h)$ have a dense range in H for each fixed $x \in H$. Then A is surjective. If A is injective and A and A^{-1} are continuous, then A is a homeomorphism.*

2.2 Constructive Homeomorphisms and Their Perturbations

In this subsection, following [27], we shall study A-proper homeomorphisms that are strongly approximation solvable and show that they are stable under condensing perturbations. The results of this subsection are based on the open mapping theorem for A-proper maps (cf. [17]) and the following result.

Theorem 22. *Let $T : X \to Y$ be a continuous, locally injective and open map that is closed, and in particular proper, on bounded and closed subsets of X. Then T is a homeomorphism if and only if T satisfies condition $(+)$.*

Theorem 23. *Let $T : X \to Y$ be continuous, locally injective, locally A-proper with respect to a projection scheme $\Gamma = \{X_n, Y_n, Q_n\}$ and for each $x_0 \in X$ let there be an $r>0$ such that $\deg(Q_nT, B(x_0,r) \cap X_n, Q_nTx_0) \neq 0$ for all large n. Then T is a homeomorphism if and only if it satisfies condition $(+)$, in which case the equation $Tx = f$ is strongly approximation solvable in a neighborhood $B(x_0,r)$ of its solution x_0.*

Replacing condition $(+)$ by the weaker closedness of the range condition, we get

Theorem 24. *Let $T : X \to Y$ be locally A-proper with respect to a projection scheme $\Gamma = \{X_n, Y_n, Q_n\}$, have closed range, and for each $x_0 \in X$ let there be an $r > 0$ such that* $\deg(Q_nT, B(x_0,r) \cap X_n, Q_nTx_0) \neq 0$ *for all large n. Then either*

(i) *T is surjective, or*

(ii) *T is continuous and injective, in which case it is a homeomorphism and the equation $Tx = f$ is strongly approximation solvable in a neighborhood $B(x_0,r)$ of its solution x_0.*

Recall that $T : X \to X^*$ is said to be of type (S_+) if $x_n \rightharpoonup x$ and $\limsup(Tx_n, x_n - x) \leq 0$ imply that $x_n \to x$.

Corollary 12. *Let X be reflexive, $T : X \to X^*$ be bounded, locally injective, and locally of type (S_+) and have closed range. Then either*

(i) *T is demicontinuous, in which case the equation $Tx = f$ is strongly approximation solvable w.r.t $\Gamma = \{X_n, P_n, X_n^*, P_n^*\}$ in a neighborhood $B(x_0,r)$ of each of its solution for each $f \in X^*$, or*

(ii) *T is continuous and injective, in which case the equation $Tx = f$ is strongly approximation solvable w.r.t $\Gamma = \{X_n, P_n, X_n^*, P_n^*\}$ in a neighborhood $B(x_0,r)$ of its unique solution x_0 for each $f \in X^*$.*

Fréchet differentiability implies the degree assumption in Theorem 23. Hence, we have

Corollary 13. *Let $T : X \to Y$ be Fréchet differentiable and A-proper with respect to a projection scheme $\Gamma = \{X_n, Y_n, Q_n\}$. Suppose that $T'(x)$ is A-proper and injective. Then T is a homeomorphism if and only if it satisfies condition $(+)$, in which case the equation $Tx = f$ is strongly approximation solvable in a neighborhood $B(x_0,r)$ of its solution x_0 for each $f \in Y$. Moreover, if $c_0 = \|(T'(x_0))^{-1}\|$, then for every $\varepsilon \in (0, c_0)$ there is an $n_0 \geq 1$ such that the approximate solutions $x_n \in B(x_0,r) \cap X_n$ of $Q_nTx = Q_nf$ satisfy*

$$\|x_n - x_0\| \leq (c_0 - \varepsilon)^{-1} \|Tx_n - f\| \quad \text{for all } n \geq n_0. \tag{10}$$

If condition $(+)$ is relaxed to the range of T being closed, then we get the following result.

Theorem 25. *Let $T : X \to Y$ be Gateaux differentiable, have closed range and $T'(x)$ be A-proper w.r.t. $\Gamma = \{X_n, Y_n, Q_n\}$ and injective for each $x \in X$. Then*

(i) *T is surjective.*

(ii) *If T is Fréchet differentiable and A-proper w.r.t. Γ, then the equation $Tx = f$ is strongly approximation solvable in a neighborhood $B(x_0,r)$ of each of its solution x_0 for each $f \in Y$ and the error estimate (10) for T holds. Moreover,* $\operatorname{card} T^{-1}(f)$ *is constant and finite on each connected component of the open set $Y \setminus T(\Sigma)$.*

(iii) *If T is continuously Fréchet differentiable and A-proper, then T is a Fredholm homeomorphism of index zero, the equation $Tx = f$ is uniquely approximation solvable for each $f \in Y$, and the unique solutions $x_n \in B(r,x_0) \cap X_n$ of $Q_nTx = Q_nf$ satisfy*

$$\|x_n - x_0\| \le k\|P_nx_0 - x_0\| \le c\,\mathrm{dist}\,(x_0,X_n)$$

for each large n, some k depending on $c_0 = \|(T'(x_0))^{-1}\|$, $\varepsilon \in (0,c_0)$, $\delta = \sup\|Q_n\|$, where $c = 2k\sup\|P_n\|$.

We have the following homeomorphism result for maps of type (S_+).

Theorem 26. *Let X be reflexive, $T : X \to X^*$ be bounded, continuous, locally injective and locally of type (S_+). Then T is a homeomorphism if and only if it satisfies condition $(+)$, in which case the equation $Tx = f$ is strongly approximation solvable in a neighborhood $B(x_0,r)$ of its solution x_0 for each $f \in X^*$.*

2.3 Nonlinear Alternatives for Perturbations of A-Proper Homeomorphisms

In this subsection, following [27], we shall now look at various nonlinear alternatives for A-proper maps and their perturbations. Finite solvability results are based on the following basic theorem on the number of solutions of nonlinear equations for A-proper maps (see [24]).

Theorem 27. *Let $T : X \to Y$ be a continuous A-proper map that satisfies condition $(+)$. Then*

(a) *The set $T^{-1}(f)$ is compact (possibly empty) for each $f \in Y$.*
(b) *The range $R(T)$ of T is closed and connected.*
(c) *Σ and $T(\Sigma)$ are closed subsets of X and Y, respectively, and $T(X \setminus \Sigma)$ is open in Y.*
(d) $\mathrm{card}\,T^{-1}(f)$ *is constant and finite (it may be 0) on each connected component of the open set $Y \setminus T(\Sigma)$.*
(e) *If T is locally injective, then* $\mathrm{card}\,T^{-1}(f)$ *is finite (it may be 0) for each $f \in Y$.*

Theorem 28. *Let $T : X \to Y$ be continuous, A-proper with respect to a projection scheme $\Gamma = \{X_n, Y_n, Q_n\}$, satisfy condition $(+)$, and for each $x_0 \in X$ let there be an $r > 0$ such that* $\deg(Q_nT, B(x_0,r) \cap X_n, Q_nTx_0) \neq 0$ *for all large n. Then either*

(i) *T is locally injective, in which case it is a homeomorphism and the equation $Tx = f$ is strongly approximation solvable in a neighborhood $B(x_0,r)$ of each of its solution x_0 for each $f \in Y$, or*
(ii) *T is not locally injective, in which case it is surjective, $T^{-1}(f)$ is compact and the cardinal number* $\mathrm{card}\,T^{-1}(f)$ *is positive, constant, and finite on each connected component of the set $Y \setminus (T)(\Sigma)$.*

We continue with k-set contractive perturbations of some A-proper homeomorphisms.

Theorem 29. *Let $T : X \to Y$ be continuous, locally injective A-proper with respect to a projection scheme $\Gamma = \{X_n, Y_n, Q_n\}$, satisfy condition $(+)$, and for each $x_0 \in X$ let there be an $r > 0$ such that $\deg(Q_nT, B(x_0, r) \cap X_n, Q_nTx_0) \neq 0$ for all large n and $C : X \to Y$ be such that $\alpha(C) < \beta(T)$ and $T + C$ satisfies condition $(+)$. Then either*

(i) *$T + C$ is injective, in which case it is a homeomorphism, or*

(ii) *$T + C$ is not injective, in which case, assuming additionally that $T + tC$ satisfies condition $(+)$, it is surjective, $T^{-1}(f)$ is compact and the cardinal number* $\operatorname{card} T^{-1}(f)$ *is positive, constant, and finite on each connected component of the set $Y \setminus (T)(\Sigma)$.*

Theorem 30. *Let $T : X \to Y$ be continuously Fréchet differentiable, A-proper w.r.t. Γ and have closed range. Let $T'(x)$ be injective and A-proper w.r.t. Γ for each $x \in X$. Suppose that $C : X \to Y$ is such that $\alpha(C) < \beta(T)$ and $T + C$ satisfies condition $(+)$. Then either*

(i) *$T + C$ is locally injective, in which case it is a homeomorphism, or*

(ii) *$T + C$ is not locally injective, in which case, assuming additionally that $T + tC$ satisfies condition $(+)$, the equation $Tx + Cx = f$ is solvable for each $f \in Y$ with $(T + C)^{-1}(f)$ compact, and the cardinal number* $\operatorname{card}(A - N)^{-1}(f)$ *is positive, constant, and finite on each connected component of the set $Y \setminus (A - N)(\Sigma)$.*

Using Theorem 26, we get the following perturbation result.

Theorem 31. *Let X be reflexive, $T : X \to X^*$ be bounded, continuous, locally injective and locally of type (S_+) and satisfy condition $(+)$, and let $C : X \to X^*$ be such that $\alpha(C) < \beta(T)$ and $T + C$ satisfies condition $(+)$. Then either*

(i) *$T + C$ is injective, in which case it is a homeomorphism, or*

(ii) *$T + C$ is not injective, in which case, assuming additionally that $T + tC$ satisfies condition $(+)$, it is surjective, $T^{-1}(f)$ is compact and the cardinal number* $\operatorname{card} T^{-1}(f)$ *is positive, constant, and finite on each connected component of the set $X^* \setminus (T)(\Sigma)$.*

Next, we give a nonconstructive extension of Theorem 17 involving general $k - \phi$-contractive perturbations.

Theorem 32. *Let $T : X \to Y$ be continuous, surjective and for some $c > 0$ and for each large n*

$$\|Q_nTx - Q_nTy\| \geq c\|x - y\| \quad \text{for all } x, y \in X_n,$$

and let $C : X \to Y$ be $k - \phi$-contractive with $k < \beta(T)$. Then either

(i) *$T + C$ is locally injective, in which case $T + C$ is a homeomorphism if and only if $T + C$ satisfies condition $(+)$, or*

(ii) $T+C$ *is not locally injective, in which case, assuming additionally that* $T+tC$ *satisfies condition* $(+)$, *it is surjective,* $T^{-1}(f)$ *is compact, and the cardinal number* $\operatorname{card} T^{-1}(f)$ *is positive, constant, and finite on each connected component of the set* $X^* \setminus (T)(\Sigma)$.

A partially constructive version of Theorem 32 and its special cases will be discussed next.

Theorem 33. *Let* $A: X \to Y$ *be continuous, surjective, and satisfy for some* $c > 0$

$$\|Q_n Ax - Q_n Ay\| \geq c\|x-y\| \quad \text{for all } x,y \in \overline{B}(0,r) \cap X_n,$$

and let $N: X \to Y$ *be quasibounded with* $|N| < c$, *where*

$$|N| = \limsup_{\|x\| \to \infty} \|Nx\| / \|x\|,$$

and be $k-\phi$*-contractive with* $k < c$ *if* ϕ *is the set measure of noncompactness, and if* ϕ *is the ball measure of noncompactness, either* A *is bounded or* $\delta k < c$, *where* $\delta = \max \|Q_n\|$. *Then either*

(a) $A-N$ *is locally injective, in which case it is a homeomorphism, or*
(b) $A-N$ *is not locally injective, in which case, the equation* $Ax - Nx = f$ *is solvable and* $(A-N)^{-1}(f)$ *is compact for each* $f \in Y$. *Moreover, the cardinal number* $\operatorname{card}(A-N)^{-1}(f)$ *is positive, constant, and finite on each connected component of the set* $Y \setminus (A-N)(\Sigma)$.
(c) *If* $\Gamma = \{P_n, X_n, Y_n, Q_n\}$ *is a projection scheme for* H, *and* ϕ *is the ball measure of noncompactness, then* $A-N$ *is A-proper w.r.t.* Γ *and the equation* $Ax - Nx = f$ *is strongly approximation solvable for each* $f \in Y$ *if* $A-N$ *is injective. Otherwise, it is strongly approximation solvable in a neighborhood* $B(r,x_0)$ *of each of its solution* x_0 *for each* $f \in Y \setminus (A-N)(\Sigma)$.

Corollary 14. *Let* $A: X \to Y$ *be a continuous surjective c-strongly* K*-monotone map and* $N: X \to Y$ *be* $k-\phi$*-contractive with* $k < c$ *if* ϕ *is the set measure of noncompactness, and if* ϕ *is the ball measure of noncompactness, either* A *is bounded or* $\delta k < c$ *and* $Q_n^* Kx = Kx$ *on* X_n, *where* $\delta = \max \|Q_n\|$. *Suppose that* $A-N$ *satisfies condition* $(+)$, *or, in particular,* N *is quasibounded with the quasinorm* $|N| < c$. *Then either*

(a) $A-N$ *is locally injective, in which case it is a homeomorphism, or*
(b) $A-N$ *is not locally injective, in which case the equation* $Ax - Nx = f$ *is solvable and* $(A-N)^{-1}(f)$ *is compact for each* $f \in Y$. *Moreover, the cardinal number* $\operatorname{card}(A-N)^{-1}(f)$ *is positive, constant, and finite on each connected component of the set* $Y \setminus (A-N)(\Sigma)$.
(c) *If* $\Gamma = \{P_n, X_n, Y_n, Q_n\}$ *is a projection scheme for* H, *and* ϕ *is the ball measure of noncompactness, then* $A-N$ *is A-proper w.r.t.* Γ *and the equation* $Ax - Nx = f$ *is strongly approximation solvable in a neighborhood* $B(r,x_0)$ *of each of its solution* x_0 *for each* $f \in Y$ *if* $A-N$ *is injective. Otherwise, it is strongly approximation solvable for each* $f \in Y \setminus (A-N)(\Sigma)$.

Next, we shall look at perturbations of c-strongly $(K^{-1})^*$-monotone maps A discussed in Corollary 8.

Corollary 15. *Let $A : X \to X$ be a c-strongly $(K^{-1})^*$-monotone map, $A : H \to H$ be Lipschitz continuous and $\Gamma = \{X_n, P_n\}$ satisfy (7). Suppose that $N : X \to Y$ is $k-\phi$-contractive with $k < c$ if ϕ is the set measure of noncompactness, and $\delta k < c$ if ϕ is the ball measure of noncompactness, where $\delta = \max \|P_n\|$. Suppose that $A+N$ satisfies condition $(+)$, or, in particular, N is quasibounded with the quasinorm $|N| < c$. Then either*

(a) *$A - N$ is locally injective, in which case it is a homeomorphism, or*
(b) *$A - N$ is not locally injective, in which case the equation $Ax - Nx = f$ is solvable and $(A-N)^{-1}(f)$ is compact for each $f \in H$. Moreover, the cardinal number $\mathrm{card}(A-N)^{-1}(f)$ is positive, constant, and finite on each connected component of the set $H \setminus (A-N)(\Sigma)$.*
(c) *If $\Gamma = \{P_n, X_n\}$ is a projection scheme for H, and ϕ is the ball measure of noncompactness, then $A - N$ is A-proper w.r.t. Γ and the equation $Ax - Nx = f$ is strongly approximation solvable in a neighborhood $B(r, x_0)$ of each of its solution x_0 for each $f \in H$ if $A - N$ is injective. Otherwise, it is strongly approximation solvable for each $f \in H \setminus (A-N)(\Sigma)$.*

Finally, we shall look at perturbations of the results from Sect. 2.1. We begin with perturbations of Gateaux differential homeomorphisms. They are proved by combining the results of Sect. 1.1 in Part I and of Sect. 2.1 in Part II.

Theorem 34. *Let $N : H \to H$ be Gateaux differentiable, $N'(x)$ be selfadjoint and $B_1 \le N'(x) \le B_2$ for each $x \in H$ and some selfadjoint regular pair $\{B_1, B_2\}$. Assume that $C : H \to H$ is quasibounded $k-\phi$-contractive with the sufficiently small quasinorm $|C|$ and k. Then either*

(i) *$I - N - C$ is locally injective, in which case it is a homeomorphism, or*
(ii) *$I - N - C$ is not locally injective, in which case the equation $x - Nx - Cx = f$ is solvable and $(I-N-C)^{-1}(f)$ is compact for each $f \in H$. Moreover, the cardinal number $\mathrm{card}(I-N-C)^{-1}(f)$ is positive, constant, and finite on each connected component of the set $H \setminus (I-N-C)(\Sigma)$.*
(iii) *If $\Gamma = \{H_n, P_n\}$ is a projection scheme for H, and ϕ is the ball measure of noncompactness, then $I - N - C$ is A-proper w.r.t. Γ and the equation $x - Nx - Cx = f$ is strongly approximation solvable for each $f \in H$ if $I - N - C$ is injective. Otherwise, it is strongly approximation solvable in a neighborhood $B(r, x_0)$ of each of its solution x_0 for each $f \in H \setminus (I-N-C)(\Sigma)$.*

Theorem 35. *Let $A : H \to H$ be a gradient map and $N : H \to H$ be $k-\phi$-contractive with k and $|N|$ sufficiently small. Suppose that there are closed subspaces H_1 and H_2 of H such that $H = H_1 + H_2$, selfadjoint maps $B_1, B_2 : H \to H$ and $q > 0$ such that*
(i) *$A - B_1$ and $B_2 - A$ are monotone maps and*
(ii) *$(B_1 x, x) \ge q\|x\|^2$ on H_1 and $(B_2 x, x) \le -q\|x\|^2$ on H_2.*

Then either

(a) $A-N$ *is locally injective, in which case, it is a homeomorphism, or*
(b) $A-N$ *is not locally injective, in which case, the equation* $Ax-Nx=f$ *is solvable and* $(A-N)^{-1}(f)$ *is compact for each* $f\in H$. *Moreover, the cardinal number* $\mathrm{card}(A-N)^{-1}(f)$ *is positive, constant, and finite on each connected component of the set* $H\setminus(A-N)(\Sigma)$.
(c) *If* $\Gamma=\{H_n,P_n\}$ *is a projection scheme for* H, *and* ϕ *is the ball measure of noncompactness, then* $A-N$ *is A-proper w.r.t.* Γ *and the equation* $Ax-Nx=f$ *is strongly approximation solvable for each* $f\in H$ *if* $A-N$ *is injective. Otherwise, it is strongly approximation solvable in a neighborhood* $B(r,x_0)$ *of each of its solution* x_0 *for each* $f\in H\setminus(A-N)(\Sigma)$.

In what follows, we give a couple of special cases.

Corollary 16. *Let* $B_1,B_2:H\to H$ *be selfadjoint, bounded invertible maps, and* $A:H\to H$ *be Gateaux differentiable on* H *such that* $B_1\le A'(x)\le B_2$ *and* $A'(x)$ *is selfadjoint for each* $x\in H$ *and*
(i) $(B_1x,x)\ge q\|x\|^2$ *on* H_1 *and* $(B_2x,x)\le -q\|x\|^2$ *on* H_2.
Assume that $N:H\to H$ *is a quasibounded* $k-\phi$*-contractive map with* $|N|$ *and* k *sufficiently small. Then the conclusions of Theorem 35 hold.*

Corollary 17. *Let* $B_1,B_2:H\to H$ *be selfadjoint, bounded invertible maps,and* $A:H\to H$ *be Gateaux differentiable on* H *such that* $B_1\le A'(x)\le B_2$ *and* $A'(x)$ *is selfadjoint for each* $x\in H$ *and*
(i) $(B_1x,x)\ge q\|x\|^2$ *on* H_1 *and* $(B_2x,x)\le -q\|x\|^2$ *on* H_2.
Assume that $N:H\to H$ *is a* k *contractive map. Then the equation* $Ax+Nx=f$ *is uniquely solvable for each* $f\in H$ *if* $qk<1$. *It is solvable if* $qk=1$.

Remark 10. The results of Part II can be used to give constructive solvability and error estimates for BVP's discussed in Sect. 1.3 of Part I, assuming additional suitable conditions ([27]).

References

1. Benevieri, P., Furi, M.: A simple notion of orientability for Fredholm maps of index zero between Banach manifolds and degree theory. Ann. Sci. Math. Quebec, **22**, no 2, 131–148 (1998)
2. Benevieri, P., Furi, M.: A degree theory for locally compact perturbations of nonlinear Fredholm maps of index zero. Abstr. Appl. Anal. 1–20 (2006)
3. Benevieri, P., Calamai, A., Furi, M.: A degree theory for a class of perturbed Fredholm maps. Fixed Point Theory Appl. **2**, 185–206 (2005)
4. Browder, F.E.: Covering spaces, fiber spaces and local homeomorphisms. Duke Math. J. **21**, 329–336 (1954)
5. Calamai, A.: The invariance of domain theorem for compact perturbations of nonlinear Fredholm maps of index zero. Nonlinear Funct. Anal. Appl. **9**, 185–194 (2004)

6. Chang, K.C., Shujie, L.: A remark on expanding maps. Proc. Am. Math. Soc. **85**, 583–586 (1982)
7. Chia-Chuan, T., Ngai-Ching, W.: Invertibility in infinite-dimensional spaces. Proc. Am. Math. Soc. **128**(2), 573–581 (1999)
8. Coatabel, M.: Boundary integral operators on Lipschitz domain: elementary results. SIAM J. Math. Anal., 613–630 (1988)
9. Dancer, N.: Order intervals of selfadjoint linear operators and nonlinear homeomorphisms. Pacific J. Math. **115**, 57–72 (1984)
10. Efendiev, M.A., Schmitz, H., Wendland, W.L.: On some nonlinear potential problems. Electronic J. Diff. Eqs. **1999**(18), 1–17 (1999)
11. Fitzpatrick, P.M., Pejsachowisz, J., Rabier, P.J.: The degree of proper C^2 Fredholm mappings I. J. Reine Angew. Math. **247**, 1–33 (1992)
12. Hernandez, J.E., Nashed, M.Z.: Global invertibility of expanding maps. Proc. Am. Math. Soc. **116**(1), 285–291 (1992)
13. Hess, P.: On nonlinear mappings of monotone type homotopic to odd operators. J. Funct. Anal. **11**, 138–167 (1972)
14. John, F.: On quasi-isometric mappings, I. Comm. Pure Appl. Math. **21**, 77–110 (1968)
15. Krasnoselskii, M.A., Zabreiko, P.O.: Geometrical Methods of Nonlinear Analysis. Springer-Verlag, Berlin, New York (1984)
16. Milojević, P.S.: A generalization of the Leray-Schauder theorem and surjectivity results for multivalued A-proper and pseudo A-proper mappings. Nonlinear Anal. TMA, **1**(3), 263–276 (1977)
17. Milojević, P.S.: Some generalizations of the first Fredholm theorem to multivalued A-proper mappings with applications to nonlinear elliptic equations. J. Math. Anal. Appl. **65**(2), 468–502 (1978)
18. Milojević, P.S.: Fredholm alternatives and surjectivity results for multivalued A-proper and condensing mappings with applications to nonlinear integral and differential equations. Czech. Math. J. **30**(105), 387–417 (1980)
19. Milojević, P.S.: Fredholm theory and semilinear equations without resonance involving noncompact perturbations, I, II. Applications, Publications de l'Institut Math. **42**, 71–82 and 83–95 (1987)
20. Milojević, P.S.: Solvability of semilinear operator equations and applications to semilinear hyperbolic equations. In: Nonlinear Functional Analysis, Marcel Dekker (Ed. P.S. Milojević), vol. 121, 95–178 (1989)
21. Milojević, P.S.: Approximation-solvability of nonlinear equations and applications. In: Fourier Analysis (Eds. W. Bray, P.S. Milojević, C.V. Stanojević), Lecture Notes in Pure and Applied Mathematics, vol. 157, 311–373, Marcel Dekker Inc., NY (1994)
22. Milojević, P.S.: Approximation-solvability of semilinear equations and applications. In: Theory and Applications of Nonlinear Operators of Accretive and Monotone Type (Ed. A.G. Kartsatos), Lecture Notes in Pure and Applied Mathematics, vol.178, 149–208, Marcel Dekker Inc., NY (1996)
23. Milojević, P.S.: Implicit function theorems, approximate solvability of nonlinear equations, and error estimates. J. Math. Anal. Appl. **211**, 424–459 (1997)
24. Milojević, P.S.: Existence and the number of solutions of nonresonant semilinear equations and applications to boundary value problems. Math. Comput. Model. **32**, 1395–1416 (2000)
25. Milojević, P.S.: Homeomorphisms and finite solvability of their perturbations for Fredholm maps of index zero with applications. Facta Universitatis. **22**(2), 123–153 (2007) (CAMS Technical Report No 11, 2007, NJIT.)
26. Milojević, P.S.: Homeomorphisms and Fredholm theory for perturbations of nonlinear Fredholm maps of index zero with applications. Electron. J. Diff. Eqs. **113**, 1–26 (2009)
27. Milojević, P.S.: Constructive homeomorphisms, error estimates and Fredholm theory for perturbations of A-proper maps with applications. (in preparation)
28. Necas, J.: Sur l'alternative de Fredholm pour les operateurs non-lineares avec applications aux problemes aux limites. Ann. Scoula Norm. Sup. Pisa. **23**, 331–345 (1969)

29. Pejsachowisz, J., Rabier, P.J.: Degree theory for C^1 Fredholm mappings of index 0. J. d'Anal. Math. **76**, 289–319 (1998)
30. Perov, A.I.: On the principle of the fixed point with two-sided estimates. Doklady Akad. Nauk SSSR. **124**, 756–759 (1959)
31. Rabier, P.J., Salter, M.F.: A degree theory for compact perturbations of proper C^1 Fredholm mappings of index zero. Abstr. Appl. Anal. **7**, 707–731 (2005)
32. Ruotsalainen, K.: Remarks on the boundary element method for strongly nonlinear problems. J. Austral. Math. Soc. Ser B. **34**, 419–438 (1993)
33. Ruotsalainen, K., Wendland, W.L.: On the boundary element method for nonlinear boundary value problems. Numer. Math. **53**, 299–314 (1988)
34. Saranen, J.: Projection methods for a class of Hammerstein equations. SIAM J. Numer. Anal. **27**(6), 1445–1449 (1990)
35. Tromba, A.J.: Some theorems on Fredholm maps. Proc. Am. Math. Soc. **34**(2), 578–585 (1972)
36. Verchota, G.: Layer potentials and regularity for the Dirichlet problem for equations in Lipschitz domains. J. Funct. Anal. **59**, 572–611 (1984)
37. Webb, J.R.L.: On the property of duality mappings and the A-properness of accretive operators. Bull. Lond. Math. Soc. **13**, 235–238 (1981)

Singular Support and $\mathscr{F}L^q$ Continuity of Pseudodifferential Operators

Stevan Pilipović, Nenad Teofanov and Joachim Toft

This article is dedicated to Professor Gradimir V. Milovanović on the occasion of his 60th birthday

1 Introduction

The content of this paper was presented at the plenary lecture of the conference "Approximation & Computation APP & COM 2008" dedicated to Professor G. V. Milovanović. Although this material is not strictly related to the main topic of the conference, it shows possible directions for numerical mathematicians interested in the approximation of different types of singular supports, wave front sets and of pseudodifferential operators in the framework of Fourier-Lebesgue spaces.

The paper, an expository one, presents some of the authors' recent results which are scattered in several other papers, [15–18]. Moreover, it contains new results on singular supports in Fourier-Lebesgue spaces, Theorem 1, and on the continuity properties of certain pseudodifferential operators, Theorem 8 and Proposition 5.

The recent study of pseudodifferential and Fourier integral operators in Fourier-Lebesgue spaces, as well as their connection with modulation spaces in different

Stevan Pilipović
Faculty of Sciences, Department of Mathematics and Informatics, University of Novi Sad, Serbia
e-mail: stevan.pilipovic@dmi.uns.ac.rs

Nenad Teofanov
Faculty of Sciences, Department of Mathematics and Informatics, University of Novi Sad, Serbia
e-mail: nenad.teofanov@dmi.uns.ac.rs

Joachim Toft
Department of Computer Science, Mathematics and Physics Linnæus University Växjö, Sweden
e-mail: joachim.toft@lnu.se

W. Gautschi et al. (eds.), *Approximation and Computation: In Honor of Gradimir V. Milovanović*, Springer Optimization and Its Applications 42,
DOI 10.1007/978-1-4419-6594-3_23,

contexts, increased the interest in such spaces, cf. [2, 5, 8, 13, 19–24]. In particular, we refer to [15, 16] for the micro-local analysis of Fourier Lebesgue spaces.

The paper is organized as follows.

In Sect. 2 we collect basic notions and notation and define (weighted) Fourier-Lebesgue spaces and their localized versions. In Sect. 3 we introduce wave-front sets with respect to Fourier Lebesgue spaces and recall some basic facts from [15] and [17]. Section 4 contains new results on the propagation of singularities with respect to the convolution operators in Fourier Lebesgue spaces. We proceed with the study of local and micro-local properties of multiplication in Sect. 5. As an application we discuss semilinear equations in the same context. The results of Sect. 5 can be found in [16]. Sections 6 and 7 are devoted to continuity properties of certain pseudodifferential operators. In Sect. 6 we prove the continuity of operators in the Fourier-Lebesgue spaces by observing the localized version of a slightly more general class $S^0_{0,0}$, Theorem 8. The same statement can be found in [17], with a different proof. In the same section we show a continuity result for a class of elliptic operators. In Sect. 7 we present some results from [15] related to the propagation of singularities for more general classes of pseudodifferential operators. To that end, we define the characteristic set of an operator, which is, in the case of operators with polyhomogeneous symbols, smaller than the usual one [11]. Finally, Theorem 11 shows that hypoelliptic operators preserve the wave-front sets, as they should.

2 Notions and Notation

In this section we collect basic notions and notation which will be used further on.

When $x,\xi \in \mathbb{R}^d$, their scalar product is denoted by $\langle x,\xi\rangle$. As usual, $\langle\cdot\rangle = (1+|\cdot|^2)^{1/2}$. The open ball centered at $x_0 \in \mathbb{R}^d$ with radius $r>0$ is denoted by $L(x_0,r)$. We denote by Γ an open cone in $\mathbb{R}^d\setminus 0$ and by X we always denote an open set in $\mathbb{R}^d$. A conic neighborhood of a point $(x_0,\xi_0)\in\mathbb{R}^d\times(\mathbb{R}^d\setminus 0)$ is the product $X\times\Gamma$, where X is a neighborhood of x_0 in $\mathbb{R}^d$ and Γ an open cone in $\mathbb{R}^d$ which contains ξ_0. Sometimes this cone is denoted by Γ_{ξ_0}. For $q\in[1,\infty]$ we let $q'\in[1,\infty]$ denote the conjugate exponent, i.e., $1/q+1/q'=1$.

The Fourier transform $\mathscr{F}$ is the linear and continuous mapping on $\mathscr{S}'(\mathbb{R}^d)$ which takes the form:

$$(\mathscr{F}f)(\xi) = \widehat{f}(\xi) \equiv (2\pi)^{-d/2}\int_{\mathbb{R}^d} f(x)e^{-i\langle x,\xi\rangle}\,dx, \quad \xi\in\mathbb{R}^d,$$

when $f\in L^1(\mathbb{R}^d)$. We recall that $\mathscr{F}$ is a homeomorphism on $\mathscr{S}'(\mathbb{R}^d)$ which restricts to a homeomorphism on $\mathscr{S}(\mathbb{R}^d)$ and to a unitary operator on $L^2(\mathbb{R}^d)$.

Assume that $a\in\mathscr{S}(\mathbb{R}^{2d})$. Then the pseudodifferential operator $a(x,D)$, defined by the Kohn-Nirenberg representation

$$(a(x,D)f)(x) = (2\pi)^{-d}\int_{\mathbb{R}^d}\int_{\mathbb{R}^d} a(x,\xi)f(y)e^{i\langle x-y,\xi\rangle}\,dy\,d\xi, \tag{1}$$

is a linear and continuous operator on $\mathscr{S}(\mathbb{R}^d)$. We say that a is the symbol of the operator $a(x,D)$. For general $a \in \mathscr{S}'(\mathbb{R}^{2d})$, the pseudodifferential operator $a(x,D)$ is defined as the continuous operator from $\mathscr{S}(\mathbb{R}^d)$ to $\mathscr{S}'(\mathbb{R}^d)$ with the distribution kernel

$$K_a(x,y) = (2\pi)^{-d/2}(\mathscr{F}_2^{-1}a)(x,y-x).$$

Here $\mathscr{F}_2F$ is the partial Fourier transform of $F(x,y) \in \mathscr{S}'(\mathbb{R}^{2d})$ with respect to the y-variable. This definition makes sense, since the mappings $\mathscr{F}_2$ and

$$F(x,y) \mapsto F(x,y-x)$$

are homeomorphisms on $\mathscr{S}'(\mathbb{R}^{2d})$. We also note that this definition of $a(x,D)$ agrees with the operator in (1) when $a \in \mathscr{S}(\mathbb{R}^{2d})$.

Assume that $m \in \mathbb{R}$. Then we recall that the Hörmander symbol class

$$S^m_{1,0} = S^m_{1,0}(\mathbb{R}^d \times \mathbb{R}^d) = S^m(\mathbb{R}^{2d})$$

consists of all smooth functions a such that for each pair of multi-indices α, β there are constants $C_{\alpha,\beta}$ such that

$$|\partial_\xi^\alpha \partial_x^\beta a(x,\xi)| \le C_{\alpha,\beta}\langle\xi\rangle^{m-|\alpha|}, \quad x,\xi \in \mathbb{R}^d.$$

We also set $S^{-\infty}_{1,0} = \cap_{m\in\mathbb{R}} S^m_{1,0}$, and

$$\operatorname{Op}(S^m_{1,0}) = \{\, a(x,D) \,:\, a \in S^m_{1,0}(\mathbb{R}^d \times \mathbb{R}^d)\,\}.$$

The symbol $a \in S^m_{1,0}(\mathbb{R}^{2d})$ is called *noncharacteristic* at $(x_0,\xi_0) \in \mathbb{R}^d \times (\mathbb{R}^d \setminus 0)$ if there is a neighborhood X of x_0, a conical neighborhood Γ of ξ_0, and constants c and R such that

$$|a(x,\xi)| > c|\xi|^m, \quad \text{if } |\xi| > R, \tag{2}$$

and $\xi \in \Gamma$. Then one can find $b \in S^{-m}_{1,0}(\mathbb{R}^{2d})$ such that

$$a(x,D)b(x,D) - Id \in \operatorname{Op}(S^{-\infty}_{1,0}) \quad \text{and} \quad b(x,D)a(x,D) - Id \in \operatorname{Op}(S^{-\infty}_{1,0})$$

in a conical neighborhood of (x_0,ξ_0) (cf. [11,15]). The point $(x_0,\xi_0) \in \mathbb{R}^d \times (\mathbb{R}^d \setminus 0)$ is called characteristic for a if it is not a noncharacteristic point of $a(x,D)$. The set of characteristic points (the characteristic set) of $a(x,D)$ is denoted by $\operatorname{Char}(a(x,D))$. We shall identify operators with their symbols when discussing characteristic sets.

The operator $a(x,D) \in \operatorname{Op}(S^m_{1,0})$ is called *elliptic* if the set of characteristic points is empty. This means that for each bounded neighborhood X of x_0, there are constants $c,R > 0$ such that (2) holds when $x \in X$.

Assume that ω and v are positive and measurable functions on $\mathbb{R}^d$. Recall that ω is called a v-moderate weight if

$$\omega(x+y) \le C\omega(x)v(y) \tag{3}$$

for some constant C which is independent of $x,y \in \mathbb{R}^d$. If v in (3) is a polynomial, then ω is called polynomially moderated. We let $\mathscr{P}(\mathbb{R}^d)$ be the set of all polynomially moderated functions on $\mathbb{R}^d$.

Let there be given $\omega \in \mathscr{P}(\mathbb{R}^{2d})$ and let $q \in [1,\infty]$. The (weighted) Fourier-Lebesgue space $\mathscr{F}L^q_{(\omega)}(\mathbb{R}^d)$ is the Banach space which consists of all $f \in \mathscr{S}'(\mathbb{R}^d)$ such that

$$\|f\|_{\mathscr{F}L^q_{(\omega)}} = \|f\|_{\mathscr{F}L^q_{(\omega),x}} \equiv \|\widehat{f} \cdot \omega(x,\cdot)\|_{L^q} < \infty. \tag{4}$$

We say that $f \in \mathscr{D}'(X)$ is locally in $\mathscr{F}L^q_{(\omega)}(\mathbb{R}^d)$, if $\chi f \in \mathscr{F}L^q_{(\omega)}(\mathbb{R}^d)$ for every $\chi \in C_0^\infty(X)$ and in that case we use the notation $f \in \mathscr{F}L^q_{(\omega),\mathrm{loc}}(X)$. We say that $f \in \mathscr{F}L^q_{(\omega),\mathrm{loc}}(X)$ at $x_0 \in X$ if there exists a function $\chi \in C_0^\infty(X)$, $\chi(x_0) \neq 0$, such that $\chi f \in \mathscr{F}L^q_{(\omega)}(\mathbb{R}^d)$. If, however, $f \notin \mathscr{F}L^q_{(\omega),\mathrm{loc}}(X)$ at $x_0 \in X$, then we say that x_0 belongs to the singular support of f, $x_0 \in \mathrm{singsupp}_{\mathscr{F}L^q_{(\omega)}} f$.

The weights $\omega(x,\xi)$ in (4) depend on both x and ξ, although $\widehat{f}(\xi)$ only depends on ξ. However, since ω is v-moderate for some $v \in \mathscr{P}(\mathbb{R}^{2d})$, different choices of x give rise to equivalent norms. Therefore, the condition $\|f\|_{\mathscr{F}L^q_{(\omega),x}} < \infty$ is independent of x in the sense that for $x_1,x_2 \in \mathbb{R}^d$ there exists a constant $C_{x_1,x_2} > 0$ such that

$$C^{-1}_{x_1,x_2}\|f\|_{\mathscr{F}L^q_{(\omega),x_2}} \leq \|f\|_{\mathscr{F}L^q_{(\omega),x_1}} \leq C_{x_1,x_2}\|f\|_{\mathscr{F}L^q_{(\omega),x_2}}.$$

In the remaining part of the paper we study weighted Fourier-Lebesgue spaces with weights which depend on ξ. Thus, with $\omega_0(\xi) = \omega(0,\xi) \in \mathscr{P}(\mathbb{R}^d)$,

$$f \in \mathscr{F}L^q_{(\omega)}(\mathbb{R}^d) = \mathscr{F}L^q_{(\omega_0)}(\mathbb{R}^d) \quad \Longleftrightarrow \quad \|f\|_{\mathscr{F}L^q_{(\omega_0)}} \equiv \|\widehat{f}\omega_0\|_{L^q} < \infty.$$

We usually assume that the underlying weight function $\omega_0(\xi)$ is given by $\omega_0(\xi) = \omega(x_0,\xi) = \langle\xi\rangle^s$, for some $x_0 \in \mathbb{R}^d$ and $s \in \mathbb{R}$. In this case we use the notation $\mathscr{F}L^q_s(\mathbb{R}^d)$ instead of $\mathscr{F}L^q_{(\omega_0)}(\mathbb{R}^d)$. If $\omega = 1$ then the notation $\mathscr{F}L^q(\mathbb{R}^d)$ is used instead of $\mathscr{F}L^q_{(\omega)}(\mathbb{R}^d)$.

3 Wave-Front Sets in $\mathscr{F}L^q$

In this section we define wave-front sets with respect to Fourier Lebesgue spaces and recall some facts from [15] and [17].

We say that a distribution $f \in \mathscr{D}'(X)$ is *micro-locally smooth* at $(x_0,\xi_0) \in X \times (\mathbb{R}^d \setminus 0)$ if there exists $\chi \in C_0^\infty(X)$ such that $\chi(x_0) \neq 0$ and an open cone Γ_{ξ_0} such that for every $N \in \mathbb{N}$ there exists $C_N > 0$ such that

$$|\mathscr{F}(\chi f)(\xi)| \leq C_N\langle\xi\rangle^{-N/2}, \quad \xi \in \Gamma_{\xi_0}.$$

The *wave-front set* of f, $\mathrm{WF}(f)$, is the complement of the set of points $(x_0,\xi_0) \in X \times (\mathbb{R}^d \setminus 0)$ where f is micro-locally smooth.

Let $\omega_0 \in \mathscr{P}(\mathbb{R}^d)$, $\Gamma \subseteq \mathbb{R}^d \setminus 0$ be an open cone and $q \in [1,\infty]$ be fixed. For any $f \in \mathscr{S}'(\mathbb{R}^d)$, let

$$|f|_{\mathscr{F}L^{q,\Gamma}_{(\omega_0)}} := \left(\int_\Gamma |\widehat{f}(\xi)\omega_0(\xi)|^q \,\mathrm{d}\xi \right)^{1/q}$$

(with obvious interpretation when $q=\infty$). We note that $|\cdot|_{\mathscr{F}L^{q,\Gamma}_{(\omega_0)}}$ defines a semi-norm on $\mathscr{S}'(\mathbb{R}^d)$ which might attain the value $+\infty$. If $\Gamma = \mathbb{R}^d \setminus 0$, $f \in \mathscr{F}L^q_{(\omega)}(\mathbb{R}^d)$ and $q<\infty$, then $|f|_{\mathscr{F}L^{q,\Gamma}_{(\omega)}}$ agrees with the Fourier Lebesgue norm $\|f\|_{\mathscr{F}L^q_{(\omega)}}$ of f.

We let $\Theta_{\mathscr{F}L^q_{(\omega_0)}}(f)$ be the set of all $\xi \in \mathbb{R}^d \setminus 0$ such that $|f|_{\mathscr{F}L^{q,\Gamma}_{(\omega_0)}} < \infty$ for some $\Gamma = \Gamma_\xi$. Its complement in $\mathbb{R}^d \setminus 0$ is denoted by $\Sigma_{\mathscr{F}L^q_{(\omega_0)}}(f)$.

The proof of the following result can be found in [15].

Proposition 1. *Assume that $q \in [1,\infty]$, $\chi \in \mathscr{S}(\mathbb{R}^d)$, and that $\omega_0 \in \mathscr{P}(\mathbb{R}^d)$. Also assume that $f \in \mathscr{E}'(\mathbb{R}^d)$. Then*

$$\Sigma_{\mathscr{F}L^q_{(\omega_0)}}(\chi f) \subseteq \Sigma_{\mathscr{F}L^q_{(\omega_0)}}(f).$$

Next we define wave-front sets in the framework of Fourier Lebesgue spaces.

Definition 1. Assume that $q \in [1,\infty]$, X is an open subset of $\mathbb{R}^d$, $f \in \mathscr{D}'(X)$ and $\omega_0 \in \mathscr{P}(\mathbb{R}^d)$. The wave-front set $\mathrm{WF}_{\mathscr{F}L^q_{(\omega_0)}}(f)$ with respect to $\mathscr{F}L^q_{(\omega_0)}(\mathbb{R}^d)$ consists of all pairs $(x_0,\xi_0) \in \mathbb{R}^d \times (\mathbb{R}^d \setminus 0)$ such that $\xi_0 \in \Sigma_{\mathscr{F}L^q_{(\omega_0)}}(\chi f)$ holds for each $\chi \in C_0^\infty(X)$ with $\chi(x_0) \neq 0$.

From Definition 1 it follows that if $(x_0,\xi_0) \in \mathrm{WF}_{\mathscr{F}L^q_{(\omega_0)}}(f)$ then

$$x_0 \in \operatorname{singsupp}_{\mathscr{F}L^q_{(\omega_0)}} f.$$

The wave-front set $\mathrm{WF}_{\mathscr{F}L^q_{(\omega_0)}}(f)$ decreases with respect to the parameter q and increases with respect to the weight function ω_0, when $f \in \mathscr{D}'(X)$ is fixed. More precisely, the following is true, cf. [15].

Proposition 2. *Assume that X is an open subset of $\mathbb{R}^d$, $f \in \mathscr{D}'(X)$, $q_j \in [1,\infty]$ and $\omega_j \in \mathscr{P}(\mathbb{R}^d)$ for $j=1,2$ satisfy $q_1 \le q_2$ and $\omega_2(\xi) \le C\omega_1(\xi)$ for some constant C which is independent of $\xi \in \mathbb{R}^d$. Then*

$$\mathrm{WF}_{\mathscr{F}L^{q_2}_{(\omega_2)}}(f) \subseteq \mathrm{WF}_{\mathscr{F}L^{q_1}_{(\omega_1)}}(f).$$

The following proposition gives a characterization of $\mathrm{WF}_{\mathscr{F}L^q_{(\omega_0)}}(f)$ via the classical wave-front set, cf. [16, 17].

Proposition 3. *Let $q \in [1,\infty]$, $X \subseteq \mathbb{R}^d$ be open, $f \in \mathscr{D}'(X)$, $\omega_0 \in \mathscr{P}(\mathbb{R}^d)$ and $(x_0,\xi_0) \in \mathbb{R}^d \times (\mathbb{R}^d \setminus 0)$. Then the following conditions are equivalent:*

(1) *there exists a $g \in \mathscr{F}L^q_{(\omega_0),\mathrm{loc}}(X)$ such that $(x_0,\xi_0) \notin \mathrm{WF}(f-g)$;*

(2) $(x_0,\xi_0) \notin \mathrm{WF}_{\mathscr{F}L^q_{(\omega_0)}}(f)$.

4 Convolution in $\mathscr{F}L^q$

In this section we show that the singular support of the convolution $f_1 * f_2$ is contained in a set described by singular supports of f_1 or f_2 when one of them is compactly supported and an additional assumption holds. First we show that if f_1 and f_2 belong to Fourier-Lebesgue spaces, then $f_1 * f_2$ belongs to a space of the same type.

Lemma 1. *Assume that $q,q_1,q_2 \in [1,\infty]$ satisfy*

$$\frac{1}{q_1}+\frac{1}{q_2}=\frac{1}{q}. \tag{5}$$

Then the convolution map $(f_1,f_2) \mapsto f_1 * f_2$ from $\mathscr{S}(\mathbb{R}^d) \times \mathscr{S}(\mathbb{R}^d)$ to $\mathscr{S}(\mathbb{R}^d)$ extends to a continuous mapping from $\mathscr{F}L^{q_1}_{(\omega_1)}(\mathbb{R}^d) \times \mathscr{F}L^{q_2}_{(\omega_2)}(\mathbb{R}^d)$ to $\mathscr{F}L^{q}_{(\omega)}(\mathbb{R}^d)$, where $\omega(\xi) \le C\omega_1(\xi)\omega_2(\xi)$, $\xi \in \mathbb{R}^d$, for some $C>0$, and $\omega,\omega_1,\omega_2 \in \mathscr{P}(\mathbb{R}^d)$. This extension is unique if $q_1<\infty$ or $q_2<\infty$.

Proof. Although the proof of Lemma 1 can be found in [16], we give here a sketch of the proof for the sake of completeness.

First we assume that $q_2<\infty$, and we let $f \in \mathscr{F}L^{q_1}_{(\omega_1)}(\mathbb{R}^d)$ and $f_2 \in \mathscr{S}(\mathbb{R}^d)$. Then $q<\infty$, $f_1 * f_2$ is well-defined as an element in $\mathscr{S}'(\mathbb{R}^d)$, and Hölder's inequality implies

$$\|f_1 * f_2\|_{\mathscr{F}L^q_{(\omega)}} \le (2\pi)^{d/2} C \|f_1\|_{\mathscr{F}L^{q_1}_{(\omega_1)}} \|f_2\|_{\mathscr{F}L^{q_2}_{(\omega_2)}}.$$

The result in this case now follows from the fact that $\mathscr{S}(\mathbb{R}^d)$ is dense in $\mathscr{F}L^{q_2}_{\omega_2}(\mathbb{R}^d)$. By similar arguments, the result follows if we assume that $q_1<\infty$.

It remains to consider the case $q_1=q_2=\infty$. We note that

$$f_1 * f_2 = (2\pi)^{d/2}\mathscr{F}(\widehat{f_1}\cdot\widehat{f_2}) \tag{6}$$

when $f_1,f_2 \in \mathscr{S}(\mathbb{R}^d)$. It remains to show that the right-hand side of (6) exists for $f_j \in \mathscr{F}L^\infty_{(\omega_j)}(\mathbb{R}^d)$, $j=1,2$, and defines an element in $\mathscr{F}L^\infty_{(\omega)}(\mathbb{R}^d)$.

In fact, from the assumptions it follows that $\widehat{f_1}$ and $\widehat{f_2}$ are measurable and essentially bounded by some polynomials. Hence, the product $\widehat{f_1}\widehat{f_2}$ is well-defined as a measurable function and essentially bounded by an appropriate polynomial. Hence, it follows from (6) that $f_1 * f_2$ is well-defined as an element in $\mathscr{S}'(\mathbb{R}^d)$. Furthermore,

$$\|f_1 * f_2\|_{\mathscr{F}L^\infty_{(\omega)}} = (2\pi)^{d/2}\|\widehat{f_1}\,\widehat{f_2}\omega\|_{L^\infty} \le (2\pi)^{d/2}C\|(\widehat{f_1}\omega_1)(\widehat{f_2}\omega_2)\|_{L^\infty}$$

$$\le (2\pi)^{d/2}C\|\widehat{f_1}\omega_1\|_{L^\infty}\|\widehat{f_2}\omega_2\|_{L^\infty} = C\|f_1\|_{\mathscr{F}L^\infty_{(\omega_1)}}\|f_2\|_{\mathscr{F}L^\infty_{(\omega_2)}},$$

and the proof is completed. □

Next we study the singular support of the convolution $f_1 * f_2$.

Theorem 1. *Let $f_1, f_2 \in \mathscr{D}'(\mathbb{R}^d)$ so that one of them is compactly supported. Assume that $q, q_1, q_2 \in [1,\infty]$ satisfy (5) and $\omega(\xi) \le C\omega_1(\xi)\omega_2(\xi)$, $\xi \in \mathbb{R}^d$ for some $C > 0$, and $\omega, \omega_1, \omega_2 \in \mathscr{P}(\mathbb{R}^d)$.*

(1) *If f_2 is compactly supported, we assume that there exists a closed set $Z \in \mathbb{R}^d$ such that for any cutoff function ψ of Z we have $(1-\psi)f_1 \in \mathscr{S}(\mathbb{R}^d)$. Then we have*

$$\mathrm{singsupp}_{\mathscr{F}L^q_{(\omega)}}(f_1 * f_2) \subseteq Z + \left(\mathrm{singsupp}_{\mathscr{F}L^{q_1}_{(\omega_1)}} f_2 \cap \mathrm{singsupp}_{\mathscr{F}L^{q_2}_{(\omega_2)}} f_2\right).$$

(2) *If instead f_1 is compactly supported, we assume that there exists a closed set $Z \in \mathbb{R}^d$ such that for any cutoff function ψ of Z we have $(1-\psi)f_2 \in \mathscr{S}(\mathbb{R}^d)$. Then we have*

$$\mathrm{singsupp}_{\mathscr{F}L^q_{(\omega)}}(f_1 * f_2) \subseteq Z + \left(\mathrm{singsupp}_{\mathscr{F}L^{q_1}_{(\omega_1)}} f_1 \cap \mathrm{singsupp}_{\mathscr{F}L^{q_2}_{(\omega_2)}} f_1\right).$$

Proof. (1) Assume that $f_2 \in \mathscr{E}'(\mathbb{R}^d)$. Let $\psi \in C^\infty(\mathbb{R}^d)$ such that $\psi \equiv 1$ on Z and $\overline{Z_\delta} = \mathrm{supp}\,\psi$ for some $\delta > 0$, where $\overline{Z_\delta}$ is the closure of the set $Z_\delta = \{x+\delta : x \in Z\}$. Put $f_1 = g_1 + h_1$, where $g_1 = \psi f_1$ and $h_1 = (1-\psi)f_1$. Since $\hat{h}_1 \in \mathscr{S}(\mathbb{R}^d)$ it follows that

$$\mathrm{singsupp}_{\mathscr{F}L^q_{(\omega)}}(h_1 * f_2) = \emptyset.$$

So we consider $g_1 * f_2$. Let $x_0 \in \mathbb{R}^d$ and let $r > 0$ be chosen so that

$$A = \{x - t : x \in L(x_0, r),\ t \in \overline{Z_\delta}\} \subseteq \complement\,\mathrm{singsupp}_{\mathscr{F}L^{q_2}_{(\omega_2)}}(f_2).$$

Let $\varphi \in C^\infty(\mathbb{R}^d)$ so that $\varphi \equiv 1$ on A and $\mathrm{supp}\,\varphi \subseteq \complement\,\mathrm{singsupp}_{\mathscr{F}L^{q_2}_{(\omega_2)}}(f_2)$. Then $g_1 * f_2 = g_1 * \varphi f_2$ in $\mathscr{D}'(L(x_0, r))$, so we have

$$x_0 \notin \mathrm{singsupp}_{\mathscr{F}L^q_{(\omega)}}(g_1 * f_2) \Longleftrightarrow x_0 \notin \mathrm{singsupp}_{\mathscr{F}L^q_{(\omega)}}(g_1 * \varphi f_2).$$

By construction,

$$\int_{\mathbb{R}^d} |\widehat{\varphi f_2}(\xi)|^{q_2}\omega_2^{q_2}(\xi)\,\mathrm{d}\xi < \infty.$$

Thus, if $x_0 \notin \operatorname{singsupp}_{\mathscr{F}L^{q_1}_{(\omega_1)}} f_1 + A$, then $x_0 \notin \operatorname{singsupp}_{\mathscr{F}L^{q}_{(\omega)}} (g_1 * \varphi f_2)$ and

$$\operatorname{singsupp}_{\mathscr{F}L^{q}_{(\omega)}} (g_1 * \varphi f_2) \subseteq \operatorname{singsupp}_{\mathscr{F}L^{q_2}_{(\omega_2)}} f_2 + \overline{Z_\delta}.$$

Letting $\delta \to 0$ we obtain

$$\operatorname{singsupp}_{\mathscr{F}L^{q}_{(\omega)}} (f_1 * f_2) \subseteq Z + \operatorname{singsupp}_{\mathscr{F}L^{q_2}_{(\omega_2)}} f_2.$$

Now, if we choose $x_0 \in \mathbb{R}^d$ and $r > 0$ such that

$$A = \{ x - t \, : \, x \in L(x_0, r), \, t \in \overline{Z_\delta} \} \subseteq \complement \operatorname{singsupp}_{\mathscr{F}L^{q_1}_{(\omega_1)}} (f_2),$$

the same arguments as above give

$$\int_{\mathbb{R}^d} |\widehat{\varphi f_2}(\xi)|^{q_1} \omega_1^{q_1}(\xi) \, \mathrm{d}\xi < \infty.$$

Thus, if $x_0 \notin \operatorname{singsupp}_{\mathscr{F}L^{q_2}_{(\omega_2)}} f_1 + A$ then

$$x_0 \notin \operatorname{singsupp}_{\mathscr{F}L^{q}_{(\omega)}} (g_1 * \varphi f_2)$$

which implies that

$$\operatorname{singsupp}_{\mathscr{F}L^{q}_{(\omega)}} (g_1 * \varphi f_2) \subseteq \operatorname{singsupp}_{\mathscr{F}L^{q_1}_{(\omega_1)}} f_2 + \overline{Z_\delta}.$$

Letting $\delta \to 0$ we obtain

$$\operatorname{singsupp}_{\mathscr{F}L^{q}_{(\omega)}} (f_1 * f_2) \subseteq Z + \operatorname{singsupp}_{\mathscr{F}L^{q_1}_{(\omega_1)}} f_2.$$

This proves (1).

(2) The proof is the same as the proof of (1). □

Theorem 2. *Let the assumptions of Theorem 1 (1) hold. Then we have*

$$\begin{aligned} \mathrm{WF}_{\mathscr{F}L^{q}_{(\omega)}} (f_1 * f_2) \subseteq \Big\{ (x-y, \xi) \in \mathbb{R}^d \times \mathbb{R}^d \setminus 0 \mid (x, \xi) \in Z \times \mathbb{R}^d \setminus 0 \\ \textit{and } (y, \xi) \in \mathrm{WF}_{\mathscr{F}L^{q_1}_{(\omega_1)}} (f_2) \cap \mathrm{WF}_{\mathscr{F}L^{q_2}_{(\omega_2)}} (f_2) \Big\}. \end{aligned}$$

If instead the assumptions of Theorem 1 (2) *hold, then we have*

$$\begin{aligned} \mathrm{WF}_{\mathscr{F}L^{q}_{(\omega)}} (f_1 * f_2) \subseteq \Big\{ (x-y, \xi) \in \mathbb{R}^d \times \mathbb{R}^d \setminus 0 \mid (x, \xi) \in Z \times \mathbb{R}^d \setminus 0 \\ \textit{and } (y, \xi) \in \mathrm{WF}_{\mathscr{F}L^{q_1}_{(\omega_1)}} (f_1) \cap \mathrm{WF}_{\mathscr{F}L^{q_2}_{(\omega_2)}} (f_1) \Big\}. \end{aligned}$$

Proof. The proof follows from Theorem 1. □

5 Multiplication in $\mathscr{F}L^q_s(\mathbb{R}^d)$

In this section we discuss the problem of multiplication in $\mathscr{F}L^q_s(\mathbb{R}^d)$ and in $\mathscr{F}L^q_{s,\mathrm{loc}}(X)$ from both the local and micro-local point of view. As an application we discuss semilinear equations in the same context. The statements of this section are proved in [16].

In the next theorem and its corollary we assume that s, s_1, and s_2 satisfy

$$0 \le s_1 + s_2 \quad \text{and} \quad s \le s_1 + s_2 - d/q', \tag{7}$$

Theorem 3. *Assume that $q \in [1,\infty]$, and let $r \ge 0$ be such that $r > d(1-2/q)$ when $q > 2$, and $r = 0$ when $1 \le q \le 2$. Also assume that $s, s_j \in \mathbb{R}$ satisfy $s \le s_j$ for $j = 1,2$ and (7), where the former inequality in (7) is strict when $s = -\min(d/q, d/q')$, and the latter inequality is strict when $s_1 = d/q'$ or $s_2 = d/q'$. Then the map $(f_1, f_2) \mapsto f_1 f_2$ from $\mathscr{S}(\mathbb{R}^d) \times \mathscr{S}(\mathbb{R}^d)$ to $\mathscr{S}(\mathbb{R}^d)$ extends uniquely to a continuous mapping from $\mathscr{F}L^q_{s_1}(\mathbb{R}^d) \times \mathscr{F}L^q_{s_2+r}(\mathbb{R}^d)$ to $\mathscr{F}L^q_s(\mathbb{R}^d)$. Furthermore,*

$$\|f_1 f_2\|_{\mathscr{F}L^q_s} \le C \|f_1\|_{\mathscr{F}L^q_{s_1}} \|f_2\|_{\mathscr{F}L^q_{s_2+r}},$$

for some constant C which is independent of $f_1 \in \mathscr{F}L^q_{s_1}(\mathbb{R}^d)$ and $f_2 \in \mathscr{F}L^q_{s_2+r}(\mathbb{R}^d)$.

Since products are defined locally, the following result is an immediate consequence of Theorem 3.

Corollary 1. *Let the assumptions of Theorem 3 hold and assume that $X \subseteq \mathbb{R}^d$ is an open set.*

Then the map $(f_1, f_2) \mapsto f_1 \cdot f_2$ from $C_0^\infty(X) \times C_0^\infty(X)$ to $C_0^\infty(X)$ extends uniquely to a continuous mapping from $\mathscr{F}L^q_{s_1,\mathrm{loc}}(X) \times \mathscr{F}L^q_{s_2+r,\mathrm{loc}}(X)$ to $\mathscr{F}L^q_{s,\mathrm{loc}}(X)$.

Note that Theorem 3 agrees with [12, Theorem 8.3.1] when $q = 2$.

We proceed with the micro-local characterization of products $f_1 \cdot f_2$ when $f_j \in \mathscr{F}L^q_{s_j}(\mathbb{R}^d)$, $j = 1,2$.

Theorem 4. *Assume that $q \in [1,\infty]$ and let $f_j \in \mathscr{F}L^q_{s_j,\mathrm{loc}}(X)$, $j = 1,2$. Then the following is true:*

(1) *if $s_1 - |s_2| \ge 0$ when $q = 1$, and $s_1 - |s_2| > d/q'$ otherwise, then*

$$\mathrm{WF}_{\mathscr{F}L^q_{s_2}}(f_1 f_2) \subseteq \mathrm{WF}_{\mathscr{F}L^q_{|s_2|}}(f_1);$$

(2) *if instead $s_1 + s_2 \ge s \ge 0$ when $q = 1$, and $s_1 + s_2 - d/q' > s \ge 0$ otherwise, and $s_2 - s \ge d/q'$, then*

$$\mathrm{WF}_{\mathscr{F}L^q_s}(f_1 f_2) \subseteq \mathrm{WF}_{\mathscr{F}L^q_{|s_2|}}(f_1).$$

We note that $f_1 f_2$ in Theorem 4 makes sense as an element in $\mathscr{F}L^q_{s,\mathrm{loc}}(X)$, for some $s \in \mathbb{R}$.

In the next theorem we consider the "critical case" $s = s_1 + s_2 - \min(d/q, d/q')$ compared to Theorem 4 (2).

Theorem 5. *Assume that $q \in [1,\infty]$, $r = 0$ when $1 \le q \le 2$, $r > d(1-2/q)$ when $q > 2$, and that $s, s_j, N_j \in \mathbb{R}$, $j = 1,2$, satisfy*

$$s_1 + s_2 > 0, \quad s = s_1 + s_2 - \min(d/q, d/q'),$$

$$N_1 \ge s_1 + |s_2| + \max(0, d(1-2/q)), \ N_2 \ge s_2 + |s_1| + \max(0, d(1-2/q)), \tag{8}$$

with strict inequalities in (8) *when $q < \infty$. If $f_1 \in \mathscr{F}L^q_{s_1,\mathrm{loc}}(X)$ and $f_2 \in \mathscr{F}L^q_{s_2+r,\mathrm{loc}}(X)$, then*

$$\mathrm{WF}_{\mathscr{F}L^q_s}(f_1 f_2) \subseteq \mathrm{WF}_{\mathscr{F}L^q_{N_1}}(f_1) \cup \mathrm{WF}_{\mathscr{F}L^q_{N_2}}(f_2).$$

Remark 1. By reasons of symmetry, it follows that Theorem 5 holds if the assumptions on f_1 and f_2 are replaced by $f_2 \in \mathscr{F}L^q_{s_2}(\mathbb{R}^d)$ and $f_1 \in \mathscr{F}L^q_{s_1+r}(\mathbb{R}^d)$.

The next result is needed later on when studying semilinear equations. Here we consider distributions of the form:

$$\mathcal{G}(x) \equiv G(x, f_1(x), \dots, f_N(x)) \tag{9}$$

when $f_1, \dots, f_N$ are appropriate distributions and $G(x,y)$ is an appropriate polynomial in the y-variable, i.e., G is of the form

$$G(x,y) = \sum_{0<|\alpha|\le m} a_\alpha(x) y^\alpha, \tag{10}$$

where a_α are appropriate distributions and $y = (y_1, \dots, y_N)$.

In the following result we assume that the a_α, $0 < |\alpha| \le m$, locally belong to appropriate classes of Fourier Lebesgue spaces.

Theorem 6. *Assume that $X \subseteq \mathbb{R}^d$ is open, $q \in [1,\infty]$ and $r \ge d/q'$ with strict inequality when $q = \infty$. Let*

$$s \ge d/q', \quad s \le \sigma \le 2s - d/q',$$

and let G and $\mathcal{G}$ be given by (9) *and* (10) *for some integer $m > 0$, where $a_\alpha \in \mathscr{F}L^1_{\sigma,\mathrm{loc}}(X)$, $0 < |\alpha| \le m$, and $f_j \in \mathscr{F}L^q_{s+(m-1)r,\mathrm{loc}}(X)$, $j = 1, \dots, N$. Then the following is true:*

(1) $\mathcal{G}$ *in* (9) *makes sense as an element in $\mathscr{F}L^q_{s,\mathrm{loc}}(X)$;*

(2) $\mathrm{WF}_{\mathscr{F}L^q_\sigma}(\mathcal{G}) \subset \cup_{j=1}^N \mathrm{WF}_{\mathscr{F}L^q_{\sigma+(m-1)r}}(f_j)$.

Let $J_k f$ denote the array of all derivatives of order α, $|\alpha| \leq k$, of f (the so called $k-jet$ of f):

$$J_k f = \{(\partial^\alpha f)\}_{|\alpha| \leq k}.$$

We denote the elements of $J_k f$ with $f_1, \ldots, f_N$, where N is the number of elements in $J_k f$. We also consider $G(x, J_k f)$, where G is the same as in Theorem 6. We also let $P(x,D)$ be the partial differential operator whose symbol $P(x,\xi)$ is of the form

$$P(x,\xi) = \sum_{|\alpha| \leq n} b_\alpha(x) \xi^\alpha, \quad \text{where } b_\alpha \in C^\infty(X), |\alpha| \leq n, \tag{11}$$

where $X \subseteq \mathbb{R}^d$ is open.

Theorem 7. *Let $X \subseteq \mathbb{R}^d$ be open, $q \in [1,\infty]$, $r, s \geq d/q'$, and consider the semilinear differential equation*

$$P(x,D)f = G(x, J_k f), \tag{12}$$

where G is the same as in Theorem 6, $a_\alpha \in \mathscr{F}L^1_{2s-d/q',\mathrm{loc}}(X)$, and P is given by (11).

Assume that $(x_0,\xi_0) \notin \mathrm{Char}(P)$, $f \in \mathscr{D}'(X)$ is a solution of (12), *and that one of the following conditions holds:*

(1) $f \in \mathscr{F}L^q_{s+k+(m-1)r,\mathrm{loc}}(X)$, *and* $s+n \geq d/q' + k + (m-1)r$;
(2) $f \in \mathscr{F}L^1_{s+k,\mathrm{loc}}(X)$.

Then $(x_0,\xi_0) \notin \mathrm{WF}_{\mathscr{F}L^1_{2s+n-d/q'}}(f)$.

Note that Theorem 7 (2) gives a hypoellipticity result, which does not depend on the order of G. If $q = 1$, and if we assume that G is real analytic, then Theorem 7 (2) can be improved as follows.

Proposition 4. *Let X be an open set in $\mathbb{R}^d$, $s \geq 0$, $G(y_1,\ldots,y_N)$ be a real analytic function, $f_j \in \mathscr{F}L^1_{s,\mathrm{loc}}(X)$, $j = 1,\ldots,N$, and let $\mathcal{G}$ be the same as in* (9). *Then the following is true:*

(1) $\mathcal{G} \in \mathscr{F}L^1_{s,\mathrm{loc}}(X)$;
(2) *If* $(x_0,\xi_0) \notin \mathrm{WF}_{\mathscr{F}L^1_\sigma}(f_j)$, $j = 1,\ldots,N$, $s \leq \sigma \leq 2s$, *then* $(x_0,\xi_0) \notin \mathrm{WF}_{\mathscr{F}L^1_\sigma}(\mathcal{G})$;
(3) *Let k and n be the same as in Theorem 7, $(x_0,\xi_0) \notin \mathrm{WF}_{\mathscr{F}L^1_{s+k}}(f)$, where f is a solution of* (12). *If P is noncharacteristic at (x_0,ξ_0), then it follows that* $(x_0,\xi_0) \notin \mathrm{WF}_{\mathscr{F}L^1_{2s+n}}(f)$.

6 Continuity of Pseudodifferential Operators on $\mathscr{F}L^q$

It is well known that an operator $a(x,D)$ whose symbol lies in the Hörmander class $S^0_{1,0}$ is continuous from L^q to L^q, $1 < q < \infty$, [25]. In this section we study the continuity of localized version of pseudodifferential operators of the Hörmander type, Theorem 8, and Proposition 5.

Note that in order to prove the continuity on $\mathscr{F}L^q(\mathbb{R}^d)$, it is not sufficient to assume that the symbol a of an operator is bounded together with its derivatives with respect to the x variable, because for such a symbol and $f \in \mathscr{F}L^q(\mathbb{R}^d)$ the convolution $\widehat{a} * \widehat{f}$ does not belong to L^q, in general. However, it is possible to show the continuity if the symbol enjoys a certain decay with respect to the x variable.

Assume that $m \in \mathbb{R}$. The class of Hörmander symbols $S^m_{\rho,\delta} = S^m_{\rho,\delta}(\mathbb{R}^d \times \mathbb{R}^d)$, $0 \le \delta \le \rho \le 1$ consists of smooth functions a such that, for all multi-indices α, β we have

$$|\partial_\xi^\alpha \partial_x^\beta a(x,\xi)| \le C_{\alpha,\beta} \langle \xi \rangle^{m-\rho|\alpha|-\delta|\beta|}, \quad x,\xi \in \mathbb{R}^d.$$

We also set

$$\operatorname{Op} S^m_{\rho,\delta} = \left\{ a(x,D) \,:\, a \in S^m_{\rho,\delta}(\mathbb{R}^d \times \mathbb{R}^d) \right\},$$

that is $a \in S^m_{\rho,\delta}$ is the symbol of the operator $a(x,D)$ given by (1).

We use the notation $E_s(D) = (1-\triangle)^{s/2}$, for the operator with the symbol $E_s(\xi) = \langle \xi \rangle^s$, $s \in \mathbb{R}$.

Theorem 8. *Assume that $q \in [1,\infty]$ and $\chi \in C_0^\infty(\mathbb{R}^d)$. Then the following is true:*

(1) *If $a \in S^0_{0,0}$, then the mapping $\chi(x)a(x,D) : \mathscr{F}L^q(\mathbb{R}^d) \to \mathscr{F}L^q(\mathbb{R}^d)$ is continuous. In particular, if $a(x,\xi) = a(\xi) \in S^0_{0,0}$, then the mapping $a(D) : \mathscr{F}L^q(\mathbb{R}^d) \to \mathscr{F}L^q(\mathbb{R}^d)$ is continuous.*
(2) *If $a \in S^m_{0,0}$ and $s \in \mathbb{R}$, then the mapping $\chi(x)a(x,D) : \mathscr{F}L^q_s(\mathbb{R}^d) \to \mathscr{F}L^q_{s-m}(\mathbb{R}^d)$ is continuous. In particular, if $a(x,\xi) = a(\xi) \in S^m_{0,0}$, then the mapping $a(D) : \mathscr{F}L^q_s(\mathbb{R}^d) \to \mathscr{F}L^q_{s-m}(\mathbb{R}^d)$ is continuous.*

To prove Theorem 8, we need some preliminaries.

First note that the compactly supported functions are dense in $\mathscr{F}L^q_s(\mathbb{R}^d)$, $1 \le q < \infty$. Thus, we look at the cases $1 \le q < \infty$ and $q = \infty$ separately.

For a given $f \in \mathscr{F}L^q_s(\mathbb{R}^d)$, $1 \le q < \infty$ we consider a sequence $\{f_n\}_{n\in\mathbb{N}}$, $f_n \in \mathscr{F}L^q_s(\mathbb{R}^d)$, $n \in \mathbb{N}$, given by $\hat{f}_n(\xi) = \chi_n(\xi)\hat{f}(\xi)$, where χ_n is a smooth function such that

$$\chi_n(\xi) = \begin{cases} 1, & |\xi| \le n - \varepsilon, \\ 0, & |\xi| \ge n + \varepsilon, \end{cases}$$

for some fixed $\varepsilon \in (0,1)$. Then

$$\lim_{n\to\infty} \|f_n - f\|_{\mathscr{F}L^q_s} = 0.$$

If, instead, $f \in \mathscr{F}L^\infty_s(\mathbb{R}^d)$, then $\hat{f}(\xi)\langle\xi\rangle^{-1}$ tends to zero at infinity, and the sequence

$$\hat{f}_n(\xi)\langle\xi\rangle^{-1} = \chi_n(\xi)\hat{f}(\xi)\langle\xi\rangle^{-1}$$

converges in the norm of $L^\infty_s(\mathbb{R}^d)$.

Let $a \in S^0_{0,0}$ and $\chi \in C_0^\infty(\mathbb{R}^d)$. We denote the support of χ by K. The oscillatory integral (1) is well defined for the symbol χa and $f_n \in \mathscr{F}L^q(\mathbb{R}^d)$, $1 \le q < \infty$, $n \in \mathbb{N}$.

Namely, applying the pseudodifferential operator $E_p(D_x)$, $p \geq 0$, we obtain

$$\chi(x)a(x,D)f(x) = \lim_{n\to\infty}(2\pi)^{-d}\int \mathrm{e}^{\mathrm{i}\langle x,\theta\rangle}E_p(D_x)\chi(x)a(x,\theta)\langle\theta\rangle^{-p}\widehat{f_n}(\theta)\,\mathrm{d}\theta,$$

which, by the Hölder inequality, gives

$$|\chi(x)a(x,D)f(x)| \leq \lim_{n\to\infty}(2\pi)^{-d}\|E_p(D_x)\chi(x)a(x,\theta)\langle\theta\rangle^{-p}\|_{L^{q'}}\|\widehat{f_n}(\theta)\|_{L^q} \leq C\|\widehat{f}\|_{L^q}$$

for $p > d/q'$. If $q = \infty$, instead, then we have

$$\begin{aligned}|\chi(x)a(x,D)f(x)| &\leq \lim_{n\to\infty}(2\pi)^{-d}\|E_{2p}(D_x)\chi(x)a(x,\theta)\langle\theta\rangle^{-p+1}\|_{L^1}\|\widehat{f_n}(\theta)\langle\theta\rangle^{-1}\|_{L^\infty}\\ &\leq C\|\widehat{f}(\theta)\langle\theta\rangle^{-1}\|_{L^\infty}\end{aligned}$$

for $p > d/q' + 1$.

Denote the Fourier transform of $a_\chi(x,\theta) - \chi(x)a(x,\theta)$ with respect to the first variable by $\mathscr{F}_1 a_\chi(\eta,\theta)$. Then, for arbitrary $N \in \mathbb{N}$, there exists $C_N > 0$ such that

$$|\mathscr{F}_1 a_\chi(\eta,\theta)| \leq C_N\langle\eta\rangle^{-N}, \quad \theta \in \mathbb{R}^d. \tag{13}$$

Now we are ready to prove Theorem 8.

Proof of Theorem 8.

(1) We have

$$\begin{aligned}|\mathscr{F}(\chi(x)a(x,D)f(x))(\xi)| &= \lim_{n\to\infty}(2\pi)^{-d}\left|\int_K \mathrm{e}^{-\mathrm{i}\langle x,\xi\rangle}\left(\int_{\mathbb{R}^d}\mathrm{e}^{\mathrm{i}\langle x,\theta\rangle}a_\chi(x,\theta)\widehat{f_n}(\theta)\,\mathrm{d}\theta\right)\mathrm{d}x\right|\\ &= \lim_{n\to\infty}(2\pi)^{-d}\left|\int_{\mathbb{R}^d}\widehat{f_n}(\theta)\left(\int_K \mathrm{e}^{-\mathrm{i}\langle x,\xi-\theta\rangle}a_\chi(x,\theta)\,\mathrm{d}x\right)\mathrm{d}\theta\right|.\end{aligned}$$

By the change of variables $\tilde{\theta} = \xi - \theta$ we obtain

$$\begin{aligned}&\|\mathscr{F}(\chi(x)a(x,D)f(x))(\xi)\|_{L^q}\\ &\quad= \lim_{n\to\infty}(2\pi)^{-d}\left(\int_{\mathbb{R}^d}\left|\int_{\mathbb{R}^d}\widehat{f_n}(\xi-\tilde{\theta})\mathscr{F}_1 a_\chi(\tilde{\theta},\xi-\tilde{\theta})\,\mathrm{d}\tilde{\theta}\right|^q\mathrm{d}\xi\right)^{1/q}.\end{aligned}$$

Now, (13) and the Minkowski inequality imply that

$$\begin{aligned}&\left(\int_{\mathbb{R}^d}\left|\int_{\mathbb{R}^d}\widehat{f_n}(\xi-\tilde{\theta})\mathscr{F}_1 a_\chi(\tilde{\theta},\xi-\tilde{\theta})\,\mathrm{d}\tilde{\theta}\right|^q\mathrm{d}\xi\right)^{1/q}\\ &\qquad\leq \int_{\mathbb{R}^d}\left(\int_{\mathbb{R}^d}\left|\widehat{f_n}(\xi-\tilde{\theta})\mathscr{F}_1 a_\chi(\tilde{\theta},\xi-\tilde{\theta})\right|^q\mathrm{d}\xi\right)^{1/q}\mathrm{d}\tilde{\theta}\\ &\qquad\leq C\int_{\mathbb{R}^d}\langle\tilde{\theta}\rangle^{-d-1}\left(\int_{\mathbb{R}^d}|\widehat{f_n}(\xi-\tilde{\theta})|^q\,\mathrm{d}\xi\right)^{1/q}\mathrm{d}\tilde{\theta}\\ &\qquad\leq C\|\widehat{f_n}\|_{L^q}.\end{aligned}$$

Therefore, we conclude that

$$\|\mathscr{F}(\chi(x)a(x,D)f(x))(\cdot)\|_{L^q} \le C\|\widehat{f}\|_{L^q},$$

which gives the proof for $1 \le q < \infty$.

If $q = \infty$, we modify the proof to obtain

$$\|\mathscr{F}(\chi(x)a(x,D)f_n(x))(\cdot)\|_{L^\infty} \le C\int_{\mathbb{R}^d} \langle\tilde{\theta}\rangle^{-d-1}\left(\sup_{y\in\mathbb{R}^d} \widehat{f_n}(y)\langle y\rangle^{-1}\,\mathrm{d}y\right)\mathrm{d}\tilde{\theta},$$

which implies

$$\|\mathscr{F}(\chi(x)a(x,D)f(x))(\xi)\|_{L^\infty} \le C\|\widehat{f}(\xi)\langle\xi\rangle^{-1}\|_{L^\infty} \le C\|\widehat{f}(\xi)\|_{L^\infty}.$$

If, instead, we assume that the symbol is a Fourier multiplier $a = a(\xi) \in S^0_{0,0}$, then it is obvious that

$$\|\mathscr{F}(a(D)f)\|_{L^q} \le C\|\mathscr{F}f\|_{L^q}.$$

(2) In the proof we use the Peetre inequality:

$$\langle\xi\rangle^t \le \langle\xi-\theta\rangle^t\langle\theta\rangle^{|t|}, \quad \xi,\theta\in\mathbb{R}^d,\ t\in\mathbb{R}.$$

For every $p \in \mathbb{N}$ we have

$$\begin{aligned}
&|\mathscr{F}(\chi(x)a(x,D)f(x))(\xi)| \\
&\quad = (2\pi)^{-d}\left|\int_{\mathbb{R}^d} \mathrm{e}^{-\mathrm{i}\langle x,\xi\rangle}\left(\int_{\mathbb{R}^d} \mathrm{e}^{\mathrm{i}\langle x,\theta\rangle} E_p(D_x)a_\chi(x,\theta)\langle\theta\rangle^{-p}\widehat{f}(\theta)\,\mathrm{d}\theta\right)\mathrm{d}x\right| \\
&\quad = (2\pi)^{-d}\left|\int_{\mathbb{R}^d} \widehat{f}(\theta)\langle\theta\rangle^{-p}\left(\int_{\mathbb{R}^d} \mathrm{e}^{-\mathrm{i}\langle x,\xi-\theta\rangle} E_p(D_x)a_\chi(x,\theta)\,\mathrm{d}x\right)\mathrm{d}\theta\right|.
\end{aligned}$$

Let $p > |s-m|+d$. The change of variables $\tilde{\theta} = \xi - \theta$, the Minkowski inequality, and the assumption $a \in S^m_{0,0}$ give

$$\begin{aligned}
&\|\chi(x)a(x,D)f(x)\|_{\mathscr{F}L^q_{s-m}} \\
&\quad = (2\pi)^{-d}\left(\int_{\mathbb{R}^d} \langle\xi\rangle^{s-m}\left|\int_{\mathbb{R}^d} I(\tilde{\theta},\xi)\widehat{f}(\xi-\tilde{\theta})\langle\tilde{\theta}\rangle^{-p}\,\mathrm{d}\tilde{\theta}\right|^q \mathrm{d}\xi\right)^{1/q} \\
&\quad \le (2\pi)^{-d}\left(\int_{\mathbb{R}^d} \left|\langle\xi\rangle^{s-m}\langle\xi-\tilde{\theta}\rangle^m\widehat{f}(\xi-\tilde{\theta})\right|^q \mathrm{d}\xi\right)^{1/q}\langle\tilde{\theta}\rangle^{-p}\,\mathrm{d}\tilde{\theta} \\
&\quad \le (2\pi)^{-d}\left(\int_{\mathbb{R}^d} \left|\langle\xi-\tilde{\theta}\rangle^s\widehat{f}(\xi-\tilde{\theta})\right|^q \mathrm{d}\xi\right)^{1/q}\langle\tilde{\theta}\rangle^{|s-m|-p}\,\mathrm{d}\tilde{\theta} \le C\|f\|_{\mathscr{F}L^q_s},
\end{aligned}$$

where

$$I(\tilde{\theta},\xi) = \int_{\mathbb{R}^d} \mathrm{e}^{-\mathrm{i}\langle x,\tilde{\theta}\rangle} E_p(D_x) a_\chi(x,\xi-\tilde{\theta})\,\mathrm{d}x.$$

If, instead, we have that $a(x,\xi) = a(\xi) \in S^m_{0,0}$, then it is obvious that

$$a(D) : \mathscr{F}L^q_s(\mathbb{R}^d) \to \mathscr{F}L^q_{s-m}(\mathbb{R}^d).$$

The proof is complete. □

As an application of Theorem 8 we give a result which concerns elliptic operators.

Proposition 5. *Assume that* $q \in [1,\infty]$, $a(x,D) \in \mathrm{Op}\, S^m_{0,0}$ *is elliptic and that* $f \in \mathscr{F}L^q_{t,\mathrm{loc}}(X)$ *for some* $t \in \mathbb{R}$. *If* $a(x,D)f \in \mathscr{F}L^q_{s,\mathrm{loc}}(X)$, *then* $f \in \mathscr{F}L^q_{s+m,\mathrm{loc}}(X)$ *and for every* $\chi \in C_0^\infty(X)$ *we have*

$$\|\chi f\|_{\mathscr{F}L^q_{s+m}} \leq C_{s,t}(\|\chi a(x,D)f\|_{\mathscr{F}L^q_s} + \|\chi f\|_{\mathscr{F}L^q_t}). \tag{14}$$

In particular, if $a(x,D) = a(D) \in S^m_{0,0}$ *is elliptic, then* (14) *holds without* χ.

Proof. By ellipticity, there exists an operator $b(x,D) \in \mathrm{Op}\ S^{-m}_{0,0}$ such that

$$\chi f = b(x,D)a(x,D)\chi f + c(x,D)\chi f,$$

for some $c \in S^{-\infty}$. Theorem 8 implies that $b(x,D)$ is continuous from $\mathscr{F}L^q_s(\mathbb{R}^d)$ to $\mathscr{F}L^q_{s+m}(\mathbb{R}^d)$ and that $c(x,D)$ is continuous from $\mathscr{F}L^q_t(\mathbb{R}^d)$ to $\mathscr{F}L^q_{s+m}(\mathbb{R}^d)$. Therefore, we have

$$\begin{aligned}\|\chi f\|_{\mathscr{F}L^q_{s+m}} &\leq \|b(x,D)a(x,D)\chi f\|_{\mathscr{F}L^q_{s+m}} + \|r(x,D)\chi f\|_{\mathscr{F}L^q_{s+m}}\\ &\leq C_{s,t}(\|\chi a(x,D)f\|_{\mathscr{F}L^q_s} + \|\chi f\|_{\mathscr{F}L^q_t}).\end{aligned}$$

The proof is complete. □

Remark 2. Theorem 8 and Proposition 5 can be formulated in the language of the symbol class $S^m_{\mathrm{loc}}(X \times \mathbb{R}^d)$, where X is an open set in $\mathbb{R}^d$. This class is introduced in [11] as the starting point in the study of pseudodifferential operators on manifolds.

A result parallel to Theorem 8 can be formulated for operators whose symbols satisfy the following condition:

$$|\partial_\xi^\alpha \partial_x^\beta a(x,\xi)| \leq C_{\alpha,\beta}\langle\xi\rangle^{m_1-\rho|\alpha|}\langle x\rangle^{m_2-\rho|\beta|}, \quad x,\xi \in \mathbb{R}^d, \tag{15}$$

for all multi-indices α,β, where $m_1,m_2 \in \mathbb{R}$ and $\rho > 0$.

Such operators belong to the class of SG pseudodifferential operators or symbol-global type operators, see [6, 10, 14]. More recently, symbol-global type operators are studied in different context in [3,4,7,9].

An extension of Proposition 5 to operators whose symbols satisfy (15) is not straightforward. For such operators a new notion of ellipticity instead of (2) should be introduced, cf. [9].

7 Pseudodifferential Operators, an Extension

In this section we present a part of our results from [15] related to the action of more general classes of pseudodifferential operators.

Here we consider appropriate subclasses of $\mathscr{P}$. More precisely, we let $\mathscr{P}_0(\mathbb{R}^d)$ be the set of all $\omega \in \mathscr{P}(\mathbb{R}^d)\bigcap C^\infty(\mathbb{R}^d)$ such that $\partial^\alpha \omega/\omega \in L^\infty$ for all multi-indices α. By Lemma 1.2 in [22] it follows that for each $\omega \in \mathscr{P}(\mathbb{R}^d)$, there is an element $\omega_0 \in \mathscr{P}_0(\mathbb{R}^d)$ such that

$$C^{-1}\omega_0 \le \omega \le C\omega_0, \tag{16}$$

for some constant C.

Assume that $0 \le \rho$. Then we let $\mathscr{P}_\rho(\mathbb{R}^{2d})$ be the set of all $\omega(x,\xi)$ in $\mathscr{P}_0(\mathbb{R}^{2d})$ such that

$$\langle \xi \rangle^{\rho|\beta|} \frac{\partial_x^\alpha \partial_\xi^\beta \omega(x,\xi)}{\omega(x,\xi)} \in L^\infty(\mathbb{R}^{2d}),$$

for every multi-indices α and β. Note that in contrast to $\mathscr{P}_0$, we do not have an equivalence between $\mathscr{P}_\rho$ and $\mathscr{P}$ when $\rho > 0$ in the sense of (16). On the other hand, if $s \in \mathbb{R}$ and $\rho \in [0,1]$, then $\mathscr{P}_\rho(\mathbb{R}^{2d})$ contains $\omega(x,\xi) = \langle \xi \rangle^s$, which seem to be the most important classes in the applications.

Assume that $\omega_0 \in \mathscr{P}_\rho(\mathbb{R}^{2d})$. Then $S^\rho_{(\omega_0)}(\mathbb{R}^{2d})$ consists of all $a \in C^\infty(\mathbb{R}^{2d})$ such that

$$|\partial_x^\alpha \partial_\xi^\beta a(x,\xi)| \le C_{\alpha,\beta}\, \omega_0(x,\xi) \langle \xi \rangle^{-\rho|\beta|}.$$

(Cf. Sects. 18.4–18.6 in [11].) Clearly, $S^\rho_{(\omega_0)} = S^m_{\rho,0}(\mathbb{R}^{2d})$ when $\omega_0(x,\xi) = \langle \xi \rangle^m$.

For later convenience we set

$$\omega_{s,\rho}(x,\xi,\eta,z) = \omega(x,\xi,\eta,z) \langle x \rangle^{s_4} \langle \eta \rangle^{s_3} \langle \xi \rangle^{\rho s_2} \langle z \rangle^{s_1}, \tag{17}$$

when $\rho \in \mathbb{R}$ and $s \in \mathbb{R}^4$.

The following definition of the characteristic set is different from the one given in the introduction (see also [11, Sect. 18.1]). In fact, here it is defined for symbols which are not polyhomogeneous, while in the case of polyhomogeneous symbols our sets of characteristic points are smaller than the set of characteristic points in [11].

Definition 2. Assume that $\rho \in (0,1]$ and $\omega \in \mathscr{P}_\rho(\mathbb{R}^{2d})$. For each open cone $\Gamma \subseteq \mathbb{R}^d \setminus 0$, open set $U \subseteq \mathbb{R}^d$ and real number $R > 0$, let

$$\Omega_{U,\Gamma,R} \equiv \{\, (x,\xi) \, : x \in U, \ \xi \in \Gamma, \ |\xi| > R \}.$$

The pair $(x_0,\xi_0)\in\mathbb{R}^d\times(\mathbb{R}^d\setminus 0)$ is called *noncharacteristic* for $a\in S^{\rho}_{(\omega)}(\mathbb{R}^{2d})$ (with respect to ω), if there is a conical neighborhood Γ of ξ_0, a neighborhood U of x_0, a real number $R>0$, and elements $b\in S^{\rho}_{(\omega^{-1})}(\mathbb{R}^{2d})$, $c\in S^{0}_{\rho,0}(\mathbb{R}^{2d})$, $c\equiv 1$ on $\Omega_{U,\Gamma,R}$ and $h\in S^{-\rho}_{\rho,0}(\mathbb{R}^{2d})$ such that

$$b(x,\xi)a(x,\xi)=c(x,\xi)+h(x,\xi),\quad (x,\xi)\in\mathbb{R}^{2d}.$$

The pair (x_0,ξ_0) in $\mathbb{R}^d\times(\mathbb{R}^d\setminus 0)$ is called *characteristic* for a (with respect to $\omega\in\mathscr{P}_\rho(\mathbb{R}^{2d})$), if it is *not* noncharacteristic for a with respect to $\omega\in\mathscr{P}_\rho(\mathbb{R}^{2d})$. The *set of characteristic points* (the characteristic set), for $a\in S^{\rho}_{(\omega)}(\mathbb{R}^{2d})$ with respect to ω, is denoted by $\mathrm{Char}(a)=\mathrm{Char}_{(\omega)}(a)$.

Assume now that the underlying weight functions satisfy:

$$\frac{\omega_2(x,\xi)}{\omega_1(x,\xi)}\le C\omega_0(x,\xi),\tag{18}$$

for some constant C.

Theorem 9. *Assume that* $0<\rho\le 1$, $\omega_1,\omega_2\in\mathscr{P}(\mathbb{R}^{2d})$ *and* $\omega_0\in\mathscr{P}_\rho(\mathbb{R}^{2d})$ *satisfy* (18). *If* $a\in S^{\rho}_{(\omega_0)}$ *and* $q\in[1,\infty]$, *then*

$$\mathrm{WF}_{\mathscr{F}L^q_{(\omega_2)}}(a(x,D)f)\subseteq\mathrm{WF}_{\mathscr{F}L^q_{(\omega_1)}}(f),\quad f\in\mathscr{S}'(\mathbb{R}^d).$$

We also have the following counter result to Theorem 9. Here, it is natural to assume that the underlying weight functions satisfy

$$C^{-1}\omega_0(x,\xi)\le\frac{\omega_2(x,\xi)}{\omega_1(x,\xi)},\tag{19}$$

for some constant C, instead of (18).

Theorem 10. *Assume that* $0<\rho\le 1$, $\omega_1,\omega_2\in\mathscr{P}(\mathbb{R}^{2d})$ *and* $\omega_0\in\mathscr{P}_\rho(\mathbb{R}^{2d})$ *satisfy* (19). *If* $a\in S^{\rho}_{(\omega_0)}$ *and* $q\in[1,\infty]$ *then*

$$\mathrm{WF}_{\mathscr{F}L^q_{(\omega_1)}}(f)\subseteq\mathrm{WF}_{\mathscr{F}L^q_{(\omega_2)}}(a(x,D)f)\cup\mathrm{Char}_{(\omega_0)}(a),\quad f\in\mathscr{S}'(\mathbb{R}^d).$$

Next we show that the statements in Theorems 9 and 10 are not true if the assumption $\rho>0$ is replaced by $\rho=0$. We assume here that $\omega_0=1$ and $\omega_1=\omega_2$, and leave the general case for the reader.

Let $a(x,\xi)=\mathrm{e}^{-\mathrm{i}\langle x_0,\xi\rangle}$ for some fixed $x_0\in\mathbb{R}^d$ and choose α in such way that $f_\alpha(x)=\delta_0^{(\alpha)}$ does not belong to $\mathscr{F}L^q_{(\omega_1)}$. Since

$$(a(x,D)f_\alpha)(x)=f_\alpha(x-x_0),$$

straightforward computations imply that, for some closed cone $\Gamma \in \mathbb{R}^d \setminus 0$,

$$\mathrm{WF}_{\mathscr{F}L^q_{(\omega_1)}}(f) = \{\,(0,\xi) : \xi \in \Gamma\,\};$$

$$\mathrm{WF}_{\mathscr{F}L^q_{(\omega_1)}}(a(x,D)f) = \{\,(x_0,\xi) : \xi \in \Gamma\,\},$$

which are not overlapping when $x_0 \neq 0$.

Finally, we apply Theorems 9 and 10 to hypoelliptic operators. Assume that $a \in C^\infty(\mathbb{R}^{2d})$ is bounded by a polynomial. Then $a(x,D)$ is called *hypoelliptic*, if there are positive constants C, $C_{\alpha,\beta}$, N, ρ, and R such that

$$|\partial_x^\alpha \partial_\xi^\beta a(x,\xi)| \le C_{\alpha,\beta}|a(x,\xi)|\langle\xi\rangle^{-\rho|\beta|},$$

and $C\langle\xi\rangle^{-N} \le |a(x,\xi)|$ when $x \in \mathbb{R}^d$ and $|\xi| > R$ (see, e.g., [1, 11]). We note that if $a(x,D)$ is hypoelliptic and $\chi \in C_0^\infty(\mathbb{R}^d)$, then $\chi(x)a(x,\xi) \in S^\rho_{(\omega)}(\mathbb{R}^{2d})$, where

$$\omega(x,\xi) = \omega_a(x,\xi) = (\langle\xi\rangle^{-2N} + |a(x,\xi)|^2)^{1/2} \in \mathscr{P}_\rho(\mathbb{R}^{2d}).$$

Furthermore, as $\mathrm{Char}_{(\omega_a)}(a) = \emptyset$, as a consequence of Theorems 9 and 10, we have the following.

Theorem 11. *Assume that $a \in C^\infty(\mathbb{R}^{2d})$ is such that $a(x,D)$ is hypoelliptic, $q \in [1,\infty]$, and that $\omega_1, \omega_2 \in \mathscr{P}(\mathbb{R}^{2d})$ satisfy*

$$C^{-1}\frac{\omega_2(x,\xi)}{\omega_1(x,\xi)} \le \omega_a(x,\xi) \le C\frac{\omega_2(x,\xi)}{\omega_1(x,\xi)}$$

for some constant C which is independent of $(x,\xi) \in \mathbb{R}^{2d}$. If $f \in \mathscr{S}'(\mathbb{R}^d)$, then

$$\mathrm{WF}_{\mathscr{F}L^q_{(\omega_2)}}(a(x,D)f) = \mathrm{WF}_{\mathscr{F}L^q_{(\omega_1)}}(f).$$

Note that for any hypoelliptic operator, we may choose the symbol class which contains the symbol of the operator in such way that the corresponding set of characteristic points is empty. Consequently, in the view of Theorem 11, it follows that hypoelliptic operators preserve the wave-front sets, as they should.

Acknowledgements This paper was supported by the Serbian Ministry of Science and Technological Development (Project # 144016).

References

1. Boggiatto, P., Buzano, E., Rodino, L.: Global hypoellipticity and spectral theory. Mathematical Research, 92, Springer- Verlag, Berlin (1996)
2. Boulkhemair, A.: Remarks on a Wiener type pseudodifferential algebra and Fourier integral operators. Math. Res. L. **4**, 53–67 (1997)

3. Cappiello, M., Rodino, L.: SG-pseudo-differential operators and Gelfand-Shilov spaces. Rocky Mountain J. Math. **36**, 1117–1148 (2006)
4. Cappiello, M., Gramchev, T., Rodino, L.: Exponential decay and regularity for SG-elliptic operators with polynomial coefficients. In: Hyperbolic problems and regularity questions (L. Zanghirati, M. Padula eds.), 49–58, Trends Math., Birkhäuser, Basel (2007)
5. Cordero, E., Nicola, F., Rodino, L.: Boundedness of Fourier integral operators on $\mathscr{F}L^p$ spaces. Trans. Am. Math. Soc. **361**, 6049–6071 (2009)
6. Cordes, H.O.: The Technique of Pseudodifferential Operators. Cambridge University Press, Cambridge (1995)
7. Coriasco, S., Rodino, L.: Cauchy problem for SG-hyperbolic equations with constant multiplicities. Ricerche Mat. **48**, 25–43 (1999)
8. Concetti, F., Toft, J.: Schatten-von Neumann properties for Fourier integral operators with non-smooth symbols, I. Ark. Mat. **47**(2), 295–312 (2009)
9. Dasgupta, A., Wong, M.W.: Spectral Theory of SG pseudo-diffeential operators on $L^p(\mathbb{R}^n)$. Studia Math. **187**(2), 185–197 (2008)
10. Egorov, Yu.V., Schulze, B.-W.: Pseudo-Differential Operators, Singularities, Applications. Birkhäuser (1997)
11. Hörmander, L.: The Analysis of Linear Partial Differential Operators. Vol III, Springer-Verlag, Berlin (1994)
12. Hörmander, L.: Lectures on Nonlinear Hyperbolic Differential Equations. Springer-Verlag, Berlin (1997)
13. Okoudjou, K.A.: A Beurling-Helson type theorem for modulation spaces. J. Funct. Spaces Appl. **7**(1), 33–41 (2009)
14. Parenti, C.: Operatori pseudo-differentiali in $\mathbb{R}^n$ e applicazioni. Ann. Mat. Pura Appl. **93**, 359–389 (1972)
15. Pilipović, S., Teofanov, N., Toft, J.: Micro-local analysis in Fourier Lebesgue and modulation spaces. Part I. preprint (2008) available at arXiv:0804.1730
16. Pilipović, S., Teofanov, N., Toft, J.: Micro-local analysis in Fourier Lebesgue and modulation spaces. Part II. preprint (2009) available at arXiv:0805.4476
17. Pilipović, S., Teofanov, N., Toft, J.: Wave-front sets in Fourier-Lebesgue space. Rend. Sem. Mat. Univ. Politec. Torino **66**(4), 41–61 (2008)
18. Pilipović, S., Teofanov, N., Toft, J.: Some applications of wave-front sets of Fourier Lebesgue types, I. In: 3rd Conference on Mathematical Modelling of Wave Phenomena, Växjö, Sweden, 9 – 13 June 2008, AIP Conference Proceeding 1106 (B. Nillson, L. Fishman, A. Karlsson, S. Nordebo, eds.), 26–35 (2009)
19. Ruzhansky, M., Sugimoto, M., Tomita, N., Toft, J.: Changes of variables in modulation and Wiener amalgam spaces. preprint (2008) available at arXiv:0803.3485v1
20. Strohmer, T.: Pseudo-differential operators and Banach algebras in mobile communications. Appl. Comput. Harmon. Anal. **20**, 237–249 (2006)
21. Toft, J.: Continuity properties for modulation spaces with applications to pseudo-differential calculus. I. J. Funct. Anal. **207**, 399–429 (2004)
22. Toft, J.: Continuity properties for modulation spaces with applications to pseudo-differential calculus, II. Ann. Global Anal. Geom. **26**, 73–106 (2004)
23. Toft, J.: Continuity and Schatten properties for pseudo-differential operators on modulation spaces. In: Modern Trends in Pseudo-Differential Operators, (J. Toft, M. W. Wong, H. Zhu, eds.), Birkhäuser Verlag, Basel, 173–206 (2007)
24. Toft, J., Concetti, F., Garello, G.: Trace ideals for Fourier integral operators with non-smooth symbols III. preprint, (2008) available at arXiv:0802.2352
25. Wong, M.W.: An Introduction to Pseudo-differential Operators, 2nd edn. World Scientific, Singapore (1999)

On a Class of Matrix Differential Equations with Polynomial Coefficients

Boro M. Piperevski

Dedicated to Professor Gradimir V. Milovanović on the occasion of his 60th birthday

1 Introduction

The relation between orthogonal polynomials and differential equations is well known.

Polynomials $\{\pi_n\}$ orthogonal on a semicircle with respect to the complex inner product

$$(f,g) = \int_0^\pi f(e^{i\theta})g(e^{i\theta})w(e^{i\theta})\,d\theta,$$

have been introduced by Walter Gautschi, Henry J. Landau, and Gradimir V. Milovanović [2].

In that paper, in the Gegenbauer case

$$w(z) = (1-z)^{\lambda-1/2}(1+z)^{\lambda-1/2}, \quad \lambda > -\frac{1}{2}, \tag{1}$$

a linear second-order differential equation for $\{\pi_n\}$ is obtained.

Polynomials $\{\pi_n^R\}$ orthogonal on a circular arc with respect to the complex inner product

$$(f,g) = \int_\varphi^{\pi-\varphi} f_1(\theta)g_1(\theta)w_1(\theta)\,d\theta,$$

where $\varphi \in (0,\pi/2)$, and the function $f_1(\theta)$ in terms of $f(z)$ is defined by:

$$f_1(\theta) = f\left(-iR + e^{i\theta}\sqrt{R^2+1}\right), \quad R = \tan\varphi,$$

have been introduced by M. de Bruin [1].

Boro M. Piperevski
Faculty of Electrical Engineering and Information Technologies, Ss Cyril and Methodius University, P.O. Box 574, Skopje, Macedonia, e-mail: borom@feit.ukim.edu.mk

W. Gautschi et al. (eds.), *Approximation and Computation: In Honor of Gradimir V. Milovanović*, Springer Optimization and Its Applications 42,
DOI 10.1007/978-1-4419-6594-3_24,

In the Jacobi case

$$w(z) = (1-z)^{\alpha}(1+z)^{\beta}, \quad \alpha,\beta > -1, \tag{2}$$

Gradimir V. Milovanovic and Predrag M. Rajkovic [3] obtained a linear second-order differential equation for $\{\pi_n^R\}$.

In [4] a class of systems

$$ax_1' + bx_2' + Ax_1 = 0,$$
$$cx_1' + dx_2' + Bx_2 = 0,$$

or in matrix form

$$\mathbf{P}\cdot\mathbf{X}' + \mathbf{M}\cdot\mathbf{X} = \mathbf{0}, \tag{3}$$

where

$$\mathbf{P} = \begin{bmatrix} a\ b \\ c\ d \end{bmatrix}, \quad \mathbf{M} = \begin{bmatrix} A\ 0 \\ 0\ B \end{bmatrix}, \quad \mathbf{X} = \begin{bmatrix} x_1(t) \\ x_2(t) \end{bmatrix}, \quad \mathbf{X}' = \begin{bmatrix} x_1'(t) \\ x_2'(t) \end{bmatrix},$$

$a = a_1t + a_2$, $b = b_1t + b_2$, $c = c_1t + c_2$, $d = d_1t + d_2$, $A, B, a_i, b_i, c_i, d_i \in \mathbb{R}$, $i = 1, 2$,

$$x_1'(t) = \frac{\mathrm{d}x_1}{\mathrm{d}t}, \quad x_2'(t) = \frac{\mathrm{d}x_2}{\mathrm{d}t},$$

is considered.

In that article, the necessary conditions $b' = 0$, $B + nd' = 0$, $ka' + A \neq 0$, $k < n$, k a positive integer, for a polynomial solution of degree n, $x_1(t) = P_{n-1}(t)$, $x_2(t) = Q_n(t)$, or in matrix form

$$\mathbf{X}_n = \begin{bmatrix} P_{n-1}(t) \\ Q_n(t) \end{bmatrix},$$

where $P_{n-1}(t)$ is a polynomial of degree $n-1$ and $Q_n(t)$ a polynomial of degree n, is obtained.

In this case, the second component of the matrix X_n is a polynomial solution of degree n of the differential equation

$$(z^2 + Qz + R)(Sz + T)w'' + (\beta_2 z^2 + \beta_1 z + \beta_0)w' + (\gamma_1 z + \gamma_0)w = 0, \tag{4}$$

and another polynomial solution of degree k, $k < n$, does not exist, if the conditions

$$\begin{aligned} &n^2 S + (\beta_2 - S)n + \gamma_1 = 0, \\ &S^2(\beta_0 + SR - QT) + T^2(S + \beta_2) - T\beta_1 S = 0, \\ &S^2(\beta_1\gamma_0 + \gamma_0^2 - \beta_0\gamma_1) + T(\gamma_1 + \beta_2)(T\gamma_1 - 2S\gamma_0) = 0, \end{aligned} \tag{5}$$

(n is the smaller one if both roots of the first condition are natural numbers), are satisfied.

The complex polynomial orthogonal on the semicircle or on the circular arc [2,3] is the unique polynomial solution of the differential equation of type (4), satisfying the conditions (5) [5].

In particular, classes of complex polynomials $\pi_n^\lambda(z)$, in the case of the Gegenbauer weight function (1), are the unique polynomial solutions of the differential equation (4), where

$$
\begin{aligned}
&Q=0, \quad R=-1, \quad S=2(2n+2\lambda-1)(n+\lambda-1)\mathrm{i}\theta_{n-1},\\
&T=-\frac{1}{(2n+2\lambda-1)^2}S,\\
&\beta_2=2\lambda S, \quad \beta_1=(2\lambda+1)T, \quad \beta_0=S, \quad \gamma_1=-n(n+2\lambda-1)S,\\
&\gamma_0=(n+2\lambda-1)[n(2n+2\lambda-1)-(n-1)T],\\
&\theta_k=\frac{k(k+2\lambda-1)}{4(k+\lambda)(k+\lambda-1)\theta_{k-1}}, \quad k\in\mathbb{N}, \quad \theta_0=\frac{\Gamma(\lambda+1/2)}{\sqrt{\pi}\,\Gamma(\lambda+1)}.
\end{aligned}
$$

Therefore, the system

$$
\left(z-\frac{S}{n(2n+2\lambda-1)}\right)\frac{\mathrm{d}G_{n-1}^\lambda(z)}{\mathrm{d}z}-\frac{1}{n(n+2\lambda-1)}\frac{\mathrm{d}\pi_n^\lambda(z)}{\mathrm{d}z}+(n+2\lambda-1)G_{n-1}^\lambda(z)=0,
$$

$$
(Sz+T)\frac{\mathrm{d}G_{n-1}^\lambda(z)}{\mathrm{d}z}+\left(z-\frac{S}{(n+2\lambda-1)(2n+2\lambda-1)}\right)\frac{\mathrm{d}\pi_n^\lambda(z)}{\mathrm{d}z}-n\pi_n^\lambda(z)=0,
$$

where $G_{n-1}^\lambda(z)$ is the Gegenbauer polynomial of degree $n-1$, is obtained.

Also, we obtain the formula

$$
\begin{aligned}
\pi_n^\lambda(z)&=(n+2\lambda-1)^2(z^2-1)^{-\lambda+1/2}\\
&\times\frac{\mathrm{d}^{n-1}}{\mathrm{d}z^{n-1}}\left[\left(z-\frac{S}{(n+2\lambda-1)(2n+2\lambda-1)}\right)(z^2-1)^{n+\lambda-3/2}\right].
\end{aligned}
$$

More generally, classes of complex polynomials $\pi_n^{\alpha,\beta}(z)$, in the case of the Jacobi weight function (2), are the unique polynomial solutions of the differential equation (4), where

$$
\begin{aligned}
&Q=0, \quad R=-1, \quad S=(2n+\alpha+\beta)(2n+\alpha+\beta-1)\mathrm{i}\theta_{n-1},\\
&T=-\frac{4n(n+\alpha)(n+\beta)(n+\alpha+\beta)+(\beta^2-\alpha^2)S+S^2}{(2n+\alpha+\beta)^2},\\
&\beta_2=(\alpha+\beta+1)S, \quad \beta_1=-(\beta-\alpha)S+(\alpha+\beta+2)T, \quad \beta_0=-(\beta-\alpha)T+S,\\
&\gamma_1=-n(n+\alpha+\beta)S, \quad \gamma_0=-n(n+\alpha+\beta+1)T+\frac{n(\beta-\alpha)S-S^2}{2n+\alpha+\beta},
\end{aligned}
$$

$$
\theta_{k-1}=\frac{1}{\mathrm{i}}\frac{\rho_k(-\mathrm{i}R)}{\rho_{k-1}(\mathrm{i}R)}, \quad k\geq 1, \quad \rho_k(z)=\int_{-1}^{1}\frac{P_k^{\alpha,\beta}(x)}{z-x}(1-x)^\alpha(1+x)^\beta\,\mathrm{d}x.
$$

Here, $P_k^{\alpha,\beta}(z)$ is the Jacobi polynomial of degree k.

Therefore, we obtain the system

$$\left(z+\frac{n(\beta-\alpha)-S}{n(2n+\alpha+\beta)}\right)\frac{\mathrm{d}P_{n-1}^{\alpha,\beta}(z)}{\mathrm{d}z}-\frac{1}{n(n+\alpha+\beta)}\frac{\mathrm{d}\pi_n^{\alpha,\beta}(z)}{\mathrm{d}z}+(n+\alpha+\beta)P_{n-1}^{\alpha,\beta}(z)=0,$$

$$(Sz+T)\frac{\mathrm{d}P_{n-1}^{\alpha,\beta}(z)}{\mathrm{d}z}+\left(z-\frac{(\beta-\alpha)(n+\alpha+\beta)+S}{(n+\alpha+\beta)(2n+\alpha+\beta)}\right)\frac{\mathrm{d}\pi_n^{\alpha,\beta}(z)}{\mathrm{d}z}-n\pi_n^{\alpha,\beta}(z)=0.$$

where $P_{n-1}^{\alpha,\beta}(z)$ is the Jacobi polynomial of degree $n-1$.

Also, we obtain the following formula

$$\pi_n^{\alpha,\beta}(z)=\frac{n+\alpha+\beta}{2n+\alpha+\beta}(z-1)^{-\alpha}(1+z)^{-\beta}\frac{\mathrm{d}^{n-1}}{\mathrm{d}z^{n-1}}\Big\{[(n+\alpha+\beta)(2n+\alpha+\beta)z$$
$$-(\beta-\alpha)(n+\alpha+\beta)-S](z-1)^{n+\alpha-1}(z+1)^{n+\beta-1}\Big\}.$$

2 Main Result

We consider a class of matrix differential equations (3).

Definition 1. We say that (3) has a polynomial solution of degree n if polynomials $P_n(t)$, $Q_n(t)$ of degree n exist such that

$$\mathbf{X}_n=\begin{bmatrix}P_n(t)\\ Q_n(t)\end{bmatrix}$$

is the solution (3).

Lemma 1. *Using the substitution*

$$\mathbf{X}=\mathbf{T}\cdot\mathbf{Y},\quad \mathbf{T}=\begin{bmatrix}f(t)&0\\0&f(t)\end{bmatrix},\quad f(t)=\mathrm{e}^{-\int(aB+Ad)/(ad-bc)\,\mathrm{d}t},$$

in (3), *we get* $\mathbf{P}_1\mathbf{Y}'+\mathbf{M}_1\mathbf{Y}=\mathbf{0}$, *where*

$$\mathbf{P}_1=\begin{bmatrix}-Ad&bB\\Ac&-aB\end{bmatrix},\quad \mathbf{M}_1=\begin{bmatrix}AB&0\\0&AB\end{bmatrix}.$$

Theorem 1. *The equation* (3) *with condition* $a\cdot b\cdot c\cdot d\cdot(ad-bc)\cdot A\cdot B\neq 0$ *has a polynomial solution of degree n, and another polynomial solution of degree k, $k<n$, does not exist, if there exists a positive integer n such that the conditions*

$$r(\mathbf{M}+n\cdot\mathbf{P}')=1,\quad r(\mathbf{M}+k\cdot\mathbf{P}')=2$$

(r the rank of a matrix), $k < n$, k a positive integer, $b' \neq 0$, $c' \neq 0$, $A + na' \neq 0$, $B + nd' \neq 0$ are satisfied.

The polynomial solution will be given by the formula

$$\mathbf{X}_n = \mathbf{T} \cdot [\mathbf{T}_1 \cdot \mathbf{U}_1]^{(n-1)}, \tag{6}$$

where

$$\mathbf{T} = \begin{bmatrix} f(t) & 0 \\ 0 & f(t) \end{bmatrix}, \quad \mathbf{T}_1 = \begin{bmatrix} B(kb+a) & 0 \\ 0 & A(kd+c) \end{bmatrix}, \quad \mathbf{U}_1 = \frac{(ad-bc)^{n-1}}{f(t)} \begin{bmatrix} 1 \\ 1 \end{bmatrix}$$

and

$$f(t) = \mathrm{e}^{-\int (aB+Ad)/(ad-bc)\,\mathrm{d}t}, \quad k = -\frac{nc'}{B+nd'} = -\frac{A+na'}{nb'}.$$

Proof. Suppose that the conditions of the theorem are satisfied and consider the algebraic matrix equation $(\mathbf{M} + n\mathbf{P}') \cdot \mathbf{W} = \mathbf{0}$. Using the conditions, we write the solution of this equation in the form

$$\mathbf{W} = \mathbf{K}, \quad \mathbf{K} = \begin{bmatrix} k_1 \\ k_2 \end{bmatrix}, \quad k_1 \neq 0, k_2 \neq 0.$$

If $\mathbf{K} = \mathbf{X}_n^{(n)}$ then $\mathbf{P} \cdot \mathbf{X}_n^{(n+1)} + (\mathbf{M} + n\mathbf{P}') \cdot \mathbf{X}_n^{(n)} = \mathbf{0}$, and $\mathbf{X} = \mathbf{X}_n$ is a polynomial solution of (3).

Now, consider the matrix differential equation

$$\mathbf{P}^* \cdot \mathbf{Z}' + \mathbf{M}^* \cdot \mathbf{Z} = \mathbf{0}, \tag{7}$$

where

$$\mathbf{Z} = \begin{bmatrix} z_1(t) \\ z_2(t) \end{bmatrix}, \quad \mathbf{P}^* = \begin{bmatrix} -Ad & bB \\ Ac & -aB \end{bmatrix}, \quad \mathbf{M}^* = \begin{bmatrix} AB + (n-1)d'A & -(n-1)b'B \\ -(n-1)c'A & AB + (n-1)a'B \end{bmatrix}.$$

With the substitution $\mathbf{Z} = \mathbf{T}_1 \cdot \mathbf{U}$,

$$\mathbf{U} = \begin{bmatrix} u_1(t) \\ u_2(t) \end{bmatrix}, \quad \mathbf{T}_1 = \begin{bmatrix} B(kb+a) & 0 \\ 0 & A(kd+c) \end{bmatrix}, \; k = -\frac{nc'}{B+nd'} = -\frac{A+na'}{nb'},$$

this equation is transformed to the equation $\mathbf{P}_1^* \cdot \mathbf{U}' - \mathbf{M}_1^* \cdot \mathbf{U} = \mathbf{0}$, where

$$\mathbf{P}_1^* = \begin{bmatrix} ad-bc & 0 \\ 0 & ad-bc \end{bmatrix},$$

$$\mathbf{M}_1^* = \begin{bmatrix} aB + Ad + (n-1)(ad-bc)' & 0 \\ 0 & aB + Ad + (n-1)(ad-bc)' \end{bmatrix}.$$

A particular solution is

$$\mathbf{U}_1 = \frac{(ad-bc)^{n-1}}{f(t)} \begin{bmatrix} 1 \\ 1 \end{bmatrix}.$$

On the other hand, if we differentiate (7) $n-1$ times and use the substitution $\mathbf{Z}^{(n-1)} = \mathbf{T}^{-1} \cdot \mathbf{X}$, where

$$\mathbf{T} = \begin{bmatrix} f(t) & 0 \\ 0 & f(t) \end{bmatrix}, \quad f(t) = \mathrm{e}^{-\int (aB-Ad)/(ad-bc)\,\mathrm{d}t},$$

then we easily obtain the same equation (3). Similarly, applying the earlier substitutions, we get the particular polynomial solution by formula (6). □

References

1. De Bruin, M.G.: Polynomials orthogonal on a circular arc. J. Comput. Appl. Math. **31**(2), 253–256 (1990)
2. Gautschi, W., Landau, H.J., Milovanović, G.V.: Polynomials orthogonal on the semicircle, II. Constr. Approx. **3**, 389–404 (1987)
3. Milovanović, G.V., Rajković, P.M.: On polynomials orthogonal on a circular arc. J. Comput. Appl. Math. **51**, 1–13 (1994)
4. Piperevski, B.M.: Sur une formule de solution polynomial d' une classe d' équations différentielles linéaires du deuxième ordre. Bulletin Mathématique de la SDM de SRM. **7–8**, 10–15 (1983/84)
5. Piperevski, B.M.: On complex polynomials orthogonal on a circular arc. (in Macedonian) Proceedings of the 7MSDR Ohrid, Republic of Macedonia 21–26 (2002) http://www.cim.feit.ukim.edu.mk

Part V
Applications

Optimized Algorithm for Petviashvili's Method for Finding Solitons in Photonic Lattices

Raka Jovanović and Milan Tuba

Dedicated to Professor Gradimir V. Milovanović
on the occasion of his 60th birthday

1 Introduction

Optically induced photonic lattices [3] have attracted much interest recently, owing to their intriguing wave guiding possibilities. Optical solitons [4] in photonic lattices are of interest because of their dynamical and structural stability during beam propagation. Solitons are found by solving the corresponding wave equation. We designed a software system for this problem based on a modified version of the well-known and tested Petviashvili's method [7], the modification consisting in the introduction of new stabilizing factors. Our software implementation also includes a heuristic definition of convergence and divergence for this type of numerical problem. Owing to the intensive calculations required for finding each solitonic solution [6] and the large number of input parameters of interest, the need for such optimization has arisen. The optimizations have been implemented through early recognition of divergent cases, achieved with a number of numerical criteria that analyze the behavior of a function representing the precision of the result at the current iteration. The use of this simulation is illustrated by solving this problem for a vortex shape input beam and a circular photonic lattice with defects.

Raka Jovanović
Institute of Physics, Pregrevica 118, 11080 Zemun, Serbia, e-mail: rakabog@yahoo.com

Milan Tuba
Faculty of Computer Science, Megatrend University, Bulevar umetnosti 29, 11070 Novi Beograd, Serbia, e-mail: tuba@ieee.org

W. Gautschi et al. (eds.), *Approximation and Computation: In Honor of Gradimir V. Milovanović*, Springer Optimization and Its Applications 42,
DOI 10.1007/978-1-4419-6594-3_25,

2 Physical Model

The behavior of counter-propagating (CP) beams in photonic lattices is described by a time-independent model for the formation of self-trapped CP optical beams, based on the theory of photo-refractive effect. The model consists of two wave equations in the paraxial approximation for the propagation of CP beams. The model equations in the computational space are of the form:

$$\mathrm{i}\partial_z F = -\Delta F - \Gamma F \frac{I+I_g}{1+I+I_g}, \quad -\mathrm{i}\partial_z B = -\Delta B - \Gamma B \frac{I+I_g}{1+I+I_g}, \tag{1}$$

where F and B are the forward and backward propagating beam envelopes, Δ the transverse Laplacian, and Γ is the dimensionless coupling constant. The quantity $I = |F|^2 + |B|^2$ is the laser light intensity. A scaling $x/x_0 \leftarrow x$, $y/x_0 \leftarrow y$, $z/L_D \leftarrow z$, is utilized in the formulation of the dimensionless propagation equations, where x_0 is the typical full width at half maximum (FWHM) beam waist and L_D is the diffraction length. The quantity I_g is the transverse intensity distribution of the optically induced lattice array, formed by positioning Gaussian beams at the sites of the lattice. Different geometries of the lattice can be considered, such as hexagonal, cylindrical, and square lattices.

3 Modified Petviashvili's Method for Finding Solitonic Solutions in Photonic Lattices

For finding the solitonic solutions in photonic lattices, we will use the well-known and tested Petviashvili method [7], modified to significantly improve stability.

Because of the rotational symmetry of the problem, (1) suggests the existence of a soliton $u(x,y)$ of the form:

$$F = u(x,y)\cos\theta \exp(\mathrm{i}\mu z), \quad B = u(x,y)\sin\theta \exp(-\mathrm{i}\mu z),$$

where μ is the propagation constant and θ an arbitrary projection angle. When this solution is substituted into (1), they both transform into one, degenerate equation:

$$-\mu u + \Delta u + \Gamma u \frac{|u|^2 + I_g}{1+|u|^2+I_g} = 0. \tag{2}$$

Solutions of (2) in the form of basic, dipole, or vortex solitons can be calculated in the inverse Fourier space. We first define the linear and nonlinear elements:

$$P = \frac{\Gamma I_g}{1+|u|^2+I_g}, \quad Q = -\frac{\Gamma |u|^2 u}{1+|u|^2+I_g}.$$

After applying the Fourier transformation to (2), we obtain

$$\hat{u} = \frac{1}{|\mathbf{k}|^2 + \mu} \left[\widehat{Pu} - \hat{Q} \right], \tag{3}$$

where $\hat{}$ stands for the $2D$ Fourier transformation. The linear (naive) iteration on (3) does not find the needed solitons, nor does it converge. Following an idea from [10] (first advanced by Petviashvili), we introduce the stabilizing factors (or projectors) to (3):

$$\alpha = \int \left[(|\mathbf{k}|^2 + \mu)\, \hat{u} - \widehat{Pu} \right] u^* \, \overrightarrow{\mathrm{d}k}, \quad \beta = - \int \hat{Q} u^* \, \overrightarrow{\mathrm{d}k},$$

and construct the following iterative scheme:

$$\hat{u}_{m+1} = \frac{1}{|\mathbf{k}|^2 + \mu} \left[\left(\frac{\alpha_m}{\beta_m} \right)^{1/2} \widehat{P_m u_m} - \left(\frac{\alpha_m}{\beta_m} \right)^{3/2} \widehat{Q_m} \right].$$

Convergence is achieved when $\alpha_m = \beta_m$, theoretically, in practice when they are close enough: $|\alpha_m - \beta_m| < \varepsilon$, and then solitonic solutions can be found.

4 Software Simulator

The possibility of solving partial differential equations in an analytical way exists for a very small class of problems [9]; (2) is not in that class. Indeed, it is very hard to find solitonic solutions analytically. We will simulate [2, 8] this problem in a finite part of space $((-xmax, xmax) \times (-ymax, ymax))$. For the lattice intensity I_g we choose a circular arrangement of beams, with missing first and third rings (Fig. 1).

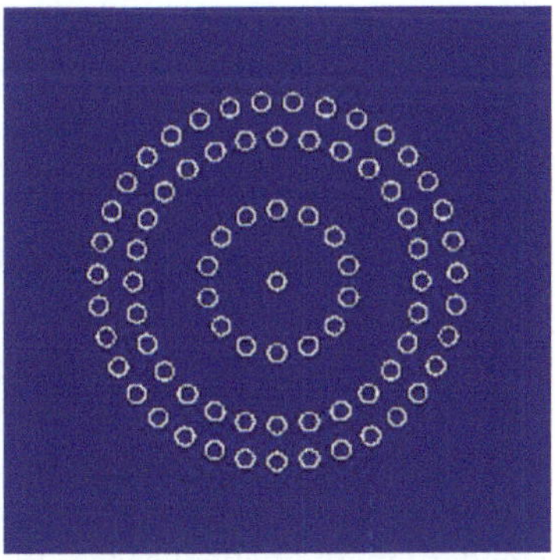

Fig. 1 Lattice intensity I_g: A circular arrangement of beams, with missing first and third rings

The value at each point of the grid is calculated in the following way:

```
for i = 0,2,4,5
  for j = 1,...,i*7
```

$$x_c = d * i * \sin\frac{2j\pi}{7i}, \quad y_c = d * i * \cos\frac{2j\pi}{7i},$$

$$I_g(x,y) := I_g(x,y) + I_0 \exp\left\{-\frac{(x-x_c)^2 + (y-y_c)^2}{\sigma^2}\right\}$$

```
  end
end
```

where I_0 is the lattice peak intensity, d the lattice spacing, and σ is the FWHM of lattice beams. For the input (initial function for the first iteration) we will be using a vortex probe defined by the formula:

$$u(x,y) = A\left(\frac{r}{\sigma}\right) \mathrm{e}^{-r^2/2\sigma^2} \mathrm{e}^{\mathrm{i}\varphi T_c},$$

where r and φ are the corresponding polar coordinates, A the amplitude of the vortex, σ the vortex width, and T_c is the topological charge (1 in our simulations).

In most simulations we are interested in the effect of changing the input parameters [1, 5]. In our problem these are μ, Γ, I_g, and u for the first iteration. The parameters I_g and u are set as in the previous section. The software implementation should be able to have different inputs for μ and Γ.

On the other hand, a criterion for convergence must be defined. We first define:

$$S1(n) = \text{Sum1} = \int_{-xmax}^{xmax} \int_{-ymax}^{ymax} |u_n(x,y) - u_{n-1}(x,y)| \,\mathrm{d}x\,\mathrm{d}y,$$

$$S2(n) = \text{Sum2} = \int_{-xmax}^{xmax} \int_{-ymax}^{ymax} |u_n(x,y)| \,\mathrm{d}x\,\mathrm{d}y.$$

Then the convergence criterion is:

$$\text{Sum1} < (\text{Sum2} * \varepsilon),$$

where ε is the level of precision we need for solitons to be stable in our experiments (we choose $\varepsilon = 5 \times 10^{-15}$).

A different problem is to decide what should be considered divergence, and to find methods that will be able to recognize these cases with the fewest number of iterations. The basic idea is to define a maximum number of iterations that will be calculated ($n_{\max}$), and if the convergence criterion has not been satisfied, the case will be considered as divergence. We used $n_{\max} = 30{,}000$ in our experiments. It is important to understand that this divergence is not the same as divergence in the standard mathematical sense. We attempt to run our simulation for a large number of different input parameter values, and in most cases no solitons will exist, or in other words, convergence will not arise. If we recognize these cases with a small number of iteration, the overall calculation time will be significantly reduced. The criteria used in our simulations are:

Absence of Stabilization Criterion

$$(\text{Sum1}/\text{Sum2} > 1) \wedge (\text{IterationNumber} > \text{MaxUnstableIteration})$$

This iteration method for finding solitons has to be stabilized, and should start leading towards the solution of (2). If $(\text{Sum1}/\text{Sum2} > 1)$, the criterion for instability is satisfied. Empirically it has been found that MaxUnstableIteration=150 is the optimal value.

Slow Convergence Criterion

Another divergence criterion used is recognition of very slowly converging simulations. These simulations may converge in a mathematical sense, but will not meet the basic convergence criterion:

$$\mathit{mod}(\text{IterationNumber}, \text{TestFrequancy}) = 0, \tag{4}$$

$$\text{TestPrecision}_i = S1(\text{IterationNumber})/S2(\text{IterationNumber}),$$

$$\frac{\text{TestPrecision}_{i-1}}{\text{TestPrecision}_i} < \text{MinConvergenceSpeed}, \tag{5}$$

$$i := i+1.$$

If (4) and (5) are both true at the same iteration, we will consider the simulation to be very slowly converging. The idea is to check every $TestFrequancy$ (in our simulation 600) iteration if $S1(IterationNumber)/S2(IterationNumber)$ has improved by more than some minimal relative threshold $MinConvergenceSpeed$ (in our simulation 1.05).

Long Interval of Instability Criterion

A high level of instability of the u, over a large number of iterations, also indicates that the simulation is divergent. The criterion for this case, checked at every iteration, is best described through pseudo code:

```
If (Sum1/Sum2) > MaxStability
    then begin
         IndicatorDiv = IndicatorDiv +1
         If (IndicatorDiv > MaxPeriodOfInstability)
```

```
            then divergence = true
        end
    else IndicatorDiv = 0
```

In our simulation we choose $MaxPeriodOfInstability = 650$, and $MaxStability = 0.05$.

Seesaw Criterion

In the seesaw criterion we analyze the behavior of the following functions:

$$\text{DirectionOfPrecision}(i) = \frac{S1(i-1)}{S2(i-1)} - \frac{S1(i)}{S2(i)},$$
$$\text{Switch}(i) = \text{DirectionOfPrecision}(i) * \text{DirectionOfPrecision}(i-1). \tag{6}$$

The condition $DirectionOfPrecision(i) > 0$ (< 0) tells us that u_i, u_{i-1} are nearer (farther) apart than u_{i-1}, u_{i-2} in the norm:

$$\text{norm}(f,g) = \int_{-xmax}^{xmax} \int_{-ymax}^{ymax} |f(x,y) - g(x,y)| \,\mathrm{d}x\mathrm{d}y.$$

$Switch(i) < 0$ tells us that *DirectionOfPrecision* has changed sign at the ith iteration, and in this case the slow convergence criterion must be reset, so we need a new criterion to find possible divergence. If the $DirectionOfPrecision$ changes sign quickly (in less than 35 iterations), and this happens *MinimalRow* times in a row, we will call this a seesaw row. Let $SeesawStartIteration/SeesawEndIteration$ be the first/last iteration in that row where $Switch(i) < 0$:

$$\text{SeesawPrecisionStart} = \frac{S1(\text{SeesawStartIteration})}{S2(\text{SeesawStartIteration})},$$
$$\text{SeesawPrecisionEnd} = \frac{S1(\text{SeesawEndIteration})}{S2(\text{SeesawEndIteration})},$$
$$\frac{|\text{SeesawPrecisionStart} - \text{SeesawEndPrecision}|}{\text{SeesawEndPrecision}} < \text{SeesawSpeed}. \tag{7}$$

If (7) is true, divergence has occurred.

Each of these criteria can sometimes be too strict, because of the chosen values of the parameters, and make incorrect judgments about convergence (some convergent cases will be thought of as divergent).

Simulation experiments have been done for the problem presented in Sect. 4, for parameter values $\Gamma \in (0,16)$ and $\mu \in (0,14)$ with a step of 0.25 and smaller for some areas. Different types of solitons have been found for various values of the parameters Γ and μ. Typical examples are presented in Figs. 2 and 3.

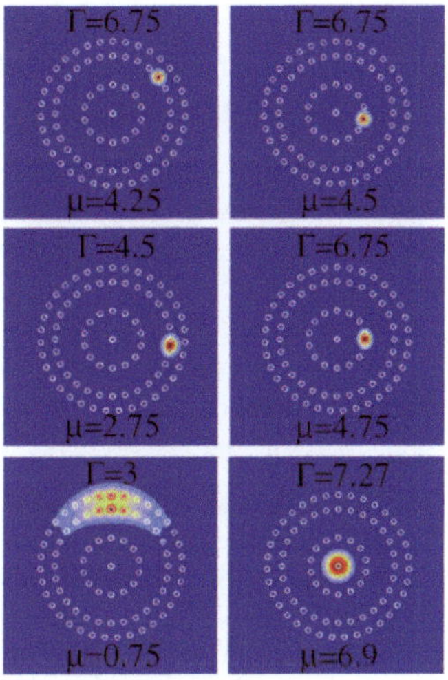

Fig. 2 Intensity distribution for all typical solitonic solutions. The on-site solitons (*first row*), off-site solitons (*second row*), the discrete solitons (*third row, left*), and the lattice central beam soliton (*right, third row*)

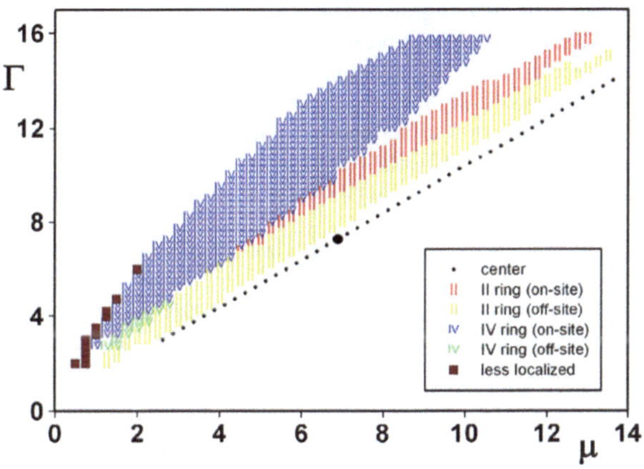

Fig. 3 Different types of solitons in the $\mu - \Gamma$ plane

5 Conclusion

The use of the modified Petviashvili's method for finding solitonic solutions in photonic lattices has been proven to be an efficient method for solving this type of problem. With the use of optimization techniques based on discovering that simulations diverge, or converge very slowly and do not find solitonic solutions in a reasonable time, in a small number of iterations (through the analysis of the precision function), the overall calculation time has improved greatly. This is best shown through the simulation results presented in Sect. 4. Only in a small portion of the investigated $\mu - \Gamma$ plane area (18–23%), solitons have been found. In these cases the average number of iterations needed to find solitons was about 5,000. Without the use of criteria for recognizing divergent simulations, the maximum number of iterations (30,000) would have been calculated, but with the use of these criteria the number of iterations calculated for most of these points has dropped to 150 (in the absence of stabilization criterion) and 600 (in the slow convergence criterion). These criteria managed to lower the overall calculation time 10–20 times, which shows their efficiency.

Acknowledgements This research was supported by the Ministry for Science and Technical Development of the Republic of Serbia, Projects 144007 and 141031.

References

1. Banks, J. et al.: Discrete-event system simulation. 4th ed. Prentice-Hall, Upper Saddle River, N. J. (2005)
2. Chung, C.: Simulation Modeling Handbook. CRC Press, West Palm Beach, FL, USA, 608 (2003)
3. Fleisher, J.W. et al.: Observation of two-dimensional discrete solitons in optically induced nonlinear photonic lattices. Nature **422**, 147–150 (2003)
4. Kivshar, Y.S., Agrawal, G.P.: Optical Solitons. Academic Press, San Diego (2003)
5. Law, A.: How to build valid and credible simulation models. Proceedings of the 37th conference on Winter simulation, 24–32 (2005)
6. Pelinovsky, D., Stepanyants, Y.: Convergence of Petviashvili's iteration method for numerical approximation of stationary solutions of nonlinear wave equations. SIAM J. Numer. Anal. **42**(3), 1110–1127 (2004)
7. Petviashvili, V. I.: Equation for an extraordinary soliton. Sov. J. Plasma Phys. **2** 257–258 (1976)
8. Seila, A. et al.: Applied Simulation Modeling, 494. Duxbury Press, North Scituate, MA, USA (2003)
9. Tian, B., Gao, Y.: Solutions of a variable-coefficient Kadomtsev-Petviashvili equation via computer algebra. Appl. Math. Comput. **84**(2–3), 125–130 (1997)
10. Yang, J. et al.: Dipole and quadrupole solitons in optically induced two-dimensional photonic lattice: theory and experiment. Stud. Appl. Math. **113**, 389 (2004)

Explicit Method for the Numerical Solution of the Fokker-Planck Equation of Filtered Phase Noise

Dejan Milić

1 Introduction

Phase noise may impair seriously a number of communication systems ranging from wireless transmission systems to coherent optical communications. The Fokker-Planck (FP) equation for the phase-noisy lightwave has been introduced by Foschini and Vannucci [1] as a rigorous method for the analysis of optical systems impaired by phase noise. A numerical solution of the equation, with integrator-filter impulse response, has been reported by Garret et al. in [2]. The authors use the operator-splitting technique with a flux-limiting adaptive scheme in radial steps to obtain very accurate results that showed excellent agreement with numerical simulation. Other methods for solving the equation have also been investigated (for example [3, 4]), with an aim to simplify the computation without compromising the accuracy. However, the majority of papers focus on the integrator-filter impulse response, and the methods are not applicable to the general case without some modifications.

In general, explicit difference schemes are significantly simpler and require less effort to implement. On the other hand, their stability is often poor and thus is the limiting factor for their use. Implicit schemes are more stable than the explicit ones, but they are more complex and more demanding. In this paper, we present a stable explicit scheme that is based on an integral representation. At the same time, it is readily applicable to a more general case with arbitrary bounded impulse response of the filter.

Dejan Milić
Faculty of Electronic Engineering, Department of Telecommunications, University of Niš, 18000 Niš, Serbia, e-mail: dmilic@elfak.ni.ac.rs

W. Gautschi et al. (eds.), *Approximation and Computation: In Honor of Gradimir V. Milovanović*, Springer Optimization and Its Applications 42,
DOI 10.1007/978-1-4419-6594-3_26,

2 Filtering Model

The response of a filter with arbitrary equivalent impulse response $h(t)$, to a phase-noisy input pulse with duration T, at any given time τ is represented by [2]:

$$z(\tau) = \int_0^T h(\tau - x)\mathrm{e}^{\mathrm{j}\varphi(x)}\,\mathrm{d}x, \quad \mathrm{j} = \sqrt{-1}.$$

The phase-noise $\varphi(t)$ is considered a continuous-path Brownian motion process with zero mean and variance $E(\varphi^2(t)) = 2\pi\Delta\nu t$, with $\Delta\nu$ being the laser linewidth [1]. For the rest of this paper, we will consider filters with bounded impulse response whose maximum duration is T, although the results may easily be extended to more general cases.

For the considered class of filters, the response is represented by:

$$z(\tau) = \int_0^\tau h(\tau - x)\mathrm{e}^{\mathrm{j}\varphi(x)}\,\mathrm{d}x,$$

having in mind that $h(\tau) = 0$, for $\tau < 0$ or $\tau > T$. By substituting $\tau - x = y$, we get

$$z(\tau) = \int_0^\tau h(y)\mathrm{e}^{\mathrm{j}\varphi(\tau - y)}\,\mathrm{d}y.$$

If we assume the initial phase to be zero, $\varphi(0) = 0$, without loss of generality, the phase process $\varphi(\tau - y)$ will have same statistical properties as the process $\varphi(\tau) - \varphi(y)$, and the filter response will be statistically equivalent to [2,5]:

$$s(\tau) = \int_0^\tau h(y)\mathrm{e}^{\mathrm{j}(\varphi(\tau) - \varphi(y))}\,\mathrm{d}y.$$

Now, we evaluate this integral numerically using, for example, the trapezoidal rule. Consider the division of the bit interval into N equal segments, each one of length $\Delta\tau = T/N$, so that $\tau_i = i\Delta\tau$. Using the trapezoidal rule, we have

$$\widehat{s}(T) = \widehat{s}_N = \Delta\tau\mathrm{e}^{\mathrm{j}\varphi(\tau)}\sum_{i=0}^{N-1} h\left(i\frac{T}{N}\right)\mathrm{e}^{-\mathrm{j}\varphi(iT/N)} = \Delta\tau\mathrm{e}^{\mathrm{j}\varphi_N}\sum_{i=0}^{N-1} h_i\mathrm{e}^{-\mathrm{j}\varphi_i}.$$

We denote $\widehat{s}_k$ to be the approximate value of the function z at time $k\Delta\tau$, using the trapezoidal rule with exactly k segments. It should be noted that k relates to both – the time and the number of segments in the trapezoidal rule. At an arbitrary moment $k\Delta\tau$, the trapezoidal rule gives

$$\widehat{s}_k = X_k + \mathrm{j}Y_k = \Delta\tau\mathrm{e}^{\mathrm{j}\varphi_k}\sum_{i=0}^{k-1} h_i\mathrm{e}^{-\mathrm{j}\varphi_i}.$$

It is easily verified that the relation of approximate function values at two consecutive moments, $k\tau$ and $(k+1)\tau$, is

$$\widehat{s}_{k+1} = \mathrm{e}^{\mathrm{j}(\varphi_{k+1}-\varphi_k)}(\widehat{s}_k + h_k \Delta\tau). \tag{1}$$

If we let $N \to \infty$, we get the exact result reported by Garret et al. [2]. The exact approach is also useful for obtaining the exact moments [6, 7] of the process in the case of integrator-filter impulse response where $h(t) = 1/T$, $0 \le t \le T$. In general, the exact moments method is limited to symbolic manipulation, which effectively excludes arbitrary impulse response of the filter. However, (1) can be used to construct a relatively simple method for describing the process dynamics. Moreover, by using sufficiently large N, the results become very close to the solution of the original FP equation, effectively providing a new method for its numerical solution.

3 Application of FP Equation

The simplified model of (1) allows the joint probability density function (pdf) of real and imaginary parts of $\widehat{s}_{k+1}$, to be determined using the known pdf of $\widehat{s}_k$. Denote the phase difference in (1), between the ends of two consecutive segments k and $k+1$, to be: $\Delta\varphi_{k+1} = \varphi_{k+1} - \varphi_k$. Being the phase drift over time T/N, the phase difference is a zero-mean Gaussian variable with variance $\sigma^2 = E(\Delta\varphi_{k+1}^2) = 2\pi\Delta\nu T/N$.

A set of equations that uniquely map real and imaginary parts X, Y of $\widehat{s}_k$, and phase difference $\Delta\varphi_{k+1}$, to the real and imaginary parts of $\widehat{s}_{k+1}$ is

$$\begin{aligned}
X_{k+1} &= (X_k + u_k\Delta\tau)\cos\Delta\varphi_{k+1} - (Y_k + v_k\Delta\tau)\sin\Delta\varphi_{k+1},\\
Y_{k+1} &= (X_k + u_k\Delta\tau)\sin\Delta\varphi_{k+1} + (Y_k + v_k\Delta\tau)\cos\Delta\varphi_{k+1},\\
Z &= \Delta\varphi_{k+1},
\end{aligned}$$

where u_k and v_k represent the real and imaginary parts of the impulse response h_k. The Jacobian of the transformation is

$$|J| = \left|\frac{D(x_{k+1}, y_{k+1}, \Delta\varphi_{k+1})}{D(x_k, y_k, z)}\right| = \begin{vmatrix} \cos\Delta\varphi_{k+1} & -\sin\Delta\varphi_{k+1} & -y_{k+1} \\ \sin\Delta\varphi_{k+1} & \cos\Delta\varphi_{k+1} & x_{k+1} \\ 0 & 0 & 1 \end{vmatrix} = 1.$$

Thus, the joint pdf functions p_k and p_{k+1} satisfy the relation:

$$p_{k+1}(x_{k+1}, y_{k+1}, z) = p_k(x_k, y_k, \Delta\varphi_{k+1}).$$

By eliminating the phase difference z from the last result, we get

$$p_{k+1}(x_{k+1}, y_{k+1}) = \int p_k(x_k, y_k, z)\,\mathrm{d}z.$$

As $\widehat{s}_{k+1}$ and $\Delta\varphi_{k+1}$ are mutually independent random variables, the last equation can be simplified to:

$$p_{k+1}(x_{k+1},y_{k+1}) = \int p_k(x_k,y_k)p_{\Delta\varphi}(z)\,\mathrm{d}z.$$

For the inverse mapping, we have

$$\begin{aligned} X_k &= X_{k+1}\cos z + Y_{k+1}\sin z - u_k\Delta\tau, \\ Y_k &= -X_{k+1}\sin z + Y_{k+1}\cos z - v_k\Delta\tau. \end{aligned}$$

Finally, the FP equation of the process $\widehat{s}_k$ is

$$p_{k+1}(x,y) = \int p_k(x\cos z + y\sin z - u_k\Delta\tau, -x\sin z + y\cos z - v_k\Delta\tau)p(z)\,\mathrm{d}z. \quad (2)$$

The result may also be stated in the form of a line integral,

$$p_{k+1}(x_0,y_0) = \oint_L p_k(x,y)p_{2\pi}\left(\frac{l-l_0}{r}\right)\frac{\mathrm{d}l}{r}, \quad (3)$$

where the contour of integration L is a circle with radius $r = \sqrt{r^2 + r_0^2}$ and center at $(-u_k\Delta\tau, -v_k\Delta\tau)$; the point $l_0 = (x_0 - u_k\Delta\tau, y_0 - v_k\Delta\tau)$ represents the point where the pdf $p_{2\pi}(z)$ has a maximum (see Fig. 1 (left)). As the pdf $p(z)$ is Gaussian, and therefore defined on the interval $(-\infty,\infty)$, (3) formally holds for the pdf $p_{2\pi}(z)$ of the phase z folded to the interval $(-\pi,\pi)$.

4 Numerical Procedure

Numerical integration of (2) can be effectively reduced to an application of the Gauss-Hermite quadrature rule with sufficiently high order of algebraic accuracy. In general, the rule is given by:

$$\frac{1}{\sqrt{2\pi}\sigma}\int_{-\infty}^{\infty} f(x)\mathrm{e}^{-x^2/(2\sigma^2)}\,\mathrm{d}x = \frac{1}{\sqrt{\pi}}\sum_{k=1}^{n} A_k f(\sqrt{2}\sigma x_k) + R_n(f).$$

The values x_k are the zeros of the nth degree Hermite polynomial H_n, and the coefficients A_k are given by:

$$A_k = \frac{2^{n-1}(n-1)!\sqrt{\pi}}{nH_{n-1}(x_k)^2}, \quad k = 1,\ldots,n. \quad (4)$$

The remainder can be estimated by:

$$R_n(f) = \frac{n!\sqrt{\pi}}{2^n(2n)!} f^{(2n)}(\zeta), \quad -\infty < \zeta < \infty.$$

Finally, after neglecting the remainder, the numerical solution of the Fokker-Planck equation is reduced to an application of the following formulae:

$$p_{k+1}(x,y) = \frac{1}{\sqrt{\pi}} \sum_{i=1}^{n} A_i p_k(x_i,y_i), \tag{5}$$

$$x_i = x\cos(\sqrt{2}\sigma z_i) + y\sin(\sqrt{2}\sigma z_i) - u_k \Delta\tau, \tag{6}$$
$$y_i = -x\sin(\sqrt{2}\sigma z_i) + y\cos(\sqrt{2}\sigma z_i) - v_k \Delta\tau,$$

where z_i are the zeros of the nth degree Hermite polynomial H_n, and the coefficients A_i are given by (4). One starts the numerical procedure by specifying a mesh in the (x,y) plane. Then, the initial condition $p_0(x,y)$ should be set to approximate the required initial condition at $t = 0$, namely $p(x,y) = \delta(x,y)$. After that, p_1, p_2, ... are found, respectively, by using (5). In general, the values x_i, y_i in (6) will not be at intersection points on the mesh, and in this case the appropriate values of the function $p_k(x_i,y_i)$ must be determined by means of interpolation.

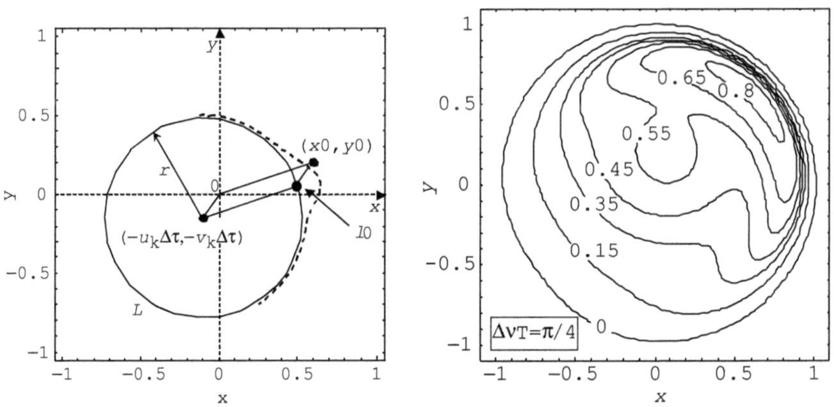

Fig. 1 Illustration of the contour of integration L. The *dashed line* indicates the pdf of the phase difference whose maximum is at the point l_0 (*left*). Contour plot of the numerical solution $p(x,y)$ of the FP equation for impulse response given in (8) (*right*)

Because of the property $|s(t)| \leq \int_0^t |h(t)|\,dt$, the solution is always confined within the disc of radius $\int_0^t |h(t)|\,dt$, thus suggesting the use of polar coordinates. After the formal changes $x = r\cos\theta$, $y = r\sin\theta$, we get

$$p_{k+1}(r,\theta) = \frac{1}{\sqrt{\pi}} \sum_{i=1}^{n} A_i p_k(r_i,\theta_i), \tag{7}$$

with

$$r_i = \sqrt{(x_i - u_k T/N)^2 + (y_i - v_k T/N)^2}, \quad \theta_i = \arctan \frac{y_i - v_k T/N}{x_i - u_k T/N},$$

where the following notation is used for simplicity,

$$x_i = r\cos\varphi_i, \quad y_i = r\sin\varphi_i, \quad \varphi_i = \theta - 2\sqrt{\frac{\pi\Delta v T}{N}} z_i.$$

By means of (7), the solution $p(x,y)$ can be found in polar coordinates. It should be noted that the solution does not represent the pdf $p(r,\theta)$. However, the pdf $p(r,\theta)$ is easily obtained by multiplying the solution by the Jacobian of the coordinate transformation, which in this case is r.

The pdf of the signal envelope $p_r(r)$, which is relevant in the performance analysis of telecommunication systems with envelope detection, may be obtained by:

$$p_r(r) = \oint_l p_{XY}(x,y)\,\mathrm{d}l,$$

where the contour of integration l is defined to be the circle $l : x^2 + y^2 = r$. The pdf of the signal phase $p_\theta(\theta)$ is obtained similarly, with the appropriate line of integration $l\colon y = x\tan\theta$.

5 Numerical Results

The method was applied in order to solve the FP equation for the case of detuned integrator-filter. The equivalent baseband impulse response that was used is

$$h(t) = \begin{cases} \exp(\mathrm{j}\pi t/(3T)), & 0 < t \leq T, \\ 0, & \text{elsewhere.} \end{cases} \tag{8}$$

This corresponds to an equivalent integrator-filter with frequency offset of $1/6$ of the bit rate. The joint pdf of the real and imaginary parts of a phase-noisy signal after the filter is shown in Fig. 1 (right). Initially, the number of segments in the time division is set to $N = 40$ in order to insure that the result is highly accurate and representative. For the same reason, an adaptive mesh in both the radial and angular directions is used. However, it has been observed that the result is practically indistinguishable from the result obtained with the number of segments set to $N = 20$. This indicates good accuracy of the model (1) and very good stability of the method. The solution of the FP equation for detuned filtering can be used to predict the system requirements regarding frequency stabilization and to provide bounds for parameter tolerances.

6 Conclusion

The presented method for the numerical solution of the Fokker-Planck equation is shown to be accurate and relatively simple to implement. Additionally, it is applicable to arbitrary bounded impulse response of the filter, without the need for significant modification. The method is well suited for use in the rigorous analysis of phase-noise influenced communication systems.

References

1. Foschini, G.J., Vannucci, G.: Characterizing filtered light waves corrupted by phase noise. IEEE Trans. Inf. Theory. **34**, 1437–1448 (1988)
2. Garret, I., Bond, D.J., Waite, J.B., Lettis, D.S.L., Jacobsen, G.: Impact of phase noise in weakly coherent optical systems: A new and accurate approach. IEEE J. Lightwave Technol. **8**, 329–337 (1990)
3. Zhang, X.: Analytically solving the Fokker-Planck equation for the statistical characterization of the phase noise in envelope detection. IEEE J. Lightwave Technol. **13**(8), 1787–1794 (1995)
4. Stefanović, M., Milić, D.: Comparison of certain methods for performance evaluation of coherent optical FSK systems. J. Opt. Commun. **20**(5), 183–187 (1999)
5. Jacobsen, G.: Noise in Digital Optical Transmission Systems. The Artech House Library, London (1994)
6. Pierobon, G.L., Tomba, L.: Moment characterization of phase noise in coherent optical systems. IEEE J. Lightwave Technol. **9**, 996–1005 (1991)
7. Monroy, I.T., Hooghiemstra, G.: On a recursive formula for the moments of phase noise. IEEE Trans. Commun. **48**(6), 917–920 (2000)

Numerical Method for Computer Study of Liquid Phase Sintering: Densification Due to Gravity-Induced Skeletal Settling

Zoran S. Nikolić

Dedicated to the 60-th anniversary of Professor Gradimir V. Milovanovic

1 Introduction

The phenomenon of liquid phase sintering (LPS) has been studied extensively not only because of its wide applicability to engineering materials but also because of the presence of a liquid phase simultaneously increases both the density of the resulting compacts and the rate of particle coarsening. During the last 50 years, many attempts [1, 5, 8, 13, 26, 27] have been made to elucidate the underlying kinetics and resultant microstructures of liquid phase sintering.

The influence of gravitational effects on grain coarsening during LPS is of both fundamental and practical interest in materials science. Settling of solid grains in a two-phase liquid–solid is a phenomenon common to several metallurgical processes including LPS, where solid–liquid segregation in liquid-phase sintered structures is related to the density difference between the solid and liquid phases. The experiments of Heaney et al. [10] have shown that the settled region was determined by the solid skeleton formation and is related to the density difference that is believed to influence grain growth and grain interaction, which alters the microstructure.

Theoretical analysis [12] has shown that in the LPS systems having significant density differences between solid and liquid, the mean coordination number is sensitive to gravity, and it varies systematically with the distance from the "top" of the specimen [14]. Therefore, it is of interest to measure the evolution of the coordination number distribution in the liquid phase sintered specimens processed in microgravity and normal gravity.

Zoran S. Nikolić
Faculty of Electronic Engineering, Department of Microelectronics, University of Niš, Aleksandra Medvedeva 14, 18000 Niš, Serbia
e-mail: zoran.nikolic@elfak.ni.ac.rs

W. Gautschi et al. (eds.), *Approximation and Computation: In Honor of Gradimir V. Milovanović*, Springer Optimization and Its Applications 42, DOI 10.1007/978-1-4419-6594-3_27,

Niemi and Courtney [16] were the first to document quantitatively this solid–liquid segregation phenomenon. They investigated the skeletal settling phenomenon and concluded that a solid skeleton formation prevents further settling when a critical solid-volume fraction is reached. As it was proposed by Courtney [3], the formation of solid skeleton is the result of interparticle collisions caused by Brownian motion and/or the density difference between solid and liquid.

Tewari and Gokhale [25] investigated the microstructure of a W–Ni–Fe alloy liquid-phase sintered in the microgravity environment of the space shuttle Columbia. Using the three-dimensional (3-D) reconstruction, they clearly revealed that, although the LPS was performed in the microgravity environment, the tungsten grains that make up the solid phase at the sintering temperature formed an almost completely connected skeletal network, and both the solid and liquid phases were co-continuous. They concluded that absence of gravity did not produce a microstructure consisting of discrete isolated tungsten grains uniformly dispersed in the liquid Ni–Fe alloy matrix at the sintering temperature.

The determination of time-dependent grain shape and grain size (i.e., grain coarsening), as well as that of coordination number distribution is presently the most challenging task in the application of solid skeleton (network) models for predicting the mass transport during LPS. Recently, results of computer simulation of skeletal settling combined with solid phase extrication during LPS in normal gravity [18] and in microgravity environments [19] have been reported.

In this paper we will be concerned with the formulation of a 3-D numerical method for simulation of microstructural evolution including densification due to gravity-induced 3-D skeletal settling during LPS. This method based on a 3-D domain topology (no shape restriction) and skeletal settling based on a settling procedure, will be developed through modification and generalization of previously defined two-dimensional methodology [20].

2 Simulation Model

For the simulation of LPS it is convenient to use multiparticle models of regular shape, because they need to store only the position, orientation and size of each particle. However, after a simulation time t ($t > 0$) most of the initially circular particles will no longer be circular because of the highly asymmetric diffusion field around and between them.

2.1 Initial Configuration

Generally speaking, it is very difficult to assume a specific particle shape. Even when all grains in a microstructure are convex and equiaxed, all of them are not spherical, or ellipsoidal, or of any specific shape, and some grains even have flat facets. Therefore, in this study we will use domains of arbitrary (convex, concave, or mixed) shape for 3-D particle representation.

To simulate the time–temperature dependent evolution of the domains during LPS (material microstructures are mostly 3-D in nature), the first step is to discretize a given cubic experimental space into small identical cubic elements, *voxels* (a 3-D equivalent of pixel), of finite size (i.e., finite resolution) defined by grid spacings Δx, Δy, and Δz of Cartesian coordinates x, y, and z, respectively. Now the domain is a volume that is equally divided into boundary and internal voxels halfway between adjacent grid points. Inasmuch as all voxels have the same shape and the same orientation, and if the mesh spacing is constant (i.e., $\Delta x = \Delta y = \Delta z = h$), the internal grid point (x_i, y_j, z_k) will be at the center of the voxel $V_{i,j,k}$. The discrete 3-D representation of domains is now defined as a set of voxels stacked along three orthogonal directions and subject to periodic boundary conditions (the experimental cube was surrounded by its translated self images), where the simulation domain is completely defined by its n_ℓ boundary voxels that lie at the surface of the domain, i.e.,

$$D^\ell = \{\ V^m_{i,j,k}\} = \{V^m(x_i, y_j, z_k)\} \quad (m = 1, 2, \ldots, n_\ell). \tag{1}$$

The advantage of this discretization method is that all elements have the same shape, size, number of neighbors, and share the same interface between them. Note that the resultant sharp surface of the domain (1) is composed of a series of small cubic elements (voxels), the volume of which can be made arbitrarily small by refining the underlying voxel grid.

A microstructure consisting of S solid-phase domains immersed in a liquid matrix can now be completely described by a binary phase matrix $\|e_{ijk}\|$, where the value of the element e_{ijk} indicates the phase present at $V_{i,j,k}$, so that

$$e_{ijk} \begin{cases} > 0 & \text{when } V_{i,j,k} \text{ belongs to the solid phase (i.e., } 1, 2, \ldots, S\text{)}, \\ = 0 & \text{when } V_{i,j,k} \text{ belongs to the liquid phase.} \end{cases}$$

Contiguous voxels of the same e_{ijk} form solid-phase domains. Solid-phase domain boundaries exist between neighboring voxels of different e_{ijk}, whereas solid-phase interfaces exist between neighboring voxels with $e_{ijk} > 0$ and liquid phase.

2.2 Settling Procedure

The settling procedure is a well-known simulation method in which solid-phase domains are subjected to a simulated gravity field: they fall under gravity over the already settled solid-phase domains (Model A). This procedure should be applied to each solid-phase domain starting with a domain having the lowest position in the vertical (z) direction inside the experimental space.

If and only if the new position of the ℓ-th domain undergoing settling, D_{T}^{ℓ}, is not already occupied by other solid-phase domains, i.e.,

$$D_{\mathrm{T}}^{\ell} \cap D^{s} = \emptyset \quad (s = 1,2,\ldots,S;\ \ell \neq s), \tag{2}$$

can the settling procedure be modeled by domain translation by a distance q along the gravitational (z) direction. The solid–liquid segregation problem in liquid-phase sintered structures is related to the density difference between the solid and liquid phases, $\Delta\rho = \rho_{\mathrm{S}} - \rho_{\mathrm{L}}$, where a density difference greater than zero causes the solid-phase domain to settle in the liquid phase, i.e.,

$$D^{\ell}(x_c^{\ell}, y_c^{\ell}, z_c^{\ell}) \mapsto D_{\mathrm{T}}^{\ell}(x_c^{\ell}, y_c^{\ell}, z_c^{\ell} - q), \tag{3}$$

where $(x_c^{\ell}, y_c^{\ell}, z_c^{\ell})$ is its center of mass and D_{T}^{ℓ} the translated domain topology. A density difference less than zero causes the solid-phase domains to float in the liquid phase, where the translation distance $(-q)$ in (3) will be replaced by $(+q)$. When $\Delta\rho$ is equal to zero, the solid-phase domains remain suspended in the liquid phase. Hereafter, we will consider only the case defined as LPS with $\Delta\rho > 0$.

However, the new settled domain can continue its motion over surfaces of domains, which have already settled, trying to reach a position of local equilibrium (Model B). If and only if condition (2) is fulfilled, can this procedure be modeled with combined translations in horizontal (x and y, sliding over settled domains) and in vertical (z, gravity) dircctions, i.e.,

$$D^{\ell}(x_c^{\ell}, y_c^{\ell}, z_c^{\ell}) \mapsto D_{\mathrm{T}}^{\ell}(x_c^{\ell} \pm q, y_c^{\ell} \pm q, z_c^{\ell} - q), \tag{4}$$

where the signs "+" or "−" can be taken at random. During this procedure, already settled domains hit by a settling domain can also move horizontally owing to accommodation with the arrived domain.

Let D^a and D^b be two solid-phase domains, the Euclidean distance between their centers

$$d(D^a, D^b) = \sqrt{(x_c^b - x_c^a)^2 + (y_c^b - y_c^a)^2 + (z_c^b - z_c^a)^2},$$

and their average effective domain radii

$$\langle r_a \rangle = \sqrt[3]{\frac{3}{4}\frac{D^a}{\pi}} \quad \text{and} \quad \langle r_b \rangle = \sqrt[3]{\frac{3}{4}\frac{D^b}{\pi}},$$

respectively. The separation distance between two domains is now

$$\lambda_{ab} = d(D^a, D^b) - (\langle r_a \rangle + \langle r_b \rangle),$$

thus the average separation distance will be

$$\langle \lambda \rangle = \langle \lambda_{ab} \rangle \quad (a, b = 1,2,\ldots,S;\ a \neq b). \tag{5}$$

If and only if $\Delta\rho > 0$, under gravity condition does Stokes's law settling usually dominate microstructure formation [3], in which the settling velocity can be calculated according to the following equation

$$v_{\mathrm{settl}} = \frac{2g\langle r\rangle^2(\rho_{\mathrm{s}} - \rho_{\mathrm{L}})}{9\eta}, \tag{6}$$

where g is the gravitational acceleration, $\langle r\rangle$ the average domain radius, and η is the liquid viscosity. Thus, the settling time for the ℓ-th solid-phase domain to travel the average separation distance $\langle\lambda\rangle$ between domains in a liquid matrix can be derived from (5) and (6),

$$\tau^{\ell}_{\mathrm{settl}} = \frac{9\eta\langle\lambda\rangle}{2g\langle r_\ell\rangle^2(\rho_{\mathrm{s}} - \rho_{\mathrm{L}})}.$$

For a given time interval Δt, the ℓ-th solid-phase domain should move a distance

$$q^{\ell}_{\Delta t} = \frac{\Delta t}{\tau^{\ell}_{\mathrm{settl}}}\langle\lambda\rangle \tag{7}$$

with the direction of movement being defined according to the selected model (A or B). Therefore, the previously defined gravity-induced domain translations (3) and (4) must be replaced by the time-dependent domain translation

$$D^{\ell}(x^{\ell}_c, y^{\ell}_c, z^{\ell}_c) \mapsto D^{\ell}_{\mathrm{T}}(x^{\ell}_c, y^{\ell}_c, z^{\ell}_c - q^{\ell}_{\Delta t}), \tag{8}$$

$$D^{\ell}(x^{\ell}_c, y^{\ell}_c, z^{\ell}_c) \mapsto D^{\ell}_{\mathrm{T}}(x^{\ell}_c \pm q^{\ell}_{\Delta t}, y^{\ell}_c \pm q^{\ell}_{\Delta t}, z^{\ell}_c - q^{\ell}_{\Delta t}). \tag{9}$$

The settling procedure defined by (8) or (9) starts with an ordered arrangement of finite-size solid-phase domains (i.e., initial randomly generated 3-D microstructure) in space, and then allows gravity-induced movements of these domains in the simulation space. A domain displacement is initiated by selecting (at random) the ℓ-th domain from S solid-phase domains and displacing it from its original position $D^{\ell}(x^{\ell}_c, y^{\ell}_c, z^{\ell}_c)$ by a trial distance (7). The trial displacement D^{ℓ}_{T} is accepted if and only if condition (2) is fulfilled, i.e., if it does not cause any two domains to overlap; otherwise it is rejected and a trial displacement of another solid-phase domain selected at random is taken. This procedure will be repeated for all (S) solid-phase domains. At the same time, the domains under the procedure (9) also are allowed to diffuse randomly in the simulation space. Such combined gravity induced and random domain movements, when repeated infinitely often (a very large number of times in real simulations), result in a microstructure that can be used (rather than (8)) for the theoretical investigation of gravity-induced LPS ordered structures.

2.3 Solid Skeleton Topology

Intergrain contacts exist in all liquid-phase sintered microstructures. German and Liu's theoretical analysis [7] has shown that such a connected grain structure is favored. Even more, Tewari and Gokhale [25] have used a 3-D reconstruction technique and have shown that in the liquid-phase sintered microstructure of a tungsten heavy alloy, a connected network of tungsten grains was produced even when LPS was performed in the microgravity environment. Therefore, it can be concluded that both, the normal gravity and microgravity environments, produce microstructures containing completely connected tungsten grains, although the mechanisms of evolution of grain connectivity are quite different in the two environments.

During LPS, solid-phase domains (grains) will be settled due to gravity. As the domains arrive at already settled domains, they make point contacts with each other and necks between them form. Neck growth will terminate when the equilibrium dihedral angle between the domain boundaries and the liquid is established. Thus, a solid skeleton forms.

A solid skeleton (*SS*) may be generally defined as a series of connected solid-phase domains (interconnected cluster of domains) arranged in a long chain, i.e.,

$$SS^{\ell}_{K(\ell)} = \bigcup_{k=1}^{K(\ell)} D^{\nu(\ell,k)} = \bigcup_{k=1}^{K(\ell)} \{V^{\nu(\ell,k)}\}, \tag{10}$$

where $K(\ell)$ and $\nu(\ell,k)$ are the vector of number of solid-phase domains included in the ℓ-th solid skeleton and the vector of their ordinal numbers, respectively. The initial microstructure consisting of S solid-phase domains immersed in a liquid matrix can now be completely described by S isolated solid skeletons of unit length using the definition (10), i.e.,

$$SS^s_1 = D^s = \{V^s\} \quad (s = 1,2,\ldots,S).$$

It is easy to understand that domains in a solid skeleton, according to their role in a set of connected domains, can be categorized into three kinds of domains: *End domain* (with exactly one neck), *Link domain* (with exactly two necks), and *Junction domain* (with more than two necks).

A skeleton network may be defined as a unique, interconnected set of connected solid-phase domains, where each domain included in the skeleton structure (10) has to have a set of data about the first-, second-, and higher-order nearest neighbors, i.e.,

$$D^{\nu(\ell,k)} = D^{\nu(\ell,k)}(m_1,\ldots,m_{k-1},m_{k+1},\ldots,m_{n_k}) \quad (k = 1,2,\ldots,K(\ell)),$$

where n_k and $m_1,\ldots,m_{n_k}$ are the number of neighboring solid-phase domains and their ordinal numbers for k-th domain, respectively. Alternatively, a skeleton network may be defined as a set of points (i.e., centers of mass of connected domains), some of which have exactly one neighbor (End points), whereas the remaining points are either points with exactly two neighbors (Link points) or points with more than two neighbors (Junction or nodal points). The skeleton network, given as a system of functions of some topological parameters, changes monotonically with time by adding new solid-phase domains during skeleton settling or by losing some smaller domains through dissolution. At the same time, solid skeleton domains will change their status from End to Link domain and from Link to Junction domain or vice versa.

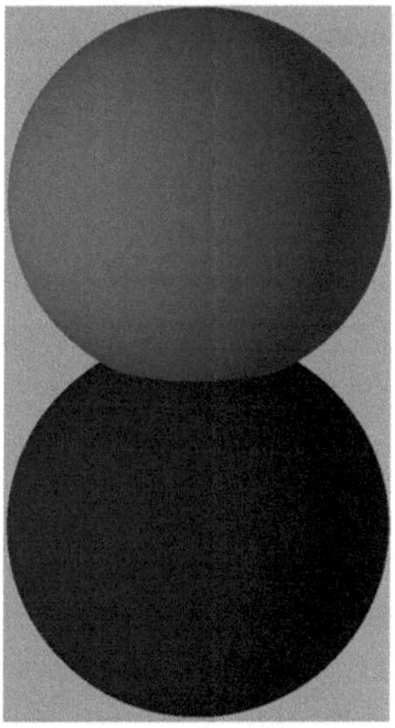

Fig. 1 Idealized – smooth model of two contacting domains with a neck between them

Let two solid-phase domains of type (1), D^a and D^b, be in contact through settling, as shown in Fig. 1. If both domains are discretized, their resultant equivalent sharp neck surface is composed of a series of small faces of contacting voxels.

V^a and V^b are said to be 6-adjacent if $d(V^a,V^b) = h$, and diagonally adjacent if $d(V^a,V^b) = h\sqrt{2}$. If the area of discrete neck surface (S_N), defined as the sum of the contacting faces (S_{CF}) of 6-adjacent and/or diagonally adjacent voxels, fulfills the inequality

$$S_N = \sum S_{CF} \geq S, \tag{11}$$

where S is the minimal neck surface area, then the domains D^a ($\equiv SS_1^a$) and D^b ($\equiv SS_1^b$) will form *a solid skeleton unit* consisting of two solid-phase domains with a discretized neck between them, where part of their boundaries are replaced by domain boundaries, i.e.,

$$SS_2^a = D^a \cup D^b = \{V^a\} \cup \{V^b\}, \quad SS_0^b = \emptyset.$$

The minimal surface S used in (11) is an empirical parameter.

During LPS some solid-phase domains included in different solid skeletons $SS^a_{K(a)}$ and $SS^b_{K(b)}$ can make new connections with each other that can be followed by neck formation. This change of current skeleton structure is defined by:

$$SS^a_{K(a)+K(b)} = \bigcup_{k=1}^{K(a)+K(b)} D^{v(a,k)\cup v(b,k)}, \quad SS_0^b = \emptyset.$$

As was proposed by Niemi and Courtney [16], the settling of the skeletal structure during LPS will be controlled by extrication of domains from the skeleton and their subsequent settling (as isolated domains) within the liquid phase, where such solid-phase domains may be extricated presumably by capillarity-driven mass transfer processes and the similarity-driven processes of Ostwald ripening. After extrication of some solid-phase domains the skeleton structure has to be replaced by at least two new solid skeletons. This skeleton structure transformation depends on domains undergoing extrication.

Let there be a solid skeleton $SS^s_{K(s),v(s,k)}$ in which ℓ-th solid-phase domain will be extricated, because the condition (11) is not fulfilled. The update of this skeleton depends on the type of the extricated domain D^ℓ:

If D^ℓ is an *End domain*, the s-th skeleton will be updated by excluding the domain D^ℓ, so that the vector $v(s,k)$ will be replaced by the vector $v'(s,k)$ $(\ell \notin v'(s,k))$ and $K(s)$ replaced by $K(s)-1$, i.e.,

$$SS^s_{K(s)} = \bigcup_{k=1}^{K(s)-1} D^{v'(s,k)}, \quad SS^\ell_1 = D^\ell.$$

If D^ℓ is a *Link domain*, the s-th skeleton will be divided into two new skeletons defined by corresponding topological parameters $(v'(s,k), K'(s))$ $(\ell \notin v'(s,k))$ and $(v''(s,k), K''(s))$ $(\ell \in v''(s,k))$, where $v(s,k) = v'(s,k) \cup v''(s,k)$ and $K(s) = K'(s) + K''(s)$.

If D^ℓ is a *Junction domain*, the s-th skeleton will be divided into several shorter skeletons defined with corresponding topological parameters $v(\bullet,k)$ and $K(\bullet)$.

Generally speaking, gravity-induced settling can be roughly separated into two stages [4]: (i) free settling of isolated solid-phase domains and (ii) skeletal settling of a connected solid structure. Free settling implies that isolated domains sink under gravity toward the experimental space bottom and slide down over the already settled domains. During their settling they can make contacts with other domains and form either new solid skeleton or be included in an exiting skeleton. Solid skeleton as a connected solid structure can also settle owing to the gravity force (skeletal settling). However, this complex time–temperature dependent settling will be realized by simultaneous translation (of type (8) or (9) depending on applied settling model) of each solid-phase domain included in the skeleton. Thus, if and only if the new position of the ℓ-th skeleton undergoing settling, $SS^\ell_{K(\ell)}\big|_T$, is not already occupied by other solid skeletons or isolated solid-phase domains (i.e., satisfying the extended condition (2)) can the movement of the ℓ-th skeleton be realized in a similar way by simultaneous (time-dependent) translation (according to selected model A or B) of all domains included in this skeleton, i.e.,

$$SS^\ell_{K(\ell)} = \bigcup_{k=1}^{K(\ell)} D^{v(\ell,k)}(x^\ell_c, y^\ell_c, z^\ell_c) \mapsto SS^\ell_{K(\ell)}\big|_T = \bigcup_{k=1}^{K(\ell)} D^{v(\ell,k)}(x^\ell_c, y^\ell_c, z^\ell_c - q^\ell_{\Delta t}),$$

$$SS^\ell_{K(\ell)} = \bigcup_{k=1}^{K(\ell)} D^{v(\ell,k)}(x^\ell_c, y^\ell_c, z^\ell_c) \mapsto SS^\ell_{K(\ell)}\big|_T = \bigcup_{k=1}^{K(\ell)} D^{v(\ell,k)}(x^\ell_c \pm q^\ell_{\Delta t}, y^\ell_c \pm q^\ell_{\Delta t}, z^\ell_c - q^\ell_{\Delta t}),$$

where $SS^{\ell}_{K(\ell)}\big|_T$ is the topology of the translated ℓ-th solid skeleton. The same sign "+" or "−" in $(x^{\ell}_c \pm q^{\ell}_{\Delta t})$ taken at random must be applied for translation of all domains included in the ℓ-th skeleton. The same methodology must be applied for generated sign in $(y^{\ell}_c \pm q^{\ell}_{\Delta t})$.

The solid skeleton gives rigidity to the structure that resists rearrangement, and the isolated domains can settle with restricted packing density. Even more, the solid skeleton can also densify through solid-skeleton densification. Generally speaking, densification through gravity-induced skeletal settling will be the result of settling of skeletons of unit length (isolated solid-phase domains) and settling of skeletons of different length $K(s)$.

It should be noted that Model B is very similar to the assumed scenario of Courtney [3], in which he proposed solid skeleton formation as a result of interparticle contacts owing to Brownian motion and/or settling owing to solid–liquid density difference.

2.4 Solution-Reprecipitation

We will assume that liquid penetrates the domain boundaries of the solid phase. Therefore, domains of solid phase are entirely surrounded by liquid, a thin film of which is present between neighboring solid-phase domains.

For a system consisting of a dispersion of (discretized) solid-phase domains in a liquid in which the solid phase has some solubility, the concentration of the dissolved solid, c, around a solid-phase domain of average effective radius $\langle r \rangle$ is given by the Gibbs-Thomson equation [6],

$$\ln\left(\frac{c}{c_o}\right) = \frac{2\gamma_{sl}\Omega}{kT}\frac{1}{\langle r \rangle}, \tag{12}$$

where c_o is the equilibrium concentration of liquid in contact with the flat solid, γ_{sl} the solid/liquid interfacial energy, Ω the molecular volume of the solid, kT has its usual meaning, and $\kappa = 1/\langle r \rangle$ is the local boundary curvature. This equation, that relates the concentration of solute at a matrix/particle boundary to the interfacial curvature, is the starting point for almost all the numerical treatments of precipitate coarsening. If $\Delta c = c - c_o$ is small, the governing (12) becomes

$$\Delta c = c_o \frac{2\gamma_{sl}\Omega}{kT}\frac{1}{\langle r \rangle}. \tag{13}$$

It can be seen that LPS is driven primarily by local differences in curvature throughout the microstructure. Therefore, the simulation model for solution-reprecipitation that operates on digitized microstructure requires a numerical method for computing curvature along digitized surfaces. Thus, the most important numerical consideration in performing an accurate computation is the determination of the local curvatures of a domain for all boundary voxels.

The mean curvature plays an important role in governing the thermodynamics of interfacial phenomena. For an infinitesimal element along a condensed phase interface the mean curvature can be defined as a ratio $\delta A/\delta V$, where δA is the incremental change in the element's area when it is normally displaced by local addition of material of volume δV [23]. For a discretized domain structure a better way would be to calculate the local curvature numerically, for example, using the interpolation functions at each boundary voxel separately. If one needs a smooth function for describing the domain boundary, then a cubic polynomial could be the simplest function of this type, as recognized by Saetre and Ryum [22] and applied by Cocks and Gill [2]. To estimate the local interfacial curvature for each boundary voxel of a solid-phase domain, the following construction will be used. The curvature at each boundary voxel lying on a domain/matrix interface (as continuous 3-D space) will be computed by fitting a quadratic polynomial to the point and its two neighbors (Fig. 2). Then the value of the local curvature can be used to determine the interfacial composition through the Gibbs-Thomson equation (12). Notice that a sharp curvature on the domain boundary requires a very fine mesh. When curvature effects were simulated, the nodal concentrations at the domain/matrix interfaces were updated at the end of each step in time according to (12).

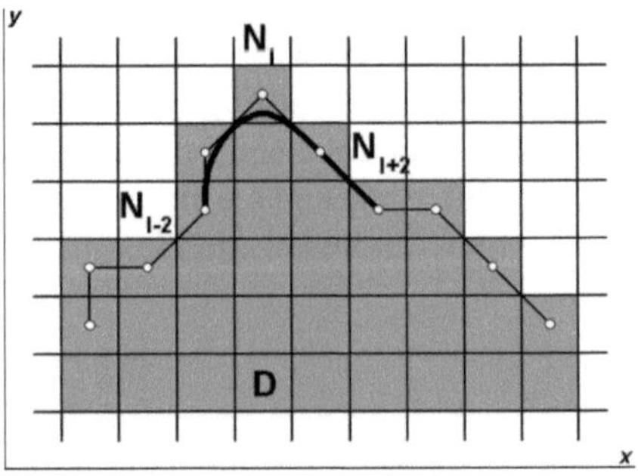

Fig. 2 Schematic illustration of geometrical method used to estimate the local interfacial curvature along the surface of a digitized solid-phase domain in 2-D

It can be seen from (13) that the concentration at an interface solid/liquid with high curvature will be above that at an interface with low curvature; thus a higher concentration around a smaller solid-phase domains gives rise to a net flux of matter from the smaller to the larger one. If there is an ensemble of domains with a size variation, it is therefore expected that larger domains will grow at the expense of smaller ones. In the process, domain coarsening occurs by mass transport through the liquid surrounding domains, with a corresponding driving force determined by the solubility of the solid-phase domains. If D_L is the concentration-independent diffusivity of the solid in the liquid, then the flux vector is $\mathbf{J} = -D_L \nabla C$. The 3-D flux for the boundary voxel $V_{i,j,k}$ from the solid-phase toward the liquid phase (through dissolution) and/or vice versa (through precipitation) is now

$$J_{i,j,k} = J^x_{i,j,k} + J^y_{i,j,k} + J^z_{i,j,k},$$

where the flux components along the x, y, and z directions are given by:

$$J^x_{i,j,k} = -D_L \cdot \left(\frac{c_{i,j,k} - c_{i+1,j,k}}{\Delta z} \cdot \delta_{i+1,j,k} + \frac{c_{i,j,k} - c_{i-1,j,k}}{\Delta z} \cdot \delta_{i-1,j,k} \right), \tag{14}$$

$$J^y_{i,j,k} = -D_L \cdot \left(\frac{c_{i,j,k} - c_{i,j+1,k}}{\Delta z} \cdot \delta_{i,j+1,k} + \frac{c_{i,j,k} - c_{i,j-1,k}}{\Delta z} \cdot \delta_{i,j-1,k} \right),$$

$$J^z_{i,j,k} = -D_L \cdot \left(\frac{c_{i,j,k} - c_{i,j,k+1}}{\Delta z} \cdot \delta_{i,j,k+1} + \frac{c_{i,j,k} - c_{i,j,k-1}}{\Delta z} \cdot \delta_{i,j,k-1} \right),$$

where the function

$$\delta_{i+1,j,k} = \begin{cases} 1, & \text{when } e_{ijk} \text{ belongs to the solid phase and} \\ & e_{i+1,j,k} \text{ to the liquid phase,} \\ 0, & \text{when } e_{ijk} \text{ and } e_{i+1,j,k} \text{ belong to the solid phase,} \end{cases} \tag{15}$$

provided that the flux (14) will be computed between solid and liquid phases and not between solid-phase domains within the neck region. The rest of the δ functions can be defined in a similar way as in (15).

Domain topology is also concerned with geometrical change, because during sintering, solid-phase domains co-evolve: grow and/or shrink. This means that the model topology is time–temperature dependent. Finally, the solid-phase evolution due to gravity-induced settling (defined by displacement of the center of mass) and due to mass transport, which takes place by dissolution and/or precipitation at the interfaces between solid-phase domains and matrix (defined by change in average effective domain radius), will be modeled by transformations

$$D^\ell\{(x^\ell_c(t),\, y^\ell_c(t),\, z^\ell_c(t)),\, \langle r_\ell(t)\rangle\} \mapsto D^s\{(x^\ell_c(t),\, y^\ell_c(t),\, z^\ell_c(t) - q^\ell_{\Delta t}),\, \langle r_\ell(t+\Delta t)\rangle\},$$

$$D^\ell\{(x^\ell_c(t),\, y^\ell_c(t),\, z^\ell_c(t)),\, \langle r_\ell(t)\rangle\} \mapsto$$
$$D^\ell\{(x^\ell_c(t) \pm q^\ell_{\Delta t},\, y^\ell_c(t) \pm q^\ell_{\Delta t},\, z^\ell_c(t) - q^\ell_{\Delta t}),\, \langle r_\ell(t+\Delta t)\rangle\},$$

for models A and B, respectively.

3 Result and Discussion

For the simulation of gravity-induced skeletal settling during LPS of tungsten heavy alloy we will use the numerical models defined earlier, as well as the previously defined simulation method for grain growth by grain boundary migration during LPS [17]. For simulations, we will use periodic microstructures to minimize edge effects in the final results. In this calculation the following data will be used: the equilibrium concentration of liquid in contact with the solid: 35 at.% W [9]; the diffusion coefficient in the liquid W–Ni alloy is estimated within an order of magnitude to be about $10^{-9}\,\mathrm{m^2s^{-1}}$ [21]; the sintering temperature: 1750 K; liquid viscosity: $5 \cdot 10^{-3}$ Pas (for liquid nickel); the acceleration: $9.81\,\mathrm{ms^{-2}}$; the liquid–solid density difference: $9\,\mathrm{gcm^{-3}}$, and the interfacial energy $0.8\,\mathrm{Jm^{-2}}$ [11].

For simulation we will apply an initial randomly generated 3-D model containing non-uniformly distributed domains of different radii with all domain centers placed in a plane. This model was obtained by applying a random generation method given elsewhere [17]: the domains (of regular or arbitrary shapes) are placed in a sequence, one at a time, at random coordinates in the simulated volume (experimental space) without overlapping any previously placed domains. If an overlap (in the simulation volume) is detected, then the random coordinate is discarded and another random coordinate is chosen, and the process repeated. All domains of the computer-generated microstructure model (that contains 72 spherical tungsten domains with a mean size of 34 μm, with the largest domain being 54 μm, and the smallest one 19 μm; area fractions of solid and liquid in a plane across the domains' centers were 75% and 25%, respectively) are immersed in liquid matrix (liquid Ni) inside the experimental space (Fig. 3). For simulation of microstructural evolution we will apply translation Model B, because it seems that this model allows for better densification. Note that Courtney [3] has also proposed solid skeleton formation as a result of interparticle contacts due to Brownian motion.

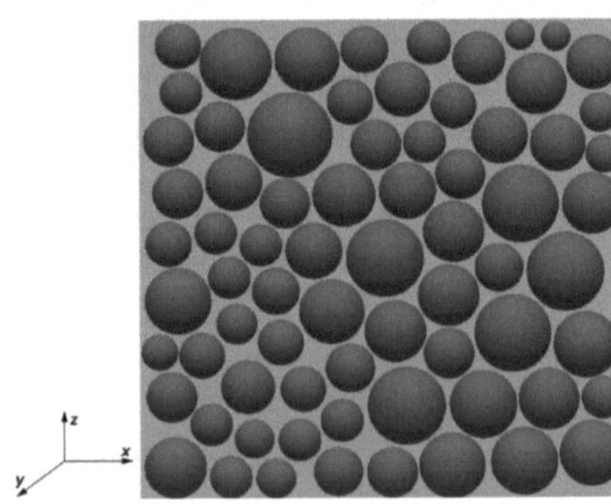

Fig. 3 Initial microstructure (smooth model instead of discretized model) with randomly distributed solid-phase domains. Domains are *dark-gray* colored and liquid is *light-gray* colored

During the initial stage, the liquid thickness between solid-phase domains remains nearly constant, because dissolution and precipitation simultaneously take place over short distances. The smaller dissolving domains give way to a new packing of small and large domains before the necks between contacting domains will be formed. The smaller domains tend to be preferentially located near the large domains, as suggested in [15]. Computed microstructures showed that after a very short time, pure settling of isolated domains was almost finished with already formed necks between contacting domains. Due to combined translation (Model B), some domains have formed bonds with neighboring domains prior to finishing complete settling and producing chain-like clusters. This means that densification inside these regions was stopped, although further densification through solid-skeleton densification was possible, too. Figure 4a shows 3-D computer-simulated microstructural evolution of liquid phase sintered W–Ni after 30 min. Note that even for short sintering time (30 min), many solid-phase domains are connected, forming three solid skeletons. It can be seen that only shorter skeletons (domain clusters) can move according to the model (Model B) inside the experimental region. Even more, there are also few still isolated domains that can move between skeletons and/or isolated domains.

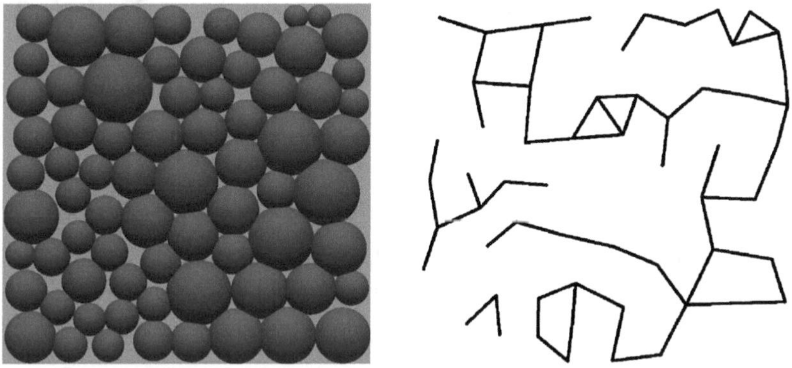
(a) (b)

Fig. 4 Computer-simulated microstructure of liquid phase sintered W–Ni after 30 min. (a) 3-D model (*Dark-gray* colored are domains and *light-gray* colored is liquid.). (b) Skeleton network with three solid skeletons

Figure 5 shows skeleton network evolutions with the skeletons of different lengths. After 60 min (Fig. 5a) the microstructure consists of one short skeleton and one very long skeleton characterized by complex skeletal structure, whereas after 120 min (Fig. 5b) the microstructure is fully densified, with already finished free movement of isolated solid-phase domains. Due to combined domain displacement, some domains have formed bonds with neighboring domains prior to finishing complete displacement and producing (very long) chain-like clusters. Such a skeleton structure was not able to settle due to geometric hindrance of adjacent domains. Therefore, densification inside the experimental region was almost stopped apart from some local densification around some isolated domains.

Generally speaking, our computed microstructures obtained have shown that the settled solid-volume fraction is dictated by the formation of a solid skeleton and can be directly related to the solid–liquid density difference. This agrees with the results of Tewari et al. [24], in which they investigated 3-D microstructures of tungsten grains in 83 wt% W–Ni–Fe alloy, liquid phase sintered for 1 and 120 min. The microstructure obtained revealed that in these specimens, processed in normal gravity, the tungsten grains were completely connected, as expected, since the density of tungsten grains is much higher than that of the liquid matrix, and therefore, the gravitational settling anchors the grains and leads to the formation of a completely connected stable grain network. Even more, they concluded that both the normal gravity and microgravity environments produce microstructures containing com-

Fig. 5 Computer-simulated skeleton network after (a) 60 min (two solid skeletons), and (b) 120 min (one solid skeleton)

pletely connected tungsten grains, although the mechanisms of evolution of grain connectivity are quite different in the two environments.

4 Conclusion

In this paper we have investigated numerically gravity-induced densification during LPS. The topological analysis was accomplished by classifying the nodal points (voxels) into solid-phase domains and solid skeletons in 3-D discrete space. Solid skeleton formation due to domain settling was introduced by the formation of skeleton units and their evolution to a large solid skeleton of connected skeleton units arranged in a long chain. For the settling procedure, two submodels were defined: pure settling model (Model A), in which solid-phase domains fall under gravity over already settled domains, and the modified (extended) model (Model B), in which settled domains continue their motion till they reach a position of local equilibrium. Since it was assumed that under gravity condition, Stokes's law settling usually dominates microstructure formation, the settling time was used for computation of average migration distance during a time interval Δt. Thus, the solid-phase evolution due to gravity-induced settling was simulated by the computation of displacement of the center of mass and the computation of mass transport due to dissolution and precipitation at the interfaces between solid-phase domains and liquid matrix.

The new methodology is illustrated by an application to a regular multidomain model. Our computer-simulated microstructures substantiate previous observations that the settled solid-volume fraction can be directly related to the solid–liquid density difference but is dictated by the formation of a solid skeleton. Although the solid skeleton can also migrate and densify through solid-skeleton densification, liquid–solid separation at the top of experimental space was not so large because of the formation of very long length skeletons with very few isolated solid-phase domains inside or outside of them. Such skeletons were not able to settle due to geometric hindrance of adjacent domains.

This approach is general and can be applied for both regular and irregular solid-phase and pore-phase domain shapes. One of the advantages of the developed model and the applied domain topology is the ability to simulate evolution of arbitrary microstructures that approximate the kinds of real powder compacts used in sintering experiments. The complex geometry of real microstructures can be incorporated in the simulation model simply by embedding a digital microstructural image (field of view) in it, i.e., by replacing the binary phase matrix with a modified (recoded) digital microstructural image.

An efficient and rigorous methodology for the geometric and topological characterization of arbitrarily complex domain microstructures is expected to play an important role in elucidating the relationships among the geometric and topological attributes of liquid phase sintered materials. A systematic investigation is currently underway, and the results will be presented in a future publication.

Acknowledgements This work was performed under the project No. 142011G supported financially by the Ministry of Science and Technological Development of the Republic of Serbia.

References

1. Ardell, A.J.: The effect of volume fraction on particle coarsening: Theoretical considerations. Acta Metall. **20**, 61–71 (1972)
2. Cocks, A.C.F., Gill, S.P.: A variational approach to two dimensional grain growth - I. Theory. Acta Mater. **44** [12], 4765–4775 (1996)
3. Courtney, T.H.: Microstructural evolution during liquid phase sintering: Part I. Development of microstructure. Metall. Trans. 8A, 679–684 (1977)
4. Courtney, T.H.: Gravitational effects on microstructural development in liquid phase sintered materials. Scripta Mater. **35** [5], 567–571 (1996)
5. Dehoff, R.T.: A geometrically general theory of diffusion controlled coarsening. Acta Metall **39** [10], 2349–2360 (1991)
6. Freundlich, H.: Kapillarchemie. Leipzig (1922)
7. German, R.M., Liu, Y.: Grain agglomeration in liquid phase sintering. J. Mater. Sci. Engng. **4**, 23–34 (1996)
8. Greenwood, G.W.: The growth of dispersed precipitates in solutions. Acta Metall. **4**, 243–248 (1956)
9. Hansen, M., Anderko, K.: Constitution of Binary Alloys. McGraw-Hill, New York (1958)
10. Heaney, D.F., German, R.M., Ahn, I.S.: The gravitational effects on low solid-volume fraction liquid-phase sintering. J. Mater. Sci. **30**, 5808–5812 (1995)
11. Huppmann, W.J., Petzow, G.: The role of grain and phase boundaries in liquid phase sintering. Ber. Bunsenges. Phys. Chem. **82**, 308–312 (1978)
12. Kipphut, C.M., Bose, A., Farooq, S., German, R.M.: Gravity and configurational energy induced microstructural changes in liquid phase sintering. Metall. Trans. A, 19A, 1905–1913 (1988)
13. Lifshitz, I.M., Slyozov, V.V.: The kinetics of precipitation from supersaturated solid solutions. J. Phys. Chem. Solids. **119** [12], 35–50 (1961)
14. Liu, Y., Heaney, D.F., German, R.M.: Gravity induced solid grain packing during liquid phase sintering. Acta Metall. Mater. **43**, 1587–1592 (1995)

15. Marder, M.: Correlations and droplet growth. Phys. Rev. Lett. **55**, 2953–2956 (1985)
16. Niemi, A.N., Courtney, T.H.: Settling in solid–liquid systems with specific application to liquid phase sintering. Acta Metall. **31** [9], 1393–1401 (1983)
17. Nikolić, Z.S.: Computer simulation of grain growth by grain boundary migration during liquid phase sintering. J. Mater. Sci. **34** [4], 783–794 (1999)
18. Nikolić, Z.S.: Liquid phase sintering – I. Computer study of skeletal settling and solid phase extrication in normal gravity environment. Science of Sintering **40**, 3–12 (2008)
19. Nikolić, Z.S.: Liquid phase sintering – II. Computer study of skeletal settling and solid phase extrication in microgravity environment. Sci. Sintering **40**, 107–116 (2008)
20. Nikolić, Z.S.: Numerical simulation of gravity induced skeletal settling during liquid phase sintering. Math. Comp. Model. **51**, 1146–1153 (2010)
21. Ono, Y., Shigematsu, T.: Diffusion of Vanadium, Cobalt, and Molybdenum in molten Iron. J. Jpn. Inst. Met. **41**, 62–68 (1977)
22. Saetre, T.O., Ryum, N.: Dynamic simulation of grain boundary migration. J. Sci. Comp. **3**, 189–199 (1988)
23. Taylor, J.E.: II-mean curvature and weighted mean curvature. Acta Metall. Mater. **40** [7], 1475–1485 (1992)
24. Tewari, A., Gokhale, A.M., German, R.M.: Effect of gravity on three-dimensional coordination number distribution in liquid phase sintered microstructures. Acta Mater. **47** [13], 3721–3734 (1999)
25. Tewari, A., Gokhale, A.M.: Application of three-dimensional digital image processing for reconstruction of microstructural volume from serial sections. Mater. Charact. **44**, 259–269 (2000)
26. Voorhees, P.W., Glicksman, M.E.: Solution to the multi-particle diffusion problem with applications to Ostwald ripening – I. Theory. Acta Metall **32** [11] 2001–2011 (1984)
27. Wagner, C.: Theorie der Alterung von Niederschlägen durch Umlösen (Ostwald-Reifung). Z Elektrochem. **65**, 581–591 (1961)

Computer Algebra and Line Search

Predrag Stanimirović, Marko Miladinović, and Ivan M. Jovanović

Dedicated to Professor Gradimir V. Milovanović
on the occasion of his 60th birthday

1 Introduction

An unconstrained minimization problem requires to find $\min Q(\mathbf{x})$, $\mathbf{x} \in \mathbb{R}^n$, where $\mathbb{R}^n$ is an n-dimensional Euclidean space and $Q : \mathbb{R}^n \to \mathbb{R}$ a given objective function. A general iterative scheme in multivariate optimization, based on line search, is given by:

$$\mathbf{x}^{(k+1)} = \mathbf{x}^{(k)} + h_k \mathbf{s}^{(k)}, \tag{1}$$

where $\mathbf{x}^{(k+1)}$ is the new iterative point, $\mathbf{x}^{(k)}$ the previous iterative point, $\mathbf{s}^{(k)}$ a search direction, and h_k a positive step size. The key problem is to find the direction vector $\mathbf{s}^{(k)}$ and a suitable step size h_k.

The search direction $\mathbf{s}^{(k)}$ is generally required to satisfy the descent condition

$$\nabla Q(\mathbf{x}^{(k)})^T \mathbf{s}^{(k)} < 0,$$

where the $n \times 1$ vector $\nabla Q(\mathbf{x}^{(k)})$ is the gradient of Q at $\mathbf{x}^{(k)}$.

Predrag Stanimirović
Faculty of Science, Department of Mathematics, University of Niš, Višegradska 33, 18000 Niš, Serbia, e-mail: pecko@pmf.ni.ac.rs

Marko Miladinović
Faculty of Science, Department of Mathematics, University of Niš, Višegradska 33, 18000 Niš, Serbia e-mail: markomiladinovic@gmail.com

Ivan M. Jovanović
Technical Faculty, University of Belgrade, Vojske Jugoslavije 12, Bor, Serbia
e-mail: jmivan@ptt.rs

W. Gautschi et al. (eds.), *Approximation and Computation: In Honor of Gradimir V. Milovanović*, Springer Optimization and Its Applications 42,
DOI 10.1007/978-1-4419-6594-3_28,

The general form of an unconstrained optimization algorithm is as follows [20].

Algorithm 1 Input. *Given an initial point* $\mathbf{x}^{(0)} \in \mathbb{R}^n$ *and a tolerance* $0 \leq \varepsilon \ll 1$.

Step 1. (Verify termination criterion) *If* $\|\nabla Q(\mathbf{x}^{(k)})\| \leq \varepsilon$, *stop.*
Step 2. (Finding the direction) *Find a vector* $\mathbf{s}^{(k)}$ *which is a descent direction.*
Step 3. (Line search) *Determine the step size* h_k *such that the objective function value decreases, i.e.,*

$$Q(\mathbf{x}^{(k)} + h_k\mathbf{s}^{(k)}) < Q(\mathbf{x}^{(k)}).$$

Step 4. (Loop) *Compute the new approximation* $\mathbf{x}^{(k+1)}$ *according to (1), set* $k := k+1$, *and go to* **Step 1**.

One can think of various multidimensional optimization methods arising from Algorithm 1. These methods will differ only by how, at each stage, they choose the next direction $\mathbf{s}^{(k)}$ and step size h_k.

The exact line search (ELS) is defined as the following special case of **Step 3** from Algorithm 1 [19,20].

Step 3-ELS. (Exact line search)

(a) *Generate an unevaluated expression of the form*

$$F(h) = Q(\mathbf{x}^{(k)} + h\mathbf{s}^{(k)}). \tag{2}$$

(b) *Compute the step size* h_k *from the following univariate optimization problem*:

$$h_k = \operatorname{argmin}_{h>0} F(h). \tag{3}$$

We also recall the well-known *Approximate minimization rule* (AMR) [19] as a special case of **Step 3** from Algorithm 1.

Step 3-AMR. (Approximate minimization rule). *At each iteration,* h_k *is selected so that*

$$h_k = \min\{h \mid F'(h) = 0, h > 0\} = \min\left\{h \;\middle|\; \nabla Q\left(\mathbf{x}^{(k)} + h\mathbf{s}^{(k)}\right)^T \cdot \mathbf{s}^{(k)} = 0,\ h > 0\right\}. \tag{4}$$

In some special cases (for example quadratic problems) it is possible to compute the step length h_k analytically, but in most cases it is computed approximately, minimizing Q along the ray $\mathbf{x}^{(k)} + h\mathbf{s}^{(k)}$ or at least reducing Q sufficiently, i.e., such that the descent $Q(\mathbf{x}^{(k)}) - Q(\mathbf{x}^{(k)} + h_k\mathbf{s}^{(k)}) > 0$ is acceptable by the user. Such a line search is called inexact (approximate) line search. Since, in procedure-oriented languages the theoretically exact optimal step size generally cannot be found, and it is also expensive to find an almost exact step size, an inexact line search with less computational effort is highly popular. Therefore, in practice inexact algorithms are the most used ones. Some of the inexact line search methods are developed in [5,9,10,13,14,16,17,19,21].

For the sake of completeness we recall the definition of three inexact line search rules (see, for example, [7,19,20]) as special cases of **Step 3**:

Step 3-Ar. (Armijo rule). *Set scalars* s_k, β, *and* σ, *with* $s_k = -\nabla Q(\mathbf{x}^{(k)})^T \mathbf{s}^{(k)} / ||\mathbf{s}^{(k)}||^2$, $\beta \in (0,1)$ *and* $\sigma \in (0,1/2)$, *and set* $h_k = \beta^{m_k} s_k$, *where* m_k *is the first nonnegative integer m for which*

$$Q(\mathbf{x}^{(k)}) - Q(\mathbf{x}^{(k)} + \beta^m s_k \mathbf{s}^{(k)}) \geq -\sigma \beta^m s_k \nabla Q(\mathbf{x}^{(k)})^T \mathbf{s}^{(k)},$$

i.e., $m = 0, 1, \ldots$ *are tried successively until the inequality above is satisfied for* $m = m_k$.

Step 3-Gl. (Goldstein rule). *Select a fixed scalar* $\sigma \in (0,1/2)$, *and* h_k *such that*

$$\sigma \leq \frac{Q(\mathbf{x}^{(k)} + h_k \mathbf{s}^{(k)}) - Q(\mathbf{x}^{(k)})}{h_k \nabla Q(\mathbf{x}^{(k)})^T \mathbf{s}^{(k)}} \leq 1 - \sigma.$$

Step 3-WP. (Wolfe-Powell rule). *Select the parameter* h_k *to satisfy*

$$Q(\mathbf{x}^{(k)}) - Q(\mathbf{x}^{(k)} + h_k \mathbf{s}^{(k)}) \geq -\sigma h_k \nabla Q(\mathbf{x}^{(k)})^T \mathbf{s}^{(k)}$$

and

$$\nabla Q(\mathbf{x}^{(k)} + h_k \mathbf{s}^{(k)})^T \mathbf{s}^{(k)} \geq \beta \nabla Q(\mathbf{x}^{(k)})^T \mathbf{s}^{(k)},$$

where σ *and* β *are some scalars with* $\sigma \in (0,1/2)$ *and* $\beta \in (\sigma, 1)$.

In this paper we develop an implementation of five different variations of three unconstrained nonlinear optimization methods: the gradient descent method, the conjugate gradient method, and the Newton method with line search. Two of these variants, the exact line search and the approximate minimization rule, assume the application of a computer algebra system for their implementation. The remaining three variants are based upon three known inexact line search rules. Implementation of the exact line search is described in details. The implementation is carried out in the programming package MATHEMATICA. A comparison of the number of iterative steps and the required CPU time between these variations is presented.

In the second section we survey the main differences between the symbolic and numerical implementation of optimization methods. In Sect. 3 we describe symbolic transformations arising in the exact line search. Also, we develop an implementation of three main optimization methods which are based on line search. Some numerical examples and comparisons are given in Sect. 4, where each method is tested on a collection of 24 unconstrained optimization test functions given in generalized or extended form.

2 Preliminaries

The following optimization tools are available:

1. Compiled programming languages such as Fortran 90, C, C++;
2. Interactive mathematical software: fast to define, solve and model small problems, less efficient for large ones:

   (a) General tools for numerical analysis (Matlab, IDL),

(b)Symbolic mathematical computer systems (Mathematica, Maple, Macsyma),
(c)Modelling tools for optimization (GAMS, AMPL).

The main advantages arising from the application of a computer algebra system are well known:

1. Symbolic implementation partially avoids truncation errors inherent in the traditional numerical implementation.

2. Programs written in procedure-oriented computer languages such as C help in computations, but are of limited value in developing an understanding of the underlying algorithms, because very little information about the intermediate steps is presented. The user interface for such programs is primitive at best and typically requires definitions of user functions and data files corresponding to a specific problem. Many investigations have a need for the capability of developing a "rapid-prototype" code to test the behavior of an algorithm before investing a substantial effort in developing a code for the algorithm in a procedure-oriented language such as C. It is often the case that one can implement a mathematical algorithm almost identically to its mathematical formulation by means of a computer algebra system. This is different from, say, numerical computation where one often has to take special care during the implementation in order to ensure numerical correctness. These numerical procedures typically look very different from their original mathematical descriptions and hence add a significant element of uncertainty to the entire implementation process [12].

3. A disadvantage of traditional languages is that they support only procedural programming. This style is an important one, but it is not the only option, and it is not always the best approach. The programming style should be chosen to fit the problem to be solved, rather than vice versa [12]. Procedural languages cannot be used easily to solve nonnumerical problems. During the implementation of the main methods for unconstrained optimization we observed two nonstandard problems. The first is the capability of manipulating effectively arbitrary objective functions. The capability of the software to process arbitrary objective function makes it generally applicable. The second is the transformation of the objective function required during the exact line search.

On the other hand, the main disadvantage of a computer algebra system is that formula manipulation by a computer requires much more processor time and memory space than the traditional algorithms in procedural programming languages. In particular, the exact line search is expensive. Especially when an iterate is far from the solution of the problem, it is not effective to solve exactly the one-dimensional subproblem (3) [20].

Implementation of optimization methods in MATHEMATICA can be found in [6,8]. Loehle in [11] observed the following additional advantages which MATHEMATICA offers in optimization:

1. Very high precision math is standard.
2. An extensive library of advanced math functions is available.
3. The notebook user interface is easy to use and interactive.
4. Most critically for optimization, a symbolic manipulation of expressions is possible.

But the question which is not investigated in detail is the following: do there exist differences in the convergence rate (number of iterations) between the variations of optimization methods based on symbolic implementation of the exact line search and approximate minimization rules with respect to numerical implementation of inexact line search procedures? The answer to this question is the main goal of the present paper. In [20] it is claimed that for many optimization methods, for example Newton's method and quasi-Newton methods, their convergence rate does not depend on the exact line search. In spite of this statement, Bhati in [6] implemented an approximate minimization rule as well as Armijo's inexact line search, and his experience is that "because of the approximate nature of the computed step length", the overall convergence rate of unconstrained optimization methods based on the Armijo's line search is slower with respect to convergence rate than the same methods based on an approximate minimization rule. Theoretical investigations concerning the convergence of line search methods with seven line search rules are presented in [19]. But the difference in the convergence rate between the symbolic implementation of optimization methods based on the exact line search as well as the approximate minimization rule with respect to numerical implementation, assuming various inexact line search procedures, is still not investigated in practice and in detail.

Our intermediary goal is to improve the implementation of standard optimization methods which are mostly written in procedure-oriented programming languages. We investigate the problem of symbolic transformations of the objective function and construction of composite objective functions from their composite parts, as is required by some subalgorithms in the implementation of nonlinear optimization methods based upon line search.

3 Implementation of Exact Line Search

Our assumption is that the objective function Q is given in an unevaluated form, and not previously defined by a subroutine. Also, we assume that the step size h is given symbolically with undefined value. A respective algorithm cannot be written in procedural languages without lexical and syntax analysis.

The internal representation of the objective function Q is by means of two parameters, denoted by q and *var*. The parameter q is an unevaluated expression which contains locally undefined (symbolic) values of the variables. The parameter *var* involves a list of variables used in Q (these variables evaluate to themselves at this stage). If $x0$ is the list representing an arbitrary n-dimensional point, the value $q0 = q[x0]$ (numerical or symbolic) can be computed using the following transformation rules:

```
q0=q/.Thread[Rule[var,x0]];
```

We develop a universal function $LineFun[q, var, xk, s, h]$ for the symbolic construction of the function $F(h)$, defined in (2). The formal parameter xk denotes the point $\mathbf{x}^{(k)}$, and the list s determines the ray $\mathbf{s}^{(k)}$. Finally, the last parameter h is also unevaluated and denotes the symbol which is used in (2).

```
LineFun[q_,var_List,xk_List,s_List,h_]:=
            Simplify[q/.Thread[Rule[var,xk+h*s]]];
```

An analogue of this function is the following function in C:

```
float LineFun(float (*q)(),float *xk,float *s,int n,float h)
   { int j;
     float xt[50];
     for(j=0; j<n; j++)   xt[j]=xk[j]+h*s[j];
     return((*q)(xt));
   }
```

We observe the following main differences between the MATHEMATICA and C function *LineFun*.

1. The C function $LineFun(q,xk,s,n,h)$ must accompany external functions corresponding to the functional parameter q. Unfortunately, the user of the executable version of the program cannot make new external definitions in the program.
2. The function written in C treats only numerical values for h.
3. It is impossible to use the C function $LineFun(q,xk,s,n,h)$ as the function $F(h)$ (of arbitrary unevaluated h) in further working.

A similar function $f1dim(x)$ is described in [18], (P. 317).

An implementation of line search in the programming language MATLAB can be found in [15]. The function `[f,g]=func1(cas,x)` calculates $f = f(x)$ and $g = f'(x)$ for univariate test functions. The parameter *cas* defines the choice of f and takes values from a set called *caselist*. Without any modifications in the code, the software is applicable only to the set of presented test functions.

We now describe the implementation of **Step 2** and **Step 3** in Algorithm 1 by means of **Step 3-ELS**.

The gradient descent method with an exact line search algorithm uses for the step size h_k the solution of the following constrained univariate minimization problem:

$$h_k = \min_h Q\left(\mathbf{x}^{(k)} - h\nabla Q(\mathbf{x}^{(k)})\right) = \min_h F(h), \quad h > 0.$$

In this case, Algorithm 1 reduces to the well-known steepest descent (Cauchy) method. The gradient vector $\nabla Q(\mathbf{x0})$ can be computed using the function *Grad* from [8]:

```
Grad[f_,var_List]:=Map[Function[D[f,#] ], var];
GradIn[f_,var_List,x0_]:=Grad[f,var]/.Thread[Rule[var,x0]]
```

The standard MATHEMATICA function `D[f, x]` gives the partial derivative $\partial f/\partial x$, and application of the standard function MAP produces a list of such partial derivatives [22].

Now, the function $Gh = F(h) = Q(xk - h\nabla Q(xk))$ can be symbolically generated using the following expression:

```
Gh=LineFun[q,var,xk,-GradIn[q,var,xk],h]
```

In the Newton minimization method with line search, the search direction is chosen as $-\mathbb{H}^{-1}(\mathbf{x}^{(k)})\nabla Q(\mathbf{x}^{(k)})$. The Hessian $\mathbb{H}(\mathbf{x}^{(k)})$ can be found exactly (symbolically) rather than by numerical approximations:

```
Hes[q_,var_List]:=
  Block[{n=Length[var],i,j},
      Return[Table[D[q,var[[i]],var[[j]]],{i,n},{j,n}]];
  ]
HesIn[q_,var_List,x0_List]:= Hes/.Thread[Rule[var,x0]]];
```

We mention that the standard MATHEMATICA function D computes partial derivatives: `D[f, x, y, ... ]` produces $\partial/\partial x \partial/\partial y \cdots f$ (see, for example, [22]).

Thereafter, the function

$$F(h) = Q\left(\mathbf{x}^{(k)} - h\mathbb{H}^{-1}(\mathbf{x}^{(k)})\nabla Q(\mathbf{x}^{(k)})\right)$$

can be formed symbolically as follows:

```
s=LinearSolve[HesIn[q,var,xk],GradIn[q,var,xk]];
Gh=Simplify[LineFun[q,var,xk,-s,h]];
```

We mention that the inverse Hessian matrix $\mathbb{H}^{-1}(\mathbf{x}^{(k)})$ can be computed using the MATHEMATICA function `Inverse`; later multiplied by the vector $\nabla Q(\mathbf{x}^{(k)})$, it gives s. But the function `LinearSolve` yields the product $\mathbb{H}^{-1}(\mathbf{x}^{(k)})\nabla Q(\mathbf{x}^{(k)})$ much more directly. About the function `LinearSolve` see [22].

Conjugate gradient methods use as step length h_k the solution of the one-dimensional optimization problem (2), (3), where the search direction is of the general form $\mathbf{s}^{(k)} = -\nabla Q(\mathbf{x}^{(k)}) + \beta_{k-1}\mathbf{s}^{(k-1)}$ (see, e.g., [20]). In particular, the Fletcher–Reeves conjugate gradient method is defined by:

$$\mathbf{s}^{(k)} = -\nabla Q(\mathbf{x}^{(k)}) + \frac{[\nabla Q(\mathbf{x}^{(k)})]^T \nabla Q(\mathbf{x}^{(k)})}{[\nabla Q(\mathbf{x}^{(k-1)})]^T \nabla Q(\mathbf{x}^{(k-1)})}\mathbf{s}^{(k-1)}. \tag{5}$$

Denote by *PrevGr* and *Gr* the numerical values of $\nabla Q(\mathbf{x}^{(k-1)})$ and $\nabla Q(\mathbf{x}^{(k)})$, respectively. According to (2), (3), and (5), the function $F(h)$ can be symbolically generated by applying the following code:

```
s=-Gr+Gr.Gr/(PrevGr.PrevGr) *Prevs;
Gh=LineFun[q,var,x0,s,h];
```

Minimization of the function $F(h)$ is accomplished by using the standard MATHEMATICA function `FindMinimum` [22]. The function `FindMinimum[f,{x,x0}]` searches for a local minimum of f, starting from the point $x = x0$ and returns a list of the form `{fmin, {x->xmin}}`, where *fmin* is the minimum value of f found, and *xmin* is the value of x for which it is found. Therefore, the optimal step length *hk* is equal to the numerical value *xmin* from the output {`fmin, {x->xmin}`}, and can be generated by the following code:

```
hk=FindMinimum[Gh,{h,1},MaxIterations->100][[2,1,2]];
```

As we are concerned with a local minimum near the value $h = 1$, we use the starting value $h0 = 1$ for h.

A common framework for the steepest descent method, Newton's method with line search, and the conjugate gradient method with *Step 3* defined by **Step 3-ELS** is described in the following algorithm:

Algorithm 2 Implementation of optimization methods based on exact line search.

- *q*, *var* denote the internal representation of the mathematical model.
- *xs* denotes the starting point.
- *opts* is the list of selected options.

Step 1. *Compute initial values* `x1=x0=xs`.
Step 2. *While the precision is not achieved, perform the following steps in the loop:*
Step 2.1. *Generate a new ray* `s` *corresponding to the chosen optimization method in the parameter* `opts`.
Step 2.2. *Generate* `F(h)=LineFun[q,var,x0,s,h]`.
Step 2.3. *Compute* `hk=` $\min_h$ `F(h)` *by applying the function* `FindMinimum`.
Step 2.4. *Compute the new iteration* `x1=x0+hk*s` *and* `q1=q[x1]`.
Step 3. *Return the list* {`q1,x1`}.

An implementation of this code is given in the Appendix.

4 Numerical Results and Comparisons

In this section we report some results obtained through the use of the package MATHEMATICA in the implementation of the above described optimization methods based on five different rules for finding the step size h_k in each iteration. The first rule is the exact line search (ELS) method, which uses the MATHEMATICA function `FindMinimum` to solve the one-dimensional problem (3). The second rule is the approximate minimization rule, which uses the standard function `FindRoot` to solve equation (4), and is implemented in [6]. We also considered the well-known Wolfe-Powel (WP), Goldstein (Gl), and Armijo-backtracking (Ar) rules as inexact line search methods, whose algorithms can be found in [20]. Our intention is to show that exact line search and the approximate minimization rule give better results compared to inexact line search methods, in the sense of number of iterations necessary to obtain the desired accuracy.

We selected 24 unconstrained optimization test functions given in generalized or extended form (taken from the collection of test functions for unconstrained optimization in [2]). For each test function we have conducted numerical experiments with the number of variables $n = 3, 5, 8, 10$, and 20. The results obtained on each test problem for each method are given it Table 1. In the table we show the number of iterative steps needed to achieve a numerical precision 10^{-6}.

We mention that a slant line in the table below means that the iterative process failed, while an asterisk indicates that the global minimum is not achieved.

Table 1 Number of iterations and average CPU time for the ELS, AMR, WP, Gl, and Ar method, 24 test functions, 3 optimization methods

Function name	Dim	Method Newton					Gradient					Steepest				
		ELS	AMR	WP	Gl	Ar	ELS	AMR	WP	Gl	Ar	ELS	AMR	WP	Gl	Ar
Extended Penalty	3	6	6	9	23	41	39	11	21	45	148	36	37	35	25	58
Perturbed quadr.	3	2	2	2	5	29	4	4	4	11	106	59	22	22	23	360
Raydan-1	10	2	2	4	17	13	26	20	27	/	76	80	58	29	46	75
Raydan-2	20	2	2	6	6	6	2	2	6	6	6	2	2	6	6	6
Diagonal1	5	5	5	8	9	82	152	13	21	51	85	23	32	32	29	34
Diagonal2	10	5	6	6	32	64	27	22	21	163	82	78	55	37	45	78
Hager	5	5	5	11	15	9	18	10	14	48	66	11	15	14	10	43
Gen. Tridiagonal-1	10	5	5	6	67	77	68	28	26	712	251	/	55	43	51	168
Diagonal4	20	2	2	2	37	58	3	3	5	48	91	8	10	4	683	460
Diagonal5	20	2	2	4	4	4	2	2	4	4	4	2	2	4	4	4
Extended PSC1	10	6	*	9	19	219	31	11	12	29	50	/	16	20	24	92
Quadratic QF1	20	2	2	2	17	49	28	21	21	136	146	110	144	144	105	96
Ext. Tridiagonal-2	8	2	2	5	23	981	37	12	31	65	49	54	79	74	76	40
Tridia	10	2	2	2	9	487	11	11	11	382	360	554	675	675	322	390
Arglinb	10	2	2	2	/	74	8	6	6	/	134	276	284	284	/	145
Arwhead	8	2	2	7	9	39	11	8	11	40	80	/	10	19	51	147
Almost Perturbed	10	2	2	2	19	75	11	11	11	89	168	/	72	72	53	51
Trid. Perturbed Q.	10	2	2	2	24	28	11	11	11	65	117	108	127	127	129	78
Power	5	2	2	2	20	111	6	6	6	89	178	/	186	186	169	122
Quartc	10	2	2	9	/	1456	2	2	9	2	1456	2	2	9	2	1456
Dixon3dq	10	2	2	2	32	26	6	6	50	233	288	260	323	323	308	206
Biggsb1	10	2	2	2	31	24	6	6	46	200	242	244	307	307	297	191
Diagonal7	20	2	2	5	36	39	2	2	5	36	39	2	2	5	36	39
Diagonal8	20	2	2	5	39	40	2	2	5	39	40	2	2	5	39	40
Average no. iterations		2.8	2.7	4.8	22.4	168	21.4	9.6	16	113.3	177.6	100.6	104.9	103.2	110.1	182.5
Average CPU time		0.22	0.24	0.36	1.23	4.76	0.66	0.32	0.36	1.52	2.06	1.03	1.47	1.06	1.09	2.26

We choose low-dimensional test problems because of limitations in the MATHEMATICA computer algebra system

For each of the three methods (Newton's method, conjugate gradients and gradient descent method, in turn) the columns represent the number of iterations for exact line search (ELS), approximate minimization rule (AMR), Wolfe-Powell (WP), Goldstein (Gl) and Armijo-backtracking (Ar) rule, respectively, tested on the respective test function. The last row represents the average number of iterations for each rule applied on each method.

As one can observe from Table 1, it is perfectly clear that the number of iterations for ELS and AMR is substantially smaller than the number of iterations for the Gl and Ar rule. Therefore, it is sufficient to compare only the results for the ELS, AMR and WP methods, which are shown in Table 2. The average CPU time is displayed in the last row of Table 1.

Table 2 Summarized performance of ELS, AMR, and WP line search from Table 1

Performance criterion	Newton	Gradient	Steepest	Total
ELS	2	0	1	3
AMR	0	7	2	9
WP	0	2	4	6
ELS and AMR	13	8	4	25
ELS and WP	0	0	0	0
AMR and WP	0	2	1	3
ELS, AMR, and WP	9	5	0	14

The entries in Table 2 represent the number of test problems in which the rules specified in the rows (observed individually and jointly) achieved the minimal number of iterations for each of the applied optimization method (Newton, Gradient, and Steepest). The last column (headed by Total) is the sum of the preceding three columns.

5 Conclusion

By achieving an implementation of the exact line search we improve conventional implementations of the main optimization methods. The main advantage arising from the application of computer algebra systems in the implementation of unconstrained optimization methods over traditional implementations in procedural languages such as FORTRAN and C, is the possibility of defining an optimization problem without previous functions or subroutines which define objectives. Also, MATHEMATICA allows one to activate the merit function and constraints at run-time, as is required in some optimization methods based on line search.

From Tables 1 and 2 we can make the following observations about the convergence rate of three main optimization methods using ELS, AMR and three inexact line searches. Obviously, for each of the three optimization methods, the number of test functions for which AMR shows better performance (achieved minimum number of iterations) is greater than the number of problems for which ELS and WP gives better results. From Table 2 it is clear that the total number of test problems where AMR achieved minimal number of iterations ($9+25+3+14 = 51$) is slightly greater than the count of the best results corresponding to ELS ($3+25+14=42$). Together, AMR and ELS gives substantially better results compared to WP ($6+3+14 = 23$). Furthermore, the average number of iterations for AMR is smaller than the average number of iterations for ELS and WP (see Table 1).

The numerical results shown in Table 1 are derived for test functions of small dimensions. Greater dimensions cause difficulties in both the CPU time and the memory space requirements. Therefore, our numerical experience confirms the known fact that formula manipulation by a computer requires much more time and memory space compared to numerical implementation. If processor time or dimensions of the problem are primary goals, then the code should be written in FORTRAN or C. While the slow execution of the code (inherent to all interpretative programming languages) is a clear disadvantage, code execution is only 1% of the total research time. In any research, most of the time is spent reading, thinking and trying out ideas. In computer languages where computer algebra tools are not available, researchers are forced to spend weeks, if not months, implementing those features only at a barely functioning level. MATHEMATICA on the other hand will allow researchers to focus on research most of the time.

An inexact algorithm, based on the acceleration of the gradient descent algorithm with backtracking is developed in [1]. The idea is to modify the step length h_k (computed by backtracking) by means of a positive parameter θ_k in a multiplicative manner by such a way as to improve the behavior of the classical gradient algorithm [1]. A similar idea is used in papers [3, 4]. It seems interesting to implement analogous improvements of ELS, AMR, Gl, and WP line search and perform analogous comparisons of convergence rates. This problem will be investigated in our future research.

Appendix

The implementation of the gradient descent method, Newton's method with line search, and the conjugate gradient method based upon the ELS procedure is contained in the following function *LineMin*, where the parameters *q*, *var* denote the internal representation of the objective function and the list *xs* is a starting point. Optional arguments *ep1* and *ep2* denote default values for the minimal gradient of the objective function and the minimal function improvement, respectively. Finally, the optional parameter *Method* selects the optimization method. In order to guarantee convergence of Algorithm 1, we require the following termination criteria:

$$\|\nabla Q(\mathbf{x}^{(k)})\| \leq eps1, \text{ and } \frac{|Q(\mathbf{x}^{(k+1)}) - Q(\mathbf{x}^{(k)})|}{1+|Q(\mathbf{x}^{(k)})|} \leq eps2,$$

where $eps1$ and $eps2$ are predefined small real numbers with default values equal to $ep1 = 10^{-6}$, and $ep2 = 10^{-10}$.

```
Grad[f_, var_List] := Map[Function[D[f, #]], var];
GradIn[f_, var_List, x0_] :=
                        Grad[f, var] /. Thread[Rule[var, x0]];

Hes[f_, var_List] :=
  Block[{n, i, j}, n = Length[var];
    Return[Table[D[f, var[[i]], var[[j]]], {i, n}, {j, n}]]; ]
HesIn[f_, var_List, x0_List] :=
                        Hes[f, var] /. Thread[Rule[var, x0]];
norm[x_] := N[Sqrt[x.x]];
```

```
LineFun[q_, var_List, xk_List, s_List, h_] :=
    Simplify[q /. Thread[Rule[var, xk + h*s]]];
Options[LineMin] =
            {ep1 -> 10^-6, ep2 -> 10^-10, Method -> Newton};
LineMin[q_, var_List, xs_List, opts___Rule] :=
  Block[{x1=x0=xs, h, hk, s=Table[0, {Length[var]}], Prevs,
   k=0, q0, q1, Gh, work1, work2, eps1, eps2, met, Gr, PrevGr},
    {eps1,eps2,met} =
                {ep1,ep2,Method} /. {opts} /. Options[LineMin];
    work1 = eps1 + 1; work2 = eps2 + 1;
    q1 = q /. Thread[Rule[var, x1]]; Gr = GradIn[q, var, x0];
    While[(work1 > eps1 || work2 > eps2),  (* Step 1 *)
       k++; x0 = x1; q0 = q1; (* Step 4 *)
       (* Step 2 *)
       PrevGr = Gr;   Prevs = s;     Gr = GradIn[q, var, x0];
       If[met === Steepest, s = -Gr];
       If[met === Newton,
            s = LinearSolve[HesIn[q, var, x0], -Gr]];
       If[met === Gradient,
            s = -Gr + Gr.Gr/(PrevGr.PrevGr)*Prevs];
       (* Step 3-ELS *)
       Gh = LineFun[q, var, x0, s, h];
       hk = FindMinimum[Gh, {h, 1},
            MaxIterations -> 100][[2,1,2]];
       (*Numerical computation of new iterations*)
        x1 = N[x0 + hk*s, 20]; q1 = q /. Thread[Rule[var, x1]];
        (* Step 4 *)
        (*Define measures for stopping criterions*)
        work1 = N[norm[x1 - x0], 20]; work2 = Abs[q1 - q0];
     ];
    Return[{k, {q1, x1}}]; ]
```

The implementation of the optimization methods based on the AMR procedure differs from the previous code only in the replacement of **Step 3-ELS** by the following code:

```
(* Step 3-AMR *)
hk = FindRoot[GradIn[q, var, x0 + h*s].s == 0, {h, 0}][[1, 2]];
```

For the implementation of the Armijo rule (**Step 3-Ar**) we have the following procedure:

```
Backtracking[f_, gk_, dk_, xk_, prom_, fk_] :=
  Module[{\[Alpha], \[Beta], sk, t, fn, xn, n=Length[prom]},
    \[Alpha] = 0.0001;   \[Beta] = 0.8;
    sk := -gk.dk/(norm[dk]*norm[dk]);
    t = sk; xn = xk + t*dk;
    fn = f /. Thread[Rule[prom, xn]];
    While[fn > fk + \[Alpha]*t*gk.dk, t = t*\[Beta];
          xn = xk + t*dk;  fn = f /. Thread[Rule[prom, xn]]; ];
    Return[t]; ]
```

which we call to find step size h_k in each iteration instead of using `FindMinimum`:

```
hk = Backtracking[q,Gr,s,x0,var,q0];
```

The remainder of the algorithm is the same as in ELS.

Similarly, we use the following code to implement Gl (**Step 3-Gl**) and WP (**Step 3-WP**) inexact line search rules:

```
Goldstein[f_, gk_, dk_, xk_, prom_, fk_] :=
  Module[{Cond, t1, t2, \[Alpha], sk, t, fn, xn,
           n=Length[prom]},
    \[Rho] = 0.1;    t1 = 0; t2 = 100;
    sk := -gk.dk/(norm[dk]*norm[dk]);    t = sk;
    xn = xk + t*dk;   fn = f /. Thread[Rule[prom, xn]];
    Cond = True;
    While[Cond,
      While[fn > fk + \[Rho]*t*gk.dk,
          t2 = t;      t = (t1 + t2)/2;
          xn = xk + t*dk;     fn = f /. Thread[Rule[prom, xn]];
      ];
      If[fn < fk + (1 - \[Rho])*t*gk.dk,
        t1 = t;    t = (t1 + t2)/2;
        xn = xk + t*dk;   fn = f /. Thread[Rule[prom, xn]],
        Cond = False
      ] ];
    Return[t]; ]

Woolf[f_, gk_, dk_, xk_, prom_, fk_] :=
  Module[{Cond, \[Sigma], t1,t2,tp, \[Alpha], sk, t,
           fn,fk1,fn1, xn, n=Length[prom]},
    \[Rho] = 0.1; \[Sigma] = 0.4;
    t1 = 0; t2 = 10;
    fk1 = gk.dk;
    sk := -gk.dk/(norm[dk]*norm[dk]);    t = sk;
    xn = xk + t*dk;   fn = f /. Thread[Rule[prom, xn]];
    Cond = True;
    While[Cond,
      While[fn > fk + \[Rho]*t*gk.dk,
         t = t1 + 1/2(t - t1)/(1 + (fk - fn)/((t - t1)fk1));
         xn = xk + t*dk;    fn = f /. Thread[Rule[prom, xn]];
      ];
      gn = GradIn[f, prom, xn];
      fn1 = gn.dk;
      If[Abs[fn1] > \[Sigma]*Abs[gk.dk],
        If[fk1 \[NotEqual] fn1,
          tp = t + (t - t1)*fn1/(fk1 - fn1);    t1 = t; t = tp,
          t = 2*t; ];
        fk1 = fn1;
        xn = xk + t*dk;    fn = f /. Thread[Rule[prom, xn]],
        Cond = False
      ] ];
    Return[t]; ]
```

Calls of the functions *Goldstein* and *Woolf* are of the form:

```
hk = Goldstein[q,Gr,s,x0,var,q0]; hk = Woolf[q,Gr,s,x0,var,q0];
```

References

1. Andrei, N.: An acceleration of gradient descent algorithm with backtracking for unconstrained optimization. Numer. Algor. **42**, 63–73 (2006)
2. Andrei, N.: An Unconstrained Optimization Test Functions Collection. http://camo.ici.ro/neculai/t1.pdf (2005)
3. Andrei, N.: A scaled BFGS preconditioned conjugate gradient algorithm for unconstrained optimization. Appl. Math. Lett. **20**, 645–650 (2007)
4. Andrei, N.: A Dai-Yuan conjugate gradient algorithm with sufficient descent and conjugacy conditions for unconstrained optimization. Appl. Math. Lett. **21** 165–171 (2008)
5. Armijo, L.: Minimization of functions having Lipschitz first partial derivatives. Pac. J. Math. **6**, 1–3 (1966)
6. Bhatti, M.A.: Practical Optimization Methods with Mathematica® Applications. Springer-Verlag, New York (2000)
7. Cohen, A.I.: Stepsize analysis for descent methods. J. Optim. Theory Appl. **33**, 187–205 (1981)
8. Culioli, J.C.: Optimization with Mathematica. In: Computational Economics and Finance (H. Varian, eds.), TELOS/Springer-Verlag, Santa Clara CA (1996)
9. Fletcher, R.: Practical Methods of Optimization. Wiley, New York (1987)
10. Goldstein, A.A.: On steepest descent. SIAM J. Contrl. **3**, 147–151 (1965)
11. Loehle, C.: Global optimization using Mathematica: A test of software tools. Math. Educ. Res. **11**, 139–152 (2006)
12. Maeder, R.E.: Computer Science with Mathematica. Cambridge University Press, Cambridge, New York, Madrid, Cape Town, Singapore, Sao Paolo (2006)
13. Lemaréchal, C.: A view of line search. In: Optimization and Optimal Control (A. Auslander, W. Oetti, J. Stocr, cds.), pp. 59–78, Springer, Berlin (1981)
14. Moré, J.J., Thuente, D.J.: On line search algorithm with guaranteed sufficient decrease. Mathematics and Computer Science Division Preprint MCS-P153-0590, Argone National Laboratory, Argone (1990)
15. Neumaier, A.: Matlab line search routines. http://www.mat.univie.ac.at/~neum/software/ls (1988)
16. Potra, F.A., Shi, Y.: Efficient line search algorithm for unconstrained optimization. J. Optim. Theory Appl. **85**, 677–704 (1995)
17. Powell, M.J.D.: Some global convergence properties of a variable-metric algorithm for minimization without exact line search. AIAM-AMS Proc., Philadelphia **9**, 53–72 (1976)
18. Press, W.H., Flannery, B.P., Teukolsky, S.A., Vetterling, W.T.: Numerical Recipes in C. Cambridge University Press, New York-Melbourne-Sydney (1990)
19. Shi, Z.-Jun: Convergence of line search methods for unconstrained optimization. Appl. Math. Comput. **157**, 393–405 (2004)
20. Wenyuu, S., Ya-xiang, Y.: Optimization Theory and Methods, Series: Springer Optimization and its Application, Vol 1, Springer, Berlin (2006)
21. Wolfe, P.: Convergence conditions for ascent methods. SIAM Rev. **11**, 226–235 (1968)
22. Wolfram, S.: The Mathematica Book. 4th ed., Wolfram Media/Cambridge University Press, Cambridge (1999)

Roots of AG-bands

Nebojša Stevanović and Petar V. Protić

Dedicated to Professor Gradimir V. Milovanović on the occasion of his 60th birthday

1 Introduction

A groupoid G on which the following is true,

$$(\forall a,b,c \in G) \quad (ab)c = (cb)a,$$

is an Abel-Grassmann's groupoid (AG-groupoid), [1]. It is easy to verify that on every AG-groupoid the *medial* law holds

$$(ab)(cd) = (ac)(bd),$$

so the class of Abel-Grassmann's groupoids belongs to the class of medial groupoids.

As in Semigroup Theory, bands and band decompositions are one of the most useful methods for research on AG-groupoids.

Let G be an AG-groupoid, $a \in G$ and $a^2 = a$, then a is an idempotent in G. If on the AG-groupoid G every element is an idempotent, then G is an AG-band. We denote by $E(G)$ the set of all idempotents in an AG-groupoid G.

An AG-groupoid G is an *AG-band Y of AG-groupoids G_α* if $G = \bigcup_{\alpha \in Y} G_\alpha$, Y is an AG-band, $G_\alpha \cap G_\beta = \emptyset$ for $\alpha, \beta \in Y$, $\alpha \neq \beta$, and $G_\alpha G_\beta \subseteq G_{\alpha\beta}$.

A congruence ρ on G is called *band congruence* if G/ρ is a band.

Nebojša Stevanović
Faculty of Civil Engineering and Architecture, University of Niš, Aleksandra Medvedeva 14, 18000 Niš, Serbia, Deceased March 11, 2009

Petar V. Protić
University of Niš, Faculty of Civil Engineering and Architecture, Aleksandra Medvedeva 14, 18000 Niš, Serbia, e-mail: pvprotic@yahoo.com

W. Gautschi et al. (eds.), *Approximation and Computation: In Honor of Gradimir V. Milovanović*, Springer Optimization and Its Applications 42, DOI 10.1007/978-1-4419-6594-3_29, © Springer Science+Business Media, LLC 2011

2 Subclasses of Roots of a Band

In this section we shall consider the class of *AG*-groupoids, defined by the identity $a^2 \cdot a^2 = a^2$. There are a few interesting subclasses of this class; these are *AG*-bands, *AG*-3-bands, *AG*-4-bands and unipotent radicals.

Let us now introduce the following notion.

Definition 1. Let G be an *AG*-groupoid, $a \in G$ an arbitrary element; if $a^2 \cdot a^2 = a^2$, we say that a is a *(square) root of idempotent.*

For an *AG*-groupoid G we define the set

$$\sqrt{E(G)} = \{a \in G : a^2 \cdot a^2 = a^2\}.$$

Definition 2. The groupoid G is a *root of a band* if $G = \sqrt{E(G)}$.

The earlier definition has its motivation in the fact that on an *AG*-groupoid idempotents form a subgroupoid, an *AG*-band.

Definition 3. Let G be an *AG*-groupoid, $a \in G$ an arbitrary element; if all products of a of length 4 are equal to a^2, we say that a is a *4-potent*. The *AG*-groupoid G is an *AG*-4-band (or a 4-band) if all of its elements are 4-potents.

It is easy to see that if the element $a \in G$ is a 4-potent, then it belongs to $\sqrt{E(G)}$, consequently every *AG*-4-band is a root of a band.

Now we are going to introduce two more important subclasses of roots of idempotents.

Definition 4 ([6]). Let G be an *AG*-groupoid, $a \in G$ an arbitrary element; if $(aa)a = a(aa) = a$ (or $a^3 = a$), we say that a is a *3-potent*. The *AG*-groupoid G is an *AG*-3-band (or a 3-band) if all of its elements are 3-potents.

Lemma 1. *Let G be an AG-groupoid; if $a \in G$ is a 3-potent, then a^2 is an idempotent.*

Proof. Let $a \in G$ be a 3-potent; then $a^2a^2 = (aa)(aa) = ((aa)a)a = aa = a^2$. Whence a^2 is an idempotent. □

From Lemma 1 it follows that an *AG*-groupoid which does not have idempotents, has no 3-potents as well.

The class of *AG*-3-bands is one of the most important classes of *AG*-groupoids. For this reason we have already dealt with this class in [6]. Here we give its relation with the class of *AG*-4-bands. Let $a \in G$ be a 3-potent element, i.e., $a^3 = a$, then $aa^3 = aa = a^2$ and $a^3a = aa = a^2$. By Lemma 1 we obtain $a^2a^2 = a^2$, which, together with the above, leads to the conclusion that a is a 4-potent. Consequently, every 3-band is a 4-band, so it is a root of a band.

The next lemma describes the intersection of the class of *AG*-4-bands with the class of *AG*-3-bands.

Lemma 2. *A left canceling AG-4-band is an AG-3-band.*

Proof. Let G be a left canceling AG-4-band, $a \in G$ an arbitrary element; since all products of a of length 4 are equal to a^2, we obtain $a((a^2)a) = a(aa^2) = a^2$. By left cancelation it follows that $(a^2)a = aa^2 = a$, consequently G is an AG-3-band. □

Definition 5. Let G be a unipotent AG-groupoid and $e \in G$ a unique idempotent in G; then if $a^2 = e$ for all $a \in G$, we say that G is a unipotent radical.

It is clear from the definitions that the intersection of the class of unipotent radicals and AG-bands is the class of one-element groupoids.

The group G satisfying $a^2 = e_G$ is a *Boolean group*.

Lemma 3. *Let G be an AG-3-band; if G is a unipotent radical, then G is a Boolean group.*

Proof. Let $a \subset G$ be an arbitrary element; because a is a 3-potent by Lemma 1, there follows $a^2 = e$, and thus $ea = a^2a = a$ and $ae = aa^2 = a$. Whence e is both a left and right identity in G.

It was proved in [4] that an AG-groupoid with a right identity is a commutative semigroup. Since $aa = a^2 = e$, it follows that $a = a^{-1}$ in the group sense, i.e., G is a Boolean group. □

We are going to give some examples of unipotent radicals and roots of idempotents.

Example 1. Let G be an AG-groupoid given by the following table:

·	1	2	3	4
1	2	4	2	1
2	1	2	1	4
3	2	4	2	1
4	4	1	4	2.

By using an AG-test [5] we can easily verify that G is an AG-groupoid; but G is not a semigroup since, for example, $(1 \cdot 1) \cdot 1 = 2 \cdot 1 = 1$ and $1 \cdot (1 \cdot 1) = 1 \cdot 2 = 4$. It is a root of band, of course; moreover, from the above it follows that $(1 \cdot (1 \cdot 1)) \cdot 1 = (1 \cdot 2) \cdot 1 = 4 \cdot 1 = 4$, which is different from $1^2 = 2$; thus G is not an AG-4-band. Consequently, it is not an AG-3-band as well.

Example 2. Let G be an AG-groupoid given by the following table:

·	1	2	3	4	5
1	2	1	1	1	1
2	1	2	2	2	2
3	1	2	3	3	3
4	1	2	3	3	5
5	1	2	3	4	3.

By using an AG-test [5] we can easily verify that G is an AG-groupoid; but G is not a semigroup since, for example, $(4 \cdot 4) \cdot 5 = 3 \cdot 5 = 3$ while $4 \cdot (4 \cdot 5) = 4 \cdot 5 = 5$. It is easy to verify that G is not an AG-3-band since, for example, $(5 \cdot 5) \cdot 5 = 5 \cdot (5 \cdot 5) = 5 \cdot 3 = 3 \cdot 5 = 3$. Also, G is not a unipotent radical since it has two idempotents 2 and 3. It is obvious that G is a root of idempotents. By simple calculation we can verify that G is a 4-band since all products of length 4 of any element are equal to its square.

Some interesting properties of 3-potents in an arbitrary AG-groupoid will be given below, as well as their connection with idempotent elements.

Lemma 4. *Let G be an AG-groupoid; if $a, b \in G$ are 3-potents, then ab is a 3-potent as well.*

Proof. Let $a, b \in G$ be 3-potents; then

$$((ab)(ab))(ab) = ((aa)(bb))(ab) = (a^2b^2)(ab) = (a^2a)(b^2b) = ab$$

and similarly $(ab)((ab)(ab)) = ab$. Hence ab is a 3-potent. □

With $T_{id}(G)$ we shall denote the set of all 3-potents of an AG-groupoid G. From Lemma 4 we have that $T_{id}(G)$ is an AG-3-band.

Corollary 1. *Let G be an AG-groupoid; then the set $B = \{a^2 : a \in T_{id}(G)\}$ is an AG-band.*

Proof. Suppose that $e, f \in B$; then there exist $a, b \in T_{id}(G)$ such that $e = a^2$, $f = b^2$. Now we have $ef = a^2b^2 = (aa)(bb) = (ab)(ab) = (ab)^2$; since $a, b \in T_{id}(G)$ by Lemma 4, it follows that $ab \in T_{id}(G)$ and $(ab)^2 \in B$. From the above it follows that B is a subgroupoid of G. By Lemma 1 it follows that B is an AG-band. □

Remark 1. Let G be an AG-groupoid and $a, b \in G$ arbitrary elements. Suppose that $a^2 = b^2$ is an idempotent; then $(ab)^2 = (ab)(ab) = (aa)(bb) = a^2b^2 = a^2a^2 = a^2$, consequently $(ab)^2 = a^2 = b^2$.

It was proved in paper [7] that the relation $\leq$ defined on an AG-band by $e \leq f \iff e = ef$ is a natural partial order.

Proposition 1. *Let G be an AG-groupoid, $a \in G$ a 3-potent element, and $e \in E(G)$; then if $ae = a$, we have that $a^2 \leq e$.*

Proof. Suppose $a^2a = aa^2 = a$ and $ae = a$; then $a^2 = aa = (ae)(ae) = a^2e^2 = a^2e$, which by the definition of natural partial order means $a^2 \leq e$. □

Till the end of this paper we are going to give the band decomposition of the class of the roots of AG-band into band indecomposable components. We are also going to prove two theorems which describe unipotent radicals as important factors in band decompositions of the roots of a band.

The earlier definitions give us the motivation to introduce a relation $\mathcal{K}$ on arbitrary *AG*-groupoids as follows,

$$a\mathcal{K}b \Longleftrightarrow a^2 = b^2. \tag{1}$$

It follows immediately from its definition that $\mathcal{K}$ is reflexive, symmetric, and transitive, i.e., it is an equivalence relation.

Theorem 1. *The relation $\mathcal{K}$ defined earlier on an arbitrary AG-groupoid G is an idempotent separating congruence relation.*

Proof. We are going to prove that $\mathcal{K}$ is compatible with the operation on G. Let $a,b,c,d \in G$ be such that $a\mathcal{K}b$, $c\mathcal{K}d$; then

$$(ac)^2 = (ac)(ac) = a^2c^2 = b^2d^2 = (bb)(dd) = (bd)(bd) = (bd)^2.$$

Hence, $\mathcal{K}$ is compatible, so it is a congruence relation. From the definition of $\mathcal{K}$ it follows that from $e\mathcal{K}f$ there follows $e = f$ for all $e,f \in E(G)$, i.e., $\mathcal{K}$ is an idempotent separating congruence. □

Theorem 2. *Let G be a root of band; then $\mathcal{K}$ is a band congruence on G, and G is band indecomposable if and only if it is a unipotent radical.*

Proof. Let G be an *AG*-groupoid and suppose that G is a band indecomposable root of band.

Let $a \in G$ be an arbitrary element. Because $a^2 \in E(G)$ we obtain $(a^2)^2 = a^2$, which implies $a^2\mathcal{K}a$. From the above and Theorem 1, we have that $\mathcal{K}$ is a band congruence on G.

Let $e,f \in E(G)$. As G is band indecomposable it follows that $\mathcal{K} = G \times G$, whence $e\mathcal{K}f$, i.e., $e^2 = f^2$ and so $e = f$. Consequently, G is a unipotent radical.

Conversely, let G be a unipotent radical, ρ a band congruence on G, $a,b \in G$ arbitrary elements, and e the unique idempotent of G. As ρ is a band congruence, we have $a^2\rho a$ and $b^2\rho b$, hence $a\rho a^2 = e = b^2\rho b$, which gives $\rho = G \times G$. Consequently, G is band indecomposable. □

Theorem 3. *Let G be a root of band; then the relation $\mathcal{K}$ is the smallest band congruence on G and the $\mathcal{K}$-classes are band indecomposable unipotent radicals.*

Proof. It was proved that $\mathcal{K}$ is a band congruence. Let ρ be a band congruence on G, and $a,b \in G$ arbitrary elements. Suppose $a\mathcal{K}b$, i.e., $a^2 = b^2$. Since ρ is a band congruence, we have $a^2\rho a$ and $b^2\rho b$, hence $a\rho a^2 = b^2\rho b$, which gives $a\rho b$. From the above we conclude that $\mathcal{K} \subseteq \rho$ for any band congruence ρ, so $\mathcal{K}$ is the smallest band congruence on G.

As $\mathcal{K}$ is an idempotent separating congruence, its equivalence classes are unipotent roots of band, i.e., unipotent radicals. By Theorem 2, unipotent radicals are band indecomposable. □

Corollary 2. *Let G be a root of a band; then G is an AG-band Y of unipotent radicals $S_\alpha, \alpha \in Y$.*

Proof. By Theorems 1 and 2, the relation $\mathcal{K}$ is a band congruence relation on G. Therefore, G is an AG-band $Y = S/_{\mathcal{K}}$ of AG-groupoids; by $S_\alpha, \alpha \in Y$, we shall mean equivalence classes of $\mathcal{K}$. Let $b \in S_\alpha$ be an arbitrary element; by $e_\alpha = b^2$, we shall mean (the unique) idempotent element of the class S_α. It is now clear that $a^2 = e_\alpha$ for all $a \in S_\alpha$. The fact that e_α is the unique idempotent follows from the fact that $\mathcal{K}$ is an idempotent separating congruence. □

One of the best approaches in the research of one type of algebraic structure is to connect it with another type which is better explored. This goal is reached for AG-3-bands by the next theorem, which connects AG-3-bands with Boolean groups.

Theorem 4. *Let G be an AG-groupoid; then G is an AG-3-band if and only if it can be decomposed into an AG-band Y of Boolean groups* $S_\alpha, \alpha \in Y$.

Proof. Let G be an AG-3-band and $\mathcal{K}$ a congruence relation defined by (1). Since G is an AG-3-band, by Lemma 1 we have that a^2 is an idempotent for all $a \in G$, so $\mathcal{K}$ is a band congruence on G.

By Corollary 2 it follows that G is an AG-band Y of AG-groupoids $S_\alpha, \alpha \in Y$, satisfying $a^2 = e_\alpha, \forall a \in S_\alpha$, where e_α is the unique idempotent in S_α. Since G is an AG-3-band, we have $a^2a = aa^2 = a$ for all $a \in G$, which means that a^2 is a neutral for a. Denote $a^2 = e$, $b \in \mathcal{K}_a$; then $be = eb = b$ since $b^2 = a^2 = e$, whence e is the identity element in $\mathcal{K}_a$. Clearly $a^2 = e$ holds for all $a \in \mathcal{K}_a$. Since an AG-groupoid with the identity is a semigroup, we have that $\mathcal{K}_a$ is a semigroup with identity $a^2 = e$.

Let $a, b \in G$ be elements such that $a\mathcal{K}b$; then

$$ab = (a^2a)(bb^2) = (a^2b)(ab^2) = (b^2b)(aa^2) = ba.$$

From the above we have that $ab = ba$ for all $a, b \in \mathcal{K}_a$, whence $\mathcal{K}_a$ is a commutative semigroup.

Conversely, suppose that $G = \cup_{\alpha \in Y} S_\alpha$, where $S_\alpha, \alpha \in Y$, are commutative semigroups satisfying $a^2 = e_\alpha$ for all $a \in S_\alpha$, Y an AG-band. We are going to prove that G is an AG-3-band. Let $x \in G$ be an arbitrary element; then there exists $\beta \in Y$ such that $x \in S_\beta$. If e_β is the identity element of S_β, then $x^2 = e_\beta$, consequently $x(xx) = xe_\beta = x$ and $(xx)x = e_\beta x = x$. From the above it follows that $(xx)x = x(xx) = x$ holds for all $x \in G$, so G is an AG-3-band. □

It is interesting to mention that $\mathcal{K} = \Delta$ if G is an AG-band.

Example 3. Let G be an AG-groupoid given by the following table:

$\cdot$	1	2	3	4	5	6	7	8
1	1	2	7	8	3	4	5	6
2	2	1	8	7	4	3	6	5
3	5	6	3	4	7	8	1	2
4	6	5	4	3	8	7	2	1
5	7	8	1	2	5	6	3	4
6	8	7	2	1	6	5	4	3
7	3	4	5	6	1	2	7	8
8	4	3	6	5	2	1	8	7

By using an *AG*-test [5] we can easily verify that G is an *AG*-groupoid, but G is not a semigroup since, for example, $(3\cdot 2)\cdot 8 = 6\cdot 8 = 3$ and $3\cdot(2\cdot 8) = 3\cdot 5 = 7$. It is easy to verify that G is an *AG*-3-band as well. Consequently, G is decomposable into an *AG*-band $T = \{1,3,5,7\}$ of commutative inverse semigroups $S_\alpha = \{\alpha, \alpha+1\}$, $\alpha \in T$ with the identity element. The band T is isomorphic with the unique *AG*-band of order 4 ($\mathbf{T_4}$ [6]), and multiplication in G_α is given by $\alpha\alpha = (\alpha+1)(\alpha+1) = \alpha$ and $\alpha(\alpha+1) = (\alpha+1)\alpha = \alpha+1$, $\alpha \in T$. Obviously, the semigroups S_α satisfy $x^2 = \alpha$, $\forall x \in S_\alpha$.

Acknowledgements Supported by Grant ON 144013 of the Ministry of Science through the Math. Inst. SANU.

References

1. Deneš, J., Keedwell, A.D.: Latin Squares and Their Applications. Akadémia Kiadó, Budapest (1974)
2. Holgate, P.: Groupoids satisfying a simple invertive law. Math Student. **61**, No 1–4, 101–106 (1992)
3. Kazim, M.A., Naseeruddin, M.: On almost semigroups. Aligarh Bull. Math. **2**, 1–7 (1972)
4. Protić, P.V., Stevanović, N.: On Abel-Grassmann's groupoids (review). In: Proceedings of the mathematical conference in Pristina. 31–38 (1994)
5. Protić, P.V., Stevanović, N.: *AG*-test and some general properties of Abel-Grassmann's groupoids. PU.M.A. **6**, No. 4, 371–383 (1995)
6. Protić, P.V., Stevanović, N.: Some relations on Abel-Grassmann's 3-bands. PU. M. A. Budapest-Siena, **14**, No. 1 (2003)
7. Protić, P.V., Stevanović, N.: Abel-Grassmann's bands. Quasigroups and related systems, (Moldavia - South Corea), Kišinjev No. 11, 95–101 (2004)

Context Hidden Markov Model for Named Entity Recognition

Branimir T. Todorović, Svetozar R. Rančić, Edin H. Mulalić

Dedicated to Professor Gradimir V. Milovanović on the occasion of his 60th birthday

1 Introduction

Named entity recognition (NER) is one of the most important subtasks of Information Extraction (IE) that seeks to locate and classify tokens (words, atomic elements) in unstructured text into predefined classes such as the names of persons, organizations, locations, expressions of times, quantities, monetary values, percentages, etc. The term "Named Entity" was coined for the Sixth Message Understanding Conference (MUC-6) [3] sponsored by DARPA (US Defense Advanced Research Projects Agency). MUC-6 and the following conference MUC-7 were focused on Information Extraction tasks, where structured information of company activities and defense-related activities is extracted from unstructured text such as newspaper articles. Identifying references to these entities in text is formally called "Named Entity Recognition and Classification (NERC)."

Named entity (NE) recognition is the core technology for understanding low-level semantics of texts. Applications of NER are numerous. For instance, extracting NEs is the first step toward automated ontology building. It is shown in [5] that

Branimir T. Todorović
Faculty of Science and Mathematics, University of Niš, Višegradska 33, 18000 Niš, Serbia
e-mail: branimirtodorovic@yahoo.com

Svetozar R. Rančić
Faculty of Science and Mathematics, University of Niš, Višegradska 33, 18000 Niš, Serbia
e-mail: rancicsv@yahoo.com

Edin H. Mulalić
Accordia Group LLC, Učitelj Tasina 38, Niš, Serbia, e-mail: edinmulalic@yahoo.com

The project was fully funded by Accordia Group LLC.

W. Gautschi et al. (eds.), *Approximation and Computation: In Honor of Gradimir V. Milovanović*, Springer Optimization and Its Applications 42,
DOI 10.1007/978-1-4419-6594-3_30,

such ontology could improve the performance of a Question Answering system. NER is also very important for automatic text summarization. In [4], named entity recognition enhanced the identification of important text segments, which were used in the summary.

Named Entity Recognition systems can roughly be divided into rule-based systems, which use linguistic grammar-based techniques, and stochastic machine learning systems. Early studies were mostly based on handcrafted rules, and recent ones use supervised machine learning. Handcrafted grammar-based systems usually give good results, however they demand months of development by experienced linguists.

The idea of supervised learning is to use a collection of annotated documents to train classifiers for the given set of NE classes. The current dominant techniques for addressing the NERC problem as supervised learning are Hidden Markov Models (HMM) [1], Maximum Entropy Models (ME) [2], and Support Vector Machines (SVM) [6].

In this paper we propose the Named Entity Recognition system based on Context Hidden Markov Model whose performance have been improved by the handcrafted grammar for named entity classes DATE, TIME, MONEY, and PERCENT.

In this paper we propose the combination of two classifiers for the Named Entity recognition problem. The first one is our modification of Hidden Markov Model, which we named Context Hidden Markov Model for named entity classes PERSON, ORGANIZATION, and LOCATION, and the second one is based on carefully crafted grammar rule-based systems for DATE, TIME, MONEY, and PERCENT. In our previous paper [7] we have compared the performance of our implementation of generative classifiers based on Hidden Markov Model described in [1] and the performance of our Context Hidden Markov Model and concluded that the Context HMM has better classification accuracy as well as implementation efficiency.

Here we report on our attempt to improve the classification accuracy of the Context HMM. We consider the problem of estimating the probabilities of events that have not been seen during the training. We have represented the final probability of some event as a weighted mixture and applied the Expectation Maximization algorithm for the estimation of the mixture weights. As will be presented in the final section, the best results were obtained when we enabled the algorithm to train the different probability mixture models for each of the named entity classes.

The rest of the paper is organized as follows. In the second section we describe the representation of the atomic elements of the text: tokens, words by using a two-element vector composed of token and feature vector. In the third section we compare the equations of the generative Hidden Markov Model and our Context Hidden Markov Model for named entity recognition. The maximum likelihood estimation of the transition probabilities is given in the fourth section. In the fifth section we explain briefly the problem of data sparseness and estimation of the probability of unseen events and propose the solution in the form of a probability mixture model. We are estimating the weights of the mixture using the Expectation Maximization algorithm described in sixth section. In the last section we have described the training scenario and the corresponding results.

2 Word-Feature Pairs

Following [1], we consider tokens (words, atomic elements) to be two-element vectors composed of word and word-features vector, denoted $< w, f >$. The feature vector is determined by the morphological and some contextual properties of the token (word). In our model, feature of each token is represented as a set of two subfeatures: f_1 - simple deterministic feature such as capitalization and digitalization and f_2 - semantic feature of important triggers.

Possible values for subfeature f_1 are shown in Table 1. Considering how important capital letters and number properties are in Roman languages, one can easily understand the importance of f_1. In our implementation, subfeature f_1 is obtained (during preprocessing) by specialized grammar rules.

Some words may be good indicators that a group of words preceding (or succeeding) them are members of a certain class. For example, in the sentence "Peter Johns works for Accordia LLC," the word LLC implies that the word Accordia is an organization. Trigger feature f_2 is responsible for capturing those semantic properties by searching for a word in lexicons of special prefixes and suffixes. A complete list of those lexicons is given in Table 2.

Beside the mentioned subfeatures, it is possible to include other syntax or semantic properties. Experiments with using POS-tag as a feature did not produce any improvement. Most state-of-the-art NER systems use well-populated dictionaries of personal names and last names, as well as dictionaries of location and organization names. Our system has not been enriched with those kinds of gazetteers yet, but early experiments showed that significant improvement can be obtained, both in precision and recall.

3 Context Hidden Markov Model for Named Entity Recognition

3.1 Hidden Markov Model in NERC

The Hidden Markov Model (HMM) can be considered as a probabilistic generative model of a sequence. Formally, HMM is a double stochastic process. The first process generates the sequence of states. In our case these states are NE classes, but considering more generally Natural Language Processing tasks, they can be Part of Speech Tags or Noun Phrase tags etc.

The stochastic process, responsible for generating the sequence of states (i.e., the sequence of classes), is described using the following assumptions. The first one is that the current state depends only on the previous state (it is known as first-order Markov assumption). In that case the transition probability is given by:

$$P(q_t \,|\, q_1, \ldots, q_{t-1}) = P(q_t \,|\, q_{t-1}).$$

The second assumption states that transition probabilities are time invariant $a_{ij} = P(q_{t-1} = i)$, $t = 1, 2, \ldots$. These assumptions define the $N \times N$ transition probability

Table 1 Subfeature f_1 values

f_1	Example	Explanation / Intuition
`OneDigitNum`	7	Digital number
`TwoDigitNum`	46	Two-digit year
`FourDigitNum`	1957	Four-digit year
`Ordinal`	123rd, 4th	Ordinal number
`YearDecade`	1990s, 90s, ’90s, mid-1990s	Year decade
`ContainsDigitAndDash`	12-08	Date
`ContainsDigitAndOneSlash`	12/08, 5/7	Date or fraction
`ContainsDigitAndTwoSlashes`	23/09/2008	Date
`ContainsDigitAndComma`	100,000	Money
`ContainsDigitAndPeriod`	3.14159	Money or percents
`OtherContainsDigit`	123	Other numbers
`AllCaps`	IBM, I-B-M, USA	Organization or country
`CapPeriod`	D.	Person name (initial)
`LowPeriod`	d.	Abbreviation
`CapOtherPeriod`	Mr., Corp.	Abbreviation
`CapPeriods`	I.B.M.	Abbreviation
`LowPeriods`	a.m.	Abbreviation
`InitialCap`	Peter	Capitalized word
`LowerCase`	walking	
`MixedCapitalAndLower`	USAir	Abbreviation (organization)
`Dot`	.	Punctuation
`Comma`	,	Punctuation
`Colon`	; : -	Punctuation
`QuestionMark`	?	Punctuation
`ExclamationMark`	!	Punctuation
`Elipsis`	..	Punctuation
`Quote`	’ ’’’’ ’’	Punctuation
`OpenBrackets`	(	Punctuation
`CloseBrackets`	)	Punctuation
`PossessiveEnding`	’s	Possessive ending
`Percent`	%	Percent
`Dollar`	$	Dollar
`Hash`	♯	Hash, Number
`Determiner`	A, a, An, AN, an, The, THE, the	Determiners
`BeginOfSentence`		Special value added to enable creating context around first word in sentence
`EndOfSentence`		Special value added to enable creating context around last word in sentence

matrix $A = \{a_{ij}\}$, where N denotes the number of states in the model, $a_{ij} \geq 0$ and $\sum_{j=1}^{N} a_{ij} = 1$. An initial state distribution $\prod = \{\pi_i\}$ has to be defined as well: $\pi_i = P(q+1 = i)$, $1 \leq i \leq N$, where $\pi_i \geq 0$ and $\sum_{i=1}^{N} \pi_i = 1$.

Table 2 Trigger feature f_2 values

NE Class	f_2	Example	Explanation / Intuition
Percent	SuffixPercent	%	Percentage suffix
Money	PrefixMoney	$	Money prefix
	SuffixMoney	Dollars	Money suffix
	CountryPossesive	Australian	Country description
Undefined	OperatorEqual	Around, nearly	Comparison operator
	OperatorGreater	Above, higher	Comparison operator
	OperatorLess	Less, lower	Comparison operator
	Cardinal	Six	Cardinal Numbers
	Ordinal	Sixth	Ordinal Numbers
Date	SuffixDate	Day	Date suffix
	WeekDate	Monday	Week date
	MonthDate	July	Month date
	SeasonDate	Summer	Season date
	PeriodDate1	Month	Period date
	PeriodDate2	Quarter	Quarter/Half of year
	PrefixDate	From, until	Date preffix
	QualifierDate	Last, next	Date descriptor
	YearWord	Twenties	Years described by word
	DeterminerDateTime	Earlier, early	Date descriptor
Time	SuffixTime	a.m.	Time suffix
	PeriodTime	Morning	Time period
Person	PrefixPersonTitle	Mr.	Person title
	PrefixPersonDesignation	President	Person designation
	FirstNamePerson	Michael	Person first name
Location	SuffixLocation	River	Location suffix
Organization	SuffixOrganization	Ltd	Organization suffix

The second stochastic process in HMM is responsible for generating the sequence of observations from the sequence of states. Usually, it is assumed that observation o_t at time step t depends solely on the current state q_t, that is $b_{ij} = p(o_t = j \mid q_t = i)$. This defines the $N \times M$ emission probability matrix $B = \{b_{ij}\}$, where M denotes the number of signals that can be emitted from each state, and for the probabilities we have $b_{ij} \geq 0$ and $\sum_{j=1}^{M} b_{ij} = 1$.

When considering the NE classification, the most likely sequence of NE classes has to be found, given the sequence of words. This problem is solved using the Viterbi decoding, whose task is to find the optimal sequence of states by maximizing the conditional probability of state sequence given the sequence of words and their features $P(\mathrm{NE}_{1:T} \mid <w,f>_{1:T}, \lambda)$. In [1], Bikel et al. applied Viterbi decoding for estimating the optimal sequence for the HMM model in which the probability for generating the first word of a name-class is factored into two parts: $P(\mathrm{NE}_t \mid \mathrm{NE}_{t-1}, w_{t-1}) \cdot p(<w,f>_t \mid \mathrm{NE}_t, \mathrm{NE}_{t-1})$. The top-level model for generating all but the first word in a name class is $P(<w,f>_t \mid <w,f>_{t-1}, NE)$.

The optimal sequence of the NE classes $\mathrm{NE}^*_{1:T}$ is obtained by maximization of conditional probability:

$$P(\mathrm{NE}_{1:T} \mid <w,f>_{1:T}, \lambda) = \frac{P(\mathrm{NE}_1 : T, <w,f>_{1:T} \mid \lambda)}{P(<w,f>_{1:T} \mid \lambda)},$$

that is,

$$\begin{aligned}\mathrm{NE}^*_{1:T} &= \arg\max_{\mathrm{NE}_{1:T}} P(\mathrm{NE}_{1:T}, < w, f >_{1:T} \mid \lambda) \\ &= \arg\max_{\mathrm{NE}_{1:T}} P(\mathrm{NE}_{1:T}, < w, f >_{1:T}, \lambda).\end{aligned}$$

The auxiliary variable

$$\delta_t(i) = \max_{NC_{1:t-1}} P(\mathrm{NE}_t = i, \mathrm{NE}_{1,t-1}, < w, f >_{1:t} \mid \lambda)$$

is introduced to derive a recursive algorithm known as Viterbi decoding:

$$\begin{aligned}\delta_t(i) = \max_{1 \le j \le N} \{ & P(< w, f >_t \mid \mathrm{NE}_t = i, \mathrm{NE}_{t-1} = j) \\ & \times P(\mathrm{NE}_t = i \mid \mathrm{NE}_{t-1} = j, < w, f >_{t-1}, \lambda) \delta_{t-1}(j) \}.\end{aligned}$$

In order to retrieve the sequence of NE classes, we keep track of arguments that maximize the auxiliary variable at each time step:

$$\begin{aligned}\psi_t(i) = \arg\max_{1 \le j \le N} \{ & P(< w, f >_t \mid \mathrm{NE}_t = i, \mathrm{NE}_{t-1} = j) \\ & \times P(\mathrm{NE}_t = i \mid \mathrm{NE}_{t-1} = j, < w, f >_{t-1}, \lambda) \delta_{t-1}(j) \}.\end{aligned}$$

3.2 Context Hidden Markov Model in NERC

The key concept that led to the derivation of our modification of HMM, which we call the Context HMM, is that the transition probability between successive states depends also on the context of surrounding words and their features. The formal derivation starts with the maximization of the conditional probability

$$\begin{aligned}\mathrm{NE}^*_{1:T} &= \arg\max_{\mathrm{NE}_{1:T}} P(\mathrm{NE}_{1:T} \mid < w, f >_{1:T} \lambda) \\ &= \arg\max_{\mathrm{NE}_{1:T}} P(\mathrm{NE}_{1:T}, < w, f >_{1:T} \mid \lambda).\end{aligned}$$

The maximum of the joint probability in the second line can be obtained as:

$$\begin{aligned}&\max_{\mathrm{NE}_{1:T}} P(\mathrm{NE}_{1:T}, < w, f >_{1:T} \mid \lambda) \\ &= \max_{1 \le i \le N} \left\{ \max_{\mathrm{NE}_{1,T-1}} P(\mathrm{NE}_T = i, \mathrm{NE}_{1:T-1}, < w, f >_{1:T} \mid \lambda) \right\}.\end{aligned}$$

Similarly to the original Viterbi decoding, we are introducing the auxiliary variable:

$$\delta_t(i) = \max_{1 \le j \le N} P(\mathrm{NE}_t = i, \mathrm{NE}_{1:t-1}, < w, f >_{1:T} \mid \lambda),$$

which also can be represented recursively, but in a simpler form than in the original algorithm:

$$\delta_t(i) = \max_{1\leq j\leq N} \{P(\mathrm{NE}_t = i \,|\, \mathrm{NE}_{t-1} = j, < w, f >_{1:T}, \lambda) \\ \times \max_{\mathrm{NE}_{1:t-2}} \{P(\mathrm{NE}_{t-1} = j, \mathrm{NE}_{1:t-2}, < w, f >_{1:T})\}\Big\}$$

and finally

$$\delta_t(i) = \max_{1\leq j\leq N} \{P(\mathrm{NE}_t = i \,|\, \mathrm{NE}_{t-1} = j, \mathrm{NE}_{1:t-2}, < w, f >_{1:T}, \lambda) \times \delta_{t-1}(j)\}\,.$$

Using the assumption that the current named entity class depends on the previous class and the appropriate context, which can be represented by a window of word feature pairs around the current word feature pair, the last equation becomes:

$$\delta_t(i) = \max_{1\leq j\leq N} \{P(\mathrm{NE}_t = i|\mathrm{NE}_{t-1} = j, < w, f >_{t-k\,:\,t+k}, \lambda) \times \delta_{t-1}(j)\}\,.$$

We also keep track of arguments that maximize the auxiliary variable at each time step, and we use backtracking to extract the optimal sequence of NE classes.

4 Training

The maximum likelihood estimate for the set of conditional probabilities

$$P(\mathrm{NE}_t/C_t) = P(\mathrm{NE}_t/\mathrm{NE}_{t-1}, < w, f >_{t-k\,:\,t+k}),$$

where we have used C_t to denote the context $\mathrm{NE}_{t-1}, < w, f >_{t-k\,:\,t+k}$, is obtained by maximization of the logarithm of the likelihood over the training corpus:

$$\bigwedge\{P(NE/C)\} = \log\prod_{t=1}^{T} P(\mathrm{NE}_t/C_t) = \sum_{(NE,C)} \sharp(NE,C)\log P(NE/C). \quad (1)$$

In the second equality of (1) the summation index is changed using the count definition $\sharp(NE,C)$ for the named entity class and its context. While maximizing (1), the constraint $\sum_{NE} P(NE/C) = 1$, $\forall C$, in the training corpus has to be satisfied.

The constrained maximization problem can be solved using the method of Lagrange multipliers. The corresponding Lagrangian is

$$E\{P(NE/C)\} = \sum_{(NE,C)} \sharp(NE,C)\log P(NE/C) + \mu\left(\sum_{NE} P(NE/C) - 1\right).$$

The solution is obtained after equating the partial derivatives of $E\{P(NE/C)\}$ with respect to each of the probabilities $P(NE/C)$ and the Lagrange multiplier μ:

$$\frac{\sharp(NE,C)}{P(NE/C)} + \mu = 0 \quad \text{and} \quad \sum_{NE} P(NE/C) - 1 = 0.$$

After summation over each named entity class NE, and using the constraint, we obtain

$$\mu = -\sharp C \quad \text{and} \quad P(NE/C) = \frac{\sharp(NE,C)}{\sharp(C)}. \tag{2}$$

An estimate in this form will assign zero probability to any event not seen in the training corpus. The probability of unseen events is directly linked to the problem of data sparseness, which we have solved using linear interpolation on back off models. We shall describe this in the following section.

5 The Sparseness of Data and the Expectation Maximization

The probabilities of transitions between states depend on the context of surrounding words and their features. Actual events of transition from one specific named entity class to another, in case of given surrounding context, are very sparse. In order to estimate the probabilities of events which were not seen during the training, we have to apply some kind of probability smoothing. We adopted the linear interpolation technique to solve the data sparseness problem.

First we have limited the width of the context to five word/feature pairs. Then we have divided such context into the eight groups of subcontexts (back off models), based on the estimated frequency of their appearance in the text.

The groups of subcontexts are given later:

- $\text{NE}_{t-1} f_{t-2} f_{t-1} w_t$, $d\text{NE}_{t-1} f_{t-1} w_t f_{t+1}$, $\text{NE}_{t-1} w_t f_{t+1} f_{t+2}$;
- $\text{NE}_{t-1} f_{t-1} w_t$, $\text{NE}_{t-1} w_t f_{t+1}$;
- $\text{NE}_{t-1} w_{t-2} f_{t-1} f_t$, $\text{NE}_{t-1} f_{t-2} w_{t-1} f_t$, $\text{NE}_{t-1} w_{t-1} f_t f_{t+1}$, $\text{NE}_{t-1} f_{t-1} f_t w_{t+1}$, $\text{NE}_{t-1} f_t w_{t+1} f_{t+2}$, $\text{NE}_{t-1} f_t f_{t+1} w_{t+2}$;
- $\text{NE}_{t-1} w_{t-1} f_t$, $\text{NE}_{t-1} f_t w_{t+1}$;
- $\text{NE}_{t-1} w_t$;
- $\text{NE}_{t-1} f_{t-2} f_{t-1} f_t$, $\text{NE}_{t-1} f_{t-1} f_t f_{t+1}$, $\text{NE}_{t-1} f_t f_{t+1} f_{t+2}$;
- $\text{NE}_{t-1} f_{t-1} f_t$, $\text{NE}_{t-1} f_t f_{t+1}$;
- NE_{t-1}.

The final probability $P(\text{NE}_t/C_t)P(\text{NE}_t/\text{NE}_{t-1}, <w,f>_{t-k\,t+k})$ is the weighted mixture of the group probabilities which are obtained as simple means of maximum likelihood probability estimates for each subcontext in the group (that is, as the mixture of equally probable events). The weights are given by:

$$P(\text{NE}_t/\text{NE}_{t-1}, <w,f>_{t-k\,t+k}) = \sum_{l=1}^{8} \mu_l P_l(\text{NE}_t \,|\, \text{NE}_{t-1}, C_l).$$

In the first scenario we have used one probability mixture model for all named entity classes and applied the Expectation Maximization algorithm to estimate the weights of the mixture. Note that the actual probabilities were obtained as maximum likelihood estimates, as described in the previous section. In the second scenario we have used a different mixture model for each Named Entity class (the same groups of subcontexts and different weights of the groups) and again used the EM algorithm to estimate the weights of each mixture model. In the next section we explain the general Expectation Maximization algorithm based on the short tutorial given in [2].

6 Expectation Maximization

The Expectation Maximization algorithm is an efficient iterative procedure for computing the Maximum Likelihood (ML) estimate in the presence of missing or hidden data. By using the Maximum Likelihood estimation we actually ensure that the observed data are the most likely for the estimated value of the unknown model parameters. Each iteration of the EM algorithm consists of two processes: The E-step, and the M-step. In the expectation, or E-step, the missing data are estimated given the observed data and the current estimate of the model parameters. In the M-step, the likelihood function is maximized under the assumption that the missing data are known. The algorithm is guaranteed to increase the likelihood of the observed data at each iteration.

Let $\mathbf{x}$ be the random vector which results from a parameterized family. Our goal is to find θ such that the likelihood $P(\mathbf{x}/\theta)$ is maximized. This is known as the Maximum Likelihood (ML) estimate for θ . In order to perform the maximization, it is more convenient to introduce the criteria function as the log likelihood function defined by $L(\theta) = \ln P(\mathbf{x}/\theta)$. Since we are maximizing $L(\theta)$, the next estimate of parameters θ should satisfy $L(\theta) > L(\theta_n)$, that is, we should maximize the following difference:

$$L(\theta) - L(\theta_n) = \ln P(\mathbf{x}/\theta) - \ln P(\mathbf{x}/\theta_n). \tag{3}$$

Let us now consider the problem of the unobserved or missing discrete variable Z with given realizations $\mathbf{z}_k$. The total probability $P(\mathbf{x}/\theta)$ may be obtained by marginalization of the hidden variable:

$$P(\mathbf{x}/\theta) = \sum_{z_k} P(\mathbf{x}, \mathbf{z}_k/\theta) = \sum_{z_k} P(x/\mathbf{z}_k, \theta) P(\mathbf{z}_k/\theta),$$

giving the new form of (3):

$$L(\theta) - L(\theta_n) = \ln \left(\sum_{z_k} P(\mathbf{x}/\mathbf{z}_k, \theta) P(\mathbf{z}_k/\theta) \right) - \ln P(\mathbf{x}/\theta_n).$$

Based on Jensen's inequality, we can show that the following holds:

$$\ln \left(\sum_k \lambda_k \mathbf{x}_k \right) \geq \sum_k \lambda_k \ln(\mathbf{x}_k),$$

for $\lambda_k \geq 0$ and $\sum_k \lambda_k = 1$. In place of the constant λ_k we will use in the following the probabilities $P(\mathbf{z}_k/\mathbf{x},\theta_n)$ for which we have the conditions $P(\mathbf{z}_k/\mathbf{x},\theta_n) \geq 0$ and $\sum_k P(\mathbf{z}_k/\mathbf{x},\theta_n) = 1$ satisfied:

$$\begin{aligned} L(\theta) - L(\theta_n) &= \ln\left(\sum_{z_k} P(\mathbf{x}/\mathbf{z}_k,\theta)P(\mathbf{z}_k/\theta)\frac{P(\mathbf{z}_k/\mathbf{x},\theta_n)}{P(\mathbf{z}_k/\mathbf{x},\theta_n)}\right) - \ln P(\mathbf{x}/\theta_n) \\ &\geq \sum_{z_k} P(\mathbf{z}_k/\mathbf{x},\theta_n)\ln\left(\frac{P(\mathbf{x}_k/\mathbf{z},\theta_n)P(\mathbf{z},\theta_n)}{P(\mathbf{z}_k/\mathbf{x},\theta_n)}\right) - \ln P(\mathbf{x}/\theta_n) \\ &= P(\mathbf{z}_k,\mathbf{x},\theta)\ln\left(\frac{P(\mathbf{x},\mathbf{z}/\theta)}{P(\mathbf{x},\mathbf{z}/\theta_n)}\right) \triangleq \Delta(\theta/\theta_n). \end{aligned}$$

Based on the previous inequality: $L(\theta) \geq L(\theta_n) + \Delta(\theta/\theta_n)$ and, after introducing the new function $l(\theta/\theta_n) = L(\theta_n) + \Delta(\theta/\theta_n)$, we have: $L(\theta) \geq l(\theta/\theta_n)$ (Fig. 1).

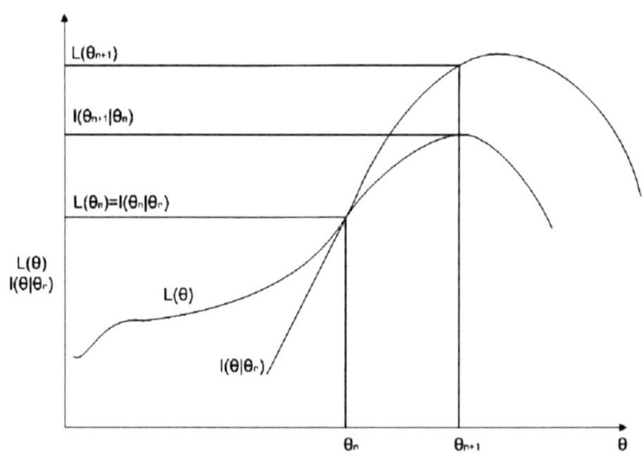

Fig. 1 Graphical interpretation of a single iteration of the EM algorithm

The function $l(\theta/\theta_n) = L(\theta_n) + \Delta(\theta/\theta_n)$ is bounded above by the likelihood function $L(\theta)$; it can be show that $l(\theta/\theta_n) = L(\theta_n)$ and therefore any θ which increases $l(\theta/\theta_n)$ also increases $L(\theta)$. The EM algorithm selects the next estimate of parameters θ such that $l(\theta/\theta_n)$ is maximized. After removing the terms which are constant with respect to θ, we obtain:

$$\theta_{n+1} = \arg\max_{\theta}\left\{\sum_{z_k} P(\mathbf{z}_k/\mathbf{x},\theta_n)\ln P(\mathbf{x},\mathbf{z}_k,/\theta_n)\right\} = \arg\max_{\theta} E_{\mathbf{z}/\mathbf{x},\theta_n}\left[\ln P(\mathbf{x},\mathbf{z}/\theta_n)\right].$$

The final equation illustrates the expectation and maximization steps of the EM algorithm: in the E step, the conditional expectation $E_{\mathbf{z}/\mathbf{x},\theta_n}[\ln P(\mathbf{x},\mathbf{z}/\theta_n)]$ is obtained as a function of the unknown parameter θ, and in the M step, this function (conditional expectation) is maximized with respect to θ.

7 Grammar Component of the NERC System

In order to improve the accuracy of our NE recognition system, we have developed a grammar with rules that define NE like TIMEX (DATE and TIME) and NUMEX (MONEY and PERCENT). This grammar is used in combination with Context HMM. TIMEX and NUMEX NE classes have a wide spectrum of possible forms to be represented. We have tried to describe and envelop a widely used and recognizable set of such expressions, which can be found in documents, in a formal way. Our intention was to describe, using grammatical rules, a small subset of natural language as a domain-specific representation of TIMEX and NUMEX. To be more precise, our grammar and parser are able to recognize parts of sentences that contain valid inscription of DATE, TIME, MONEY and PERCENT.

Internally we maintain dictionaries which contain collections of words with the same or similar meaning if it is used in context of the four mentioned categories. So we have, for example, collections for:

- Words describing numbers (NUMBER),
- Abbreviations for time zones (TIME SUFFIX),
- Time intervals equal or longer than one day (DATE PERIOD),
- Time units shorter than day (TIME PERIOD),
- Prepositions in front of Date sequences (DATE PREFIX).

In a preprocessing phase we perform a search in the collections, and if a word is found in any of them, it is substituted with the word which represents the whole collection. During parsing we try to find a grammar rule that describes any of the four categories.

Lowest-level rules define numbers, both cardinal and ordinal, written by digits or words. The simplest category is PERCENT, as it consists of number and percentage symbol or word: < percent > ::= < number > PERCENT.

Rules for MONEY contain number, currency and optionally country:
< money > ::= < number >< countryCurrency >.

Rules for TIME cover a wide range of definitions of exact time, which possibly include time zone, as well as time interval. Less precise time tokens like morning, evening, afternoon, and its starting or ending part also.

Rules for DATE recognize date appearance both as single and as date interval. They cover formal and informal date units longer than day, including weeks, quarter, fiscal year up to millennium, etc. There are also mixed rules and the parser is able to recognize TIME description wrapped around DATE, as part of it, as in the example:

- Yesterday - DATE;
- Early yesterday - TIME;
- Early yesterday morning - TIME.

Rules with qualifiers this, last, next etc., help recognize both time and date qualified description. The grammar contains more than 150 rules for TIME and more than 400 rules for DATE.

General and comprehensive rules improve the recall, but reduce the precision of the NERC system. We have solved this problem by introducing rules that handle exceptions for TIME and DATE.

8 Experimental Results

The results which we discuss here are obtained on the MUC 7 text corpus. The documents from the training part of the MUC 7 corpus were divided in two fractions. One was used for maximum likelihood estimation of the subcontext probabilities, and the other was used by the EM algorithm, which estimated the weights of the probability mixtures. The training was repeated with the roles of the training fractions changed and the final weights of the mixture were obtained by averaging.

The performance of NER systems is measured using the precision (P), recall (R) and F-measure. The precision (P) measures the number of correct NEs, obtained by the NER system, over the total number of NEs extracted by the NER system. The recall (R) measures the number of correct NEs, obtained by the NER system over the total number of NEs in a text that has been used for testing. The F-measure represents the harmonic mean of precision and recall:

$$F = \frac{RP}{R+P}.$$

The results reported in Tables 3 and 4 are obtained using only the Context Hidden Markov Model. In Table 3 we give the precision, recall and F-measure of the classifier for each class, when only one probability mixture model for all classes is estimated using the EM. Results shown in Table 4 were obtained when each named entity class had its own probability mixture model whose weights were estimated using the EM.

Results reported in Tables 5 and 6 are obtained when the Context Hidden Markov model was combined with the classifier based on handcrafted grammar rules.

Table 3 Classification results for Context HMM, without the grammar, multiple probability model

NEClass	P(%)	R(%)	F(%)
Person	94.36	95.52	94.93
Organization	92.10	88.95	90.50
Location	95.45	86.50	90.75
Date	96.73	91.83	94.22
Time	98.04	84.14	90.56
Money	98.06	96.19	97.12
Percent	100.0	100.0	100.0
Total	94.57	90.07	92.27

Table 5 is obtained with one probability mixture model for all classes and Table 6 with eight mixture models, that is, one for each class. The best results were obtained when the Context HMM was combined with the handcrafted rules, and the separate mixture models were used for each Named Entity Class.

Table 4 Classification results for Context HMM without the grammar, single probability mixture models

NEClass	P(%)	R(%)	F(%)
Person	95.04	90.95	92.95
Organization	89.95	89.34	89.65
Location	89.39	90.18	89.78
Date	95.99	97.88	96.93
Time	98.69	98.69	98.69
Money	100.0	100.0	100.0
Percent	100.0	100.0	100.0
Total	92.46	91.99	92.22

Table 5 Classification results for Context HMM combined with the grammar, multiple probability mixture model

NEClass	P(%)	R(%)	F(%)
Person	94.44	95.61	95.02
Organization	92.10	88.90	90.47
Location	95.26	86.12	90.46
Date	95.99	97.88	96.93
Time	98.69	98.69	98.69
Money	100.0	100.0	100.0
Percent	100.0	100.0	100.0
Total	94.53	91.73	93.11

Table 6 Classification results for Context HMM combined with the grammar, single probability mixture models

NEClass	P(%)	R(%)	F(%)
Person	95.05	90.29	92.61
Organization	89.71	88.50	89.10
Location	89.25	89.97	89.61
Date	95.77	87.85	91.64
Time	96.97	83.58	89.78
Money	98.04	95.24	96.62
Percent	100.0	100.0	100.0
Total	92.10	89.08	90.57

References

1. Bikel, D., Schwartz, R., Weischedel, R.: An algorithm that learns what's in a name. Machine Learning, Vol. **34**, No 1-3 211–231 (1939)
2. Borthwick, A.: A Maximum Entropy Approach to Bamed Entity Recognition. Ph.D. Thesis, New York University (1999)
3. Grishman, R., Sundheim, B.: Message Understanding Conference - 6: A brief history. In: Proc. 16th Int'l Conf. on Computational Linguistic (COLING-96), Copenhagen, 466–471 (1996)
4. Mayfield, J., McNamee, P., Piatko, C.: Named entity recognition using hundreds of thousands of features. In: Proc. of the seventh conference on Natural language learning at HLT-NAACL 2003, Volume 4, 184–187 (2003)
5. Hassel, M.: Exploitation of named entities in automatic text summarization for Swedish. In: 14th Nordic Conference of Computational Linguistics, NODALIDA (2003)
6. Mann, G.: Fine-grained proper noun ontologies for question answering. In: SemaNet'02: Building and Using Semantic Networks (2002)
7. Todorović, B., Rančić, S., Marković, I., Mulalić, E., Ilić, V.: Named entity recognition and classification using context hidden Markov model. In: Proc. of the Ninth Symposium on Neural Networks Applications in Electrical Engineering (NEUREL2008), Belgrade, Serbia, (2008)

On the Interpolating Quadratic Spline

Zlatko Udovičić

Dedicated to Professor Gradimir V. Milovanović
on the occasion of his 60th birthday

1 Introduction

Definition 1. Let $\Delta = \{a = x_0 < x_1 < \cdots < x_n = b\}$ be a partition of the interval $[a,b]$. The function $S_\Delta : [a,b] \to \mathbb{R}$, with the following properties:

1. S_Δ is an algebraic polynomial of second degree on every subinterval $[x_k, x_{k+1}]$, $0 \le k \le n-1$,
2. $S_\Delta \in C^1[a,b]$

is called a quadratic spline relative to the partition Δ. If the interpolating properties

- $S_\Delta(x_k) = f(x_k), \quad 0 \le k \le n,$

are added to the previous definition, a quadratic interpolating spline of the function f will be obtained, usually denoted by $S_{\Delta,f}$.

Much more details on splines can be found in [1].

Let the values $f(x_k)$, $0 \le k \le n$, of the function f, be given. From property 1° of the above definition, it follows that

$$x \in [x_k, x_{k+1}] \Rightarrow S_{\Delta,f}(x) = a_k x^2 + b_k x + c_k = s_k(x), \quad 0 \le k \le n-1,$$

Zlatko Udovičić
Faculty of Sciences, Department of Mathematics, Zmaja od Bosne 33-35, 71000 Sarajevo, Bosnia and Herzegovina, e-mail: zzlatko@pmf.unsa.ba

W. Gautschi et al. (eds.), *Approximation and Computation: In Honor of Gradimir V. Milovanović*, Springer Optimization and Its Applications 42,
DOI 10.1007/978-1-4419-6594-3_31,

which means that there are $3n$ unknown coefficients. To determine these coefficients, $2n$ equations will be obtained from the interpolating properties, while $n-1$ equations will be obtained from property 2° of the same definition. Thus, there is one parameter which can be freely chosen. The respective equations are:

$$a_k x_k^2 + b_k x_k + c_k = f(x_k), \quad 0 \le k \le n-1, \tag{1}$$

$$a_k x_{k+1}^2 + b_k x_{k+1} + c_k = f(x_{k+1}), \quad 0 \le k \le n-1, \tag{2}$$

$$2a_k x_{k+1} + b_k = 2a_{k+1} x_{k+1} + b_{k+1}, \quad 0 \le k \le n-2. \tag{3}$$

After subtracting (1) from (2), one obtains:

$$b_k = \frac{f(x_{k+1}) - f(x_k)}{x_{k+1} - x_k} - a_k(x_{k+1} + x_k).$$

The last equation, together with (3), after an elementary calculation, gives the recurrence formulas for calculating the coefficients a_k, $1 \le k \le n-1$. Hence, the coefficients of the quadratic interpolating spline can be calculated in the following way:

$$\begin{aligned}
&a_0 \;-\; \text{arbitrary},\\
&a_{k+1} = \frac{f(x_{k+2}) - f(x_{k+1})}{(x_{k+2} - x_{k+1})^2} - \frac{f(x_{k+1}) - f(x_k)}{(x_{k+2} - x_{k+1})(x_{k+1} - x_k)} - a_k \frac{x_{k+1} - x_k}{x_{k+2} - x_{k+1}},\\
&\qquad\qquad 0 \le k \le n-2,\\
&b_k = \frac{f(x_{k+1}) - f(x_k)}{x_{k+1} - x_k} - a_k(x_{k+1} + x_k), \quad 0 \le k \le n-1,\\
&c_k = f(x_k) - a_k x_k^2 - b_k x_k, \quad 0 \le k \le n-1.
\end{aligned}$$

If the partition of the interval $[a,b]$ is uniform, i.e., if $x_k = x_0 + kh$, $0 \le k \le n$, where $x_0 = a$ and $h = (b-a)/n$, the recurrence relations for calculating the coefficients a_k and b_k becomes much more simple:

$$\begin{aligned}
a_k &= \frac{f(x_{k+1}) - 2f(x_k) + f(x_{k-1})}{h^2} - a_{k-1}\\
&= \sum_{i=1}^{k} (-1)^{k-i} \frac{f(x_{i+1}) - 2f(x_i) + f(x_{i-1})}{h^2} + (-1)^k a_0, \quad 1 \le k \le n-1,
\end{aligned}$$

and

$$b_k = \frac{f(x_{k+1}) - f(x_k)}{h} - a_k(x_{k+1} + x_k), \quad 0 \le k \le n-1. \tag{4}$$

2 Basic Results

To obtain an error estimate, we used a technique similar to one used in [2] for the error estimate of the approximation by the cubic interpolating spline. From here on, we assume a uniform partition. We begin with two auxiliary lemmas.

Lemma 1. *Let $f \in C^4[a,b]$. Then*

$$s_k'(x_k) - f'(x_k) = \frac{h^2}{6} f'''(x_k) + \frac{h^3}{24} f^{iv}(\alpha_k) - h\left(a_k - \frac{1}{2} f''(x_k)\right)$$

and

$$s_k'(x_{k+1}) - f'(x_{k+1}) = \frac{h^2}{6} f'''(x_{k+1}) - \frac{h^3}{24} f^{iv}(\beta_k) + h\left(a_k - \frac{1}{2} f''(x_{k+1})\right),$$

for some $\alpha_k, \beta_k \in (x_k, x_{k+1})$, $0 \le k \le n-1$.

Proof. By using the relation (4) and Taylor expansion of the function f, we obtain

$$\begin{aligned} s_k'(x_k) - f'(x_k) &= 2a_k x_k + b_k - f'(x_k) \\ &= 2a_k x_k + \frac{1}{h}\left(f(x_{k+1}) - f(x_k)\right) - a_k(x_{k+1} + x_k) - f'(x_k) \\ &= -a_k h + \frac{1}{h}\left(hf'(x_k) + \frac{h^2}{2} f''(x_k) + \frac{h^3}{6} f'''(x_k) + \frac{h^4}{24} f^{iv}(\alpha_k)\right) - f'(x_k) \\ &= \frac{h^2}{6} f'''(x_k) + \frac{h^3}{24} f^{iv}(\alpha_k) - h\left(a_k - \frac{1}{2} f''(x_k)\right), \end{aligned}$$

which proves the first equality.

The second equality can be proved in a similar way. □

Lemma 2. *Let $f \in C^4[a,b]$ and let $M_4 = \max_{x\in[a,b]} \left|f^{iv}(x)\right|$. If*

$$a_0 = \frac{1}{4}\left(3f''(x_1) - f''(x_2)\right),$$

then the following inequality holds:

$$\left|2a_k - \frac{1}{2}\left(f''(x_k) + f''(x_{k+1})\right)\right| \le \frac{4k-3}{6} M_4 h^2, \quad 1 \le k \le n-1.$$

In particular, we have the following inequality

$$\left|2a_0 - \frac{1}{2}\left(f''(x_0) + f''(x_1)\right)\right| \le \frac{1}{2} M_4 h^2.$$

Proof. Because of simplicity we use the notation f''_k and f'''_k instead of $f''(x_k)$ and $f'''(x_k)$, respectively. We also recall

$$\frac{f(a+h)-2f(a)+f(a-h)}{h^2}=f''(a)+\frac{h^2}{12}f^{iv}(\theta),$$

for some $\theta\in(a-h,a+h)$.

If the number k is even, i.e., if $k=2m$ for some positive integer m, we have

$$\begin{aligned}
a_{2m}-\frac{1}{2}f''_{2m} &= \sum_{i=1}^{2m}(-1)^i\left(f''_i+\frac{h^2}{12}f^{iv}(\tau_i)\right)+a_0-\frac{1}{2}f''_{2m}\\
&= \frac{1}{2}f''_{2m}+\sum_{i=1}^{2m-1}(-1)^i f''_i+a_0+\frac{h^2}{12}\sum_{i=1}^{2m}(-1)^i f^{iv}(\tau_i)\\
&= \frac{1}{2}\sum_{i=1}^{m-1}(f''_{2i+2}-f''_{2i+1})+\frac{1}{2}\sum_{i=1}^{m-1}(f''_{2i}-f''_{2i+1})\\
&\quad+\frac{1}{2}f''_2-f''_1+a_0+\frac{h^2}{12}\sum_{i=1}^{2m}(-1)^i f^{iv}(\tau_i)\\
&= \frac{1}{2}\sum_{i=1}^{m-1}\left(hf'''_{2i+1}+\frac{h^2}{2}f^{iv}(\xi_{2i+1})-hf'''_{2i+1}+\frac{h^2}{2}f^{iv}(\eta_{2i+1})\right)\\
&\quad+\frac{1}{2}f''_2-f''_1+a_0+\frac{h^2}{12}\sum_{i=1}^{2m}(-1)^i f^{iv}(\tau_i)\\
&= a_0-f''_1+\frac{1}{2}f''_2\\
&\quad+\frac{h^2}{4}\sum_{i=1}^{m-1}\left(f^{iv}(\xi_{2i+1})+f^{iv}(\eta_{2i+1})\right)+\frac{h^2}{12}\sum_{i=1}^{2m}(-1)^i f^{iv}(\tau_i),
\end{aligned}$$

for some $\tau_i\in(x_{i-1},x_{i+1})$, $1\le i\le 2m$ and some $\xi_{2i+1}\in(x_{2i+1},x_{2i+2})$, $\eta_{2i+1}\in(x_{2i},x_{2i+1})$, $1\le i\le m-1$.

In this case (if the number k is even) we also have

$$\begin{aligned}
a_{2m}-\frac{1}{2}f''_{2m+1} &= \sum_{i=1}^{2m}(-1)^i\left(f''_i+\frac{h^2}{12}f^{iv}(\tau_i)\right)+a_0-\frac{1}{2}f''_{2m+1}\\
&= -\frac{1}{2}f''_{2m+1}+\sum_{i=1}^{2m}(-1)^i f''_i+a_0+\frac{h^2}{12}\sum_{i=1}^{2m}(-1)^i f^{iv}(\tau_i)
\end{aligned}$$

$$=-\frac{1}{2}\sum_{i=1}^{m}(f''_{2i+1}-f''_{2i})-\frac{1}{2}\sum_{i=1}^{m}(f''_{2i-1}-f''_{2i})-\frac{1}{2}f''_1+a_0+\frac{h^2}{12}\sum_{i=1}^{2m}(-1)^i f^{iv}(\tau_i)$$

$$= -\frac{1}{2}\sum_{i=1}^{m}\Big(hf_{2i}''' + \frac{h^2}{2}f^{iv}(\xi_{2i}) - hf_{2i}''' + \frac{h^2}{2}f^{iv}(\eta_{2i})\Big) - \frac{1}{2}f_1'' + a_0 + \frac{h^2}{12}\sum_{i=1}^{2m}(-1)^i f^{iv}(\tau_i)$$

$$= a_0 - \frac{1}{2}f_1'' - \frac{h^2}{4}\sum_{i=1}^{m}\left(f^{iv}(\xi_{2i}) + f^{iv}(\eta_{2i})\right) + \frac{h^2}{12}\sum_{i=1}^{2m}(-1)^i f^{iv}(\tau_i),$$

for some $\tau_i \in (x_{i-1}, x_{i+1})$, $1 \le i \le 2m$, and some $\xi_{2i} \in (x_{2i}, x_{2i+1})$, $\eta_{2i} \in (x_{2i-1}, x_{2i})$, $1 \le i \le m$.

According to the previous conclusions, we have that

$$2a_{2m} - \frac{1}{2}\left(f_{2m}'' + f_{2m+1}''\right) = 2a_0 - \frac{3}{2}f_1'' + \frac{1}{2}f_2'' + \frac{h^2}{4}\left[\sum_{i=1}^{m-1}\left(f^{iv}(\xi_{2i+1}) + f^{iv}(\eta_{2i+1})\right) - \sum_{i=1}^{m}\left(f^{iv}(\xi_{2i}) + f^{iv}(\eta_{2i})\right)\right] + \frac{h^2}{6}\sum_{i=1}^{2m}(-1)^i f^{iv}(\tau_i),$$

and with the choice $a_0 = \left(3f''(x_1) - f''(x_2)\right)/4$, we conclude that

$$\left|2a_{2m} - \frac{1}{2}\left(f_{2m}'' + f_{2m+1}''\right)\right| \le \frac{h^2}{4}\left[2(m-1)M_4 + 2mM_4\right] + \frac{h^2}{6}2mM_4 = \frac{4k-3}{6}M_4h^2.$$

The case when the number k is odd can be proved in a similar way. The last inequality can be verified directly. □

By using the previous lemmas it is easy to check that with the choice

$$a_0 = \frac{1}{4}\left(3f''(x_1) - f''(x_2)\right),$$

for $1 \le k \le n-1$, the following inequalities

$$\left|s_k'(x_{k+1}) - f'(x_{k+1}) - \left(s_k'(x_k) - f'(x_k)\right)\right| \le \frac{8k-3}{12}M_4h^3 \tag{5}$$

hold, and in particular,

$$\left|s_0'(x_1) - f'(x_1) - \left(s_0'(x_0) - f'(x_0)\right)\right| \le \frac{3}{4}M_4h^3.$$

Theorem 1. *Let $f \in C^4[a,b]$, $M_i = \max_{x\in[a,b]}|f^{(i)}(x)|$ for $i \in \{3,4\}$, and let $a_0 = \left(3f''(x_1) - f''(x_2)\right)/4$. For $x \in (x_k, x_{k+1})$, $1 \le k \le n-1$, the following inequalities hold:*

$$\left|s_k''(x) - f''(x)\right| \le M_3h + \frac{8k-3}{12}M_4h^2,$$

$$\left|s_k'(x) - f'(x)\right| \le M_3h^2 + \frac{8k-3}{12}M_4h^3,$$

$$\left|s_k(x) - f(x)\right| \le M_3h^3 + \frac{8k-3}{12}M_4h^4.$$

Especially, for $x \in (x_0, x_1)$, we have

$$\left|s_0''(x) - f''(x)\right| \le M_3 h + \frac{3}{4} M_4 h^2,$$

$$\left|s_0'(x) - f'(x)\right| \le M_3 h^2 + \frac{3}{4} M_4 h^3,$$

$$\left|s_0(x) - f(x)\right| \le M_3 h^3 + \frac{3}{4} M_4 h^4.$$

Proof. For $x \in (x_k, x_{k+1})$, $1 \le k \le n-1$, we have

$$\begin{aligned}
s_k''(x) - f''(x) &= 2a_k - f''(x) = \frac{1}{h}(2a_k x_{k+1} + b_k - 2a_k x_k - b_k) - f''(x) \\
&= \frac{1}{h}\left[s_k'(x_{k+1}) - f'(x_{k+1}) - (s_k'(x_k) - f'(x_k))\right] \\
&\quad + \frac{1}{h}\left[f'(x_{k+1}) - f'(x) - (f'(x_k) - f'(x))\right] - f''(x) \\
&= \frac{1}{h}\left[s_k'(x_{k+1}) - f'(x_{k+1}) - (s_k'(x_k) - f'(x_k))\right] \\
&\quad + \frac{1}{h}\Bigg[(x_{k+1} - x)f''(x) + \frac{(x_{k+1} - x)^2}{2} f'''(\lambda_{k+1}) \\
&\quad - (x_k - x)f''(x) - \frac{(x_k - x)^2}{2} f'''(\lambda_k)\Bigg] - f''(x) \\
&= \frac{1}{h}\left[s_k'(x_{k+1}) - f'(x_{k+1}) - (s_k'(x_k) - f'(x_k))\right] \\
&\quad + \frac{1}{2h}\left[(x_{k+1} - x)^2 f'''(\lambda_{k+1}) - (x_k - x)^2 f'''(\lambda_k)\right],
\end{aligned}$$

for some $\lambda_k \in (x_k, x)$ and $\lambda_{k+1} \in (x, x_{k+1})$. Now, the first inequality follows from (5) and the last equality.

Since

$$s_k(x_k) - f(x_k) = s_k(x_{k+1}) - f(x_{k+1}),$$

there exists $X \in (x_k, x_{k+1})$ such that $s_k'(X) - f'(X) = 0$. Now, we have that

$$s_k'(x) - f'(x) = \int_X^x \left(s_k''(t) - f''(t)\right) \mathrm{d}t,$$

from which the second inequality follows immediately.

Finally, the last inequality follows from the fact that

$$s_k(x) - f(x) = \int_{x_k}^x \left(s_k'(t) - f'(t)\right) \mathrm{d}t.$$

If $x \in (x_0, x_1)$, the corresponding inequalities can be verified directly. □

3 A Note

Since the error estimate, described by the last theorem, depends on the position of the point x, the obtained estimates are very useful if the point x belongs to the first half of the partition. If the point x belongs to the second half of the partition, after renumbering the nodes $a = x_{-n}$, $b = x_0$, $x_{-k} - x_{-k-1} = h$, $0 \le k \le n-1$, and changing the labels $x \in [x_{-k-1}, x_{-k}] \Rightarrow S_{\Delta,f}(x) = a_{-k}x^2 + b_{-k}x + c_{-k} = s_{-k}(x)$, $0 \le k \le n-1$, one can check (in the same way) that the following theorem holds.

Theorem 2. *Let $f \in C^4[a,b]$, $M_i = \max_{x\in[a,b]} |f^{(i)}(x)|$ for $i \in \{3,4\}$, and let $a_0 = (3f''(x_{-1}) - f''(x_{-2}))/4$. Then, for $x \in (x_{-k-1}, x_{-k})$, $1 \le k \le n-1$, the following inequalities hold:*

$$\left|s''_{-k}(x) - f''(x)\right| \le M_3 h + \frac{8k-3}{12} M_4 h^2,$$

$$\left|s'_{-k}(x) - f'(x)\right| \le M_3 h^2 + \frac{8k-3}{12} M_4 h^3,$$

$$\left|s_{-k}(x) - f(x)\right| \le M_3 h^3 + \frac{8k-3}{12} M_4 h^4$$

Especially, for $x \in (x_{-1}, x_0)$ we have the following inequalities

$$\left|s''_0(x) - f''(x)\right| \le M_3 h + \frac{3}{4} M_4 h^2,$$

$$\left|s'_0(x) - f'(x)\right| \le M_3 h^2 + \frac{3}{4} M_4 h^3,$$

$$\left|s_0(x) - f(x)\right| \le M_3 h^3 + \frac{3}{4} M_4 h^4.$$

Acknowledgements This research was partially supported by the grant No. 11-14-21618.1/2007, Government of the Municipality of Sarajevo, Ministry of Science and Education.

References

1. Chui, K.C.: An Introduction to Wavelets. Academic Press Inc., New York (1992)
2. Stoer, J.: Numerische Mathematik 1, 7th ed. Springer-Verlag, Berlin (1994)
3. Udovičić, Z.: Some modifications of the trapezoidal rule. Sarajevo J. Math. Vol. 2 (15) No. 2, 237–245 (2006)

Visualization of Infinitesimal Bending of Curves

Ljubica S. Velimirović, Svetozar R. Rančić, Milan Lj. Zlatanović

Dedicated to Professor Gradimir V. Milovanović on the occasion of his 60th birthday

1 Preliminaries

Infinitesimal bending of surfaces and manifolds was widely studied in [1, 2, 9, 10, 12, 17–19]. Infinitesimal bending of curves in E^3 was studied in [5, 15, 19, 20]. Such problems have physical applications (in the study of elasticity, for example) and have a long history. This work presents a follow-up of the results given in [19].

First we give some basic facts, definitions and theorems discussed in [5] and [19].

Definition 1. Consider a continuous regular curve

$$C : \mathbf{r} = \mathbf{r}(u), \tag{1}$$

included in a family of the curves

$$C_\varepsilon : \mathbf{r}_\varepsilon = \mathbf{r}(u) + \varepsilon \mathbf{z}(u), \quad \varepsilon \geq 0, \ \varepsilon \to 0,$$

Ljubica S. Velimirović
Faculty of Science and Mathematics, University of Niš, Višegradska 33, 18000 Niš, Serbia
e-mail: vljubica@pmf.ni.ac.rs

Svetozar R. Rančić
Faculty of Science and Mathematics, University of Niš, Višegradska 33, 18000 Niš, Serbia
e-mail: rancicsv@yahoo.com

Milan Lj. Zlatanović
Faculty of Science and Mathematics, University of Niš, Višegradska 33, 18000 Niš, Serbia
e-mail: zlatmilan@yahoo.com

The first and the third author were supported by Project 144032D MNTR Serbia

W. Gautschi et al. (eds.), *Approximation and Computation: In Honor of Gradimir V. Milovanović*, Springer Optimization and Its Applications 42,
DOI 10.1007/978-1-4419-6594-3_32,

where u is a real parameter and we get C for $\varepsilon = 0$ $(C \equiv C_0)$. A family of curves C_ε is an *infinitesimal bending of the curve* C if $\mathrm{d}s_\varepsilon^2 - \mathrm{d}s^2 = o(\varepsilon)$,where $\mathbf{z} = \mathbf{z}(u)$ is the *infinitesimal bending field* of the curve C.

Theorem 1 ([5]). *A necessary and sufficient condition for* $\mathbf{z}(u)$ *to be an infinitesimal bending field of the curve* C *is* $\mathrm{d}\mathbf{r} \cdot \mathrm{d}\mathbf{z} = 0$.

The next theorem is related to the determination of an infinitesimal bending field of the curve C.

Theorem 2 ([19]). *An infinitesimal bending field for the curve* C*, given by* (1)*, is*

$$\mathbf{z}(u) = \int [p(u)\mathbf{n}(u) + q(u)\mathbf{b}(u)]\,\mathrm{d}u + \mathrm{const.}, \tag{2}$$

where $p(u)$, $q(u)$ *are arbitrary integrable functions, and the vectors* $\mathbf{n}(u)$, $\mathbf{b}(u)$ *are, respectively, the unit principal normal and binormal vector field of the curve* C *and* const. *is an arbitrary constant.*

Here we exclude the trivial infinitesimal bending. Considering that the unit binormal and principal normal vector fields are

$$\mathbf{b} = \frac{\dot{\mathbf{r}} \times \ddot{\mathbf{r}}}{|\dot{\mathbf{r}} \times \ddot{\mathbf{r}}|}, \qquad \mathbf{n} = \frac{(\dot{\mathbf{r}} \cdot \dot{\mathbf{r}})\ddot{\mathbf{r}} - (\dot{\mathbf{r}} \cdot \ddot{\mathbf{r}})\dot{\mathbf{r}}}{|\dot{\mathbf{r}}||\dot{\mathbf{r}} \times \ddot{\mathbf{r}}|},$$

an infinitesimal bending field can be written in the form

$$\mathbf{z}(u) = \int \left[p(u) \frac{(\dot{\mathbf{r}} \cdot \dot{\mathbf{r}})\ddot{\mathbf{r}} - (\dot{\mathbf{r}} \cdot \ddot{\mathbf{r}})\dot{\mathbf{r}}}{|\dot{\mathbf{r}}||\dot{\mathbf{r}} \times \ddot{\mathbf{r}}|} + q(u) \frac{\dot{\mathbf{r}} \times \ddot{\mathbf{r}}}{|\dot{\mathbf{r}} \times \ddot{\mathbf{r}}|} \right] du,$$

where $p(u)$, $q(u)$ are arbitrary integrable functions, or in the form

$$\mathbf{z}(u) = \int [P_1(u)\dot{\mathbf{r}} + P_2(u)\ddot{\mathbf{r}} + Q(u)(\dot{\mathbf{r}} \times \ddot{\mathbf{r}})]\,du,$$

where $P_i(u)$, $i = 1,2$, $Q(u)$ are arbitrary integrable functions, too.
Based on the above the next theorem is almost evident.

Theorem 3. *Any closed curve is infinitesimally not rigid.*

Remark 1. Infinitesimal deformations of a special kind where considered in [15].

Example 1. Consider an ellipse

$$C: \frac{x^2}{a^2} + \frac{y^2}{b^2} = 1,$$

or, in parametric form,

$$C: r(u) = (a\cos u,\, b\sin u,\, 0),$$

for $a = 2$ and $b = 1$; we choose an infinitesimal bending field $\mathbf{z}(u)$ given by (2), where

$$p(u) = \sqrt{\cos^2 u + 4\sin^2 u}, \quad q(u) = 0.$$

Figure 1 presents both the initial curve C and the deformed curves C_ε, $\varepsilon = 0, 0.2, 0.4, 0.6, 0.8$.

Remark 2. Although one gets the impression from Fig. 1 that the deformations pass through space, the deformed curves stay in the plane.

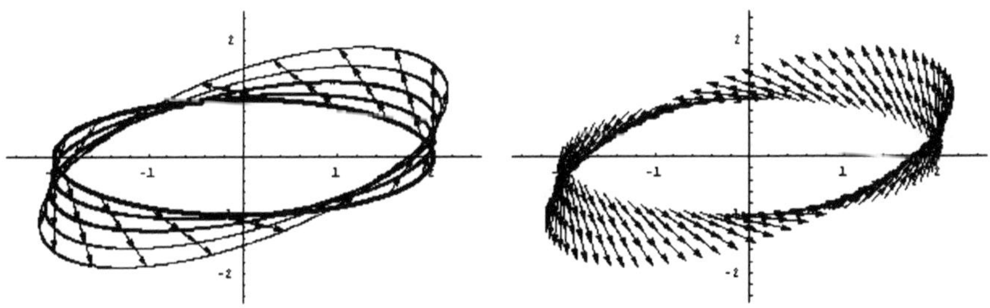

Fig. 1 Infinitesimal bending of an ellipse

2 Infinitesimal Bending in the Plane

In this section we consider infinitesimal bending of plane curves remaining plane (belonging to the same or to a different plane) after deformation. We determine an infinitesimal bending field in this case. Some examples will be given.

We shall consider a regular closed plane curve in polar coordinates, $C: \rho = \rho(\theta)$, $\theta \in [0, 2\pi]$. Under infinitesimal bending, this curve is included in the family of curves

$$C_\varepsilon : \rho_\varepsilon = \rho_\varepsilon(\theta), \quad \theta \in [0, 2\pi], \ \varepsilon \geq 0, \ \varepsilon \to 0.$$

In vector form, these curves are given as follows:

$$\begin{aligned} C: \ & \mathbf{r} = \mathbf{r}(\theta), \quad \theta \in [0, 2\pi], \\ C_\varepsilon : \ & \mathbf{r}_\varepsilon = \mathbf{r}(\theta) + \varepsilon \mathbf{z}(\theta), \quad \theta \in [0, 2\pi], \end{aligned}$$

where $\mathbf{z}(\theta)$ is the infinitesimal bending field.

Remark 3. In the case of a piecewise smooth curve, at the points of non regularity we assume the bending field to be continuous along the curve, i.e., $\mathbf{z}(\theta - 0) = \mathbf{z}(\theta + 0)$.

We next obtain a bending field in which the curve remains in its initial plane after infinitesimal bending.

Theorem 4 ([19]). *An infinitesimal bending field for a curve*

$$C : \rho = \rho(\theta) \tag{3}$$

that under infinitesimal bending remains included in a family of plane curves

$$C_\varepsilon : \rho_\varepsilon = \rho_\varepsilon(\theta), \quad \varepsilon \geq 0,\ \varepsilon \to 0, \tag{4}$$

is given in the form:

$$\mathbf{z}(\theta) = \int p(\theta)\mathbf{n}(\theta)\,\mathrm{d}\theta + c. \tag{5}$$

If the equation of the curve is given in vector form,

$$C : \mathbf{r}(\theta) = \rho(\theta)\cos\theta\mathbf{i} + \rho(\theta)\sin\theta\mathbf{j},$$

we have

$$\mathbf{z}(\theta) = \int p(\theta)\frac{(\rho\ddot{\rho} - \rho^2 - 2\dot{\rho}^2)[(\rho\cos\theta + \dot{\rho}\sin\theta)\mathbf{i} + (\rho\sin\theta - \dot{\rho}\cos\theta)\mathbf{j}]}{\sqrt{\rho^2 + \dot{\rho}^2}|\rho\ddot{\rho} - \rho^2 - 2\dot{\rho}^2|}\,\mathrm{d}\theta.$$

Using

$$\rho(\theta)\cos\theta + \dot{\rho}(\theta)\sin\theta = (\rho(\theta)\sin\theta)^{\cdot},\ \rho(\theta)\sin 0 - \dot{\rho}(\theta)\cos\theta = -(\rho(\theta)\cos\theta)^{\cdot},$$

we choose $p(\theta) = |\dot{\mathbf{r}}| = \sqrt{\rho(\theta)^2 + \dot{\rho}(\theta)^2}$, which leads to

$$\mathbf{z}(\theta) = \int [(\rho(\theta)\sin\theta)^{\cdot}\mathbf{i} - (\rho(\theta)\cos\theta)^{\cdot}\mathbf{j}]\,\mathrm{d}\theta = \rho(\theta)\sin\theta\mathbf{i} - \rho(\theta)\cos\theta\mathbf{j}.$$

Example 2. For the Cassini curve

$$C:\ \rho(\theta) = \pm\sqrt{a^2\cos 2\theta \pm \sqrt{b^4 - a^4\sin^2 2\theta}}, \quad \theta \in [0, 2\pi],$$

for graphical representation we choose $a = 1.9$, $b = 2$. We obtain two vector fields:

$$\begin{aligned}\mathbf{z}_{1,2}(\theta) = &\pm \sqrt{a^2\cos 2\theta + \sqrt{b^4 - a^4\sin^2 2\theta}}\sin\theta\mathbf{i}\\ &\mp \sqrt{a^2\cos 2\theta + \sqrt{b^4 - a^4\sin^2 2\theta}}\cos\theta\mathbf{j}.\end{aligned}$$

The deformed curves, shown in Fig. 2, are given by the equation

$$\begin{aligned}\rho_\varepsilon(\theta) &= \sqrt{a^2\cos 2\theta + \sqrt{b^4 - a^4\sin^2 2\theta}}(\cos\theta \pm \varepsilon\sin\theta)\mathbf{i}\\ &+ \sqrt{a^2\cos 2\theta + \sqrt{b^4 - a^4\sin^2 2\theta}}(\sin\theta \mp \varepsilon\cos\theta)\mathbf{j}.\end{aligned}$$

Finally, we get $\rho_\varepsilon(\theta) = \rho(\theta)\sqrt{1 + \varepsilon^2}$, $\theta \in [0, 2\pi]$.

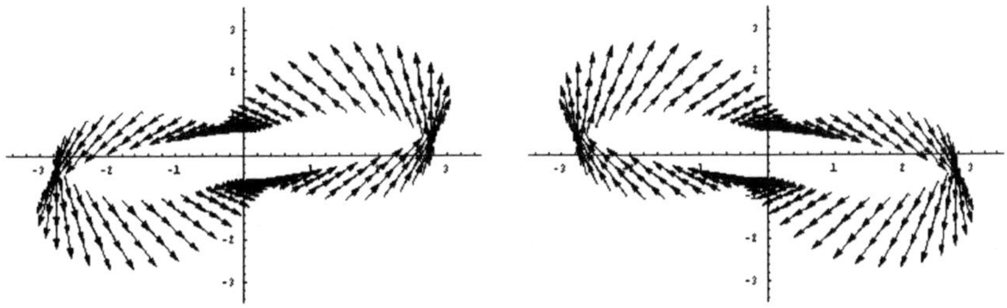

Fig. 2 Infinitesimal bending field for Cassini Curve

3 Variation of Torsion and Curvature

Under infinitesimal bending geometric magnitudes change. Some geometric magnitudes were discussed in [19]. We will here discuss curvature and torsion in more detail.

Theorem 5. *For a regular plane curve of class C^3, given by* (3)*, the variation of torsion τ under infinitesimal bending given by* (2) *is independent of the function $p(\theta)$, but depends on the function $q(\theta)$.*

Proof. Let us find the unit normal vector $\mathbf{n}(\theta)$ and the unit binormal $\mathbf{b}(\theta)$ of the curve C:

$$\mathbf{n}(\theta) = \left(-\frac{\rho\cos\theta + \dot\rho\sin\theta}{\sqrt{\rho^2+\dot\rho^2}}, \frac{-\rho\sin\theta + \dot\rho\cos\theta}{\sqrt{\rho^2+\dot\rho^2}}, 0\right), \quad \mathbf{b}(\theta) = (0,0,1).$$

Now we have the family of curves $C_\varepsilon : \rho_\varepsilon(u) = \rho(\theta) + \varepsilon\mathbf{z}(\theta)$, $\varepsilon \geq 0$, $\varepsilon \to 0$, i.e.,

$$C_\varepsilon : \rho_\varepsilon(u) = \left(\rho\cos\theta - \varepsilon\int\frac{p(\theta)(\rho\cos\theta + \dot\rho\sin\theta)}{\sqrt{\rho^2+\dot\rho^2}}\mathrm{d}\theta,\right.$$
$$\left.\rho\sin\theta + \varepsilon\int\frac{p(\theta)(-\rho\sin\theta + \dot\rho\cos\theta)}{\sqrt{\rho^2+\dot\rho^2}}\mathrm{d}\theta, \varepsilon\int q(\theta)\mathrm{d}\theta\right).$$

The variation of torsion under an infinitesimal bending is

$$\delta\tau = \left.\frac{\partial\tau_\varepsilon}{\partial\varepsilon}\right|_{\varepsilon=0} = \frac{1}{\rho^2+2\dot\rho^2-\rho\ddot\rho}q''(\theta) + \frac{-\dot\rho(2\rho+3\ddot\rho)+\rho\rho^{(3)}}{(\rho^2+2\dot\rho^2-\rho\ddot\rho)^2}q'(\theta)$$
$$+\frac{\rho^2+6\dot\rho^2-4\rho\ddot\rho+3\ddot\rho^2-2\dot\rho\rho^{(3)}}{(\rho^2+2\dot\rho^2-\rho\ddot\rho)^2}q(\theta). \quad (6)$$

We can see that the variation of torsion depends only on the function $q(\theta)$. □

In order to keep the curve in the initial plane, we require $\delta\tau = 0$. In this case, based on (6), if $\rho^2+2\dot\rho^2-\rho\ddot\rho \neq 0$, i.e., $\rho(\theta) \neq \cos(\theta + c_2)$, and denoting

$$a(\theta)=1,\ b(\theta)=\frac{-\dot\rho(2\rho+3\ddot\rho)+\rho\rho^{(3)}}{(\rho^2+2\dot\rho^2-\rho\ddot\rho)},\ c(\theta)=\frac{\rho^2+6\dot\rho^2-4\rho\ddot\rho+3\ddot\rho^2-2\dot\rho\rho^{(3)}}{(\rho^2+2\dot\rho^2-\rho\ddot\rho)},$$

we get from (6)

$$a(\theta)q''(\theta)+b(\theta)q'(\theta)+c(\theta)q(\theta)=0. \tag{7}$$

Note that b is the negative logarithmic derivative of $\rho^2+2\dot\rho^2-\rho\ddot\rho$.
Let us discuss (7).

1. For $c(\theta)=0$, we have as solution of (7)

$$q(\theta)=\int c_1\left(\rho^2+2\dot\rho^2-\rho\ddot\rho\right)\mathrm{d}\theta+c_2.$$

2. For $b(\theta)=0$, we have two special solutions of (7):

2.1. For $c(\theta)=C=\text{const.}$, the solution is $q(\theta)=c_1\cos\sqrt{C}\theta+c_2\sin\sqrt{C}\theta$.
2.2. For $c(\theta)=C/(k\theta+n)^2$, we get Euler's differential equation with solution

$$q(\theta)=c_1(k\theta+n)^{(k-\sqrt{k^2-4C})/(2k)}+c_2(k\theta+n)^{(k+\sqrt{k^2-4C})/(2k)}.$$

3. For $c(\theta)=0$ and $b(\theta)=0$, the solution is $q(\theta)=c_1+c_2\theta$.
4. For $c(\theta)\neq 0$ and $b(\theta)\neq 0$, we solve the differential equation (7) by functions $u(\theta),v(\theta)$ where $q(\theta)=u(\theta)v(\theta)$. We choose

$$v(\theta)=\mathrm{e}^{-1/2\int b(\theta)\,\mathrm{d}\theta}=\rho^2+2\dot\rho^2-\rho\ddot\rho,$$

and we obtain for u the equation

$$u''(\theta)+\left(c(\theta)-\frac{b'(\theta)}{2}-\frac{b^2(\theta)}{4}\right)u(\theta)=0,$$

with solution of the same type as one that was given in **case 2**. The solution of (7) is

$$q(\theta)=u(\theta)v(\theta)=u(\theta)(\rho^2+2\dot\rho^2-\rho\ddot\rho).$$

Example 3. Consider a circle defined by $\rho(\theta)=a$, $\theta\in[0,2\pi]$. The variation of the torsion is

$$\delta\tau=\left.\frac{\partial\tau_\varepsilon}{\partial\varepsilon}\right|_{\varepsilon=0}=a^{-2}(q''(u)+q(u)).$$

If the curve undergoing infinitesimal deformation remains a plane curve, we have $q''(u)+q(u)=0$, from where one gets the solution $q(u)=c_1\cos(u)+c_2\sin(u)$, $c_1,c_2\in\mathbb{R}$. For $p(u)=1$, $c_1=1$ and $c_2=1$, the family of deformed curves is given in Fig. 3

Theorem 6. *For a regular plane curve of class C^3 the variation of the curvature κ under bending defined by the field* (2) *is independent of the function $q(\theta)$, but depends on the function $p(\theta)$.*

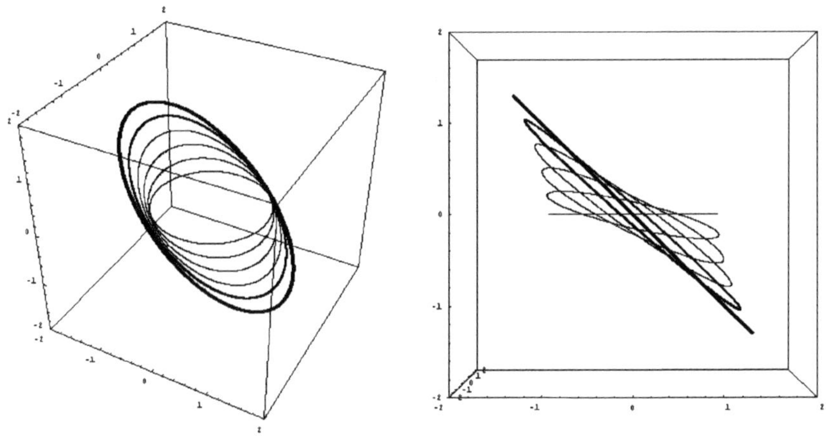

Fig. 3 Infinitesimal deformation of a circle

Proof. In a similar manner as in Theorem 5, we get an expression for the variation of curvature:

$$\delta\kappa = \left.\frac{\partial \kappa_\varepsilon}{\partial \varepsilon}\right|_{\varepsilon=0} = \frac{1}{\rho^2+\dot{\rho}^2}p'(\theta) - \frac{\dot{\rho}(\rho+\ddot{\rho})}{(\rho^2+\dot{\rho}^2)^2}p(\theta).$$

The variation of curvature vanishes if and only if

$$p'(\theta) - \frac{\dot{\rho}(\rho+\ddot{\rho})}{\rho^2+\dot{\rho}^2}p(\theta) = 0. \tag{8}$$

Note that the coefficient of p is half the logarithmic derivative of $\rho^2+\dot{\rho}^2$.

Under the condition $\rho \neq c_1 e^{\pm i\theta}$ we discuss (8) as follows:

1. If $\dot{\rho}(\rho+\ddot{\rho}) = 0$, i.e., $\rho_1(\theta) = c_1$, $\rho_2(\theta) = c_2\cos\theta + c_3\sin\theta$, solving (8), we get $p(\theta) = c$, where $c, c_1, c_2, c_3 \in \mathbb{R}$.
2. If $\dot{\rho}(\rho+\ddot{\rho}) \neq 0$, the solution of (8) is

$$p(\theta) = c\sqrt{\rho^2+\dot{\rho}^2}. \tag{9}$$

Remark 4. For a regular plane curve of class C^3 given by (3) the variation of curvature κ under infinitesimal bending is zero if and only if the function $p(\theta)$ is given by (9).

Remark 5. For a regular plane curve of class C^3 given by (3) the variation of curvature κ under infinitesimal bending is zero if only if $p(\theta) = cs'(\theta)$, where $s(\theta)$ is is arclength of curve (3).

Theorem 7. *A regular plane curve of the class C^3, given by* (3), *under an infinitesimal bending is included in a family of plane curves given by* (4) *if the equation*

$$ds_\varepsilon^2 - ds^2 = \varepsilon^2 \left(p(\theta)^2 + q(\theta)^2\right)$$

holds, where the functions p,q are given by (2).

Proof. Calculating the difference of the squares of line elements of the initial and the deformed curve (3) and (5), we get

$$ds_\varepsilon^2 - ds^2 = \varepsilon^2 p(\theta)^2 + \varepsilon^2 q(\theta)^2 + \rho^2 + \dot{\rho}^2 - (\rho^2 + \dot{\rho}^2) = \varepsilon^2(p(\theta)^2 + q(\theta)^2). \quad \square$$

4 InfBend

InfBend (formerly SurfBend) is our visualization tool devoted to visual representation of infinitesimally bent curves and rotational surfaces. The tool in the previous version, that we named SurfBend, was aimed at creating 3D presentation and visualizing applications of infinitesimal bending of flexible toroid-like surfaces, and was partialy presented at the ESI Conference Rigidity and Flexibility, Vienna, 2006 [21]. These rotational surfaces were obtained by revolution of a meridian in the shape of a polygon. We were also able to show curves formed by a polygon and their infinitesimally deformed shapes, as well as visualize surfaces created during deformation [14].

We have extended our research and widened InfBend to include the application of infinitesimal bending to a class of nonrigid curves. Our goals are to create an easy-to-use tool for:

- Infliction of curves and deformation given by (2). The ability to symbolically inflict curve **r**, also functions p and q, is given,
- Visual presentation and achievement of quick basic and 3D calculations. It is very useful to interactively examine bent curves and resulting surfaces and the influence of infinitesimal bending fields on them.

InfBend is developed in Object-Oriented language C++. It uses explicitly defined functions with n independent variables over a set of elementary functions. Each of the elementary functions is wrapped in an appropriate class, and we created a class hierarchy to support building a tree structure as an expression tree of the function. Every class in the hierarchy has overridden abstract members:

```
double evaluate (double * arguments);
Function * derive(int argNumber);
```

The function `evaluate` is designed to calculate the value of a wrapped elementary function and as an argument it takes an array with double values of n parameters for which we want to calculate the function value.

The function `derive` is designed to build a new tree structure according to the derivation rules for elementary functions, widened with rules for composite functions, having as argument the ordinal of the ith independent variable for which the partial derivation is wanted. The return value is also of the `Function * type`, so this enables us to build the tree structure for arbitrary-order partial derivation, and calculate values of the obtained function by calling the evaluate member function.

It is possible to start from an explicitly defined function entered as input string, then parse it and check its consistency. We use a formal parsing technique, that includes the grammar describing such functions as can be found in [4], and as parsing tool we have used the GOLD Parser [3]. The GOLD Parser is a free, multi language, pseudo open source parsing system that can be used to develop programming languages, scripting languages and interpreters. After parsing, we build an internal tree structure – an expression tree of the function as described in Object-Oriented design pattern Composite [6]:

```
MainFunction * pF;
if( ManagerFunction::parse( string functionInscription )
   pF = ManagerFunction::build(string functionInscription));
```

We also used the famous OO patterns Singleton, Abstract Factory [6] in producing function objects, building trees, and evaluating functions.

According to Theorems 2 and 3 we consider infinitesimal bending fields $\mathbf{z}$ for closed curves and parameter values $u \in [0, 2\pi]$. If we calculate the definite integral

$$\mathbf{z}(a) - \mathbf{z}(0) = \int_0^a [p(u)\mathbf{n}(u) + q(u)\mathbf{b}(u)]\,\mathrm{d}u, \tag{10}$$

we have

$$\mathbf{z}(a) = \int_0^a [p(u)\mathbf{n}(u) + q(u)\mathbf{b}(u)]\,\mathrm{d}u + \mathbf{z}(0).$$

The additive constant const. mentioned in (2) is here $\mathbf{z}(0)$. Considering that we can find partial derivatives, and calculate their values for passed arguments, we can apply numerical methods for calculating (10) and produce values of the bending field for discrete values of $a \in [0, 2\pi]$. We use the well-known generalized Simpson formula [13] for numerical integration, and a user is able to supply the number of division points, as well as the number of points inside every segment bounded by successive division points. Further, we can calculate values of the supplied curve $\mathbf{r}$ in the division points, and knowing ε we have points of the bent curve

$$C_\varepsilon : \mathbf{r}_\varepsilon = \mathbf{r}(\theta) + \varepsilon \mathbf{z}(\theta), \quad \theta \in [0, 2\pi].$$

Visualization of bent curves $\mathbf{r}_\varepsilon$ is obtained by using the OpenGL [7,11] standard. It should therefore be portable, although it has only been tested on the Microsoft Windows platform. Rising control to interactive level has been done using MFC [16]. We are able to rotate a 3D object and see it from different angles and points of view. It is possible to use sliders, to easily, interactively, adjust important parameters for bending calculations like ε, the number of segments and the number of inner points inside segments for numerical integration, as well as additive constants in each of the three dimensions. During deformation, curves describe surfaces and we are able to give visualization of these surfaces in fill or wire mode with the ability to adjust semi transparency of hidden lines.

InfBend is a free software and is available from http://tesla.pmf.ni.ac.rs/people/Ljubica%20Velimirovic/

The following examples were treated using our visualization tool InfBend.

Example 4. Consider an astroid given by the parametric equation

$$C:\ r(u) = (\cos^3(u), \sin^3(u), 0);$$

we choose the infinitesimal bending field $\mathbf{z}(u)$ given by (10), where $p(u) = 1$, $q(u) = \sin(u)$. Figure 4 presents both the initial curve C and the deformed curves C_ε, $\varepsilon = 0, (0.08), 0.4$.

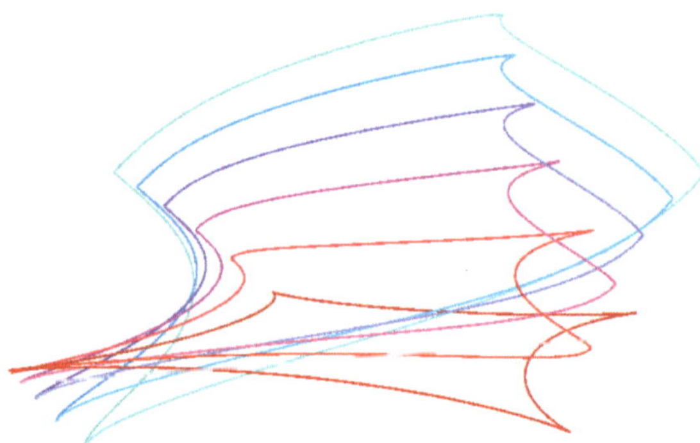

Fig. 4 Infinitesimal bending of an astroid

Example 5. Consider a cardioid given by the parametric equation

$$C:\ r(u) = ((1+\sin(u))\cos(u), (1+\sin(u))\sin(u), 0);$$

we choose the infinitesimal bending field $\mathbf{z}(u)$ given by (10), where

$$p(u) = \sqrt{2(1+\sin(u))}, \quad q(u) = 2(\pi - u).$$

Fig. 5 presents both the initial curve C and the deformed curves C_ε, $\varepsilon = 0, (0.03), 0.15$.

Example 6. Consider a circle given by the parametric equation

$$C:\ r(u) = (\cos(u), \sin(u), 0);$$

we choose the infinitesimal bending field $\mathbf{z}(u)$ given by (10), where $p(u) = 0$, $q(u) = \cos(u) + \sin(u)$. Figure 6 presents both the initial curve C and the deformed curves C_ε, $\varepsilon = 0, (0.08), 0.4$. In this case the curves and formed surface are visualized in wire mode with semi transparency.

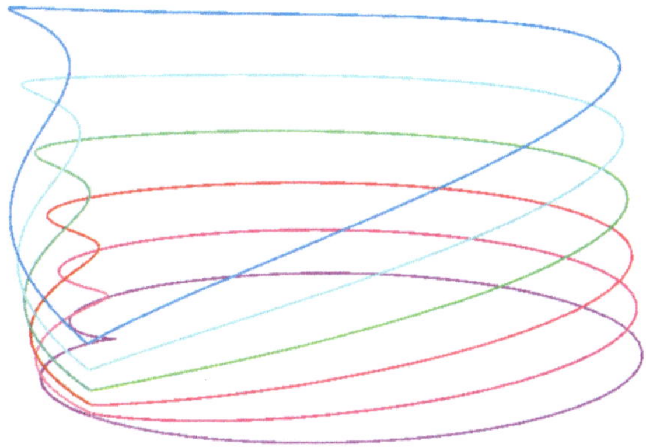

Fig. 5 Infinitesimal deformation of a cardioid

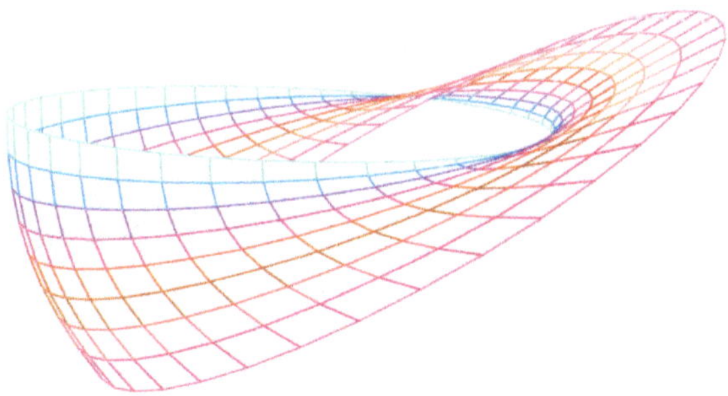

Fig. 6 Surface made by infinitesimally deformed initial circle

References

1. Aleksandrov, A. D.: O beskonechno malyh izgibaniyah neregulyarnyh poverhnostei. Matem. Sbornik. **1**(43), 3, 307–321 (1936)
2. Cohn-Vossen, S.: Unstarre geschlossene Flächen. Math. Ann. **102**, 10–29 (1930)
3. Cook D.: GOLD Parser Builder. www.devincook.com/goldparser.
4. Ćirić, M., Rančić, S. R.: Parsing in different languages. Facta Universitatis (Niš), Ser. Elec. Energ. **18** (2), 299–307 (2005)
5. Efimov, N.V.: Kachestvennye voprosy teorii deformacii poverhnostei. UMN **3.2**, 47–158 (1948)
6. Gamma, E., Helm, R., Johnston, R., Vlisides, J.: Design Patterns – Elements of Reusable Object-Oriented Software. Addison-Wesley, Reading, MA, USA (1995)
7. Glasser, G., Stachel, H.: Open Geometry: OpenGL + Advanced Geometry. Springer, Berlin (1999)
8. Ivanova-Karatopraklieva, I., Sabitov, I Kh.: Deformation of surfaces. J. Math. Sci. New York **70** (2), 1685–1716 (1994)
9. Ivanova-Karatopraklieva, I., Sabitov, I Kh.: Bending of surfaces II. J. Math. Sci. New York **74** (3), 997–1043 (1995)
10. Kon-Fossen, S.E.: Nekotorye voprosy differ. geometrii v celom. Fizmatgiz, Moskva. **9** (1959)
11. McReynolds, T., Blythe, D.: Advanced Graphics Programming Using OpenGL. Morgan Kaufmann Publishers, Los Altos, CA (2005)

12. Meziani, A.: Infinitesimal bendings of high orders for homogeneous surfaces with positive curvature and a flat point. J. Diff. Eqs. **239** (1), 16–37 (2007)
13. Milovanović, G.V.: Numerička Analiza. Naučna knjiga, Beograd (1985)
14. Rančić, S.R.; Velimirović, L.S.: Visualization of infinitesimal bending of some class of toroid. Int. J. Pure Appl. Math. **42** (4), 507–514 (2008)
15. Sabitov, I Kh.: Isometric transformations of a surface inducing conformal maps of the surface onto itself. Mat. Sb. **189** (1), 119-132 (1998)
16. Shepherd, G., Kruglinski, D.: Programming with Microsoft Visual C++.NET. 6th ed. Microsoft Press (2003)
17. Stachel, H.: Higher order flexibility of octahedra. Period. Math. Hung. **39** (1–3), 225–240 (1999)
18. Velimirović, L.S.: Change of area under infinitesimal bending of border curve. Buletins for Applied Mathematics (BAM), Hungary PC-129 (2000)
19. Velimirović, L.S.: Change of geometric magnitudes under infinitesimal bending. Facta Universitatis – Series: Mechanics, Automatic Control and Robotics, **3** (11), 135–148 (2001)
20. Velimirović, L.S.: Infinitesimal bending of curves. Matematicki bilten Skopje, Makedonija **25**(LI) 25–36 (2001)
21. Velimirović, L.S., Rančić S.R.: Higher order infinitesimal bending of a class of toroids. European J. Combin. **31** (4), 1136–1147 (2010)

MIX
Papier aus verantwortungsvollen Quellen
Paper from responsible sources
FSC® C105338

If you have any concerns about our products,
you can contact us on
ProductSafety@springernature.com

In case Publisher is established outside the EU,
the EU authorized representative is:
Springer Nature Customer Service Center GmbH
Europaplatz 3, 69115 Heidelberg, Germany

Printed by Libri Plureos GmbH
in Hamburg, Germany